Springer Series in **Materials Science** 18

Edited by Manuel Cardona

Springer Series in **Materials Science**

Advisors: M. S. Dresselhaus · H. Kamimura · K. A. Müller
Editors: U. Gonser · A. Mooradian · R. M. Osgood · M. B. Panish · H. Sakaki
Managing Editor: H. K.V. Lotsch

1 **Chemical Processing with Lasers**
By D. Bäuerle

2 **Laser-Beam Interactions with Materials**
Physical Principles and Applications
By M. von Allmen

3 **Laser Processing of Thin Films and Microstructures**
Oxidation, Deposition and Etching of Insulators
By I. W. Boyd

4 **Microclusters**
Editors: S. Sugano, Y. Nishina, and S. Ohnishi

5 **Graphite Fibers and Filaments**
By M. S. Dresselhaus, G. Dresselhaus, K. Sugihara, I. L. Spain, and H. A. Goldberg

6 **Elemental and Molecular Clusters**
Editors: G. Benedek, T. P. Martin, and G. Pacchioni

7 **Molecular Beam Epitaxy**
Fundamentals and Current Status
By M. A. Herman and H. Sitter

8 **Physical Chemistry of, in and on Silicon**
By G. F. Cerofolini and L. Meda

9 **Tritium and Helium-3 in Metals**
By R. Lässer

10 **Computer Simulation of Ion-Solid Interactions**
By W. Eckstein

11 **Mechanisms of High Temperature Superconductivity**
Editors: H. Kamimura and A. Oshiyama

12 **Dislocation Dynamics and Plasticity**
By T. Suzuki, S. Takeuchi, and H. Yoshinaga

13 **Semiconductor Silicon**
Materials Science and Technology
Editors: G. Harbeke and M. J. Schulz

14 **Graphite Intercalation Compounds I**
Structure and Dynamics
Editors: H. Zabel and S. A. Solin

15 **Crystal Chemistry of High-T_c Superconducting Copper Oxides**
By B. Raveau, C. Michel, M. Hervieu, and D. Groult

16 **Hydrogen in Semiconductors**
By S. J. Pearton, J. W. Corbett, and M. Stavola

17 **Ordering at Surfaces and Interfaces**
Editors: A. Yoshimori, T. Shinjo, and H. Watanabe

18 **Graphite Intercalation Compounds II**
Transport and Electronic Properties
Editors: H. Zabel and S. A. Solin

19 **Laser-Assisted Microtechnology**
By S. M. Metev and V. P. Veiko

20 **Microcluster Physics**
By S. Sugano

21 **The Metal-Hydrogen System**
Basic Bulk Properties
By Y. Fukai

22 **Ion Implantation in Diamond, Graphite and Related Materials**
By M. S. Dresselhaus and R. Kalish

H. Zabel S. A. Solin (Eds.)

Graphite Intercalation Compounds II

Transport and Electronic Properties

With Contributions by
G. L. Doll G. Dresselhaus M. S. Dresselhaus
P. C. Eklund M. Endo N. A. W. Holzwarth
J.-P. Issi J. T. Nicholls R. Schlögl S. A. Solin
S. Tanuma H. Zabel

With 216 Figures

Springer-Verlag
Berlin Heidelberg New York
London Paris Tokyo
Hong Kong Barcelona
Budapest

Professor Dr. Hartmut Zabel
Ruhr-Universität Bochum, Fakultät für Physik und Astronomie
Universitätsstrasse 150, W-4630 Bochum 1, Fed. Rep. of Germany

Stuart Solin, Ph. D.
NEC Research Institute, Inc.
4 Independence Way, Princeton, NJ 08540, USA

Guest Editor: Professor Dr. Manuel Cardona
Max-Planck-Institut für Festkörperforschung, Heisenbergstrasse 1
W-7000 Stuttgart 80, Fed. Rep. of Germany

ISBN 3-540-53921-2 Springer-Verlag Berlin Heidelberg New York
ISBN 0-387-53921-2 Springer-Verlag New York Berlin Heidelberg

Library of Congress Cataloging-in-Publication Data
Graphite intercalation compounds II: transport and electronic properties / H. Zabel, S. A. Solin (eds.): with contributions by G. L. Doll [et al.] p. cm. - (Springer series in materials science; v. 18) Includes bibliographical references and index. ISBN 0-387-53921-2 (U.S.) 1. Clathrate compounds. 2. Graphite. I. Zabel, H. (Hartmut), 1946-. II. Solin, S. A. (Stuart A.), 1942-. III. Doll, G. L. IV. Series. QD474.G75 1992. 541.2'2 - dc20 91-41103 CIP

Typesetting: Macmillan India Limited, Bangalore, India
54/3140 - 5 4 3 2 1 0 - Printed on acid-free paper

Preface

The research on graphite intercalation compounds often acts as a forerunner for research in other sciences. For instance, the concept of staging, which is fundamental to graphite intercalation compounds, is also relevant to surface science in connection with adsorbates on metal surfaces and to high-temperature superconducting oxide layer materials. Phonon-folding and mode-splitting effects are not only basic to graphite intercalation compounds but also to polytypical systems such as superconductors, superlattices, and metal and semiconductor superlattices. Charge transfer effects play a tremendously important role in many areas, and they can be most easily and fundamentally studied with intercalated graphite. This list could be augmented with many more examples. The important message, however, is that graphite intercalation compounds represent a class of materials that not only can be used for testing a variety of condensed-matter concepts, but also stimulates new ideas and approaches.

This volume is the second of a two-volume set. The first volume addressed the structural and dynamical aspects of graphite intercalation compounds, together with the chemistry and intercalation of new compounds. This second volume provides an up-to-date status report from expert researchers on the transport, magnetic, electronic and optical properties of this unique class of materials. The band-structure calculations of the various donor and acceptor compounds are discussed in depth, and detailed reviews are provided of the experimental verification of the electronic structure in terms of their photoemission spectra and optical properties.

The discovery of superconductivity in KC_8 many years ago raised fundamental questions about the mechanism of electronic coupling and anisotropy in these layered materials and spurred on research in this area. Although the highest superconducting transition temperature known today in intercalated graphite is only about 4 K, this nonetheless represents an exciting 30-fold increase over the original transition temperature in KC_8. A review of the superconducting properties of intercalated graphite is also included in this volume.

Thermal transport properties represent a very complex topic in condensed-matter physics, with phonon, electronic, and magnetic contributions convoluted together in a nontrivial manner. Intercalated graphite materials offer the unique opportunity to change, via the intercalate species, the relative importance of the various contributions, thus casting light on the most important scattering mechanisms.

The magnetic properties of intercalated graphite provide another exciting arena for fundamental physics. The properties are naturally quasi two dimensional, and the local anisotropy fields – determined by the choice of the magnetic ions – allow the investigation of Ising, *XY*, and Heisenberg-like systems in planar structures.

Moreover, using the freedom to change the stage of the compound, crossover effects in the transition from two- to three-dimensional interactions have been studied. These and other aspects of the magnetic properties of graphite intercalation compounds are discussed in great depth in this volume.

As in Volume I the emphasis in the present volume is on the fundamental aspects of the physics and chemistry of intercalated compounds. In addition, a chapter is included on another modification of graphite and its intercalated compounds, namely the so-called graphite fiber materials.

The editors are grateful to the contributors to this second volume for their efforts to provide comprehensive and up-to-date reviews and for their cooperation in all details. We are also grateful to the publishers and their staff for their encouragement and patience in all parts of this production, and in particular to the managing editor of the Springer Series in Materials Science, Dr. Helmut K.V. Lotsch.

Bochum *Hartmut Zabel*
Princeton *Stuart A. Solin*
February 1992

Contents

Contributors

Doll, Gary L.
Physics Department, General Motors Research Laboratories
Warren, MI 48090, USA

Dresselhaus, Gene
Massachusetts Institute of Technology, Cambridge, MA 02139, USA

Dresselhaus, Mildred S.
Massachusetts Institute of Technology, Cambridge, MA 02139, USA

Eklund, Peter C.
Department of Physics, University of Kentucky
Lexington, KY 40506-0055, USA

Endo, Morinobu
Massachusetts Institute of Technology, Cambridge, MA 02139, USA

Holzwarth, Natalie A. W.
Department of Physics, Wake Forest University
Winston-Salem, NC 27109, USA

Issi, Jean-Paul
Unité de Physique-Chimie, et de Physique des Matériaux,
Université Catholique de Louvain, 1, Place Croix-du-Sud
B-1348 Louvain-la-Neuve, Belgium

Nicholls, James T.
Massachusetts Institute of Technology, Cambridge, MA 02139, USA

Schlögl, Robert
Institut für Anorganische Chemie, Universität Frankfurt
Niederurseler Hang, W-6000 Frankfurt/M. 50, Fed. Rep. of Germany

Solin, Stuart A.
NEC Research Institute, Inc.
4 Independence Way, Princeton, NJ 08540, USA

Tanuma, Sei-ichi
Department of Materials Science, Iwaki Meisei University
Iwaki, Fukushima 970, Japan

Zabel, Hartmut
Ruhr-Universität Bochum, Fakultät für Physik und Astronomie
Universitätsstrasse 150, W-4630 Bochum 1, Fed. Rep. of Germany

1. Introduction

By Hartmut Zabel and Stuart A. Solin

This volume on the transport and electronic properties of graphite intercalation compounds (GICs) is the second in a two-part series. The first volume focused on structure and vibrations in these unusual materials. As with the subject matter of Vol. I, there have been several international, binational, and domestic meetings and conferences on GICs, where exciting new results on electronic and transport properties have been reported. However, this subject has not been reviewed in depth since the excellent treatises of the early 1980s [1.1–5]. Thus, Vol. II of this series is most timely. Although these two volumes are complementary, we intend that they be individually self-contained. Therefore, some introductory material presented in Vol. I will of necessity be presented again here.

The most noteworthy feature of GICs is their high degree of anisotropy. This anisotropy manifests itself in a variety of stacking arrangements of the host graphene layers and the intercalated layers of the guest species that occupy the galleries formed between graphene layers. Underlying the myriad of structural arrangements is the phenomenon of staging, which is characterized by a long-range ordered sequence of n graphene layers interspersing pairs of guest layers to form a stage-n GIC. Structures with stage numbers, as high as 10 have been reported in the literature. In addition, GICs are known to form fractional stages of the type n/m in which only n of the m galleries of the stacking repeat unit contain guest layers.

The above-described structural complexity of GICs relates only to the quasi-one-dimensional character of their stacking arrangements. But the guest species themselves can adopt a variety of planar structures, including those which are commensurate, incommensurate, or liquid-like with respect to the hexagonal structure of the graphene bounding layers. These in-plane structures endow GICs with a quasi-two-dimensional character, which in the presence of inter-layer correlation sometimes results in compounds with a three dimensionally ordered superlattice structure. The structures exhibited by GICs and the phase transitions between them (including order-order, order-disorder, and dimensionality crossover phenomena) were the subject of Vol. I. But clearly these aspects of structure will have a direct impact on the electronic and transport properties of GICs. It is this aspect that we address in the current volume.

In addition to their structural characterization, GICs are also characterized as donor or acceptor compounds, according to whether the guest species gives up electronic charge to, or receives it from, the host layers. They are further

Springer Series in Materials Science, vol. 18
H. Zabel · S.A. Solin (eds.): Graphite Intercalation Compounds II

identified as binary, ternary, etc., compounds based on the number of distinct guest species that occupy the galleries. This ternary GICs may contain two donor species, two acceptor species, or both a donor and an acceptor–and in each of these, the guests may be distributed either homogeneously or heterogeneously. Whereas there is a plethora of structural information available on both binary and ternary GICs, electrical and transport studies of the latter are relatively rare. Therefore, their treatment in the current volume is brief.

Since Vol. I of this series was published, the GIC field has progressed rapidly with some notable advances in the study of structure and vibrations. These most recent results, as well as those on transport and electronic properties, have been presented in part in the *Proceedings* of the Fourth, Fifth and Sixth International Symposia on GICs held in Jerusalem in May 1987 [1.6], in Berlin in May 1989 [1.7] and in Orléans in May 1991 [1.8]. Work was also presented at the biannual Symposia on Graphite Intercalation Compounds as part of the Materials Research Society Fall Meeting in Boston, which was most recently held in 1988 [1.9]. This volume on transport and electronic properties discusses most of the relevant material mentioned in those proceedings.

The present volume contains eight chapters including the introduction. Chap. 2 Holzwarth reviews the theoretical aspects of the electronic band structure of both pristine graphite and of GICs, which for the purpose of analysis are grouped as low-stage and high-stage compounds. The reader should find particularly useful her concise discussion of the theoretical techniques commonly used to determine the band structures of GICs, since these methods are also applicable to many other layered solid-state systems. Pristine graphite and stage-1 compounds with alkali metals exhibit a well-defined three-dimensional structure, and the band structure of these materials can therefore be calculated rigorously using first-principles pseudopotentials and self-consistent band energies. In this context, Holzwarth describes the various approaches for treating the effective one-electron potentials and wavefunctions for the computation of the electronic band structure.

After a detailed discussion of the electronic band structure of graphite, including its famous π and σ bands, Chap. 2 addresses the three major questions that occur most frequently in the literature:

1) How much charge is transferred between the guest and the host species?
2) How are the graphite and guest bands altered and how big is their mutual hybridization?
3) What is the nature of the "interlayer" band in GICs? (It has been talked about so many times but still remains somewhat nebulous.)

Holzwarth also reviews for the first time band-structure calculations of some ternary compounds, which are of interest because of their superconducting properties and metal–insulator transitions. The band structure of the high-stage compounds is more difficult to describe, as we already mentioned, and often the effect of intercalation is treated in a parameterized manner as an energy shift of the graphite bands. For stages higher than 2 an additional question is discussed thoroughly; namely, the theoretical justification for the charge distribution over the various inequivalent graphite layers from the bounding layers next to the guest layers to the interior layers.

Chapter 3 by Schlögl complements Chap. 2 by cataloging experimentally measured electronic properties of GICs determined by a variety of modern spectroscopic techniques ranging from ultraviolet photoelectron spectroscopy (UPS), and X-ray photoelectron spectroscopy (XPS), to soft X-ray emission spectroscopy (SXS). Schlögl discusses the experimentally determined band structure and density of states of pristine graphite and compares these with the corresponding properties of the intercalated compounds. He points out that even in pristine graphite the position and shape of spectral lines may be obscured because of chemical contamination at the surface and/or structural defects. This effect may be enhanced in intercalated samples by defects introduced upon intercalation and by the air sensitivity of most intercalation compounds. After reviewing briefly relevant theoretical models treating the charge transfer between guest and host layers, Schlögl describes in detail the experimental requirements for measuring the charge transfer via spectroscopic techniques for donor and acceptor compounds.

In Chap. 4 Eklund and Doll examine again the intriguing issue of charge transfer in GICs via their optical properties as manifested in optical reflectance measurements and in Raman spectroscopy. In Vol. I, Solin reviewed the effect of charge transfer on the phonon frequencies. The review in the present volume extends the previous review and includes a more quantitative analysis and comparison with theoretical models. However, the main emphasis remains on reflectance measurements of donor and acceptor compounds for probing quantitatively the consequences of subtle changes in the spatial dependence of the charge density.

A limited number of donor GICs exhibit superconductivity, which was first observed in a stage-1 potassium graphite binary compound. Because neither potassium nor graphite exhibit superconducting properties in their bulk form, this discovery spurred much activity in the field. Moreover, it was shown that superconducting GICs exhibit either type-I or type-II characteristics, depending on whether the magnetic field is directed parallel or perpendicular to the c axis. In Chap. 5, Tanuma addresses this work, as well as the pressure dependence of the anisotropy and the more recently discovered superconductors, such as the potassium-mercury ternary compound.

In Chap. 6, Issi details aspects of the thermal and electrical transport of the nonsuperconducting acceptor GICs, including those which are magnetic. Because of the pronounced structural anisotropy of GICs, the transport properties are highly anisotropic as well. Many years ago, reports on conductivity values for some acceptor compounds rivaling those of copper made some headlines. Although the promise could not be fulfilled, the transport properties of GICs remains a very intriguing subject for fundamental tests of anisotropic scattering theory as well as for applications. After a general introduction to the subject, Issi illustrates the basic ideas with experiments involving metal chloride compounds in graphite. A recent paper by *McRae* and *Mareche* [1.10] offers a complementary review of other acceptor and donor compounds.

The general subject of magnetism in GICs is treated in Chap. 7 by G. Dresselhaus et al., who review the structure and the magnetic properties of magnetic donor and acceptor compounds. In these layered magnetic materials the ratio of the interplanar to interlayer interaction can be varied by several orders of magnitude through staging, making these compounds interesting for investigations of dimensional crossover effects. Moreover, the local spin anisotropy may be varied by the choice of intercalate, providing, for instance, exciting examples for two-dimensional XY-type magnetism. Chapter 7 reviews in depth the theoretical aspects, as embodied in the magnetic Hamiltonians, and the experimental aspects associated with a variety of measurement techniques.

In Chap. 8, M.S. Dresselhaus and M. Endo review the different types of carbon fiber materials and their preparation techniques and structural characteristics, which, in turn, are related to their intercalation ability. The transport and mechanical properties of intercalated graphite fibers are of significant interest because of their large aspect ratio (length/diameter). They provide a favorable geometry for highly accurate a-axis transport measurements and excellent mechanical and chemical stability. Chapter 8 also discusses the current and potential applications of the fiber compounds.

Volumes I and II together cover most of the theoretical and experimental aspects of current research on GICs. New research opportunities, yet too early for a review, are already appearing on the horizon. Emerging fields include the application of scanning tunneling microscopy of pristine and intercalated graphite [1.11], the deposition of a few monolayers of graphite for intercalation and adsorption studies [1.12], the use of GICs for fundamental studies of two-dimensional metallic hydrogen [1.13], the study of nucleation and growth in one dimension [1.14], tunneling spectroscopy of molecular intercalates [1.15], and layer-rigidity effects [1.16], to mention just a few. In due time such topics may be at the center of interest in another review of the field.

References:

1.1 M.S. Dresselhaus, G. Dresselhaus: Adv. Phys. 30, 139 (1981)
1.2 P. Pfluger, H.J. Güntherodt: *Festkörperprobleme* (Advances in Solid State Physics), Vol. 21, ed. by J. Treusch (Vieweg, Braunschweig 1981)
1.3 S.A. Solin: Adv. Chem. Phys. **49**, 455 (1982)
1.4 R. Clarke, C. Uher: Adv. Phys. **33**, 469 (1984)
1.5 H. Zabel, P.C. Chow: Comments Cond. Mat. Phys. **12**, 225 (1986)
1.6 D. Davidov, H. Selig (eds.): Proc. Fourth Int. Symp. on Graphite Intercalation compounds, Jerusalem, Israel, 1987, Synth. Met. **23** (1988)
1.7 K. Lüders, R. Schöllhorn (eds.): Proc. Fifth Int. Symp. on Graphite Intercalation Compounds, Berlin, Germany, 1989, Synth. Met. **34** (1990)

1.8 S. Flandrois (ed.): Proc. Sixth Int. Symp. on Graphite Intercalation Compounds, Orléans, France, 1991, Synth. Met. (1992)
1.9 M. Endo, M.S. Dresselhaus, G. Dresselhaus: Symposium on Graphite Intercalation Compounds: Science and Applications, Boston, USA, 1988, Materials Research Society Extended Abstracts, Pittsburgh 1988
1.10 E. McRae, J.F. Mareche: J. Mater. Res. **3**, 75 (1988)
1.11 R. Wiesendanger, D. Anselmetti, V. Geiser, H.R. Hidber, H.J. Güntherodt: Synth. Mct. **34**, 175 (1990); D. Tomanek, G. Overney, H. Miyazaki, S.D. Mahanti, H.J. Güntherodt: Phys. Rev. Lett. **63**, 876 (1989)
1.12 N.R. Gall, S.N. Mikhailov, E.V. Rut'kov, A.Ya. Tontegode: Synth. Met. **34**, 497 (1990)
1.13 S. Miyajima, M. Kabasawa, T. Chiba, T. Enoki, Y. Maruyama, H. Inokuchi: Phys. Rev. Lett. **64**, 319 (1990)
1.14 N. Metoki, H. Suematsu, Y. Murakami, Y. Ohishi, Y. Fujii: Phys. Rev. Lett. **64**, 657 (1990)
1.15 D.A. Neumann, H. Zabel, A. Magerl, Y.B. Fan, S.A. Solin: In Ref. [1.8]
1.16 S. Lee, H. Miyazaki, S.D. Mahanti, S.A. Solin: Phys. Rev. Lett. **62**, 3066 (1989)

2. Electronic Band Structure of Graphite Intercalation Compounds

By Natalie A.W. Holzwarth

With 25 Figures

This chapter reviews the results of electronic band structure calculations for several graphite intercalation compounds. Although we have attempted to include a broad range of results in this review, there are undoubtedly omissions for which we apologize. Earlier reviews of this subject were given by *Fischer* [2.1], *Dresselhaus* and *Dresselhaus* [2.2], *Pfluger* and *Güntherodt* [2.3a], and *Kamimura* [2.3b]. In addition, collections of electronic structure studies can be found in several conference proceedings and reports, some of which are listed in the references [2.4–14].

This review is arranged as follows. In Sect. 2.1, we briefly review the general framework and methodologies for electronic structure calculations that have been used to study graphite intercalation compounds. In Sect. 2.2, we review the electronic structure of graphite itself. Section 2.3 considers "first-principles" electronic band structure results for graphite intercalation compounds that have been published during the past 15 years. These results are mainly for low-stage donor compounds. Section 2.4 includes a description of various treatments for higher-stage compounds. A brief summary and conclusions are given in Sect. 2.5.

2.1 Methods for Band Structure Calculations

The term electron energy bands refers to the one-electron eigenstates of a perfect crystal. Because of the periodicity of the crystal, these states form a continuum of levels, parameterized by wave vector $\boldsymbol{k}$. [2.15] There are two aspects of this procedure that define the relationship between the energy bands and the states of the actual many-electron materials. First, an effective one-electron Hamiltonian must be constructed. This Hamiltonian generally takes the form

$$\mathscr{H} = -\frac{\hbar^2 \nabla^2}{2m} + V(\boldsymbol{r}) . \tag{2.1}$$

Here m is the mass of an electron and $V(\boldsymbol{r})$ is an effective one-electron potential, which models the interaction of each electron with the nuclei and other electrons of the material. Second, a convenient numerical method must be devised for computing the eigenstates of the Hamiltonian:

$$\mathscr{H}\,\Psi_n(\boldsymbol{k},\boldsymbol{r}) = \varepsilon_n(\boldsymbol{k})\,\Psi_n(\boldsymbol{k},\boldsymbol{r}) . \tag{2.2}$$

Springer Series in Materials Science, vol. 18
H. Zabel · S.A. Solin (eds.): Graphite Intercalation Compounds II

Here $\Psi_n(\boldsymbol{k}, \boldsymbol{r})$ denotes the one-electron wavefunction and $\varepsilon_n(\boldsymbol{k})$ denotes the corresponding one-electron energy having band index n and wave vector $\boldsymbol{k}$. There are two common procedures for constructing the effective one-electron Hamiltonian (2.1): the Hartree-Fock theory [2.15, 16] and the density functional theory developed by *Hohenberg* and *Kohn* [2.17] and *Kohn* and *Sham* [2.18].

According to the Hartree-Fock theory, the one-electron wavefunctions $\Psi_n(\boldsymbol{k}, \boldsymbol{r})$ are factors in the antisymmetrized many-electron wavefunction, and (2.1) is derived by minimizing the energy of the system with respect to functional variations in the one-electron wave functions. The electronic charge density $\varrho(\boldsymbol{r})$ is determined from a sum of contributions of all occupied states:

$$\varrho(\boldsymbol{r}) = \sum_{\substack{n\boldsymbol{k} \\ \text{(occupied)}}} |\Psi_n(\boldsymbol{k}, \boldsymbol{r})|^2 . \tag{2.3}$$

The resultant Hartree-Fock effective potential $V(\boldsymbol{r})$ is the sum of the electrostatic interactions $V_\mathrm{N}(\boldsymbol{r})$ and $V_\mathrm{H}(\boldsymbol{r})$ of the "one" electron with the nuclei and electronic charge density of the material, respectively, as well as an integral operator $V_\mathrm{X}(\boldsymbol{r}, \boldsymbol{r}')$, which represents the exchange interaction:

$$V(\boldsymbol{r}) = V_\mathrm{N}(\boldsymbol{r}) + V_\mathrm{H}(\boldsymbol{r}) + V_\mathrm{X}(\boldsymbol{r}, \boldsymbol{r}') . \tag{2.4}$$

The electrostatic Coulomb interaction due to the electron distribution, or the Hartree potential, is a solution to the Poisson equation:

$$\nabla^2 V_\mathrm{H}(\boldsymbol{r}) = -4\pi e^2 \varrho(\boldsymbol{r}) . \tag{2.5}$$

The exchange operator is given by

$$V_\mathrm{X}(\boldsymbol{r}, \boldsymbol{r}') = -\frac{e^2}{2} \sum_{\substack{n\boldsymbol{k} \\ \text{(occupied)}}} \frac{\Psi_n(\boldsymbol{k}, \boldsymbol{r}) \Psi_n(\boldsymbol{k}, \boldsymbol{r}')}{|\boldsymbol{r} - \boldsymbol{r}'|} . \tag{2.6}$$

The Hartree-Fock equations should be solved self-consistently so that the effective potential $V(\boldsymbol{r})$ in the Schrödinger equation (2.2) is the same as that formed (2.4) from the electronic charge density (2.5) and the exchange operator (2.6). In the general literature of electronic band structure calculations, the Hartree-Fock approach has been used less frequently than the density functional approach. This is especially true for metallic materials. Part of the reason for the unpopularity of the Hartree-Fock treatment of metals is that it is possible to show [2.15] that for a noninteracting free-electron gas, the density of states vanishes at the Fermi surface. This analytical demonstration of nonphysical behavior of the Hartree-Fock equations for an idealized metal raises doubts about their ability to model metals in general. In Sect. 2.2 we compare the results of a Hartree-Fock calculation for graphite [2.19] with those of local density treatments. There has been a limited effort [2.20] to use Hartree-Fock theory to study intercalation compounds.

Historically, the development of density functional theory was based on many of the ideas of the Hartree-Fock formulation. In density functional theory, the one-electron wavefunction $\Psi_n(\boldsymbol{k}, \boldsymbol{r})$ also arises from a variational principle.

However, density functional theory is formulated upon minimization of the energy of the system with respect to variations in the electronic charge density $\varrho(\boldsymbol{r})$. This charge density can be determined by solving the effective Schrödinger equation (2.2) for noninteraction electrons moving in an effective potential $V(\boldsymbol{r})$ and summing over all occupied one-electron states as in (2.3). The effective potential contains the same electrostatic contributions as given in (2.4) for the Hartree-Fock theory. However, the exchange-integral operator is replaced by an effective exchange-correlation potential $V_{\text{XC}}(\varrho(\boldsymbol{r}))$, which results from the minimization of the exchange and correlation energy $E_{\text{XC}}(\varrho)$ with respect to the electron density:

$$V_{\text{XC}}(\varrho(\boldsymbol{r})) = \frac{\partial E_{\text{XC}}(\varrho(\boldsymbol{r}))}{\partial \varrho(\boldsymbol{r})} . \tag{2.7}$$

The exchange-correlation energy functional $E_{\text{XC}}(\varrho)$ is usually taken from calculations for a free-electron gas, and thus depends on the electron density through the Fermi wave vector $k_{\text{F}} = (3\pi^2\varrho)^{1/3}$, Several accurate calculations of $E_{\text{XC}}(\varrho)$ for a free-electron gas have been parametrized to convenient analytic form for use in this so-called local density approximation. For example, the form of *Hedin* and *Lundqvist* [2.21] has been used extensively. A recent paper by *Pickett* [2.22] comprehensively reviews the local density approximation. Thus, in place of (2.4), the effective potential in the local density approximation is given by

$$\text{V}(\boldsymbol{r}) = V_{\text{N}}(\boldsymbol{r}) + V_{\text{H}}(\boldsymbol{r}) + V_{\text{XC}}(\varrho(\boldsymbol{r})) . \tag{2.8}$$

In density functional theory using the local density approximation (2.8), the Schrödinger equation (2.2) and the effective potential (2.8) must be determined self-consistently. In both the Hartree-Fock and density functional formalisms, the *total* energy of the many-electron system is determined variationally, while the one-electron band energies $\varepsilon_n(\boldsymbol{k})$ are nominally only variational parameters. However, because they are eigenvalues of an effective Hamiltonian (2.2), and because of certain identities, such as Koopman's theorem [2.15, 16], the one-electron energy bands $\varepsilon_n(\boldsymbol{k})$ can be related to spectroscopic and Fermi-surface properties of the materials.

The second aspect of an electronic band structure calculation is the numerical methods used to solve the eigenvalue problem (2.2). These are discussed in solid-state textbooks such as [2.15]. A few of the methods that have been used for graphite intercalation compounds are discussed here. Two general approaches are based on "direct" solution and basis set expansion techniques. Examples of direct solution methods are the augmented plane-wave (APW) [2.23, 24] and the *Korringa*, *Kohn*, and *Rostoker* (KKR) [2.25, 26] methods. Special procedures must be developed to solve the Schrödinger equation in the core regions of the crystal in the presence of the strong core potential. Often this is done by dividing the one-electron potential $V(\boldsymbol{r})$ into spherical regions separated by interstitial regions. If we modify $V(\boldsymbol{r})$ to muffin-tin form so that it is strictly spherical within each atomic sphere and strictly flat (at the average potential value), we can solve the Schrödinger equation essentially exactly by either the APW or KKR formalisms. The approach has two shortcomings: the replacement of $V(\boldsymbol{r})$ by its muffin-tin form is rather unphysical for anisotropic materials such as graphite intercalation compounds, and the APW and KKR formalisms require iterative procedures for the band-structure determinations in addition to the self-consistency interations. Several authors [2.27–29] have applied muffin-tin corrections to the KKR wavefunctions to obtain reasonable results for graphite and intercalation compounds. Altmann et al. developed a more elegant approach [2.30, 31] based on a cellular method, which avoids the problem of the muffin-tin approximation and yields reasonable results for graphite. In general, however, the intensive computation involved in these methods discourages their use in self-consistent treatments.

More recent numerical methods are formulated in terms of basis set expansions of the one-electron wavefunctions:

$$\Psi_n(\boldsymbol{k},\boldsymbol{r}) = \sum_i C_n^i(\boldsymbol{k})\,\Phi_i(\boldsymbol{k},\boldsymbol{r})\,. \tag{2.9}$$

The coefficients $C_n^i(\boldsymbol{k})$ are determined in solving the eigenvalue problem (2.2). The basis functions $\Phi_i(\boldsymbol{k},\boldsymbol{r})$ can be chosen in a variety of ways ranging from plane waves,

$$\Phi_i(\boldsymbol{k},\boldsymbol{r}) = \frac{1}{\sqrt{\Omega}}\,\mathrm{e}^{\mathrm{i}(\boldsymbol{k}+\boldsymbol{G}_i)\cdot\boldsymbol{r}}\,, \tag{2.10}$$

where $\boldsymbol{G}_i$ denotes a reciprocal-lattice vector of the crystal and Ω denotes the volume of the unit cell, to linear combinations of atomic orbitals (LCAO functions):

$$\Phi_i(\boldsymbol{k},\boldsymbol{r}) = \sum_{\boldsymbol{T}} \mathrm{e}^{\mathrm{i}\boldsymbol{k}\cdot(\boldsymbol{\tau}+\boldsymbol{T})}\,\phi_i(\boldsymbol{r}-\boldsymbol{\tau}-\boldsymbol{T})\,, \tag{2.11}$$

where $\phi_i(\boldsymbol{r}-\boldsymbol{\tau}-\boldsymbol{T})$ denotes a function of type i centered at the site $\boldsymbol{\tau}$, $\boldsymbol{T}$. We use $\boldsymbol{\tau}$ to denote positions within the unit cell and $\boldsymbol{T}$ to denote a lattice translation. The plane-wave function (2.10) form a complete set for the lattice, but for many materials, the LCAO functions (2.11) are more rapidly convergent. In some cases, it is convenient to use a mixed-basis set including terms of both types. In general, the basis functions $\Phi_i(\boldsymbol{k},\boldsymbol{r})$ are not orthogonal and have overlap integrals

$$S_{ij}(\boldsymbol{k}) = \int \Phi_i(\boldsymbol{k},\boldsymbol{r})\,\Phi_j(\boldsymbol{k},\boldsymbol{r})\,d^3r\,. \tag{2.12}$$

Thus the solution of the differential eigenvalue equation (2.2) is reduced to the solutions of a generalized algebraic eigenvalue problem:

$$\sum_j [H_{ij}(\boldsymbol{k}) - \varepsilon_n(\boldsymbol{k})\,S_{ij}(\boldsymbol{k})]\,C_n^j(\boldsymbol{k}) = 0\,, \tag{2.13}$$

where the Hamiltonian matrix elements are given by

$$H_{ij}(\boldsymbol{k}) = \int \Phi_i(\boldsymbol{k},\boldsymbol{r})\,\mathscr{H}\,\Phi_j(\boldsymbol{k},\boldsymbol{r})\,d^3r\,. \tag{2.14}$$

There is a wide variety of choices for the atomic-orbital functions. Some that have been used for graphite intercalation problems range from very approximate to the "state-of-the-art" treatments. The early LCAO calculations used semiempirical schemes to evaluate matrix elements. So-called tight-binding calculations for graphite [2.32–34] used only near-neighbor terms in the matrix evaluations. Another approximate LCAO scheme successfully used for graphite and intercalation compounds is the extended Hückel approximation [2.35–38]. In this scheme, the atomic-orbital functions $\phi_i(\boldsymbol{r} - \boldsymbol{\tau} - \boldsymbol{T})$ are taken as analytic functions that approximate the atomic-valence wavefunctions, and the Hamiltonian is parameterized in terms of site energies. This approach has also been used in a Green's-function formalism to study graphite intercalation compounds [2.39].

First-principles LCAO calculations have been successfully used in many of the electronic structure calculations for graphite and intercalation compounds. Taking the atomic-orbital functions $\phi_i(\boldsymbol{r} - \boldsymbol{\tau} - \boldsymbol{T})$ to be of the Gaussian form [2.40], the overlap and Hamiltonian matrix elements can be evaluated from lattice sums of analytic integrals. Taking the atomic-orbital functions $\phi_i(\boldsymbol{r} - \boldsymbol{\tau} - \boldsymbol{T})$ to be Slater-type orbitals [2.41, 42] or numerical functions [2.43–47], the matrix elements $S_{ij}(\boldsymbol{k})$ and $H_{ij}(\boldsymbol{k})$ can be efficiently evaluated by numerical integration. Several authors have taken advantage of the layer geometry of these materials by using an orthogonalized plane-wave (OPW) expansion within a layer plane and one-dimensional Bloch functions in the perpendicular direction [2.48]. Replacing the ionic potentials with pseudopotentials permitted two-dimensional plane waves to be used as basis functions within a layer, and the charge transfer could be calculated self-consistently [2.49, 50].

The APW and KKR methods were reformulated into noniterative ("linearized") algorithms by *Andersen* [2.51], leading to a number of LCAO-like techniques that have been used for graphite and its intercalation compounds, such as the linearized muffin-tin orbital (LMTO) method [2.52] and the full-potential linearized augmented plane-wave (FLAPW) method [2.53–55]. Empirical pseudopotentials have been used [2.49, 50, 56] to study graphite and its intercalation compounds. Recently, first-principles ("norm-conserving") pseudopotentials [2.57] together with mixed-basis functions (both plane-wave and LCAO functions) [2.58] have been used in a series of studies [2.59–63].

2.2 Graphite

2.2.1 General Features

Any discussion of the electronic band structure of graphite intercalation compounds must start with a discussion of the band structure of graphite itself. As one of the crystalline forms of carbon, the basic notions that describe the electronic structure of graphite were developed in the 1930s by *Pauling* [2.64]. Pauling's notion of the directed σ bonds due to *s-p* hybridization of the states of atomic carbon explains both graphite's trigonal geometry and diamond's tetrahedral geometry. For graphite, Pauling's hybrid orbitals are of the type sp^2, forming three equal σ bonds that connect in planes.

Figure 2.1 shows the lattice structure of graphite. It is a layered structure with hexagonal planes of carbon covalently bonded in trigonal geometry having a bond length of 1.42 Å. According to *Pauling* [2.64], there is no covalent bonding between the layers; the interlayer forces are small and the interlayer separation is large (3.35 Å). These weak interlayer forces, however, are sufficiently strong to cause a regular pattern in the interlayer stacking. In Bernal graphite [2.65], this stacking is ABAB, so that the space-group symmetry is D_{6h}^4 (which happens to be the same as the space group for hexagonally close-packed structures). Using the coordinates defined in the figure, the crystal translations can be taken as [2.66]

$$\left.\begin{aligned} \boldsymbol{T}_1 &= a\left(+\frac{1}{2}\hat{x}+\frac{\sqrt{3}}{2}\hat{y}\right) \\ \boldsymbol{T}_2 &= a\left(-\frac{1}{2}\hat{x}+\frac{\sqrt{3}}{2}\hat{y}\right) \end{aligned}\right\} \quad a = 2.4612\ \text{Å}\,,$$

$$\boldsymbol{T}_3 = c\hat{z} \qquad c = 6.7079\ \text{Å}\,. \tag{2.15}$$

There are four atoms in a unit cell of graphite; these are shown in the diagram with vectors pointing from a center of inversion having the following values:

$$\begin{aligned} \boldsymbol{\tau}_{\alpha 1} &= \tfrac{1}{2}\boldsymbol{T}_3, \quad \boldsymbol{\tau}_{\alpha 2} = -\tfrac{1}{2}\boldsymbol{T}_3\,, \\ \boldsymbol{\tau}_{\beta 1} &= \tfrac{1}{3}(\boldsymbol{T}_1+\boldsymbol{T}_2)+\tfrac{1}{2}\boldsymbol{T}_3, \quad \boldsymbol{\tau}_{\beta 2} = -\tfrac{1}{3}(\boldsymbol{T}_1+\boldsymbol{T}_2)-\tfrac{1}{2}\boldsymbol{T}_3\,. \end{aligned} \tag{2.16}$$

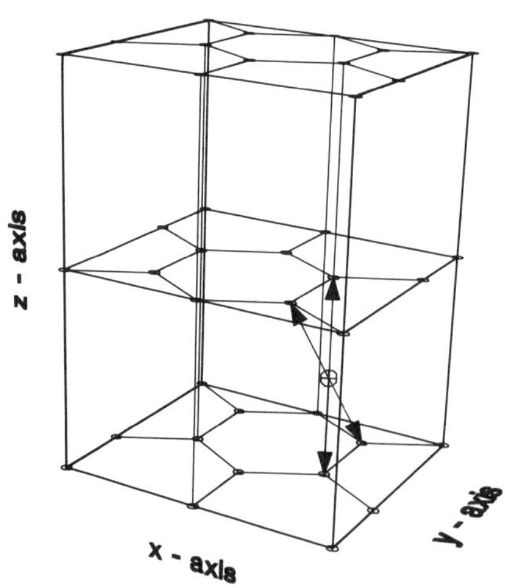

Fig. 2.1. Lattice structure of graphite shown in the coordinate system defined by (2.15, 16). Carbon sites are indicated by open circles within each layer plane. Four atomic basis sites within a unit cell are indicated by arrows pointing from the center of inversion

For a single layer of graphite, the atoms of type α and type β are equivalent. However, for three-dimensional graphite as shown in Fig. 2.1, atoms of type α have vertical neighbors in adjacent layers, while atoms of type β have no vertical neighbors in adjacent layers.

An important consequence of graphite's layered structure is its wide range of electronic change densities. Figure 2.2 shows two contour plots of the electronic charge density of graphite in the same geometry as that of the structural diagram in Fig. 2.1. These plots were constructed using mixed-basis pseudopotential techniques as described by *Holzwarth* et al. [2.59]. The maximum electron density in both planes is 2.14 electrons/Å^3, which occurs along the C–C bonds. The minimum density in the carbon plane is 0.15 electrons/Å^3, which occurs at the centers of the hexagons. On the other hand, the minimum electron density in the perpendicular direction is nearly ten times smaller (0.02 electrons/Å^3). On the basis of this geometry alone, intercalation of various chemicals between graphite layers is quite likely.

For comparison, we make a small digression to the other crystalline form of carbon – diamond. It is formed from Pauling's hybrid orbitals of the type sp^3, resulting in four equal covalent bonds that form a tetrahedral network. The tetrahedral C–C bond length is 1.54 Å, 8.5% larger than that of graphite. Figure 2.3 shows a contour plot of the diamond charge density in a plane containing two of the four carbon bonds. It is amusing to note that the peak density in diamond, which occurs along the covalent bond, is 1.90 electrons/Å^3, slightly less than that of graphite. However, the diamond structure is more than 50% more closely packed than graphite. Its minimum electron density is 0.11 electrons/Å^3, ten times larger than the minimum density in graphite.

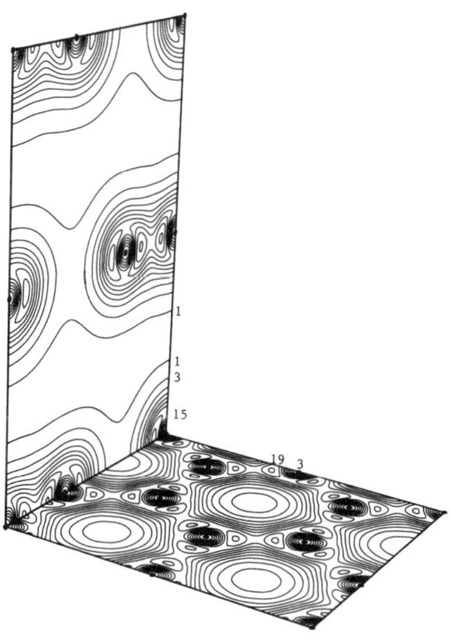

Fig. 2.2. Contour plot of the valence electron density of graphite from a self-consistent pseudopotential calculation shown in a carbon layer plane and in a perpendicular plane along the c axis using the same geometrical perspective as in Fig. 2.1. Carbon sites are indicated by filled circles. Contour labels are given in units of 0.1 electrons/Å^3; levels are spaced at 0.2 electrons/Å^3, starting at the level 0.1 electrons/Å^3. The pseudodensity differs from the actual valence density within a sphere of radius 1.0 Bohr about each carbon site

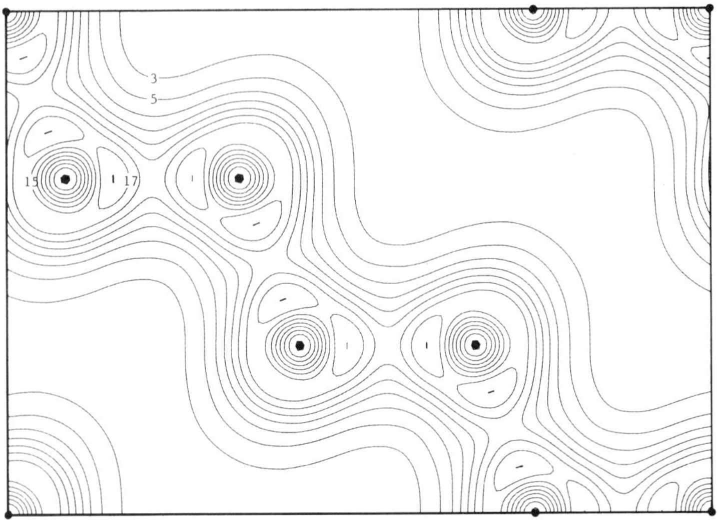

Fig. 2.3. Contour plot of valence electron density of diamond from a self-consistent pseudopotential calculation shown in a plane containing two carbon bonds. Carbon sites are indicated by filled circles. Contour labels are given in units of 0.1 electrons/Å; levels are spaced at 0.2 electrons/Å^3, starting at the level 0.1 electrons/Å^3

The relationship between the two structural forms of crystalline carbon is very interesting in itself. For example, *Fahy* et al. [2.67] have recently studied a possible mechanism for transforming graphite into diamond, involving an energy barrier of 0.33 eV; the energy of the ground-state graphite is only approximately 0.02 eV per carbon atom lower than that of diamond.

To discuss the band structure of graphite, we must define the Brillouin zone. For the unit cell defined in (2.15), the reciprocal-lattice vectors are given by

$$\begin{aligned} \boldsymbol{G}_1 &= \frac{4\pi}{\sqrt{3}a}\left(+\frac{\sqrt{3}}{2}\hat{\boldsymbol{x}} + \frac{1}{2}\hat{\boldsymbol{y}}\right), \\ \boldsymbol{G}_2 &= \frac{4\pi}{\sqrt{3}a}\left(-\frac{\sqrt{3}}{2}\hat{\boldsymbol{x}} + \frac{1}{2}\hat{\boldsymbol{y}}\right), \\ \boldsymbol{G}_3 &= \frac{2\pi}{c}\hat{\boldsymbol{z}}. \end{aligned} \tag{2.17}$$

Figure 2.4 shows the corresponding Brillouin zone together with the standard symmetry labels.

2.2.2 Electronic Band Structure

The electronic band structure of graphite has been extensively studied using a variety of calculational techniques mentioned in Sect. 2.1. Table 2.1 compiles a

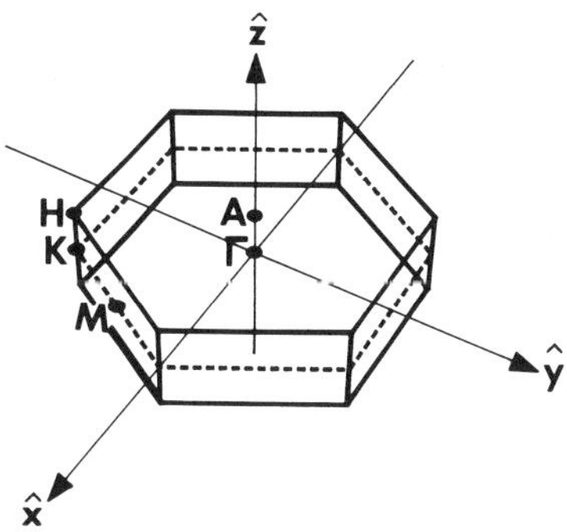

Fig. 2.4. Brillouin zone for graphite shown in the coordinate system defined by (2.17) and indicating standard symmetry labels

Table 2.1. Electronic band structure calculations for graphite. "2D" indicates calculation for a single layer. Underlined energies were quoted in the reference; nonunderlined energies were estimated from published band diagrams

Year	Reference	Method	Widths of occupied bands			Interlayer interaction[a]	Interlayer band[b]
			$sp^2\ \sigma$	π	total		
1955	[2.32]	tight binding, 2D	7.0 eV	5.6 eV	12.6 eV	– eV	– eV
1967	[2.33]	tight binding, 2D	11.4	4.1	14.1	–	–
1969	[2.56]	pseudopotential	–	9.8	–	1.7	1.8
1970	[2.41]	LCAO, 2D	14.4	7.3	19.5	–	–
1972	[2.34]	tight binding, 2D	13.6	5.3	15.9	–	–
1974	[2.35]	extended Hückel, 2D	22.0	6.8	25.8	–	–
1974	[2.42]	LCAO	16.0	8.0	20.7	1.1	?
1976	[2.48]	LCAO and OPW	16.6	8.8	23.3	1.3	4.2
1978	[2.36]	extend Hückel, 2D	17.0	5.6	21.2	–	–
1980	[2.19]	Hartree-Fock, 2D[c]	23.4	14.6	30.5	–	–
1980	[2.40]	LCAO	14.9	7.9	21.5	0.8	?
1981	[2.31]	cellular method	17.0	8.1	18.0	1.4	4.3
1982	[2.29]	corrected KKR	14.8	8.2	19.5	1.0	7.1
1982	[2.59]	mixed-basis pseudopotentials[c]	17.5	9.1	20.8	1.5	3.7
1986	[2.43]	LCAO[c]	15.3	8.5	20.5	1.0	6.7
1987	[2.54]	FLAPW[c]	15.0	8.7	19.6	?	3.8
1989	[2.47]	LCAO[c]	15.4	8.1	19.7	1.1	6.0

[a] Splitting of π bands at K point near Fermi level ($\varepsilon(K_4) - \varepsilon(K_2)$).
[b] "Interlayer" band minimum measured above E_F.
[c] Self-consistent calculation.

list of some of these calculations. For each calculation, the table lists the bandwidths of the occupied valence bands (σ, π, and total), the splitting of the π bands at the K point near the Fermi level as a measure of the interaction between layers, and the distance between the Fermi level and the lowest unoccupied σ bands, dubbed the "interlayer band" by *Posternak* et al. [2.53]. The table includes only calculations that presented the entire valence band. Early studies [2.68–72] considered only the π bands. The table also omits several calculations [2.67, 73, 74] that considered graphite in modified crystalline forms. In the 1980s computers became sufficiently powerful and computer codes more efficient, so calculations could be performed self-consistently as intended by the original Hartree-Fock and density functional theories. Even with self-consistency, however, discrepancies remain among the various calculations done with different calculational methods, with different choices of the exchange-correlation functional in local density theory [2.47], and with different k-space sampling [2.54] for calculating the self-consistent density according to (2.3). The occupied bandwidths vary by 2 eV; the position of the interlayer band varies considerably more in these calculations [2.47]. The Hartree-Fock calculation stands out with much larger bandwidths than any of the density functional theory results.

Despite the discrepancies, many of the results listed in Table 2.1 do relate to experimentally observed physical properties of graphite. For example, the calculations of *Bassani* and *Pastori Parravicini* [2.33], *Willis* et al. [2.42], and *Johnson* and *Dresselhaus* [2.72] were helpful in interpreting spectroscopic data. The calculations of *Nagayoshi* et al. [2.48] and *Tatar* and *Rabii* [2.29] were consistent with the empirical Fermi-surface parameters [2.71, 75] of graphite. The calculations of *Holzwarth* et al. [2.59] determined valence electronic charge density contours very similar to those derived from X-ray form factors. *Jansen* and *Freeman* [2.54] calculated elastic constants of graphite in good agreement with experimental values.

Another area of success is the recent explanation of the scanning tunneling micrograph (STM) results for graphite by *Tománek* and *Louie* [2.76]. Since well-ordered surfaces of graphite are easily prepared, they are often used to demonstrate and calibrate STM equipment. Somewhat surprising is that the STM picture of a graphite surface under low bias voltage shows only every other graphite atom in the hexagonal layer, revealing a triangular lattice instead of a hexagonal one. The basic idea of *Tománek* and *Louie* [2.76] comes from the fact that the STM image is formed primarily from electrons near the Fermi level of the material which, for graphite, corresponds to a very small volume of k-space near the K-H edge of the Brillouin zone. As we shall discuss in more detail below, for a single layer of graphite, the K-point eigenstates are doubly degenerate and are concentrated equally on both the α- and β-type atoms of the unit cell. However, in three-dimensional graphite as shown in Fig. 2.1, the α- and β-type atoms are no longer equivalent. The α-type atoms have closer neighbors in adjacent planes and form two bands with appreciable c-axis dispersion. The β-type atoms have farther neighbors in adjacent layers and form two bands with very small c-axis dispersion and with band edges closer to the Fermi level. Consequently, there is a significantly higher contribution of states near the Fermi level from β-type atoms than from α-type atoms. This basic structure is preserved at the surface layer of graphite and results in the tunneling probe

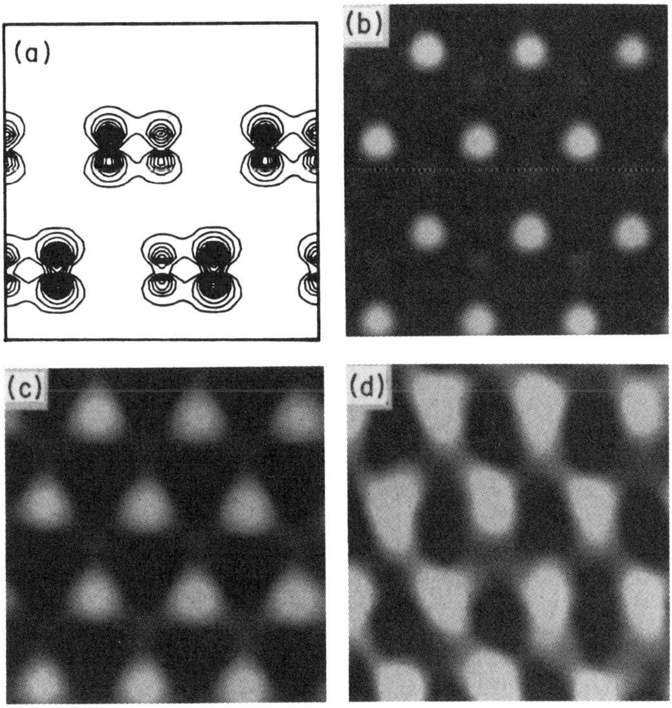

Fig. 2.5. Comparison of theoretical STM charge density to experimental tunneling current at the surface of Bernal graphite for a bias voltage of 0.25 V [2.76]: (**a**) contour plot of theoretical STM charge density in a plane perpendicular to the surface, (**b**) gray-scale plot (white corresponding to large current densities) of theoretical STM charge density in a plane parallel to the surface at a distance of 1 Å from the topmost layer, (**c**) theoretical results of plot (**b**) modified using a Gaussian function to filter out high-frequency components, (**d**) experimental STM image

detecting appreciably more current near a β-type site than an α-type site. Figure 2.5 shows the simulation of this effect. Figure 2.5a shows a contour plot of the calculated charge density, which contributes to the tunneling current in the plane perpendicular to a surface. This plot reveals the π-like nature of the tunneling current and the asymmetry in the contributions from α and β sites. The gray-scale plots in Figs. 2.5b, c are very close to the experimental results shown in Fig. 2.5d.

Figure 2.6 shows a typical band-structure diagram for Bernal graphite and the corresponding density of states. These results were generated using the method described by *Holzwarth* et al. [2.59]. It is interesting to note that Pauling's sp^2 hybrid σ orbitals, which correspond to the occupied σ bands, actually span an energy range of 15–18 eV. Each carbon atom has one remaining valence electron, which is accommodated in the lower two π bands indicated in the diagram with a dotted line. The π bands are degenerate near the Fermi

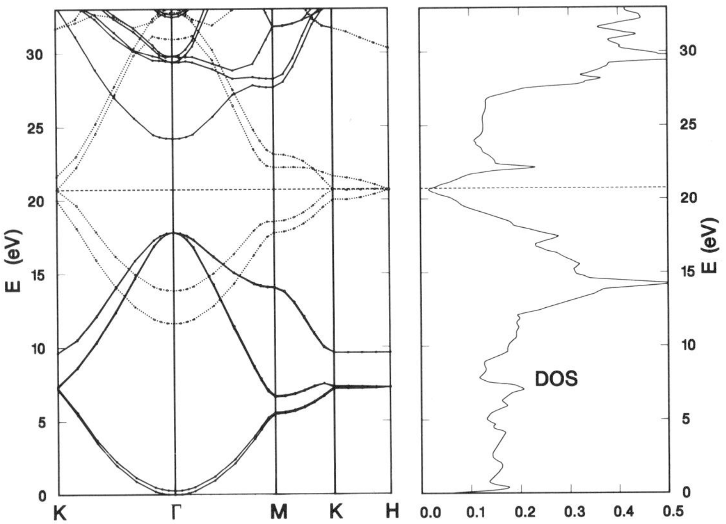

Fig. 2.6. Band structure (left panel) and corresponding density of states (DOS) (right panel) for Bernal graphite from a mixed-basis pseudopotential calculation [2.59]. In the band-structure plot, the Brillouin zone labels are those shown in Fig. 2.4, and the zero of energy is taken at the bottom of the lowest valence band. The dotted curves indicate π bands; full curves indicate σ bands. The Fermi level is indicated in both plots with a dashed line. The units of the DOS are states per eV per C atom

level, causing the famous minimum in the density of states at the Fermi level. The bandwidth for the occupied π bands is 8–9 eV; the total bandwidth of the occupied states is 20–21 eV. While the electronic structure of graphite is dominated by interactions within a layer, the interactions between layers are not negligible, having a magnitude of 1–2 eV for the π bands near the Fermi level. It is amusing to compare this situation with that of diamond shown in Fig. 2.7, generated using the same computational techniques. In the case of diamond, the occupied states can all be described by σ bonds between Pauling's sp^3 hybrid orbitals, resulting in a valence bandwidth of 23.0 eV.

Two features of the electronic band structure of graphite that are important for intercalation compounds are the unoccupied "interlayer" band and the π bands near the Fermi level. The former is important because it is spatially and energetically close to donor intercalant states; the latter is involved with charge transfer. Several recent papers [2.43, 47, 53, 77, 78] have discussed the role of the interlayer states in intercalation compounds. Figure 2.8 shows a contour plot of the wavefunction of the lowest-energy interlayer state at the Γ point. This contour plot corroborates the results of other workers [2.43, 47, 53, 78], showing the peak amplitude to be between carbon planes. An LCAO analysis of this state [2.47] shows that it has a large contribution from carbon 3s orbitals, as represented in the pseudo-wave-function plot of Fig. 2.8 by a sign change in the amplitude near the carbon planes.

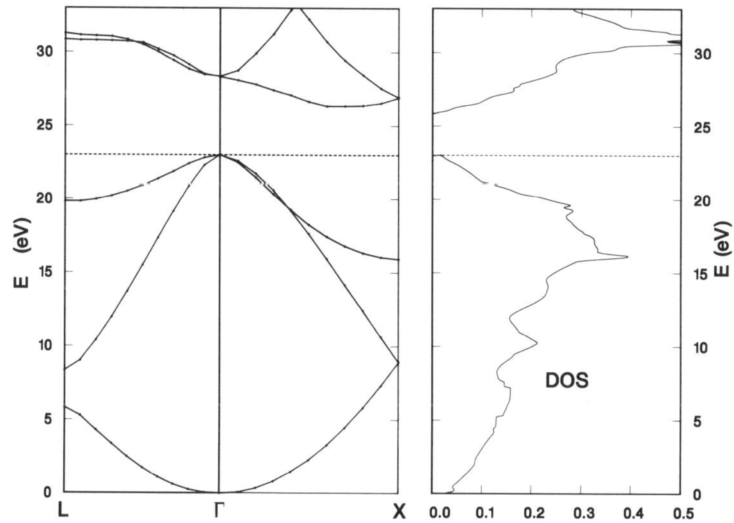

Fig. 2.7. Band structure (left panel) and corresponding density of states (DOS) (right panel) for diamond from a mixed-basis pseudopotential calculation [2.59]. In the band-structure plot, the Brillouin zone labels are the standard labels for an fcc lattice, and the zero of energy is taken at the bottom of the lowest valence band. The full curves indicate σ bands; there are no π bands for this structure. The Fermi level (at the top of the valence band) is indicated in both plots with a dashed line. The units of the DOS are states per eV per C atom

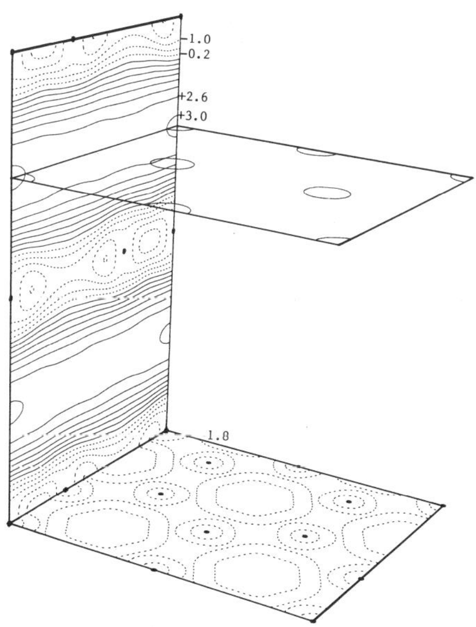

Fig. 2.8. Contour plot of the wavefunction of the lowest-energy interlayer state of graphite at the Γ point shown in the same geometry as in Fig. 2.2, but with a contour plot added in a plane midway between two carbon planes. Carbon sites are indicated by filled circles. Negative contours are indicated with a dotted line; positive contours are indicated with a full line. The wavefunction, from a self-consistent pseudopotential calculation, is normalized so that the integral of its squared magnitude is equal to the volume of the unit cell. In these units, contours are given in intervals of 0.4, starting at the level -1.8

While the interlayer states are thought to be strongly affected by intercalation, the graphite π bands near the Fermi level are thought to remain fairly rigid in the intercalation process. The cause of this rigid behavior is the antibonding form of these states and their resulting spatial distribution near the intercalant, as will be discussed in more detail below. In anticipation of this discussion, Figure 2.9b plots one of these wavefunctions – the π state at the M point of graphite, 2.4 eV above the Fermi level. Unfortunately, this wavefunction is complex so that its phase variations are not easily illustrated. Therefore, we have plotted its squared magnitude (normalized to the unit cell volume) and have shown it, for comparison, next to the corresponding plot for the bonding state at the bottom of the π band. The bonding π band in Fig. 2.9a shows a relatively small volume, located near the graphite planes, having a density smaller than 0.5; the plane midway between carbon planes has a nearly uniform density between 0.4 and 0.7. By contrast, the antibonding π band in Fig. 2.9b shows a substantial volume having a density between 0 and 0.4; the plane midway between carbon planes has density mostly between 0 and 0.4, with small regions of density between 0.4 and 0.5 located above α-type carbon sites.

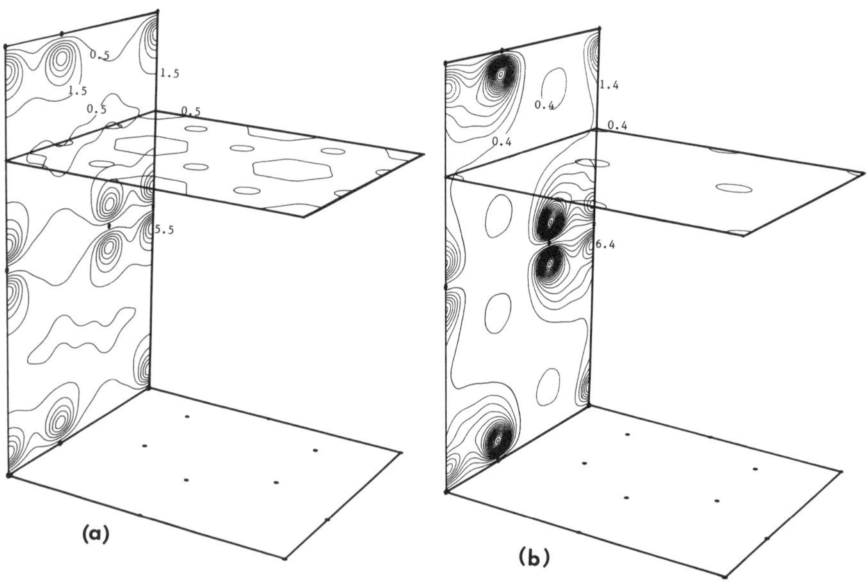

Fig. 2.9. Contour plots of squared magnitudes of wavefunctions for π bands of graphite from self-consistent pseudopotential calculations shown in same geometry as in Fig. 2.8. Carbon sites are indicated by filled circles. Squared magnitudes of wavefunctions are normalized so that their integrals are equal to the volume of the unit cell. **(a)** Lowest π state at the Γ point. Contours are given in intervals of 1.0, starting at the level 0.5. **(b)** π state located 2.4 eV above the Fermi level at the M point. Contours are given in intervals of 1.0, starting at the level 0.4

2.3 Electronic Band Structures of Low-Stage Graphite Intercalation Compounds

2.3.1 General Features

As we will discuss below, there have been a number of first-principles calculations of the electronic structure of low-stage donor graphite intercalation compounds using the techniques described in Sect. 2.1. Many of these results have been successful in explaining experimentally observed properties of graphite intercalation compounds. Before discussing the results themselves, we consider a general framework for discussing their basic features. The following questions have been studied in the literature:

(1) How much charge is transferred between the intercalant and graphite?

(2) How are the graphite and intercalant bands altered by intercalation, or how much hybridization occurs between the graphite and intercalant bands?

(3) What is the role of the "interlayer" band in graphite intercalation compounds?

In fact, these questions are difficult to address quantitatively because they imply the notion of a reference graphite system and a reference intercalant system, neither of which exist in nature. The intercalation process introduces both geometric and electronic changes in the graphite lattice. In most electronic structure studies, the geometric effects have been treated empirically, while attention has been focused on the electronic effects. A reasonable choice for the reference systems that incorporates the geometric effects could be based on the notion that the structure of an intercalation compound can be described as the sum of two sublattices, one lattice including the graphite layers and the other containing the intercalant layers. The two sublattices would then form the reference systems.

The reference "graphite" system defined in this way would thus differ from the lattice of Bernal graphite because of an increase in the spacing between layers, a change in the stacking of the layers, and a small dilation of the layers themselves. The Hamiltonian for the compound then could be formulated according to the sum of contributions from the graphite lattice, the intercalant lattice, and an interaction term that arises from the Hartree and exchange-correlation potential of the charge redistributed between the two systems:

$$\mathscr{H} = \mathscr{H}^0_{\text{graphite}} + \mathscr{H}^0_{\text{intercalant}} + \mathscr{H}_{\text{interaction}} \,. \tag{2.18}$$

Having defined the reference Hamiltonian, we can also define the one-electron eigenstates of the reference systems:

$$\begin{aligned} &\mathscr{H}^0_{\text{graphite}}\,\Psi^0_{\text{graphite},\,m}(\boldsymbol{k},\boldsymbol{r}) = \varepsilon^0_{\text{graphite},\,m}(\boldsymbol{k})\,\Psi^0_{\text{graphite},\,m}(\boldsymbol{k},\boldsymbol{r})\,,\\ &\mathscr{H}^0_{\text{intercalant}}\,\Psi^0_{\text{intercalant},\,n}(\boldsymbol{k},\boldsymbol{r}) = \varepsilon^0_{\text{intercalant},\,n}(\boldsymbol{k})\,\Psi^0_{\text{intercalant},\,n}(\boldsymbol{k},\boldsymbol{r})\,. \end{aligned} \tag{2.19}$$

The band indices m and n refer to the distinct bands of each reference system. The wave vector $\boldsymbol{k}$ is defined within the common Brillouin zone of the intercalation compounds and the reference systems. Unfortunately, the reference eigenstates $\{\Psi^0_{\text{graphite},m}(\boldsymbol{k},\boldsymbol{r}), \Psi^0_{\text{intercalant},n}(\boldsymbol{k},\boldsymbol{r})\}$ do not form a complete orthogonal basis set upon which to express the states of the intercalant system; in fact, they form an *overcomplete* basis set. This causes a small complication in the analysis, mainly in considering question (3). While this reference system framework has not been overtly used to a great extent in the literature, it does provide a well-defined background for discussing the three questions above.

The notion of charge transfer between the graphite and intercalant systems involves two considerations: identification of the bands and occupancy of these bands in the intercalation compound. The question of band identification assumes that for some bands in the intercalation compound, there is a clear correspondence with the states of the reference systems:

$$\begin{aligned} \Psi_m(\boldsymbol{k},\boldsymbol{r}) &\approx \Psi^0_{\text{graphite},m}(\boldsymbol{k},\boldsymbol{r}) \, , \\ \Psi_n(\boldsymbol{k},\boldsymbol{r}) &\approx \Psi^0_{\text{intercalant},n}(\boldsymbol{k},\boldsymbol{r}) \, , \end{aligned} \tag{2.20}$$

The assumption is that these special states remain fairly rigid upon intercalation, although their occupancies change in the charge transfer process.

As we discussed in Sect. 2.2.2, the rigidity of the graphite π bands near the Fermi level occurs as a result of their antibonding nature. The antibonding nature can be described by the following simple analysis, based on idealizing $\mathscr{H}^0_{\text{graphite}}$ to the Hamiltonian of a *single* carbon layer, or equivalently, to that of a lattice of *noninteracting* carbon layers. The main features of the electronic structure of the π bands of two-dimensional graphite can be described by the following simple tight-binding picture. The translation vectors $\boldsymbol{T}_1$ and $\boldsymbol{T}_2$ and reciprocal-lattice vectors $\boldsymbol{G}_1$ and $\boldsymbol{G}_2$ defined above apply to two-dimensional graphite. There are two basis atoms per unit cell, which we can take as $\boldsymbol{\tau}_{\alpha 1}$ and $\boldsymbol{\tau}_{\beta 1}$. (Since we are discussing a single layer, we will drop the index "1" in the following discussion.) Taking LCAO functions based on a single π orbital on each carbon site, the Hamiltonian matrix takes the form:

$$\mathscr{H}^0_{\text{graphite}} = \begin{bmatrix} H_{\alpha\alpha} & H_{\alpha\beta} \\ H_{\beta\alpha} & H_{\beta\beta} \end{bmatrix} . \tag{2.21}$$

If we analyze the matrix element using the simplest tight-binding form – that is, assuming that the basis functions are orthonormal and include only nearest-neighbor terms in the Hamiltonian – then the matrix elements are given by the expressions [2.72, 79]

$$\begin{aligned} H_{\alpha\alpha} &= H_{\beta\beta} \equiv \mathscr{E}_0 \, , \\ H_{\alpha\beta} &= H^*_{\beta\alpha} = \mathscr{E}_1 \left(\mathrm{e}^{\mathrm{i}k_y a/\sqrt{3}} + 2\mathrm{e}^{-\mathrm{i}k_y a/2\sqrt{3}} \cos(k_x a/2)\right) . \end{aligned} \tag{2.22}$$

The eigenvalues of this sample model are

$$\varepsilon^0_{\text{graphite},\pm}(k_x, k_y) = \mathscr{E}_0 \pm \mathscr{E}_1 \sqrt{1 + 4\cos^2(k_x a/2) + 4\cos(\sqrt{3}\,k_y a/2)\cos(k_x a/2)}\,, \tag{2.23}$$

with corresponding ratios of eigenstate coefficients

$$\frac{C^{\alpha}_{\pm}}{C^{\beta}_{\pm}} = \pm \frac{(e^{ik_y a/\sqrt{3}} + 2e^{-ik_y a/2\sqrt{3}}\cos(k_x a/2))}{\sqrt{1 + 4\cos^2(k_x a/2) + 4\cos(\sqrt{3}\,k_y a/2)\cos(k_x a/2)}}\,. \tag{2.24}$$

In this simple framework, $\mathscr{E}_0$ corresponds to the Fermi level of two-dimensional graphite and can be taken to be 0; $\mathscr{E}_1$ should take a value of approximately -3 eV to be consistent with the full band structure results shown in Fig. 2.6. The bands $\varepsilon^0_{\text{graphite},\pm}(k_x, k_y)$ are shown in the Brillouin zone of two-dimensional graphite in Fig. 2.10, where the lower bands corresponds to the + eigenstate and the upper band corresponds to the $-$ eigenstate. From the eigenstate coefficients (2.24), it is evident that the lower band corresponds to the adjacent carbon sites α and β having the same phase (bonding configuration), while the upper band corresponds to sites having opposite phase (antibonding configuration). Actually, the phases vary with wave vector $\boldsymbol{k}$; the maximum phase variation of ± 1 occurs at the Γ point $\{k_x = 0, k_y = 0\}$. At the K point $\{k_x = 4\pi/3a, k_y = 0\}$, the eigenstates are degenerate and have amplitude on either the α or the β site. These are indicated schematically in the band diagram of Fig. 2.10. These basic bonding and antibonding structures are preserved in more realistic models of the graphite π bands, as shown in the contour diagrams for three-dimensional graphite in Fig. 2.9.

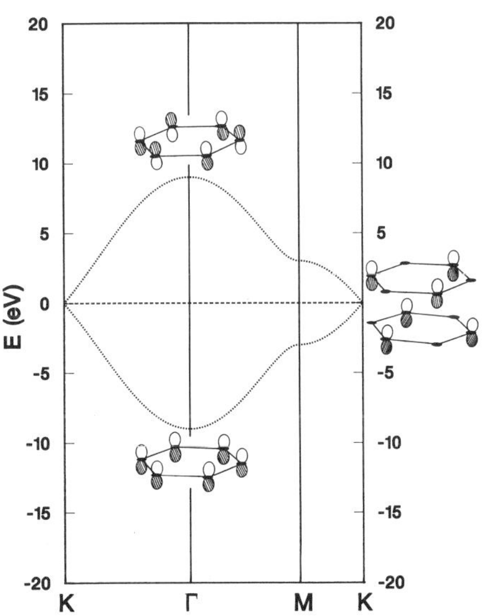

Fig. 2.10. Band-structure diagram for the simple two-dimensional model of the graphite π bands as represented by the function $\varepsilon^0_{\text{graphite},\pm}(k_x, k_y)$ given by (2.23). Corresponding eigenstates are shown schematically at the Γ and K points with π orbitals on each carbon site. Blank lobes indicate positive phases, and shaded lobes indicate negative phases

Since the structure of the graphite π bands near the K point is so important for understanding the Fermi-surface properties of graphite, it has been studied [2.71, 75] in great detail. In particular, if one analyzes the dispersion of these bands in the vicinity of the K point, as defined by $\boldsymbol{k} = \boldsymbol{k}_K + \boldsymbol{\kappa}$, a linear dispersion relation of the form

$$\varepsilon^0_{\text{graphite},\,\pm}(\boldsymbol{k}_K + \boldsymbol{\kappa}) = \mathscr{E}_0 \pm \mathscr{E}'_1 k \tag{2.25}$$

is obtained for small $\boldsymbol{k}$.

Deviations from the rigid-band behavior (2.20) depend on the magnitude of matrix elements between the two reference systems:

$$V_{mn} \equiv \int \Psi^{0*}_{\text{graphite},\,m}(\boldsymbol{k}, \boldsymbol{r}) \mathscr{H} \Psi^0_{\text{intercalant},\,n}(\boldsymbol{k}, \boldsymbol{r})\, \mathrm{d}^3 r\,. \tag{2.26}$$

In general, V_{mn} will not be small. However, in discussing the rigid-band model we consider $\Psi^0_{\text{graphite},\,m}(\boldsymbol{k}, \boldsymbol{r})$ to correspond to a graphite π state near the Fermi level, denoted by $\Psi^0_{\text{graphite, F}}(\boldsymbol{k}, \boldsymbol{r})$, and $\Psi^0_{\text{intercalant},\,n}(\boldsymbol{k}, \boldsymbol{r})$ to correspond to the lowest-energy intercalant band, denoted by $\Psi^0_{\text{intercalant},\,0}(\boldsymbol{k}, \boldsymbol{r})$.

For many systems, the intercalant wave function $\Psi^0_{\text{intercalant},\,0}(\boldsymbol{k}, \boldsymbol{r})$ is a relatively smooth function, concentrated in the intercalant layer. Figure 2.9b showed that for a representative wave function, the amplitude of $\Psi^0_{\text{graphite, F}}(\boldsymbol{k}, \boldsymbol{r})$ in the integration region of (2.26) is quite small. Above we also showed that $\Psi^0_{\text{graphite, F}}(\boldsymbol{k}, \boldsymbol{r})$ has oscillatory behavior in this spatial region. Both of these behaviors indicate that matrix elements (2.26) are likely to have small values for these states. This is a qualitative argument and not a rigorous symmetry analysis, so in general there will be a small hybridization of the graphite Fermi-level states with the intercalant states. This same analysis can be used to conclude that the graphite states at the *bottom* of the π band *do* have significant hybridization matrix elements (2.26) with $\Psi^0_{\text{intercalant},\,0}(\boldsymbol{k}, \boldsymbol{r})$.

Once the relationship between intercalant and reference states (2.20) has been established, charge transfer between the graphite and intercalant states can be described in terms of the relative alignment of their energy levels. Corresponding to (2.20), the energy bands of the compound and reference systems have a zero-order relationship of the form

$$\begin{aligned} \varepsilon_m(\boldsymbol{k}) &\approx \varepsilon^0_{\text{graphite},\,m}(\boldsymbol{k})\,, \\ \varepsilon_n(\boldsymbol{k}) &\approx \varepsilon^0_{\text{intercalant},\,n}(\boldsymbol{k})\,, \end{aligned} \tag{2.27}$$

where in reference to the rigid-band model, we have in mind that $m \equiv \mathrm{F}$ for the graphite system and $n \equiv 0$ for the intercalant system. This relationship presupposes that both reference systems have a common reference energy, which in general is a nontrivial problem. Assuming that this reference energy can be found and that the Fermi level E_F of the intercalation compound can be determined within the same energy scale, the charge transfer relations are as follows. For donor compounds (2.27) implies that charge is transferred from the intercalant band to the graphite bands if, for the intercalant valence states

$\varepsilon^0_{\text{intercalant, 0}}(\boldsymbol{k})$,

$$E_F < \varepsilon^0_{\text{intercalant, 0}}(\boldsymbol{k}) \,. \tag{2.28}$$

Similarly, for acceptor compounds, charge is transferred to the intercalant band from the graphite bands if, for the intercalant valance states $\varepsilon^0_{\text{intercalant, 0}}(\boldsymbol{k})$,

$$E_F > \varepsilon^0_{\text{intercalant, 0}}(\boldsymbol{k}) \,. \tag{2.29}$$

In this framework, the charge transfer would be complete for either of these two cases. Partial charge transfer would occur if for some of the intercalant valence states, $\varepsilon^0_{\text{intercalant, 0}}(\boldsymbol{k})$ crosses the Fermi level of the compound. This measure of charge transfer can be made quantitative by evaluating the Fermi-surface volumes of the intercalation compound compared with those of the reference systems. This measure of charge transfer, based on the Fermi-surface properties of the intercalation compound, ignores any charge redistribution associated with hybridization of the reference bands. However, as long as the correspondence of the compound and reference states of (2.20) are well-defined, the Fermi-surface charge transfer can be uniquely determined and related to the Fermi-surface properties of the intercalation compound.

Another approach to charge transfer involves the total charge transfer throughout the spectrum of states, not only at the Fermi-surface, as is important, for example, in the analysis of core-level spectroscopies. An example is *Mulliken* population analysis [2.80], which is easily implemented in LCAO treatments. Another example is a geometrical partitioning of charge or a decomposition in terms of overlapping atomic density functions [2.44]. Although none of these methods – Mulliken population analysis, geometric partitioning, or overlapping atom decompositions – gives unique numerical values, they all give helpful qualitative information about the total charge transfer. We expect the total charge transfer to be considerably different from the Fermi-surface charge transfer discussed above.

The reference graphite and intercalant systems can also be used to analyze the role of the unoccupied interlayer state of graphite in intercalation compounds. In fact, the identification of the intercalant band $\varepsilon^0_{\text{intercalant, 0}}(\boldsymbol{k})$ within the intercalation compound is complicated by the presence of the graphite interlayer band, as has been discussed by several authors. [2.43, 47, 53, 77, 78]. In this case, the relevant reference states can be denoted $\Psi^0_{\text{graphite}, I}(\boldsymbol{k}, \boldsymbol{r})$ for the graphite interlayer state and $\varepsilon^0_{\text{intercalant, 0}}(\boldsymbol{k}, \boldsymbol{r})$ for the intercalant state. To evaluate the interaction of these two states, we must consider both the Hamiltonian matrix element (2.26) and the spatial overlap matrix element:

$$S_{mn} = \int \Psi^{0*}_{\text{graphite}, m}(\boldsymbol{k}, \boldsymbol{r})\, \Psi^0_{\text{intercalant}, n}(\boldsymbol{k}, \boldsymbol{r})\, d^3r \,, \tag{2.30}$$

where the index $m \equiv I$, referring to the interlayer band, and the index $n \equiv 0$, referring to the intercalant band. Because of the spatial distribution of these two states, which is highly concentrated near the intercalant layers, we expect *both* the interaction matrix element V_{I0} and the overlap matrix element S_{I0} to be substantial. For donor compounds we also expect the reference energies

$\varepsilon^0_{\text{graphite},\,1}(\boldsymbol{k})$ and $\varepsilon^0_{\text{intercalant},\,0}(\boldsymbol{k})$ to be close (within a few eV of each other). If we ignore all the other states of the system and construct the simplest possible model for this two-state interaction, the following 2×2 determinantal equation results:

$$\begin{vmatrix} H_{11} - \varepsilon_I & V_{10} - \varepsilon_I S_{10} \\ V_{01} - \varepsilon_I S_{01} & H_{00} - \varepsilon_I \end{vmatrix} = 0 \,. \tag{2.31}$$

The eigenvalues ε_I of the intercalation compound resulting from this model exhibit a very interesting structure [2.81]. Dropping indices from the interaction and overlap matrices, and approximating $H_{11} \approx \varepsilon^0_{\text{graphite},\,1}(\boldsymbol{k}) \equiv \varepsilon^0_{\text{graphite}}$ and $H_{00} \approx \varepsilon^0_{\text{intercalant},\,0}(\boldsymbol{k}) \equiv \varepsilon^0_{\text{intercalant}}$, the intercalant-interlayer eigenvalues can be written

$$\varepsilon_{I\pm} = \frac{\varepsilon^0_{\text{graphite}} + \varepsilon^0_{\text{intercalant}} + 2SV}{2(1 - S^2)} \pm \sqrt{\left(\frac{\varepsilon^0_{\text{graphite}} + \varepsilon^0_{\text{intercalant}} + 2SV}{2(1 - S^2)}\right)^2 + \frac{V^2 - \varepsilon^0_{\text{graphite}}\varepsilon^0_{\text{intercalant}}}{1 - S^2}} \,. \tag{2.32}$$

This result demonstrates that the overlap matrix element S is important in the structure of the intercalant-interlayer eigenstates of the intercalation compound. Figure 2.11 compares the eigenvalues of the system for typical values of the parameters when the overlap S is 0 when it has a value of 0.6. In the former case, the intercalant eigenstates $\varepsilon_{I\pm}$ are symmetrically split about the mean energy of the two reference levels. In the latter case, the lower eigenvalue ε_{I-} is close to the

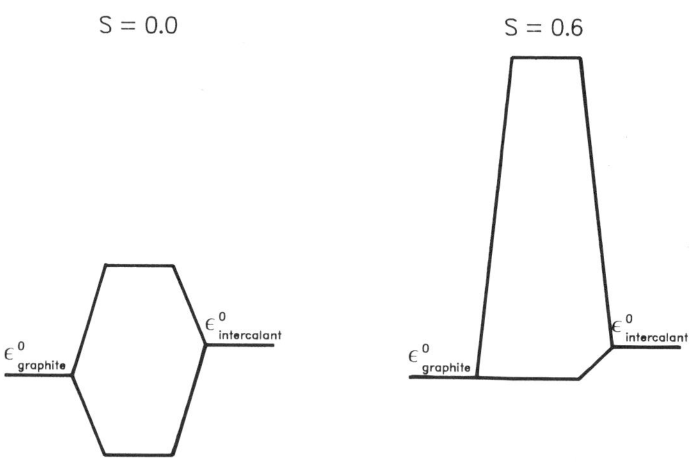

Fig. 2.11. Energy-level diagram for interaction between the interlayer graphite state and the intercalant state as modeled by (2.32). Parameters were chosen to be $\varepsilon^0_{\text{graphite}} = 1$ eV, $\varepsilon^0_{\text{intercalant}} = 2$ eV, and $V = 3$ eV. The effects of the overlap parameter $S = 0.6$ (right panel) are compared with the zero overlap case (left panel)

two reference levels, while the upper eigenvalue ε_{I+} is raised to a much higher energy. This basic structure of ε_{I-} and ε_{I+} is preserved for other reasonable approximations to the diagonal matrix elements H_{11} and H_{00}. From several of the electronic band structure results for donor graphite intercalation compounds, we can quite reasonably make the identification $\varepsilon_I \approx \varepsilon^0_{\text{intercalant},\,0}(\boldsymbol{k}) \approx \varepsilon^0_{\text{graphite},\,1}(\boldsymbol{k})$. Whether we should think of this state as $\Psi^0_{\text{intercalant},\,0}(\boldsymbol{k},\boldsymbol{r})$ or as $\Psi^0_{\text{graphite},\,1}(\boldsymbol{k},\boldsymbol{r})$ is a difficult question, since both states occupy substantially the same region of space. For simplicity, in the discussion below we will refer to this state as the intercalant-interlayer state.

To take advantage of the periodic structures of the intercalation compounds, relative to the structure of graphite, we must consider the changes in the unit cell and Brillouin zone upon intercalation. The two most common stoichiometries

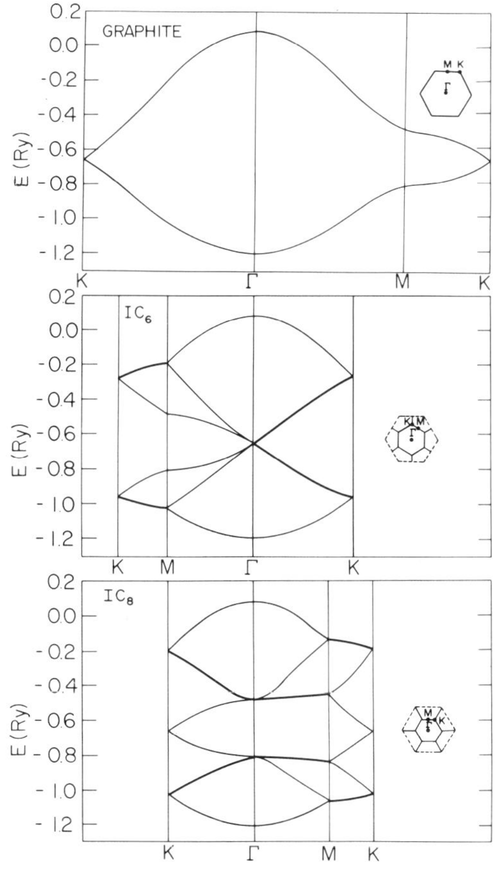

Fig. 2.12. The π bands of two-dimensional graphite, adopted from the calculation of *Painter* and *Ellis* [2.41] (reproduced from [2.27]), shown in the two-dimensional Brillouin zone of graphite (top), IC_6 (middle), and IC_8 (bottom), where I stands for intercalant. Degenerate bands are indicated with thickened lines. Insets show the relationships of two-dimensional Brillouin zones to those of graphite

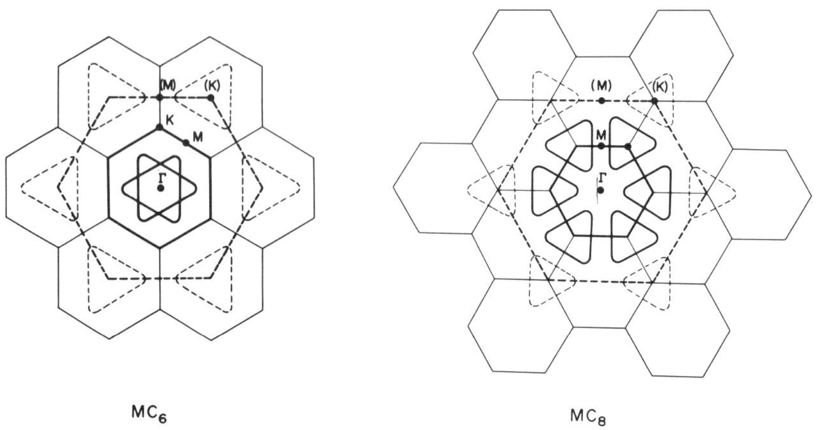

Fig. 2.13. Two-dimensional cross sections of the Fermi surfaces of graphite intercalation compounds MC_6 (left) and MC_8 (right), where M stands for intercalant as determined from the two-dimensional rigid-band model assuming full charge transfer. Dashed lines denote the Brillouin zones and Fermi surfaces in the two-dimensional graphite structures [2.27]

for the stage-1 donor intercalation compounds are MC_6 and MC_8, where M stands for an intercalant atom. Ignoring the layer stacking, we can focus on the changes to the two-dimensional Brillouin zones. The MC_6 structure has a $\sqrt{3} \times \sqrt{3}$ $R30°$ superlattice, while the MC_8 structure has a 2×2 superlattice [2.82]. To identify the graphite bands in the new Brillouin zones, we must consider "zone folding". This is shown in Fig. 2.12 for the π bands of two-dimensional graphite. In the rigid-band model, it is the graphite π bands near the Fermi level that are important in the intercalation compounds, specifically those near the M–K region of the graphite Brillouin zone. For the MC_6 compound, this feature is "folded" to the M–Γ region of the superlattice Brillouin zone. For the MC_8 compound, the feature is folded to the Γ–K region of the superlattice Brillouin zone. Furthermore, within the rigid-band model, the occupancies of these π bands change upon intercalation. The Fermi surface of two-dimensional graphite, which is a point at the K point of the Brillouin zone, expands to accommodate the transferred charge. For substantial charge transfer within the rigid-band approximation, the cross sections of the Fermi surfaces of the intercalation compounds take the shape of rounded triangles centered at the K point in the graphite Brillouin zone, as shown in Fig. 2.13. Also shown in this figure are the cross sections of the Fermi surfaces corresponding to the MC_6 and MC_8 superlattice Brillouin zones. The MC_6 structure has two interpenetrating Fermi surfaces centered at the zone center; for the MC_8 structure, the Fermi surfaces are centered at the K point corners of the intercalant Brillouin zone.

2.3.2 Lithium Graphite Intercalation Compounds

Lithium graphite compounds are in many respects the simplest of the intercalation compounds. The first-stage compound LiC_6 forms with the $\sqrt{3} \times \sqrt{3}$

$R30°$ superlattice. The stacking sequence is AαAαAαAα . . . [2.83], where the roman letters refer to the registry of the graphite planes and the Greek letters refer to the registry of the intercalant planes. Other possible structures have been suggested [2.84] for low temperatures. The symmetry of the AαAαAαAα . . . structure is D_{6h}^1 or $P6/mmm$. Aside from the change in the stacking structure, the insertion of Li into the graphite lattice causes the interlayer spacing to increase by 10 % to 3.70 Å and the intralayer bonds to dilate by 1 %.

Several groups have studied the electronic structure of LiC_6 as listed in Table 2.2. The results of these authors are more or less in agreement with each other and suggest that the intercalant band lies completely above the Fermi level of the compound corresponding to (2.28). In the sense of the rigid-band model, LiC_6 is an ideal donor compound, raising the Fermi level of the reference graphite by approximately 1 eV. On this basis, the charge transfer from the Li to the graphite is complete, although a Mulliken population analysis [2.81] shows some hybridization of the Li states with the occupied graphite bands. Figure 2.14 shows typical band structure, from the calculation of *Holzwarth* et al. [2.60]. The folded graphite π bands near the Fermi level in the M–Γ and the L–A directions are easily identified and are very slightly distorted from those of

Table 2.2. Electronic band structure calculations for lithium graphite intercalation compounds. Underlined energies were quoted in the reference; nonunderlined energies were estimated from published band diagrams

Year	Reference	Method	Widths of occupied band			c-axis dispersion[a]	Intercalant band[b]
			$sp^2\ \sigma$	π	total		
		LiC_6					
1978	[2.27]	corrected KKR	15.8 eV	9.8 eV	21.8 eV	0.8 eV	1.6 eV
1980	[2.50]	pseudopotential[c]	10.1	5.4	14.3	0.5	0.7
1980	[2.40]	LCAO	15.1	9.4	23.4	0.6	?
1982	[2.52]	LMTO[c]	16.3	9.5	21.3	1.5	1.5
1983	[2.53]	FLAPW[c]	16.9	9.3	20.4	–	2.2
1983	[2.60]	mixed-basis pseudopotentials[c]	17.1	10.0	21.6	1.8	1.3
		LiC_{12}					
1980	[2.40]	LCAO	15.1	9.1	22.8	–	?
1983	[2.60]	mixed-basis pseudopotentials[c]	17.3	10.0	21.6	–	1.7
		LiC_{18}					
1983	[2.60]	mixed-basis pseudopotentials	17.9	9.6	21.6	–	2.0

[a] Maximum c-axis dispersion of bands near the Fermi level.
[b] Intercalant-interlayer band minimum measured above E_F.
[c] Self-consistent calculation.

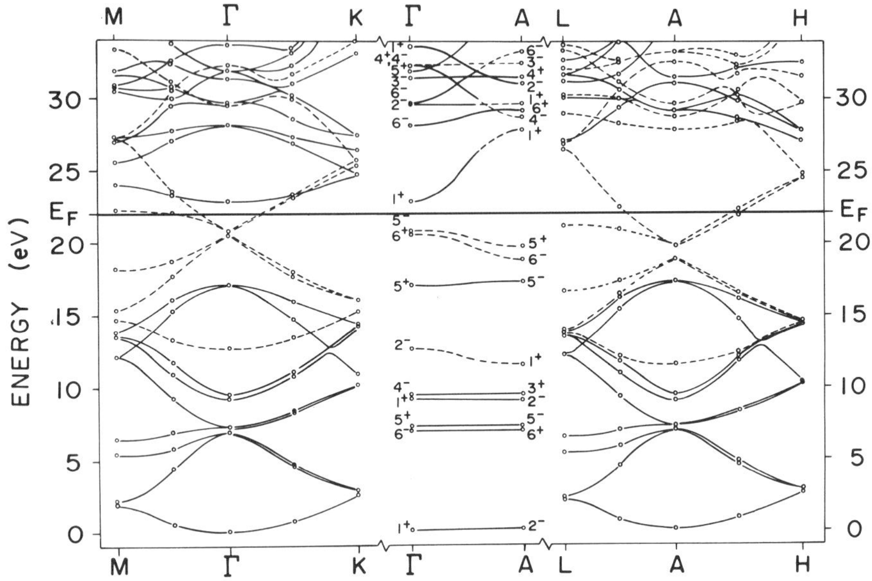

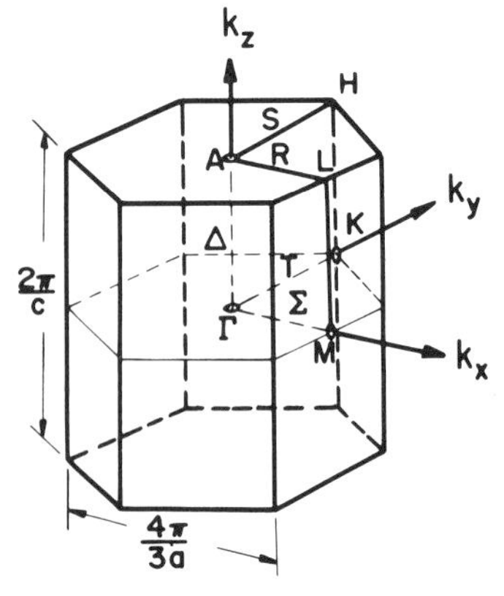

Fig. 2.14. Band structure of LiC_6 from a self-consistent mixed basis pseudopotential calculation. Dashed lines denote π bands; full lines denote σ bands. Zero of energy is taken at the bottom of the σ band. The inset on the right indicates the Brillouin zone for this structure. The indicated lattice constants are $a = 2.485$ Å and $c = 3.70$ Å [2.60]

the rigid-band model shown in Fig. 2.12. The Fermi surface of LiC_6 has also been calculated by more than one author. Figure 2.15 shows the results calculated by *Ohno* [2.50], using empirical pseudopotentials and a self-consistent determination of the charge transfer. The two sheets of the Fermi surface are shown separately. Because the c axis of LiC_6 is only 10% larger than that of graphite, the c-axis dispersion of the bands is comparable to that of graphite. As a consequence, the Fermi-surface cross section in the Γ plane is smaller and the cross section in the A plane is larger than that of the two-dimensional rigid-band model of Fig. 2.13.

The results of the band-structure calculations for LiC_6 have been used to calculate a number of experimentally accessible results. For example, *Chen* and

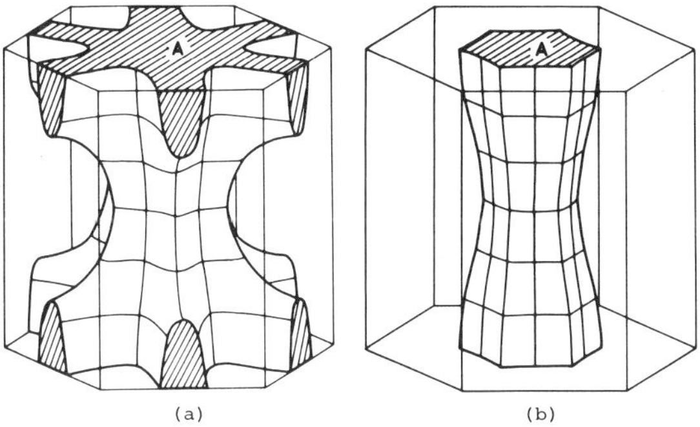

Fig. 2.15. Sketch of the (**a**) lower and (**b**) upper sheets of the Fermi surface for LiC_6 calculated self-consistently using empirical pseudopotentials and basis functions formed from two- dimensional plane waves and numerical functions along the c axis [2.50]

Rabii [2.84a] were able to use the band-structure results to calculate the optical spectrum of LiC_6 in very good agreement with experimental dielectric constants and electron-energy-loss data. *Chou* et al. [2.85] were able to calculate the Compton profiles for LiC_6 in very good agreement with experimental measurements.

In addition to the energy bands, self-consistent electronic structure calculations also yield the charge density of the material according to (2.3). The total valence density of LiC_6 is very similar to that of graphite, the charge transfer effects being small on the scale of the total valence density. To see the effects of charge transfer more clearly, it is helpful to consider the difference between the valence charge of the intercalation compound $\varrho(\boldsymbol{r})$ and that of the reference graphite compound $\varrho^0_{\text{graphite}}(\boldsymbol{r})$ defined in Sect. 2.2.2:

$$\delta\varrho(\boldsymbol{r}) \equiv \varrho(r) - \varrho^0_{\text{graphite}}(\boldsymbol{r}) . \tag{2.33}$$

Figure 2.16 shows contours of this difference density plotted in the carbon plane, the lithium plane, and a plane including the c axis. We see from this difference density that the valence charge is highly polarized to screen the intercalant ion, causing an excess charge close to the intercalant ions and a deficit of charge in the carbon planes. We can break this difference density into two contributions, one associated with the conduction electrons alone and the other associated with the lower valence states [2.60]. Then we find the conduction-electron density contribution to have a nearly undistorted π-like form, as is consistent with the rigid-band picture. Consequently, the main contribution to the polarization (or hybridization) charge comes from the lower-bonding π and σ bands of graphite.

Higher-stage lithium intercalated graphite compounds have been studied by *Samuelson* and *Batra* [2.40] and *Holzwarth* et al. [2.60]. *Samuelson* and *Batra* [2.40] calculated the electronic structure of LiC_{12} using the experimentally determined structure [2.83] of an AAαAAαAAα . . . stacking sequence and a D^1_{6h} ($P6$/mmm) space group. They used a well-converged, but non-self-consistent LCAO method. They found the electronic structure of LiC_{12} to be very similar to that LiC_6, with the additional feature of an approximately 0.5 eV

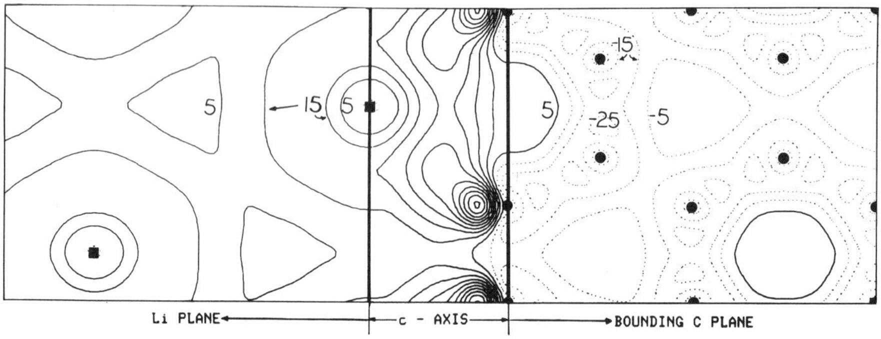

Fig. 2.16. Contour plot of difference density $\delta\varrho(\boldsymbol{r})$ for LiC_6 as defined by (2.33) from a mixed-basis pseudopotential calculation, plotted in the Li plane (left panel), in the plane containing the c axis (center panel), and in the C plane (right panel). Atomic positions are denoted by filled squares for Li and circles for carbon. Contour labels are given in units of 0.001 electrons/Å^3. Contours are given in intervals of 0.01 electrons/Å^3 starting at the level -0.025 electrons/Å^3. Negative contours are denoted with dotted lines [2.60]

splitting of the graphite π bands due to the interaction of the AA layers. The Fermi surface has four sheets with cross sections similar to (although smaller than) those shown in the rigid-band model of Fig. 2.13 and relatively little c-axis dispersion.

To focus on the general features of higher-stage compounds, *Holzwarth* et al. [2.60] carried out self-consistent mixed-basis pseudopotential calculations on second- and third-state lithium graphite, using hypothetical crystal structures with layer stacking and stoichiometry AαABβBCγCAαAB ... and LiC_{12} for the second stage, and AαABCβCABγBCAαA ... and LiC_{18} for the third stage. These compounds are not known to exist, but their structures are representative of structures found in many intercalation compounds other than Li. The electronic structure found by *Holzwarth* et al. [2.60] for the second-stage compound was similar to the results of *Samuelson* and *Batra* [2.40], having a form similar to that of LiC_6, with the additional splitting of the carbon π bands due to the interaction of the neighboring carbon layers. Figure 2.17 reproduces the electronic structure of the third-stage compound and shows additional features caused by inequivalent graphite layers. As will be discussed in more detail in Sect. 2.4, the bands corresponding to states in the bounding layers (layers closest to the intercalant planes) generally lie at lower energy than those in the interior layers. The Fermi surface has six sheets. In general, more Fermi-surface electrons are contributed by the bounding layers than are contributed by the interior layers. This effect is shown in the contour plot of Fig. 2.18b, which presents the density due to electrons in the partly filled bands of LiC_{18}. These bands near the Fermi level show the expected rigid π-like form, with more than twice as much density on the bounding layers as on the interior layers. Figure 2.18a shows the difference density of LiC_{18} as defined by (2.33). This total difference density $\delta p(\boldsymbol{r})$ demonstrates the effectiveness of the screening of the

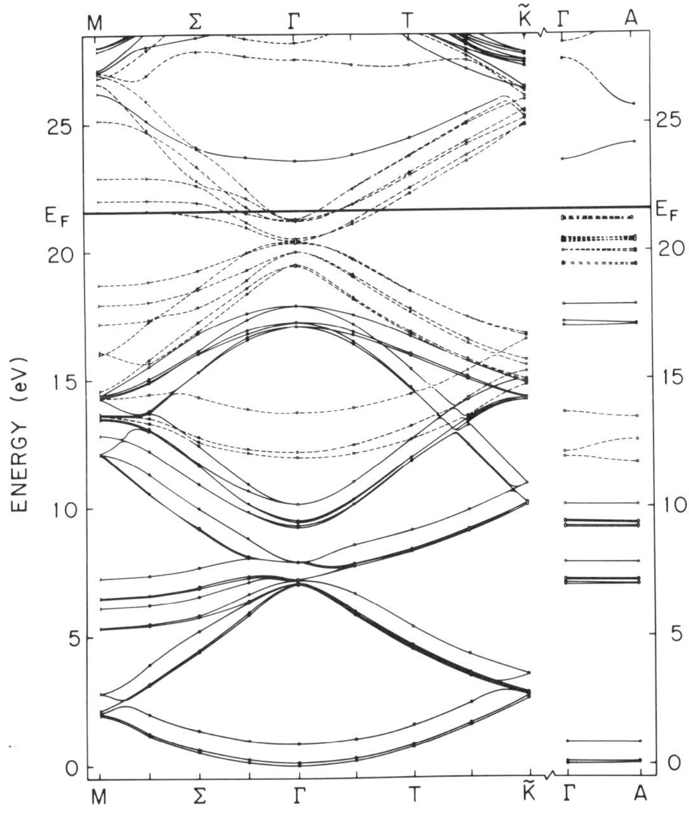

Fig. 2.17. Band-structure diagram for LiC_{18} in the hypothetical third-stage structure AαABCβCABγBCAαA . . . from mixed-basis pseudopotential calculations. Dashed lines denote π bands; full lines denote σ bands. The zero of energy was taken at the bottom of the lowest σ band. Brillouin-zone labels are the same as for Fig. 2.14 [2.60]

intercalant ions and of the donated charge by the polarization or hybridization of the lower-lying graphite bands, resulting in a highly localized difference density near the intercalant ions. The difference density near the interior layers is more than ten times smaller.

2.3.3 Alkali and Alkaline Earth-Metal Graphite Intercalation Compounds

The first published "first-principles" electronic structure study of a graphite intercalation compound was that of *Inoshita* et al. [2.37] for KC_8. This compound forms in the 2×2 superlattice structure, with a layer stacking [2.86, 87a] of AαAβAγAδAαA . . ., so that it has the space group D_{2h}^{24} (Fddd) and contains 16 carbon atoms and two potassium atoms per unit cell. The carbon bond lengths within a layer are expanded by less than 1%, but the spacing between graphite layers is expanded by 61% to 5.41 Å in KC_8. In constrast to

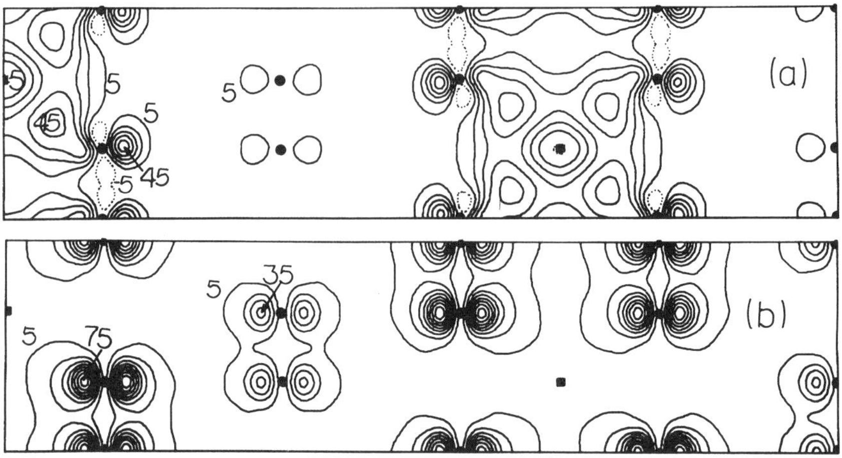

Fig. 2.18. Contour plots of valence density for hypothetical LiC_{18} plotted in a plane containing the *c* axis and passing though Li atoms and C–C bonds. Atomic positions are denoted by filled squares for Li and circles for carbon. Contour values are given in units of 0.001 electrons/$Å^3$. Contours given in intervals of 0.01 electrons/$Å^3$ starting at the level $-$ 0.005 electrons/$Å^3$. Negative contours are denoted with dotted curves: (**a**) difference density $\delta\varrho(\boldsymbol{r})$ defined by (2.33); (**b**) density due to electrons in the conduction bands only [2.60]

the calculations for LiC_6, which are all in reasonable agreement with each other, the calculations of KC_8 have been much more controversial. Part of the reason for this differences in the superlattice structures and in the corresponding foldings of the layer graphite bands shown in Fig. 2.12. In both materials the interlayer-intercalant band has a minimum at the Γ point in he Brillouinz one, which remains at the same location in the superlattice Brillouin zone. However, the folding of the layer graphite bands is different in the two cases. For IC_6, the π bands at the K point of graphite are folded to the Γ point in the IC_6 material. The separation of the interlayer-intercalant band and the graphite π bands is several eV for all points in the Brillouin zone. This qualitative relationship is relatively insensitive to the details of the band-structure calculations and is obtained in all the results listed in Table 2.2. On the other hand, for IC_8, the π bands at the M point of graphite are folded to the Γ point in the IC_8 material. In this case, the separation of the interlayer-intercalant band and the graphite π bands is small at the Γ point of the material. The close proximity of the graphite π bands and the interlayer-intercalant band in this structure creates strict precision demands on the electronic structure calculations. Since these demands are close to, if not beyond, the limits of calculational accuracy, the calculations for KC_8 have been quite variable, as indicated in Table 2.3.

The extended Hückel results of *Inoshita* et al. [2.37] placed the intercalant-interlayer band approximately 0.2 eV *below* the Fermi level of KC_8. Since this band has appreciable dispersion about the *c* axis, it gave a quasi-spherical contribution to the Fermi surface, in addition to the zone corner contributions

Table 2.3. Electronic band structure calculations for heavy alkali and alkaline earth-metal graphite intercalation compounds. Underlined energies were quoted in the reference; nonunderlined energies were estimated from published band diagrams

Year	Reference	Method	Widths of accupied band sp^2 σ	π	total	c-axis dispersion[a]	Intercalant band[c]
		KC_8					
1977	[2.37]	extended Hückel	–	5.2	–	0.2	− 0.2
1979	[2.49]	pseudopotential[c]	–	9.2	–	0.6	− 0.9
1982	[2.28]	corrected KKR	16.0	9.4	21.4	< 0.1	$\underline{+1.8}$
1985	[2.62]	mixed-basis pseudopotentials[c]	17.0	$\underline{9.6}$	22.7	0.4	+ 1.7
1986	[2.43]	LCAO[c]	–	9.0	–	< 0.1	+ 2.6
1987	[2.44]	LCAO[c]	–	9.1	–	0.5	$\underline{+2.9}$
		RbC_8					
1985	[2.62]	mixed-basis pseudopotentials[c]	17.2	$\underline{9.5}$	22.8	0.3	+ 2.0
1986	[2.43]	LCAO[c]	–	9.2	–	< 0.1	+ 2.0
		CsC_8					
1985	[2.62]	mixed-basis pseudopotentials[c]	17.2	$\underline{9.5}$	22.9	0.3	+ 2.8
1986	[2.43]	LCAO[c]	–	9.3	–	< 0.1	+ 2.6
		BaC_6					
1983	[2.61]	mixed-basis pseudopotentials[c]	$\underline{17.2}$	$\underline{8.6}$	$\underline{21.3}$	$\underline{0.7}$	$\underline{-1.4}$

[a] Maximum c-axis dispersion of bands near the Fermi level.
[b] Lowest intercalant band minimum measured above (+) or below (−) E_F.
[c] Self-consistent calculation.

anticipated from the rigid-band picture of Fig. 2.14. This result was corroborated by the self-consistent empirical pseudopotential calculations of *Ohno* et al. [2.49]. On the other hand, the non-self-consistent calculations of *DiVincenzo* and *Rabii* [2.28] and the self-consistent calculations of *Saito* and *Oshiyama* [2.43] find the intercalant-interlayer band to be well above the Fermi level, so the Fermi surface has no zone-center contributions. Finally, the self-consistent results of *Tatar* [2.62] and *Mizuno* et al. [2.44] suggest that the Fermi surface does have zone-center contributions due to small deviations from the rigid-band behavior of the graphite π bands folded to the Γ point of the Brillouin zone, with the intercalant-interlayer band well above the Fermi level. Figures 2.19, 2.20 present the results of both of these authors. The calculation of *Mizuno* et al. [2.44] used the experimental structure, while *Tatar* [2.62] used the simplified layer stacking AαAβAαAβA. . . to decrease the unit cell size by one half. The two

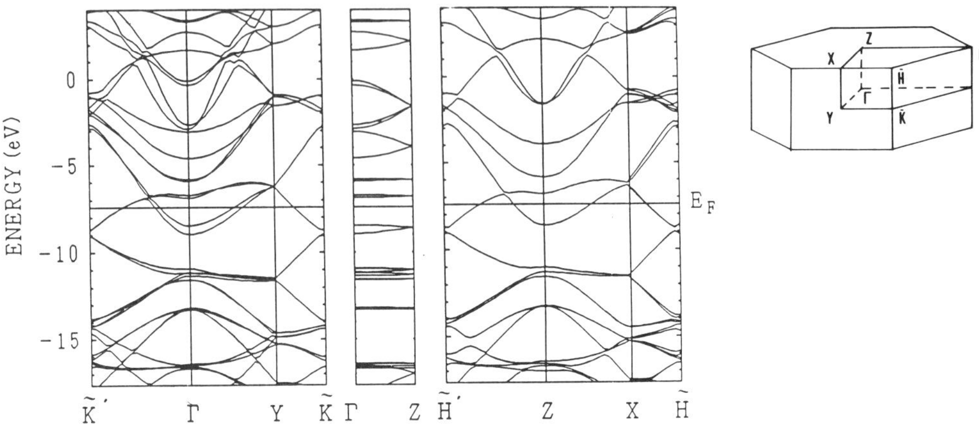

Fig. 2.19. Band-structure diagram for KC_8 in the AαAβAγAδAαA . . . structure calculated using the self-consistent LCAO method [2.44]. Lowest σ bands are not shown. Inset shows the Brillouin-zone labels

results are very similar, although they do differ in the size of the zone-center contribution to the Fermi surface. The bands in the results of *Mizuno* et al. have a minimum of 1.5 eV below the Fermi level at Γ, while *Tatar*'s bands have a minimum of 0.5 eV below the Fermi level at Γ.

Given these two results, we can again discuss the charge transfer. The charge transfer from the intercalant bands to the Fermi-level bands is essentially complete for these results, although the intercalant contributes to the spectrum of occupied states through hybridization. From an overlapping atom decomposition of the charge density, *Mizuno* et al. [2.44] estimate the total occupied potassium charge to be 0.4 electrons. In this picture, 0.6 electrons/potassium atom is distributed to carbon atoms; the inequivalent carbon atoms receive unequal amounts of charge.

The band-structure results for KC_8 have been used to calculate a number of physical properties of the material. For example, one area of interest is the superconducting properties that have been discussed by *Takada* [2.87b], *Kamimura* [2.87c], *Al-Jishi* [2.88], and *Shimizu* and *Kamimura* [2.89] on the basis of the band-structure results of *Ohno* et al. [2.49]. These authors argue that the presence of "three-dimensional" conduction electrons at the Fermi surface is essential for explaining the superconducting properties of these materials, and that there must be an appreciable involvement of the lattice vibrations of the potassium ions. Another area of interest is the interpretation of various spectroscopy experiments, recently reviewed by *Koma* et al. [2.77].

In addition to calculating the electronic band structure of KC_8, *Tatar* [2.62] and *Saito* and *Oshiyama* [2.43] have also calculated the electronic band structures of RbC_8 and CsC_8. Both authors find the bands for the heavier alkali-metal intercalation compounds to be very similar to those of KC_8. *Tatar* [2.62]

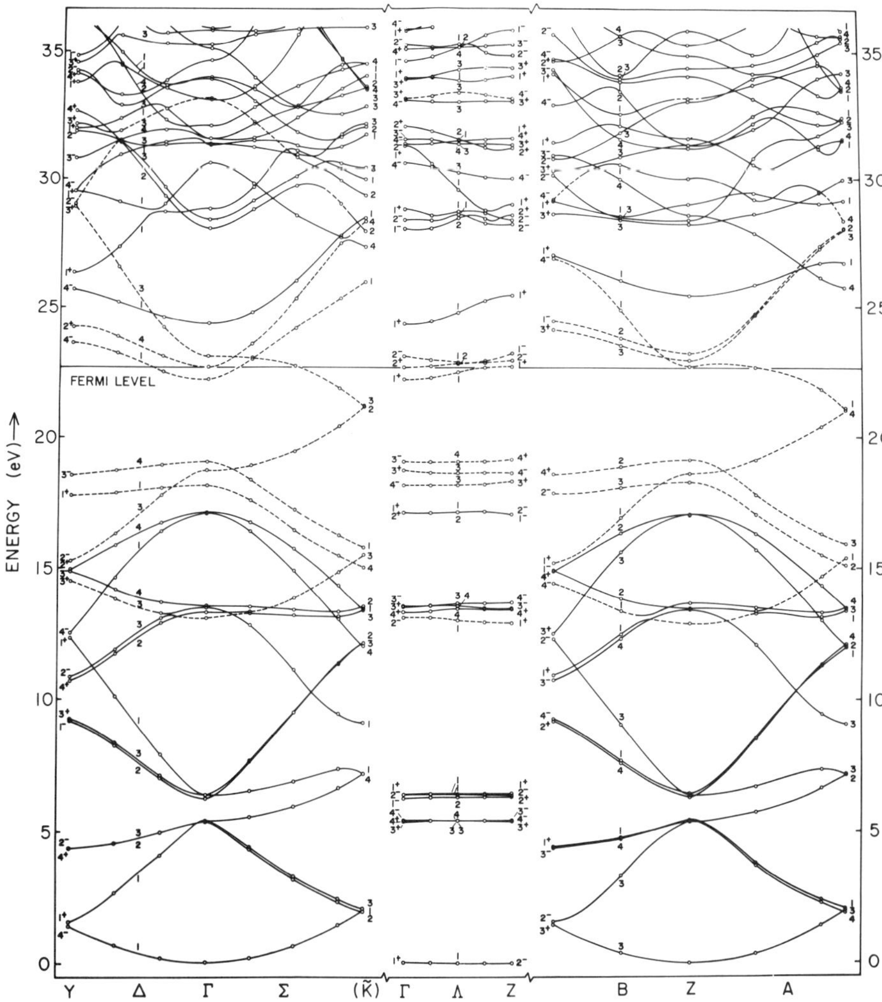

Fig. 2.20. Band-structure diagram for KC_8 in the AαAβAαAβAαA . . . structure calculated using the self-consistent mixed-basis pseudopotential method [2.62]. Dashed lines denote π bands; full lines denote σ bands. The zero of energy was taken at the bottom of the lowest σ band. Brillouin-zone labels are similar to those of Fig. 2.19

finds that the intercalant-interlayer band moves to higher energy in the sequence of increasing atomic number, $\varepsilon_I(\mathrm{K}) < \varepsilon_I(\mathrm{Rb}) < \varepsilon_I(\mathrm{Cs})$, and the size of zone-center contributions to the Fermi surface decreases in the same sequence. In the results of *Saito* and *Oshiyama* [2.43], there are no zone-center contributions to the Fermi surface for any of the alkali-metal intercalants, and the Fermi surface-charge transfer from the intercalant to the graphite π bands is 100%. However, a Mulliken population analysis shows that in CsC_8, Cs retains 6% of its charge through hybridization.

The alkaline earth-metal graphite intercalation compound BsC_6 has been studied to a small extent. It was originally hoped [2.90] that Ba with its two valence electrons might be able to donate both to the graphite layers, resulting in a highly conducting material. Unfortunately, this turned out not to be the case. The layer stacking of BaC_6 was determined by *Guérard* et al. [2.91] to be AaAβAαAβAαA ..., corresponding to the D_{6h}^4 ($P6_3/mmc$) space group. The spacing between graphite layers is 5.25 Å, and the carbon bond lengths within a graphite layer are expanded by 1% compared with graphite.

The electronic band structure was studied by *Holzwarth* et al. [2.61] using mixed-basis pseudopotential techniques and a modified layer stacking AαAβA-γAαA ... to decrease the size of the unit cell. For this material, not only does the lowest intercalant-interlayer state lie *below* the Fermi level, but the substantial 5d character in several of the low-lying intercalant states causes hybridization between these intercalant states and the graphite π states near the Fermi level, in violation of the rigid-band approximation. An estimate of the total charge transfer suggests that of the two electrons in the 6*s* shell of Ba, approximately one is donated to the graphite π bands and one is retained in the intercalant-interlayer band, in qualitative agreement with experimental X-ray photoemission results [2.90].

2.3.4 Ternary Graphite Intercalation Compounds

A number of ternary graphite intercalation compounds have been identified and studied experimentally. Apart from solid solutions of intercalation compounds studied by *Tatar* [2.62] and by *Akera* and *Kamimura* [2.92], two systems that have been the subject of recent theoretical investigations – $KHgC_{4n}$ and KH_xC_{4n} – are summarized in Table 2.4. Both materials have the intriguing trilayer form of intercalant "sandwiches" with the layer sequences generally –C–K–Hg–K–C– or –C–H–K–H–C–. These materials are not simple donor compounds; the intercalant sandwich layers generally contribute several types of states to the valence bands of the compound.

The crystal structure of the alkali-metal-amalgam graphite intercalation compounds $KHgC_4$ was determined by *Lagrange* et al. [2.93]. This first-stage compound was found to have the same layer stacking as KC_8. The distance between graphite layers is 10.16 Å, and the C–C bond length within a layer is nearly identical to that of graphite. There is some ambiguity in the trilayer structure of the intercalant. The X-ray data [2.93] are consistent with the Hg layers being either planar or buckled hexagonal networks. In the former case, the space group of the material is the same as that of KC_8, D_{2h}^{24} (Fddd); in the latter case, the space group has lower symmetry.

The first theoretical study of a ternary graphite intercalation compound was the calculation by *Senbetu* et al. [2.55] for the second-stage mercury amalgam compound $KHgC_8$. These authors carried out self-consistent FLAPW calcu-

Table 2.4. Electron band structure calculations for ternary graphite intercalation compounds. Underlined energies were quoted in the reference; nonunderlined energies were estimated from published band diagrams

Year	Reference	Method	Widths of occupied bands			c-axis dispersion[a]	Intercalant band[b]
			$sp^2\ \sigma$	π	total		
		$KHgC_4$					
1987	[2.38]	extend Hückel		5.8			− 6.5, − 0.7, − 0.1
1988	[2.63]	mixed-basis pseudopotentials[c]	$\underline{18.0}$	$\underline{9.6}$	$\underline{21.9}$	< 0.1	$\underline{-6.8}$, $\underline{-1.0}$. $\underline{-0.8}$
		$KHgC_8$					
1985	[2.55]	approximate FLAPW		9.8			−2.7, +2.1, + 2.8
		$KH_{0.5}C_8$					
1989	[2.45]	LCAO[c]		7.8			− 0.5
		$KH_{1.0}C_8$					
1989	[2.45]	LCAO[c]		7.8			− 2.2

[a] Maximum c-axis dispersion of bands near the Fermi level.
[b] Intercalant band minima measured above (+) or below (−) E_F.
[c] Self-consistent calculation.

lations for a K–Hg–K sandwich and then superposed these bands on those of two-dimensional graphite. The relative alignment of the intercalant-band with respect to graphite bands was not determined self-consistently.

Recently, first- and second-stage forms of this compound have been studied by *Kertesz* and *Guloy* [2.38] using the extended Hückel method for both possible structures suggested by the X-ray data [2.93]. On the basis of total charge transfer from Mulliken population analysis, for $KHgC_4$ they found that for each unit containing two potassium atoms, 0.1 electrons are retained by the K atoms, 1.7 electrons are donated to the Hg bands, 0.2 electrons are donated to the graphite bands. For $KHgC_8$ they found that for each unit containing two potassium atoms, 0.1 electrons are retained by the K atoms, 1.6 electrons are donated to the Hg bands, and 0.3 electrons are donated to the graphite bands.

Holzwarth et al. [2.63] have studied the first-stage compound $KHgC_4$, assuming the planar Hg structure and a slightly simplified stacking sequence, using self-consistent mixed-basis pseudopotential techniques. These results are in basic agreement with those of the previous workers, although the Hg 6p levels are generally lower in energy than those determined by *Kortesz* and *Guloy* [2.38] and closer to the band structure determined for the K–Hg–K sandwich by *Senbetu* et al. [2.55]. The alignment of the intercalant and graphite bands is such that both systems are energetically interwined. But because of the spatial separation, there is relatively little hybridization between these states. Figure 2.21 shows the band structure and corresponding density of states. The charge transfer for the Fermi-level bands (not including polarization charge) was found to be approximately 1.5 electrons donated to the amalgam bands and 0.5 electrons donated to the graphite π bands.

The results of the three theoretical studies are summarized in Table 2.4, which lists for the intercalant bands the band minima associated with the Hg 6*s*, 6*p*π, and 6*p*σ states.

The ternary graphite intercalation system involving hydrogen has recently attracted theoretical attention. *Mizuno* and *Nakao* [2.45] have studied the electronic structure of the second-stage compounds, KH_xC_8 for $x = 0.5$ and 1.0. These compounds have some structural similarities to the $KHgC_{4n}$ system,

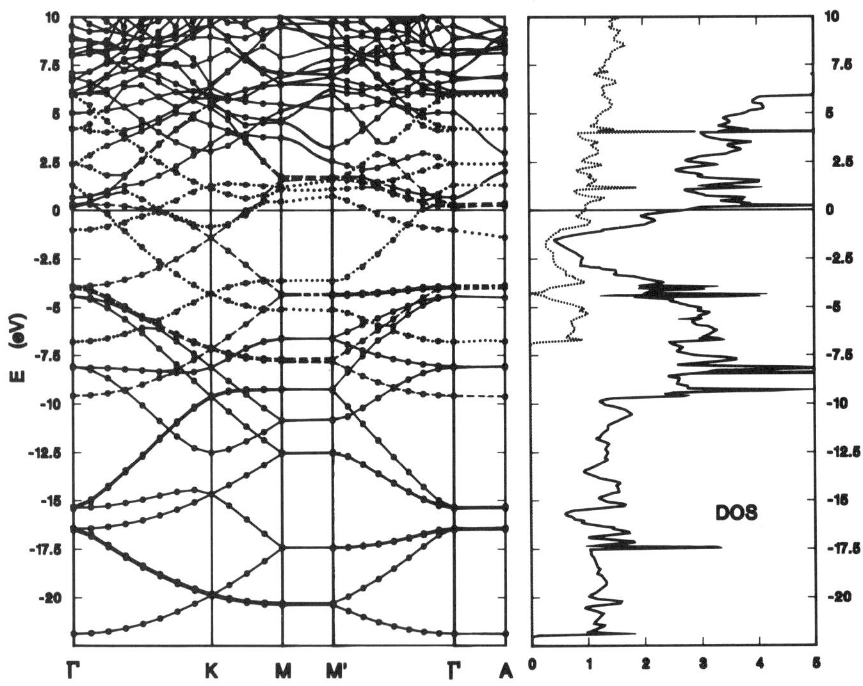

Fig. 2.21. Band-structure diagram and density of states for $KHgC_4$, calculated using the self-consistent mixed-basis pseudopotential method described [2.63]. Brillouin-zone shape is similar to that of Fig. 2.19 (using the label *M* instead of *Y*). Dotted lines denote states associated with the amalgam layer; dashed lines denote states associated with graphite π bands. The zero of energy is the Fermi level. The density of states (solid line) is given in units of states per eV per $K_2Hg_2C_8$. The partial density of Hg states is indicated in the same graphite with a dotted line

although the arrangement of the H atoms within the K–H–K sandwich is not known. Using models for the H layer inferred from magnetic resonance measurements [2.94], *Mizuno* and *Nakao* [2.45] carried out self-consistent LCAO calculations. Their band-structure results are similar to bands of KC_8, modified by interaction of the two neighboring graphite layers and with the addition of a hydrogen band partly below the Fermi-level. Table 2.4 summarizes these results, listing for the intercalant band the band minimum associated with the H states.

2.4 Electronic Band Structure of High-Stage Graphite Intercalation Compounds

2.4.1 General Features

The phenomenon of staging is unique to graphite intercalation compounds. Without venturing into the question of why staging should occur [2.95], the phenomenon of staging does introduce some additional aspects to the electronic structure. The three questions posed for low-stage graphite intercalation (Sect. 2.3.1) can also be posed for higher-stage (third and greater) graphite intercalation compounds. In addition to these questions of charge transfer, band hybridization, and the role of the interlayer band, a new question can be added: What are the effects of inequivalent graphite layer? The terminology has been developed [2.2] to reference the graphite layers nearest the intercalant layers as the *bounding* layers and the remaining graphite layers as the *interior* layers. For example, in a third-stage compound, there are two bounding layers and one interior layer of graphite in each unit cell.

The simplest model to describe the physics of inequivalent graphite layers is one that smears out the charge within each layer to make uniform sheets having magnitudes of charge per unit area of σ_b and σ_i for the bounding and interior graphite layers, respectively. Within this model, the intercalant ions themselves create no net fields because of symmetry, and therefore need not be explicitly considered. The charged graphite sheets do create electric fields within the compound. If we assume that $\sigma_b > \sigma_i$, the largest field occurs between the intercalant layer and the bounding graphite layer, while much smaller fields occur in the interior regions, as shown for donor compounds in Fig. 2.22. These fields result in different electrostatic potentials at each graphite layer, as shown in the lower panel of the figure. For donor compounds, the interior layers lie at a higher potential energy than the bounding layers. If we now consider the effects of these electrostatic potentials on the electronic band structure, we develop the picture shown in Fig. 2.23. If we ignore, for the moment, the interlayer interactions, the effect of the electrostatic potentials at each layer is to adjust the alignment of the layer bands. Figure 2.23 was constructed by taking the reference energy of each layer to be the K-point energy of the graphite π bands of

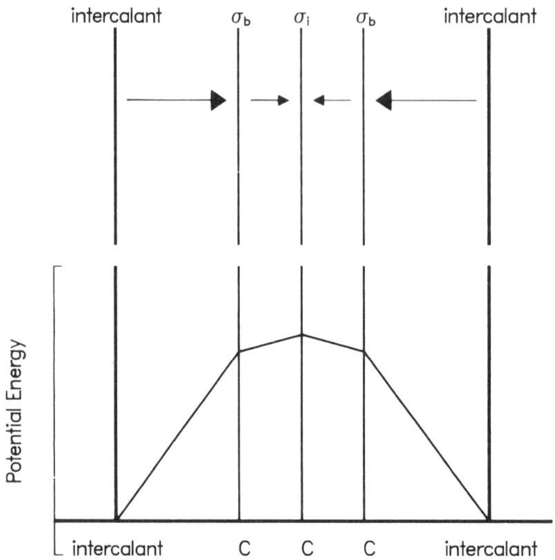

Fig. 2.22. An idealized model of a third-stage donor graphite intercalation compound. The upper panel shows geometry of bounding and interior graphite layers with corresponding layer charge density σ_b and σ_i, respectively. The resultant electric field strengths between the layers are indicated by arrows. The lower panel shows the electrostatic potential along the c axis for the same model

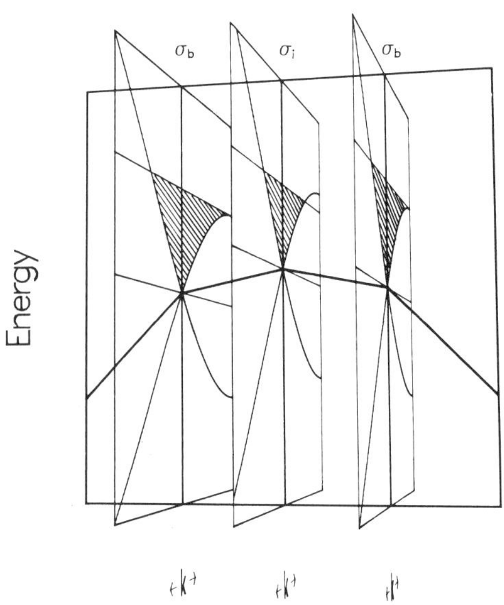

Fig. 2.23. Schematic drawing of energy bands near the Fermi level for a third-stage donor graphite intercalation compound within same model as in Fig. 2.22. Ignoring interlayer interactions, two-dimensional graphite π bands are aligned according to the electrostatic potential in each layer plane. Shaded regions indicate filled states due to charge transfer from the intercalant

each layer. Since the interior-layer bands lie at higher energy, for any given Fermi level of the compound, they can accommodate a smaller amount of charge than can the bounding layers, demonstrating that the assumption $\sigma_b > \sigma_i$ is self-consistent. For acceptor compounds, the signs of charges, the orientations of the electrostatic fields, and the signs of the energy shifts are reversed, but the self-consistent distribution of charge again has the result $\sigma_b > \sigma_i$. A more realistic treatment of these high-stage compounds must at least include the effects of the interaction between graphite layers, which is energetically comparable to the electrostatic energy shifts we have been discussing [2.96]. These systems have been studied with two different approaches – first-principles calculations and rigid-band-like models. Both are briefly described below.

2.4.2 First-Principles Calculations

In addition to the self-consistent mixed-basis pseudopotential calcultions for a hypothetical third-stage lithium intercalated graphite compound by *Holzwarth* et al. [2.60] discussed above, self-consistent local density calculations for high-stage graphite intercalation compounds have been carried out by *Ohno* and collaborators [2.46, 97] and *Chan* and collaborators [2.98].

In the work of *Ohno* et al. [2.46, 97] the superlattice introduced by the intercalant and the subsequent zone folding was neglected. The intercalant ions were treated as uniformly charged sheets. Because of this geometry, the intercalant ions thus contribute a constant potential. The region between the intercalant planes is then treated self-consistently using an LCAO basis, assuming the stoichiometry $C_{12n}X$, with each intercalant ion X transferring f units of charge. Figures 2.24, 25 show representative results for donor compounds for stages 3 and 5, respectively, for $f = 1$. The bands for the third-stage compound in Fig. 2.24 look qualitatively similar to those of the hypothetical third-stage Li graphite compound shown in Fig. 2.17, except for the zone folding. The stage-5

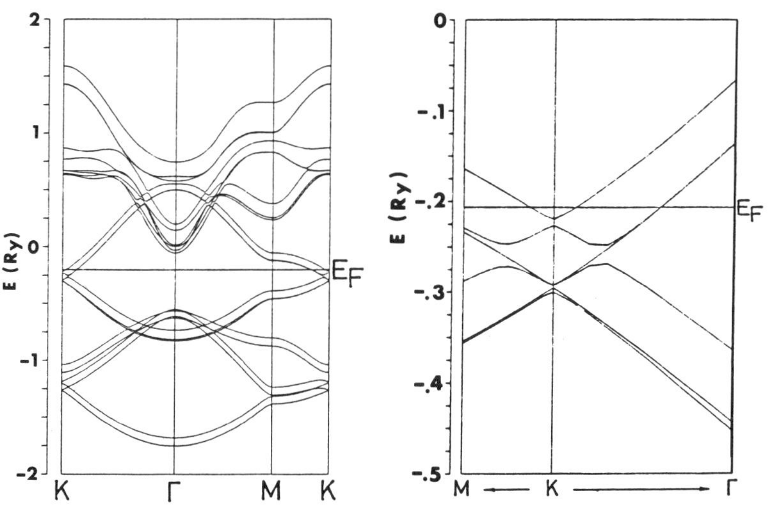

Fig. 2.24. Self-consistent band structure of a third-stage graphite intercalation compound in which intercalant layers are represented by charged sheets. The calculation was performed using a self-consistent LCAO method, for the case in which the stoichiometry of the compound is $C_{12n}X$, where $n = 3$ and where the intercalant ion X donates charge $f = 1$ electron. The left panel shows the complete band structure; the right panel shows an expanded section of the π band structure near the K point of the Brillouin zone [2.46, 97]

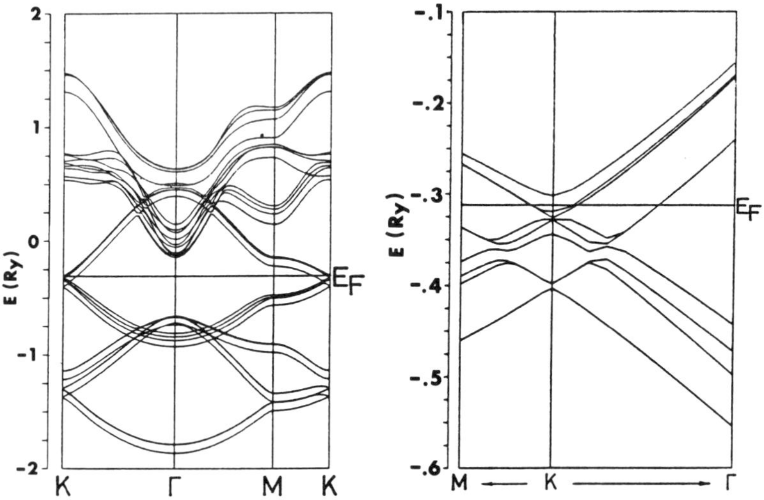

Fig. 2.25. Self-consistent band structure of a fifth-stage graphite intercalation compound in which intercalant layers are represented by charged sheets. The calculation was performed using a self-consistent LCAO method, for the case in which the stoichiometry of the compound is $C_{12n}X$, where $n = 5$ and where the intercalant ion X donates charge $f = 1$ electron. The left panel shows, the complete band structure; the right panel shows an expanded section of the π band structure near the K point of the Brillouin zone [2.46, 97]

results shown in Fig. 2.25 differ from the stage-3 results, primarily in the increased number of bands per unit cell. In general, the bands associated with the interior layers lie at higher energy and contain less charge than those associated with the bounding layers, as discussed in the simple model above. While the graphite π bands near the Fermi level participate directly in the charge transfer, the lower π and σ bands exhibit polarization effects in reponse to the electrostatic fields. In addition to studying the general behavior of the charge distribution along the c axis of these compounds, the authors discussed the orbital magnetic susceptibility and the total energy.

Chan et al. [2.98] performed self-consistent electronic structure calculations for general high-stage compounds, focusing on their total energy, specifically as it affects interlayer bond lengths and phonon frequencies. Their treatment of the intercalant was different from that of *Ohno* et al. [2.46, 97], treating the intercalants as point ions having variable charge. In fact, most of their work was performed using the geometry of a first-stage compound. The effects of high-stage configurations were inferred from the adjustable intercalant charge. They were able to show that the interlayer C–C bond length decreases with increasing concentration of acceptor charge and increases with increasing concentration of donor charge, in good agreement with experiment. Similarly, the vibrational frequencies involving interlayer motions increase with increasing concentration of acceptor charge and decrease with increasing concentration of donor charge. Part of the mechanism of the bond lengthening and weakening in donor compounds can be explained by the polarization response of the lower π and σ bands states, which causes a depletion of charge from the C–C bonds (as also seen in the calculations for lithium compounds shown in Figs. 2.16, 18).

2.4.3 Parametrized Models

There have been a number of parametrized models of high-stage graphite intercalation compounds. We describe a few here. The value of this approach is that once the model is developed, we can focus attention on evaluating various quantities [2.96–105] that relate to the Fermi-surface, transport, magnetic, optical, and other experimentally accessible properties of these materials.

Pietronero et al. [2.99] considered the distribution of charge within graphite layers of high-stage graphite intercalation compounds, using a model similar to that discussed in Sect. 2.4.1. By representing the form of the π energy bands of each noninteracting graphite layer to the linear approximation as in (2.25), and by requiring that each band be filled to the Fermi level, they were able to solve the electrostatic equations analytically. Qualitatively, they found that the linear dispersion of the graphite π bands near the Fermi level limits the amount of charge that each layer can accommodate for a given Fermi level, so that the conduction electrons in high-stage graphite intercalation compounds are distributed over many layers. The screening of this long-range charge distribution by the lower valence states of graphite was parametrized by a dielectric constant.

This approach was extended by *Strässler* and *Pietronero* [2.100] to calculations of the total energy of the system and to the study of electrical conductivity. The analysis of the total energy as a function charge transfer enabled them to estimate changes in the C–C bond length, deriving an analytic expression for the bond length as a function of charge transfer. Their results are qualitatively similar to but differ numerically from the recent first-principles results of *Chan* et al. [2.98].

Blinowski and *Rigaux* [2.101] have studied optical and magnetic properties of higher-stage graphite intercalation compounds using extensions of the *Slonczewski* and *Weiss* [2.71] and *McClure* [2.75] parametrization of the graphite π bands near the Fermi level. Included in their model is the interaction matrix element of nearest-neighbor π orbitals within a graphite layer (similar to the $\mathscr{E}_1$ and $\mathscr{E}_1'$ parameters defined in (2.22, 25), and the interaction matrix element of nearest-neighbor π orbitals in adjacent graphite layers. In addition, their model parametrizes the electrostatic energies of each layer. Because of the planar averaging of the excess charges in the intercalant layers, the intercalant enters their model only by virtue of determining the charge transfer, contributing a uniform electrostatic potential-energy shift to the system (Sect. 2.4.1). Much of the calculation could be carried out analytically in terms of the matrix element parameters. By studying the bands as a function of the electrostatic energy shifts for nonequivalent graphite layers, they could study the magnitude of the charging effects relative to the interlayer hopping matrix element. The form of the band structures was quite similar to that found by *Ohno* et al. [2.46, 97] (shown in the right-hand panels of Figs. 2.24, 25). These energy bands were used by the authors to analyze interband optical transitions.

This approach has recently been extended by *Yang* and *Eklund* [2.102] to make a detailed analysis of the optical properties of high-stage potassium graphite intercalation compounds. *Blinowski* and *Rigaux* [2.101] also used their analytic formalism to study the magnetic susceptibility of graphite intercalation compounds, extending earlier work by *Safran* and *DiSalvo* [2.103]. Since the magnetic susceptibility is such a complicated function of the band parameters, *Blinowski* and *Rigaux* [2.101] had to simplify the model and ignore the interlayer interaction parameters. Despite the complicated nature of the calculation, they were able to carry out the bulk of the calculation analytically, finding that the orbital susceptibility is a very sensitive function of the band parameters. This work has recently been extended by *Saito* [2.104], who was able to take interlayer interactions into account.

A different method for studying high-stage graphite intercalation compounds was developed by *Leung* and *Dresselhaus* [2.105]. Their work is an extension of the tight-binding parametrization of the graphite bands by *Johnson* and *Dresselhaus* [2.72]. In this work, the zone-folding effects were considered in detail within the context of the rigid-band model, while the effects of the intercalant were parametrized in terms of an energy shift parameter for the bounding graphite layers. This model was used to identify various optical transitions and de Haas-van Alphen frequencies in the high-stage intercalation compounds.

2.5 Summary and Conclusions

In this review we have attempted to focus on four basic questions and to develop a framework for addressing them. The four questions are

(1) How much charge is transferred between the intercalant and graphite?
(2) How are the graphite and intercalant bands altered by intercalation?
(3) What is the role of the graphite interlayer band?
(4) What are the effects of inequivalent graphite layers for high-stage compounds?

A convenient framework for addressing these questions is the notion of reference graphite and intercalant systems, as defined by (2.18). On the basis of the literature results and from the perspective of the reference graphite and intercalant systems, we can develop the following mental picture of the charge transfer process. First, starting with the neutral reference graphite system, the introduction of the intercalant changes the occupancy of the rigid antibonding bands near the Fermi level. Second, the lower-lying bonding bands of the reference graphite then respond by polarizing or hybridizing with the intercalant states, effectively screening both the intercalant ions and the excess charge in the graphite π bands. Third, the graphite π band occupancy and the polarization charge readjust so that they are self-consistent.

In answer to questions (1) and (2), there are two completely different measures of charge transfer: the Fermi-surface charge transfer and the total charge transfer. For many of the intercalation compounds, the Fermi-surface charge transfer involves only the charge in the rigid graphite π bands (although for highly electronegative intercalants such as Ba or for ternary graphite intercalation compounds, there are Fermi-surface contributions from the intercalant states). On the other hand, the total charge transfer takes into account contributions from the entire valence spectrum of the compound due to polarization or hybridization effects. The total charge transfer tends to be highly localized near the intercalant ions, whereas the Fermi-surface charge transfer is constrained to retain the form of the graphite π bands and tends to be more extended. The difference between the distribution of the total transferred charge and the distribution of charge transferred to states near the Fermi level is best illustrated by contour plots for the model third-stage donor intercalation compound shown in Fig. 2.18. The total charge transfer affects physical properties of the intercalation compound such as the C–C bond lengths, force constants, and core-level shifts. The Fermi-surface charge transfer affects Fermi-surface properties such as the electrical conductivity and the de Haas–van Alphen frequencies.

We discussed question (3), using a simple model based on the reference systems. The donor intercalant and interlayer states are difficult to disentangle, so the term interlayer-intercalant state describes the situation well. In answer to question (4), we discussed the electrostatic fields produced by the distribution of charge among the inequivalent carbon layers, showing that the self-consistent distribution is such that the bounding carbon layers have more excess charge than do the interior layers, as has been confirmed by detailed calculations and by experiment.

By focusing on these simple ideas, we did not mean to minimize the importance of detailed self-consistent electronic structure calculations for graphite intercalation compounds. Both detailed first-principles calculations and simple models are needed to study these materials. The detailed calculations are important for obtaining quantitative results and establishing the credibility of the simple models. The simple models are important for guiding our qualitative understanding of these materials. By adjusting parameters with the help of detailed calculations and experiment, the simple models can be used for quantitative study.

Our literature review, shows that the electronic band structure of donor graphite intercalation has been thoroughly studied. The ternary donor compounds such as KH_xC_{4n} will perhaps continue to challenge band-structure theorists in the near future. On the other hand, there has been relatively little computational modeling for the acceptor compounds [2.106]. This subject has been hampered by the lack of the detailed experimental chemical and structural information that band-structure theorists desire. First-principles techniques have progressed to the point of being able to address some of these chemical and structural questions. It may be fruitful to focus computational power on modeling the electronic, lattice, and chemical structures for some of the graphite acceptor compounds. On the other hand, since the qualitative understanding of these materials has been well advanced by the study of simple models, these efforts should also be continued. In any case, the study of graphite intercalation compounds and their electronic structure has provided a very rich example of basic solid-state phenomena for several generations of physicists, and will undoubtedly continue to do so for several future generations.

Acknowledgements. Portions of this work were supported by NSF Grant No. DMR-8501022 and by the Wake Forest University Research and Publication Fund. I would like to thank H. Kamimura, K. Nakao, T. Ohno, R. Saito, and R. Tatar for their communications regarding this review. Much appreciation is extended to J.E. Fischer, L.A. Girifalco, and S. Rabii for introducing me to graphite intercalation compounds several years ago and for their many collabrations over the years. I would also like to thank S.G. Louie for many collaborative efforts in this field.

References

2.1 J.E. Fischer: Electronic properties of graphite intercalation compounds, in *Physics and Chemistry of Materials with Layered Structures*, ed. by F. Lévy, Vol. 6 (Reidel, Dordrecht 1979) pp. 481–532

2.2 M.S. Dresselhaus, G. Dresselhaus: Adv. Phys. **30**, 139–326 (1981)

2.3a P. Pfluger and H.-J. Güntherodt: Festkörperproblem **21**, 271–311 (1981)

2.3b H. Kamimura: Physics Today **40**, 64–71 (1987)

2.4 F.L. Vogel, A. Hérold (eds.): *Intercalation compounds of Graphite*, Proc. Franco American Conf., La Napoule, France, May 23–27, 1977, Mat. Sci. Eng., Vol. 31 (Elsevier Sequoia, Lausanne 1977)

2.5 C.F. van Bruggen, C. Haas, H.W. Myron (eds.): *Layered Materials and Intercalates*, Proc. Intern. Conf., Univ. of Nijmegen, The Netherlands, Aug. 28–31, 1979, Physica, Vol. 99 B + C (North Holland, Amsterdam 1980)

2.6 Y. Nishina, S. Tanuma, H.W. Myron (eds.): *Physics and Chemistry of Layered Materials*, Proc. Yamada Conf. IV, Sept. 8–10, 1980, Sendai, Japan, Physica, Vol. 105 B + C (North Holland, Amsterdam 1981)

2.7 L. Pietronero, E. Tosatti (eds.): *Physics of Intercalation Compounds*, Proc. Int. Conf., Trieste, Italy, July 6–10, 1981, Springer Ser. Solid-State Sci. Vol. 38 (Springer, Berlin, Heidelberg 1981)

2.8 M.S. Dresselhaus, G. Dresselhaus, J.E. Fischer, M.J. Moran (eds.): *Intercalated Graphite*, Mat. Res. Soc. Symp., Boston, Mass., Nov. 1–2, 1982, Mat. Res. Soc. Symp. Proc., Vol. 20 (North Holland, New York 1983)

2.9 K. Nakao, S.A. Solin (eds.); *Graphite Intercalation compounds*, Proc. Intern. Symp., Tsukuba, Japan, May 27–30, 1985, Syn. Met., Vol. 12 (Elsevier Sequoia, Lausanne 1985)

2.10 S.-I. Tanuma, H. Kamimura (eds.): *Graphite Intercalation Compounds*, Progress of Research in Japan (World Scientific, Singapore 1985)

2.11 J. Bok, C. Rigaux, I. Rosenman, H. Kamimura (eds.);: *Composés d'Insertion du Graphite*, Coll. Franco-Jap., Paris, France, Oct. 8–11, 1985, Ann. Phys. Fr. Colloq., Vol. 11, Col. 2, Suppl. 2, 1986

2.12 M.S. Dresselhaus (ed.): Intercalation in Layered Materials, Proc. 10th Course Erice Summer School, Erice, Trapani, Sicily, Italy, July 5–15, 1986, NATO ASI Series, Series B: Physics Vol. 148 (Plenum, New York 1986)

2.13 D. Davidov, H. Selig (eds.): *Graphite Intercalation Compounds*, Proc. 4th Intern. Symp., Jerusalem, Israel, May 24–29, 1987, Syn. Met. Vol. 23 (Elsevier Sequoia, Lausanne 1988)
2.14 A.P. Legrand, S. Flandrois (eds.): *Chemical Physics of Intercalation*, Proc. NATO Adv. Study Inst., Castera Verduzan, France, June 10–19, 1987, NATO ASI Series, Series B: Physics Vol. 172 (Plenum, New York 1987)
2.15 J. Callaway: *Quantum Theory of the Solid State* (Academic, New York 1974) Chap. 4
2.16 F. Seitz: *The Modern Theory of Solids* (McGraw-Hill, New York 1940) Chap. 6
2.17 P. Hohenberg, W. Kohn: Phys. Rev. **136**, B864–871 (1964)
2.18 W. Kohn, L.J. Sham: Phys. Rev. **140**, A1133–1138 (1965)
2.19 R. Dovesi, C. Pisani, C. Roetti: Int. J. Quantum Chem. **17**, 517–529 (1980)
2.20 R. Jebli, G. Volpilhac, J. Hoarau, F. Achard: J. Chim. Phys. **81**, 873–876 (1984)
2.21 L. Hedin, B. I. Lundqvist: J. Phys. C **4**, 2064–2083 (1971)
2.22 W.E. Pickett: Computer Phys. Rep. **9**, 115–198 (1989)
2.23 J.E. Slater: Phys. Rev. **51**, 151–156 (1950)
2.24 T. Loucks: *Augmented Plane Wave Method* (Benjamin, New York 1967)
2.25 J. Korringa: Physica **13**, 392–400 (1947)
2.26 W. Kohn, N. Rostoker: Phys. Rev. **94**, 1111–1120 (1954)
2.27 N.A.W. Holzwarth, S. Rabii, L.A. Girifalco: Phys. Rev. B **18**, 5190–5205 (1978)
2.28 D.P. DiVincenzo, S. Rabii: Phys. Rev. B **25**, 4110–4125 (1982)
2.29 R.C. Tatar, S. Rabii: Phys. Rev. B **25**, 4125–4141 (1982)
2.30 S.L. Altmann, W. Barton, C.P. Mallett: J. Phys. C **11**, 1801–1812 (1978)
2.31 C.P. Mallett, J. Phys. C **14**, L213–L220 (1981)
2.32 W.M. Lomer: Proc. Roy. Soc. London **A227**, 330–349 (1955)
2.33 F. Bassani, G. Pastori Parravicini: Nuovo Cimento B **50**, 95–127 (1967)
2.34 J. Zupan: Phys. Rev. B **6**, 2477–2482 (1872)
2.35 E.-K. Kortela, R. Manne: J. Phys. C **7**, 1749–1756 (1974)
2.36 A. Zunger: Phys. Rev. B **17**, 626–641 (1978)
2.37 T. Inoshita, K. Nakao, H. Kamimura: J. Phys. Soc. Japan **43**, 1237–1243 (1987): J. Phys. Soc. Japan **45**, 689 (1978) (erratum)
2.38 M. Kertesz, A.M. Guloy: Inorg. Chem. **26**, 2852–2857 (1987)
2.39 G. Volpilhac, L. Ducasse, F. Achard, J. Hoarau: Phys. Rev. B **23**, 4236–4241 (1981); G. Volpilhac, F. Achard. L. Ducasse, J. Hoarau: J. Phys. C **18**, 1613–1625 (1985); G. Volpilhac, F. Achard: J. Chim. Phys. **86**, 553–559 (1989)
2.40 L. Samuelson, I.P. Batra: J. Phys. C **13**, 5105–5124 (1980)
2.41 G.S. Painter, D.E. Ellis: Phys. Rev. B **1**, 4747–4752 (1970)
2.42 R.F. Willis, B. Fitton, G.S. Painter: Phys. Rev. B **9**, 1926–1937 (1974)
2.43 M. Saito, A. Oshiyama: J. Phys. Soc. Japan **55**, 4341–4348 (1986)
2.44 S. Mizuno, H. Hiramoto, K. Nakao: J. Phys. Soc. Japan **56**, 4466–4476 (1987)
2.45 S. Mizuno, K. Nakao: Phys. Rev **B 40**, 5771–5773 (1989); S. Mizuno, K. Nakao: J. Phys. Soc. Japan **58**, 3679–3686 (1989)
2.46 T. Ohno: "Electronic structures of higher stage graphite intercalation compounds", Ph.D. Thesis, University of Tokyo (1981)
2.47 C. Frétigny, R. Saito, H. Kamimura: J. Phys. Soc. Japan **58**, 2098–2108 (1989)
2.48 H. Nagayoshi, K. Nakao, Y. Uemura: J. Phys. Soc. Japan **41**, 1480–1487 (1976)
2.49 T. Ohno, K. Nakao, H. Kamimura: J. Phys. Soc. Japan **47**, 1125–1133 (1979)
2.50 T. Ohno: J. Phys. Soc. Japan **49**, Suppl. A, 899–902 (1980)
2.51 O.K. Andersen: Phys. Rev. B **12**, 3060–3083 (1975)
2.52 R.V. Kasowski: Phys. Rev. B **25**, 4189–4195 (1982)
2.53 M. Posternak, A. Baldereschi, A.J. Freeman, E. Wimmer, M. Weinert: Phys. Rev. Letters **50**, 761–764 (1983)
2.54 H.J.F. Jansen, A.J. Freeman: Phys. Rev. B **35**, 8207–8214 (1987)
2.55 L. Senbetu, H. Ikezi, C. Umrigar: Phys. Rev. **32**, 750–754 (1985)
2.56 W. van Haeringen and H.-G. Junginger: Solid State Comm. **7**, 1723–1725 (1969)
2.57 D.R. Hamann, M. Schlüter, C. Chiang: Phys. Rev. Letter **43**, 1494–1497 (1979); G.P. Kerker: J. Phys. C **13**, L189–194 (1980)

2.58 S.G. Louie, K.-M. Ho, M.L. Cohen: Phys. Rev. B **19**, 1774–1782 (1979)
2.59 N.A.W. Holzwarth, S.G. Louie, S. Rabii: Phys. Rev. B **26**, 5382–5390 (1982)
2.60 N.A.W. Holzwarth, S.G. Lauie, S. Rabii: Phys. Rev. B **28**, 1013–1025 (1983)
2.61 N.A.W. Holzwarth, D.P. DiVincenzo, R.C. Tatar, S. Rabii: Int. J. Quantum Chem. **23**, 1223–1230 (1983)
2.62 R.C. Tatar: A theoretical study of the electronic structure of binary and ternary first stage alkali intercalation compounds of graphite, Ph.D. Thesis, University of Pennsylvania, 1985
2.63 N.A.W. Holzwarth, Q.S. Wang, S.D. Had: Phys. Rev. B **38**, 3722–3732 (1988)
2.64 L. Pauling: *The Nature of the Chemical Bond*, 3rd ed. (Cornell University Press, Ithaca 1960)
2.65 J.D. Bernal: Proc. Roy. Soc. (London) **A106**, 749–773 (1924)
2.66 R.C. Weast, M.J. Astle: *CRC Handbook of Chemistry and Physics*, 60th edition (CRC, Florida 1979) p. B-186
2.67 S. Fahy, S.G. Louie, M.L. Cohen: Phys. Rev. B **34**, 1191–1199 (1986)
2.68 P.R. Wallace: Phys. Rev. **71**, 622–634 (1947)
2.69 C.A. Coulson, R. Taylor: Proc. Roy. Soc. A **65**, 815–824 (1952)
2.70 D.F. Johnston: Proc. Roy. Soc. A **227**, 349–358 (1955)
2.71 J.C. Slonczewski, P.R. Weiss: Phys. Rev. **109**, 272–279 (1958)
2.72 L.G. Johnson, G. Dresselhaus; Phys. Rev. B **7**, 2275–2285 (1973)
2.73 R.R. Haering: Can. J. Phys. **36**, 352–362 (1958)
2.74 L. Samuelson, I.P. Batra, C. Roetti: Solid State Commun **33**, 817–820 (1980)
2.75 J.W. McClure: Phys. Rev. **108**, 612–618 (1957)
2.76 D. Tománek, S.G. Louie: Phys. Rev. **37**, 8327–8336 (1988)
2.77 A. Koma, K. Miki, H. Suematsu, T. Ohno, H. Kamimura: Phys. Rev. **34**, 2434–2438 (1986); H. Kamimura: in Ref. [2.11], pp. 39–48
2.78 M. Posternak, A. Baldereschi, A.J. Freeman, E. Wimmer: Phys. Rev. Letters **52**, 863–866 (1984)
2.79 P. Alstrøm: Syn. Met. **15**, 311–322 (1986)
2.80 R.S. Mulliken: J. Chem. Phys. **23**, 1833–1840 (1955)
2.81 N.A.W. Holzwarth, S.G. Louie, S. Rabii: Phys. Rev. B **30**, 2219–2222 (1984)
2.82 S.A. Solin: Adv. Chem. Phys. **49**, 455–532 (1982)
2.83 D. Guérard, A. Hérold: Carbon **13**, 337–345 (1975)
2.84 N. Kambe, M.S. Dresselhaus, G. Dresselhaus, S. Basu, A.R. McGhie, J.E. Fischer: Mater. Sci. Eng. **40**, 1–4 (1979)
2.84a N.-X. Chen, S. Rabii: Phys. Rev. Lett. **52**, 2386–2389 (1984)
2.85 M.Y. Chou, M.L. Cohen, S.G. Louie: Phys. Rev. B **33**, 6619–6626 (1986)
2.86 W. Rüdorff, E. Schultze: Z. Anorg. Chem. **227**, 156 (1954)
2.87a P. Lagrange, D. Guérard, A. Hérold: Ann. Chim. (Paris) **3**, 143 (1978)
2.87b Y. Takada: J. Phys. Soc. Japan **51**, 63–72 (1981)
2.87c H. Kamimura: In Ref. [2.7] pp. 80–89
2.88 R. Al-Jishi: Phys. Rev. B **28**, 112–116 (1983)
2.89 A. Shimizu, H. Kamimura: Syn. Met. **5**, 301–313 (1983)
2.90 M.E. Preil, J.E. Fischer, S.B. DiCenzo, G.K. Wertheim: Phys. Rev. B **30**, 3536–3538 (1984)
2.91 D. Guérard, M. Chaabouni, P. Lagrange, M. El Makrini, A. Hérold: Carbon **18**, 257–264 (1980)
2.92 H. Akera, H. Kamimura: Syn. Met. **12**, 275–280 (1985); H. Akera, H. Kamimura: J. Phys. Soc. Japan **55**, 2326–2337 (1986)
2.93 P. Lagrange, M. El Makrini, A. Hérold: Rev. Chem. Miner. **20**, 229 (1983)
2.94 T. Saito, K. Nomura, K. Mizoguchi, K. Mizuno, K. Kume, H. Suematsu: J. Phys. Soc. Japan **58**, 269–278 (1989)
2.95 S. Safran: Phys. Rev. Lett. **44**, 937 (1980)
2.96 N.A.W. Holzwarth: Phys. Rev. B **21**, 3665–3674 (1980)
2.97 T. Ohno, N. Shima, H. Kamimura: Solid State Comm. **44**, 761–765 (1982); T. Ohno, H. Kamimura: Physica **117B & 118B**, 611–613 (1983); T. Ohno, H. Kamimura, J. Phys. Soc. Japan **52**, 223–232 (1983)

2.98 C.T. Chan, W.A. Kamitakahara, K.M. Ho, P.C. Eklund: Phys. Rev. Lett. **58**, 1528–1531 (1987); C.T. Chan, W.A. Kamitakahara, K.M. Ho: Syn. Met. **23**, 327–332 (1988)

2.99 L. Pietronero, S. Strässler, H.R. Zeller, M.J. Rice: Phys. Rev. Lett. **41**, 763–767 (1978); L. Pietronero, S. Strässler, H.R. Zeller: Solid State Comm. **30**, 399–401 (1979)

2.100 S. Strässler, L. Pietronero: Festkörperproblem **21**, 313–324 (1981); L. Pietronero, S. Strässler: Phys. Rev. Lett. **47**, 593 (1981)

2.101 J. Blinowski, C. Rigaux: J. Phys. (Paris) **41**, 667–676 (1980): J. Blinowski, C. Rigaux: J. Phys. (Paris) **45**, 545–555 (1984); C. Rigaux: In Ref. [2.12] pp. 235–255

2.102 M.H. Yang, P.C. Eklund: Phys. Rev. B **38**, 3505–3516 (1988)

2.103 S.A. Safran, F.J. DiSalvo: Phys. Rev. B **20**, 4889 (1979)

2.104 R. Saito: In Ref. [2.11] pp. 189–198; R. Saito, H. Kamimura: Phys. Rev. B **33**, 7218–7227 (1986)

2.105 S.Y. Leung, G. Dresselhaus: Phys. Rev. B **24**, 3490–3504 (1981)

2.106 G. Campagnoli, E. Tosatti: J. Phys. C **15**, 1457–1480 (1982)

3. Electron Spectroscopy of Graphite Intercalation Compounds

By Robert Schlögl

With 29 Figures

This chapter deals with experiments designed to probe the electronic structure of GICs. A vast number of such experiments have been carried out since the explosion in GIC science in the late 1970s. The possible applications of GICs led to many experiments probing the electronic structure very near the Fermi edge in energy and investigating the shape of Fermi surfaces. These experiments and their comparison with various theoretical models led to the formulation of models for the charge distribution in GICs that will be described in this chapter. The experiments and corresponding background information can be found in review articles [3.1–3] and in other chapters of this book.

Turning to spectroscopic experiments probing a large portion of the energy distribution or even measuring core-level properties of GICs, the discussions in the literature concentrated on the charge distribution between host graphite and guest metals or molecules. Important experimental methods in this field are such photoemission techniques as ultraviolet photoelectron spectroscopy (UPS), X-ray photoelectron spectroscopy (XPS), and soft X-ray emission spectroscopy (SXS). This chapter deals with practical aspects in the application of these methods and discusses results published in the literature as well as new case studies. Due to practical problems only a limited number of GIC systems, largely of the donor type have been reported in the literature so far. This chapter reviews a large number of previously unpublished results on a variety of GIC systems thereby extending the experimental basis for the discussion of the electronic structure of GICs. Only a little work has been done so far to spectroscopically map out the band structure of GICs (Sect. 3.2.2c), and direct information about experimental band structures by electron spectroscopy is thus very scarce. Most work was done with angle integrating experiments, which permit extraction of only limited band-structure information. This chapter thus concentrates on the issue of charge transfer, which is adequately dealt with by electron spectroscopy because the latter can probe the electronic structures of both guest and host.

The electronic structure of GICs is studied successfully with methods other than electron spectroscopy: separate chapters in this book deal with these issues. A few results from the application of some of these methods are given in the following section because they were used in the interpretation of findings from electron spectroscopy. We attempt to derive a picture of the present state of knowledge about charge distribution.

Springer Series in Materials Science, vol. 18
H. Zabel · S.A. Solin (eds.): Graphite Intercalation Compounds II

EXAFS and XANES can be used to determine the local structure and the chemical identity of intercalated acceptor molecules [3.4]. A more general description of the method and its application to GICs is given by *Bonnin* and *Kaiser* [3.5]. Mössbauer spectroscopy has now been extended [3.6] to noniron species and successfully used to determine structural, chemical, and charge transfer information from tin [3.7] and antimony [3.8] halide acceptor intercalates. Antimony halide GICs served also as a study object for a remarkable cooperative project in applying Fermi-surface methods to elucidate relationships between geometric and electronic structures [3.9]. Electron-energy-loss spectroscopy has been applied again to study donor GICs [3.10], and Compton scattering in light-element GICs has been applied to test potential models for theoretical work [3.11]. The application of magnetic resonance of both electrons and nuclei in elucidating models of electronic structure and charge transfer has recently been reviewed by CONARD [3.12].

The vast amount of transport measurements and their interpretation in the light of structural effects has been summarized [3.1, 13], with recent updates in other chapters of this book. A very rigorous study of electrical conductivity of a wide range of materials over a large temperature regime led to some general conclusions about the structure-conductivity relationship valid for all types of GICs. A key observation was that liquid-like intercalate structures result in high specific conductivities [3.14, 15].

These few results are given to draw the reader's attention to the many facets of the "simple" fundamental problem of valence charge distribution in GICs. In the following, we review pristine graphite and the methods of photoemission for GICs. Then we formulate the problem in the framework of localized charge transfer between host and guest, i.e., as in a chemist's picture. Answers to the problem for archetypal donor and acceptor GICs will be discussed, followed by some case studies that illustrate how the chemical complexity of intercalation augments the basic charge transfer problem.

3.1 Essential Concepts

3.1.1 Pure Graphite as a Host for Intercalation

The basic concepts and theoretical models of the electronic structure of graphite as seen in the framework of band-structure theory are concisely developed by *Holzwarth* in chap. 2 of this book. Previous work is discussed in the review of *Dresselhaus* and *Dresselhaus* [3.1]. Attempts to describe the electronic structure of GICs more in detail have had to consider the problem in a full three-dimensional framework, rather than in an idealized two-dimensional picture. However, GICs are characterized by a large anisotropy in chemical bonding, which, for both graphite and intercalate, is strong in the basal plane (a, b) directions and very much weaker in the perpendicular stacking (c) direction.

The electronic interplay of host and guest, which is an overall weak interaction on the scale of chemical bond energies, is controlled by minute structural details influencing the local geometry of interactions in all three directions of space. This sensitivity, referred to as zone folding in the band-structure framework, indicates in the chemical bonding a covalent interaction by orbital overlap and hybridization rather than bond formation through electrostatic interactions of ions.

The electronic processes of intercalation can be described in the approximation of local interaction as ion-pair interaction, as alloy formation, or as a Lewis acid-base reaction. In all cases, graphite is amphoteric; i.e., it can accept electrons from or donate electrons to the intercalate, a property unique in intercalation chemistry.

The picture of ions of graphite hexagons interacting with intercalated cations (donor GICs) or anions (acceptor GICs) involves the transfer of full electrons (rather than partial charges) in redox reactions, and was frequently put forward because of its conceptual simplicity allowing quantitative treatment in theories. The procedure of electrochemical intercalation and the application of GICs as battery materials seem to offer strong experimental support for this model.

The two other models build on sharing of electrons between graphite and intercalate and require rehybridization of molecular orbitals. These effects are difficult to treat quantitatively; thus an aim of experimental electronic structure investigations is to discriminate between redox and acid-base reactions as prototype chemical interactions in intercalation. This problem is much more demanding for the three-dimensional solid GICs than for many existing molecular analogies, where ion pairs between alkali metals and aromatic radical ions with sandwich structures have been synthesized and characterized in their electronic structure [3.16]. One significant difference is the restriction of the interaction to graphite π electrons; the σ system of graphite has to remain largely unchanged to preserve the planar structure of infinite polymers of benzene rings. In some cases where the intercalate interacts with the graphite σ system (overoxidation in electrochemical intercalation, carbon fluorination or chlorination), the overall geometric arrangement is changed from the planar system, and complete loss of the metallic properties is observed. This unwanted strong interaction has been termed "chemical charge transfer" [3.13]. The rigidity of the graphene layers and the exclusion of significant σ interactions are reasons for the overall weak chemical bonding of the graphite–intercalate complex and represent the major differences to molecular analogues.

Staying with the molecular picture of aromatic sandwich complexes, we can consider intercalation as the reduction in the π bond order (acceptor) or increase of charge density in antibonding π orbitals. Both processes will be tolerated by the system to only a small extent before loss of the aromatic delocalization and geometric rearrangement occur. Consequently, the overall interaction will be limited and an intercalation equilibrium with only a shallow minimum in the total energy of the system will result, which corresponds well to the experimental situation. With this picture it is easy to understand why GICs are all so air sensitive: the aromatic carbon structure is perturbed and thus chemically activated for redox reactions, leading to thermodynamically strongly favored carbon oxygen (from oxygen or water in air) or carbon halogen bonds (from hydrolized intercalate molecules).

In the band-structure framework, graphite as a semimetal is characterized by a very low density of states near the Fermi edge and a very anisotropic distribution of this density of states in space. Intercalation will thus shift the Fermi energy to a much larger extent than in alloys of normal metals. This has severe consequences for the spectroscopy of GICs, as will be illustrated below. Furthermore, it limits the number of states in a given energy interval determined by the intercalate's chemical potential much more than in conventional metals. This in turn limits the total extent of electronic interaction. For the spectroscopic experimentalist the chance arises to depict directly the intercalation-induced changes in the energy distribution, since they will be on the order of 1 eV rather than of a fraction of an electron volt, as in many conventional alloy systems.

The valence electronic structure of graphite has recently been reinvestigated by *Takahashi* et al. [3.17] and compared with a variety of theoretical models. The agreement between the high angular resolved experimental data and the theories is good, particularly with the model of *Holzwarth* et al. [3.18]. These measurements provide for the present work essential information for the assignment of the angular integrated spectra commonly measured and are also in good agreement with the pioneering experimental work reported by *McFeely* et al. [3.19]. Angular resolved photoemission data of prototype donor GICs are also reported in the literature and will be discussed below.

Figure 3.1 shows the valence band spectra of clean single crystalline graphite (HOPG as a clean approximation of a single crystal in the stacking direction) for different experimental conditions and for an emission cone perpendicular to the basal plane. Cleanliness of the surface, which can be achieved only by thermal annealing followed by in situ cleavage, and precision in the microscopic orientation, influence the rich structure of the spectra. The dependence of the spectral shape on the analyzer mode, in particular for the spectrum excited by the He I resonance (21.2 eV), is clearly visible. This needs to be taken into account when spectra from different sources are compared: analyzer mode and transmission factors modify the overall spectral shape. For a straightforward comparison in this work many UPS data were remeasured with the appropriate instrumental parameters. The use of different parameters is justified by the different spectral sensitivities emphasizing the Fermi-edge features in the retarding ratio mode and the more strongly bound states, plus secondary electron features in the pass-energy mode.

Variation of the excitation energy helps to assign the spectral features due to the dependence of the cross section of states with different symmetry on the excitation energy. This dependence, reviewed by *Cardona* and *Ley* [3.20], permits emphasis of graphite π bands with He I (21.2 eV) radiation, σ bands

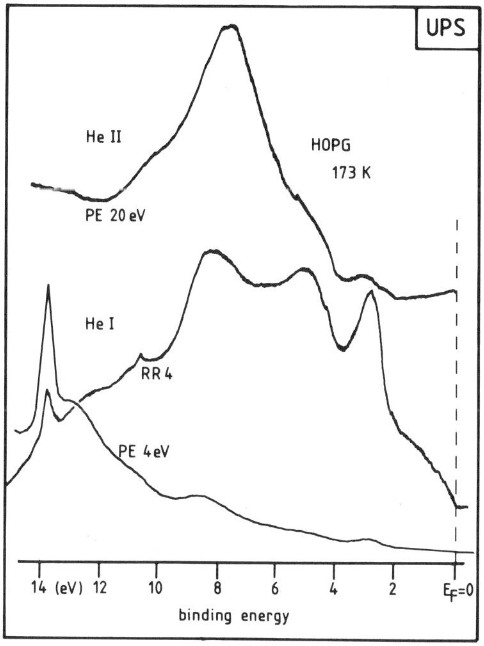

Fig. 3.1. Valence band spectra of graphite (0001). The HOPG surface was cleaved in UHV at 78 K and subsequently heated to 173 K. The two He I spectra show the influence of the analyzer operation mode on the shape of the spectrum. The finite intensity close to the Fermi energy in the He II spectrum arises from a He satellite

with the He II (40.8 eV) radiation, and intercalate states with X-rays excitation (Mg$K\alpha$ 1254.6 eV).

An unambiguous assignment of π and σ bands in angle integrated spectra can be obtained by comparing UPS data with SXS data. The fundamental work was done by *Bey reuther* et al. [3.21]. Using the angular dependence of the emitted intensity as a function of the symmetry of the initial valence state and applying polarized light from a synchrotron source, we can numerically separate the experimental DOS into σ and π bands, provided that the normalization problem is suitably solved [3.21, 22]. In this way, a π band containing the top of the valence band with a spectral width of ca. 10 eV and a σ band of ca. 15 eV width overlap in a region of ca. 5 eV and yield an experimental graphite valence band of ca. 22 eV width. This deconvolution agrees fairly well with a theoretical partial density-of-states map based upon an ab initio pseudopotential calculation [3.22].

The core-level spectrum of pure graphite consists of a single line from the C 1s emission. This spectrum is frequently compared for pristine and intercalated graphite in order to deduce charge transfer information. To interpret spectral differences in a meaningful way, some information about the reference spectrum is required. The position of the line was finally fixed to 284.45 $\pm$ 0.05 eV [3.23] after some discrepancies in the early work [3.20, 24] were resolved. Its lineshape is asymmetric with a tail to higher binding energies (b.e.) arising as a final state effect from the coupling of the core hole with the valence band [3.25]. A rigorous investigation into the asymmetry of the line based on mapping of both occupied and unoccupied valence states by XPS and bremsstrahlung isochromate spectroscopy (BIS), respectively, revealed the complexity of the coupling between core and valence states in graphite [3.26]. Thus, no detailed theory for the lineshape of the graphite C 1*s* emission exists today.

The issue is further complicated by the fact that surface contaminations and structural defects influence the position and lineshape of the peak. Heteroatoms such as oxygen or halogens result in detectable screening shifts for part of the carbon atoms, giving rise to structures at the higher b.e. side [3.27, 28]. For GICs this means that all unwanted perturbations of the σ system caused by intercalation chemistry can be easily detected by XPS, but the perturbations obscure important information, e.g., about unevenly distributed layer charges.

The position of the peak depends on the overall structural integrity of the sample as established in a linear correlation between C 1s peak position and the lattice parameter in the *c* direction [3.27]. Interpretation of small intercalation shifts may thus be ambiguous because of the inevitable structural modifications of pristine graphite upon intercalation. Introduction of many localized defects, as with mild ion sputtering, strongly affects both lineshape and line position [3.28]. These drastic effects can be related to the fundamental changes in the corresponding valence band spectra, which indicate the generation of a high density of states after chemical or ion etching. These states are, however, not additional π states but must be considered as strongly localized σ states, into which the former π states have been rehybridized as a consequence of bond breaking at defect sites.

For GICs we can conclude that creation of a limited number of defects during intercalation can strongly influence the graphite spectrum and so strongly perturb the identification of true intercalation-induced changes of the overall electronic structure. Mild intercalation conditions and structurally well-defined samples of GICs are thus essential prerequisites for useful spectroscopic work. Finally, work reported in the literature [3.24, 25, 27, 28] shows that the linewidths of spectra from different graphites differ from 1.7 eV down to ca. 1.1 eV. Unresolved chemical shifts due to defects are important, as is lifetime broadening, i.e., the efficiency of core-hole relaxation from weakly bound valence electrons. This valence band relaxation will sensitively depend on the density and symmetry of states available and thus reflect both defect formation and modification of the overall electronic structure.

In summary, we can conclude that the easy way of comparing spectra between pristine graphite and GICs and interpreting the spectral differences as intercalation-induced chemical shifts is rather dangerous. All parameters that influence the graphite spectrum are changed simultaneously upon intercalation, and not only the charge transfer is an unknown quantity among these parameters. Such effects are well known in many other systems studied by photoemission, including metals, but there they generally obscure the spectroscopic information to a much lesser extent than in the low density of states semimetal graphite.

3.1.2 The Charge Transfer Problem

The strong anisotropy in chemical bonding of graphite allows us to treat the modifications of its electronic structure upon intercalation within the framework of a perturbation of the parent electron distribution. The intercalate serves as a source (donor) or sink (acceptors) for electrons. In complex systems such as ternary donor GICs some interactions between the intercalated elements need to be considered to account for the reduced electron donation of the alloy intercalated.

The anisotropy led also to the simplified treatment of intercalation as a two-dimensional effect. A continuum of a charged layer, donors or acceptors, is placed between two symmetrically equivalent graphene layers, the electronic structure of the latter being represented by a perturbed two-dimensional calculation of graphite [3.1, 13, 29]. Such models served well for the semiempirical description of GICs with a small concentration of intercalates, i.e., for the dilute limit. Within this limit it is appropriate to consider intercalation-induced shifts of the Fermi edge by modifying the occupation of states derived from rigid graphite bands. For donor GICs, the Fermi level of the synthetic metal moves closer to the vacuum level; for acceptor GICs it moves further away relative to the Fermi edge of pristine graphite. Its position is given in the two-dimensional limit [3.29] by the degeneracy of the π valence and π conduction bands (near the edge of the Brillouin zone, point P in [3.29]). Graphite is thus considered as a zero-gap semiconductor.

In the three-dimensional electronic structure calculation of *Slonzewski* and *Weiss* [3.30], this degeneracy is lifted by the weak interlayer interaction, leading to an overlap of the valence and conduction bands of ca. 0.04 eV to render graphite a semimetal. The density of states is for ca. 2 eV above and below the overlap region, rather symmetric with very low occupation. The symmetry of the states is exclusively π-like. Only ca. 5 eV below the Fermi edge, the π states overlap with the tops of the σ bands, which dominate ca. 10 eV below the Fermi edge. The edge jump in the DOS cannot be measured with spectroscopic methods, because of its extreme low occupancy.

This view of graphite was the basis for the theoretical concept of moving the energy distribution of rigid bands for dilute intercalates [3.31], frequently confirmed by experimental studies. Soon it was realized that this method was inadequate for concentrated GICs (stage index of 1 to ca. 4). For more rigorous treatments, a full three-dimensional picture is required, taking into account the modified geometry of the Brillouin zones as well as suitable potentials for the carbon atom and the intercalates. Up to now, this problem has been treated only for first-stage donor GICs, in particular for LiC_6 and KC_8. All acceptor GICs are still treated in semiempirical theories within the rigid-band limit.

In all theoretical concepts, the GICs are viewed as an idealized structure with infinite perfect layers for both graphite and intercalates (referred to as the *Rüdorff* model [3.32]. But it is now well established that most GICs are built from islands of intercalate atoms separated by domain boundaries commonly known as *Daumas-Herold* structures [3.33]. Furthermore, a regular homogeneous structural relationship between graphite and intercalate is assumed to construct the Brillouin zone of the GIC from the graphite structure with a modified reciprocal-lattice parameter. This technique allows us to fold the bands of pristine graphite back into the smaller Brillouin zone of the intercalation compound keeping symmetry and band parameters from the graphite structure [3.1, 13]. Such idealizations, disregarding the real structure and the dynamics of the solid, are presently unavoidable. Although they hamper the comparison between theory and experiment and lead to controversies in the literature, these theoretical models have contributed significantly to the development of the experimental spectroscopic understanding of GICs.

Recently the band-structure model of graphite was augmented by a new so-called interlayer band crossing the Fermi energy, which is believed to be empty in pristine graphite but to have a finite population in GICs [3.34].

The sandwich approximation of the GIC structure with complete screening of the transferred charge by the adjacent graphene layers and the rigid-band approximation can be well supported for many GICs using NMR, a technique probing the electronic ground state of matter. Only first-stage compounds of the heavy alkali metals were found to deviate clearly from the rigid-band model [3.12].

The degree of deviation between a given experiment and the rigid-band concept depends on the method used for probing the electronic structure of the GIC. The experiments either relate to only a window in energy of the DOS or suffer from technical problems such as cross-section modulation or resolution. In both cases, experiments yield only perturbed information about band structures.

In summary, the charge transfer problem can be described as follows: For all acceptor GICs and for dilute donor GICs the electronic structure of the resulting synthetic metal can be derived from the parent graphite structure by shifting the Fermi level relative to the vacuum level in rigid bands. For concentrated donor GICs this model is inappropriate, and a completely new band-structure calculation in a modified Brillouin zone is necessary. The observed changes in the DOS of GICs can be achieved in two ways:

(1) By redox reaction of the intercalate, leading to solvate stabilized or free ions, with transfer of complete charges (ionic model).

(2) By a covalent bonding interaction, e.g., between alkali-metal valence s states and graphite π states and/or the graphite interlayer band (covalent model).

3.1.3 Concepts of Charge Transfer

The concepts of charge transfer may be grouped into two categories – those assuming the transfer of full electronic charges (electrons), and those assuming the transfer of partial charges. Concepts in the latter category require covalent interactions between host and guest, i.e., an orbital overlap between graphite states and molecular/atomic states of the intercalate. Concepts with full electron transfer are well known to chemists as redox reactions. Partial charge transfer occurs in chemistry with alloys and coordination compounds. The chemistry of these two concepts is vastly different, but this discrimination may appear rather artificial to a physicist, since band-structure theories can accommodate any degree of charge transfer.

It seems immediately obvious that electron transfer processes with non-*d* elements should involve redox reactions, i.e., the formation of ions intercalated in the macrocounterion graphite. This notion is exemplified by a band-structure calculation of LiC_6 (stage-1, concentrated limit) suggesting a charge transfer formula $Li^+C_6^-$ [3.35] or a spectroscopic investigation of EuC_6, leading *Kaindl* and coworkers [3.36] to a formulation of $Eu^{2+}C_6^{2-}$, which reminds the chemist of metal carbides rather than of GICs.

Some acceptor GICs called graphite salts are clearly the result of a redox reaction which can be quantitatively studied in electrochemical cells. In these reactions graphite is oxidized and protons are reduced to hydrogen. A typical example is a sulfuric acid GIC with $C_{24}^+\ HSO_4^- \times nH_2SO_4$, consisting of graphite cations and solvated intercalate anions.

The case of the metal halide GIC is more complex. Formulations like $C_{27}^+AlCl_4^- \times nAlCl_3$ suggest also a redox reaction with a reduced oxidizing agent such as chlorine gas not being present within the GIC. Such reactions may, on the other hand, be regarded as Lewis acid-base reactions between the Lewis base Cl^- and the Lewis acid $AlCl_3$, with the graphite being oxidized in a side reaction resulting in covalent CCl_x groups. The problem cannot be resolved by accurate chemical analysis (determination of the chlorine excess), as implied by the sum formula given above, because the interpretation of the chemical analysis requires knowledge of the chemical structure of the intercalate. In the case of aluminium chloride, the intercalate is a complex structure based on a sandwich of chloride ions with a distribution of sites for aluminium ions [3.37]. In this way local structures with terminating and bridging chlorine ions occur, in which the charges at the anion sites are different (formal: Cl terminal -1. Cl bridging -0.5).

Such a macroanionic structure of the intercalate, which may be regarded as a defect structure of a block of the parent metal chloride, obscures the relation between a chemical analysis and the electronic structure, because the self-reaction of the intercalate has produced the excess charge and not the charge transfer from the graphite. Aluminium chloride has reacted with Cl^- ions generated from oxidation of graphite with chlorine gas (unwanted side reaction of intercalation) to form $AlCl_4^-$ ions embedded in the macroanion "intercalated $AlCl_3$". None of the excess charges is thus a consequence of the reversible charge transfer guest-host, which is a collective interaction between the graphene layer and the macroanion leading to a small additional negative charge on the intercalate. This may be stored in polarized Al–Cl chemical bonds, rather than in individual anions (Sect. 3.4).

A discussion of this effect, which has generated many controversies about the charge transfer, can be found in the literature [3.4, 8, 38]. It is the covalent bonding in the molecular structure of the intercalate that introduces this complication of polyvalent anions. The ionic character of the charge transfer in most acceptor GICs is thus difficult to assess and not a clearcut example, as assumed in former models of GICs. Such a model based on optical studies was formulated by *Dzurus* and *Henning* [3.39] and had a strong influence on theoretical studies of acceptor GICs. It explained the metallic character of GICs on the basis of a structure $C^{n+,-}$ Intercalate$^{n-,+}$, with the excess carriers in the graphite responsible for the transport properties.

In donor GICs the idea of metal cations and graphite anions has frequently been the basis of discussions. Some chemical arguments illustrate the difficulties with this ionic model. First, the interlayer distance of donor GICs is too large for metal cations and slightly too small for free metal atoms. Consequently, the characteristic distance between atoms in free alkali metals is slightly larger than in first-stage GICs. This reduction in molar volume has been interpreted in terms of ionization of the alkali metal [3.40], but is in contradiction to the notion of free cations of alkali-metal atoms, the repulsive Coulomb force of which would increase the interionic distance. *Dresselhaus* and *Dresselhaus* [Ref. 3.1, p. 187] have shown that these GICs may be well described with a soft sphere model of metal and carbon atoms with the metal spheres overlapping the carbon spheres. Such a structural model is consistent with the covalent charge transfer model requiring orbital overlap.

The fact that the in-plane density of alkali metal decreases between stage-1 and stage-2 compounds is also inconsistent with the view that ions in the first-stage GICs should have their equilibrium positions determined by the repulsive forces. It is generally assumed [3.1, 2, 13] that the charge transfer per intercalated atom increases from stage-1 to stage-2, which again is inconsistent with an ionic formulation of first-stage GICs, but agrees well with a partial charge transfer and a dependence of the charge transfer fraction on the density of the intercalate (stage index).

A variety of ternary GICs with alkali-metal and main-group-metal alloys have been synthesized [3.2, 41, 42]. These sandwiches of alloy compounds with strong intrametallic interaction [3.42, 43] exchange significantly less charge with graphite than in the corresponding binary alkali GIC.

Alkali-metal GICs can react with a variety of organic molecules to form ternary compounds [3.44]. Some of these reactions involve the reduction of the organic material, which is incompatible with an ionic model of the alkali intercalate. There is, however, little doubt that many of these interactions are ion-dipole interactions and may therefore be described as solvation of alkali metals by organic donor solvents. Significantly, the solvation of KC_{24} with tetrahydrofurane did not modify the interaction between graphite and potassium [3.45], indicating that only part of the alkali-metal valence charge is donated to graphite, even in a stage-2 compound.

The chemical reactivity of concentrated donor GICs against many acceptor molecules is identical to that of a highly dispersed free alkali metal [3.46–48] and gives little evidence that the electron acceptors react with graphite anions.

In summary, there is evidence that the charge transfer in some GICs is adequately described by the ionic model. In many acceptor compounds, however, the degree of ionic charge transfer (or ionicity) is often overestimated, due to the neglect of self-reactions of the intercalate or oxidation of the intercalate by auxiliary oxidation agents such as free halogens. In concentrated donor GICs the ionic model is not adequate to describe the charge transfer, as follows from some chemical arguments. These GICs may be better described in a covalent model with localized orbital overlap. This model should be regarded as adequate for most GICs, with the ionic model being an extreme boundary case.

3.2 The Situation in the Literature

3.2.1 Charge Transfer in Theories of KC_8

In this section we review a few details about band-structure calculations. A more detailed analysis is provided by *Holzwarth* in Chap. 2 of this volume. The present discussion focuses on chemical implications from the theoretical work that may be observed in electron spectroscopy. Most of the analyses described for potassium–graphite have been extended to other heavy alkali metals and to intercalated barium without many differences from the results described here.

The first calculation was done in a tight-binding approximation neglecting all graphite σ orbitals [3.49]. The GIC is described as a metal with two conduction bands that are degenerate at the edges of the Brillouin zone. The Fermi surface was found to consist of a three-dimensional central feature and at the edges of the Brillouin zone two-dimensional features that resemble those of pristine graphite. Intercalation thus produces a three-dimensional new Fermi surface in the formerly empty center of the Brillouin zone.

These results were confirmed by a self-consistent calculation of *Ohno* et al. [3.50] using atomic pseudopotentials. According to this calculation, the top of the graphite π bands are placed below the potassium 4*s* states, leading to partial charge transfer of 0.6 states per potassium atom, which would be consistent with the covalent model.

However, the band-structure calculation of *Ohno* et al. is in complete contradiction to another calculation by *DiVincenzo* and *Rabii* [3.51] using muffin-tin potentials in the framework of the KKR theory [3.51]. Here the potassium 4*s* band is located 1.8 eV above the Fermi level and thus is completely empty, leading to a complete charge transfer in agreement with the ionic model. The lowest π band in the GIC is of the same shape as in pure graphite, but shifted relative to the σ states by 2 eV to lower energies.

In a further self-consistent calculation by *Tartar* and *Rabii* [3.52] also using atomic pseudopotentials in the density functional approximation, the three-dimensional feature in the center of the Brillouin zone was reproduced as a pocket, and large trigonal prisms along the graphite planes were found at the edges of the Brillouin zone.

All these calculations agree that the Fermi surface consisting of two elements – a central three-dimensional feature and a structure at the edges of the Brillouin zone. Depending on the theory used, this is more or less similar to the Fermi surface of free graphite. Controversy exists about the origin of the central feature in the calculation of *Tartar* and *Rabii* [3.52]. Evaluation of the charge-density distribution led to the conclusion that graphite π states from bands running along the in-plane directions constitute the Fermi-surface feature. A similar analysis of the calculation of *Ohno* et al. [3.50] led to a rescaling of the charge transfer in favor of the ionic model and to a population analysis ascribing the central Fermi surface to the manifestation of the interlayer band [3.34] running along the c direction perpendicular to the graphite plane [3.53]. The electrons in this band are expected [3.34] to behave like free electrons without much dispersion.

In summary, these theories agree now in the complete charge transfer from the potassium atom. Their Fermi surfaces are made up exclusively from graphite-derived states, but differ in the location of the states, which should be found either in the graphite planes [3.52] or in the space between the layers [3.53].

Recently, the issue of charge transfer again became controversial with a calculation by *Mizuno* et al. using a self-consistent LCAO-based method in the framework of the local density functional formalism [3.54]. These authors found two inequivalent types of carbon atoms (closer or further from the potassium atom), the interaction of which influences the graphite π bands to form a central three-dimensional Fermi surface in the Brillouin zone. The charge transfer was found to be 0.6 electrons per potassium atom. Their calculation illustrated how the degeneracy of graphite π bands is lifted by allowing the electronic configuration of the three atoms involved to deviate from unity. The final self-consistent configuration is given with

- C type A: $2s$ 1.25 $2p$ 3.05 (sp^2 + 1.30 delocalized electrons)
- C type B: $2s$ 0.98 $2p$ 2.87 (sp^2 + 0.85 delocalized electrons)
- potassium: $4s$ 0.40

Intercalation influences all valence states of the covalent host material graphite, including the σ states. Therefore, it is a phenomenon that requires covalent interactions leading to rehybridization of all valence states of graphite. Although potassium states are not involved in the Fermi-edge feature in this calculation, covalent interactions may have occurred with lower-lying states of suitable symmetry. Otherwise, it is difficult to rationalize the partial charge transfer from the potassium atom. The conduction band should exhibit a modified π symmetry resulting from the above-mentioned interaction of K 4s states with graphite states below the Fermi edge.

3.2.2 The Charge Transfer Problem in Spectroscopic Experiments

The following sections consider the experimental situation concerning the charge transfer in GICs for several methods. Only a few experiments with photon or electron emission techniques have been performed for acceptor GICs. The discussion will concentrate on the heavy alkali metal GICs, for which most information is available.

The general impression from the literature is that there are only minor differences in the results for various intercalated alkali metals. We refer also to Chap. 4 in this volume by *Eklund* and *Doll*, which discusses the charge transfer problem from the point of view of optical reflectance measurements.

(a) Soft X-Ray Emission Spectroscopy (SXS)

Soft X-ray emission investigates the energy distribution from photons created in the relaxation of a core level by valence-band electrons. The technique is element specific and has been applied to the investigation of both the graphite valence band and the alkali-metal valence states. The large energy difference between initial and final states excludes complicating selection rule effects [3.20]. The narrow linewidth of the initial core level allows an energy resolution better than 1 eV. The symmetry of the valence states can be determined directly from angle-resolved measurements. Excitation can be achieved either by gentle photon irradiation or by a more drastic electron beam irradiation. The technique is only moderately surface sensitive, with information ranging from a depth of ca. 1–3 μm. An introduction to the specific application of this technique to graphite is provided in by *Beyreuther* et al. [3.21].

The first C–K emission study of stage-1 and stage-2 potassium GICs was reported by *Eisberg* et al. [3.55]. The key result of this synchrotron excitation study was the assignment of the Fermi-edge feature to π symmetry. The intensity of this feature, as all other deviations from the spectrum of pristine graphite is strongly stage dependent; i.e., it depends directly on the concentration of donor atoms. The investigation shows clearly that the rigid-band approximation is inappropriate for these GICs, since all spectral features change upon intercalation. There is a pronounced redistribution of spectral intensity within the π band: the distance of the maxima between, the π and σ bands reduces by 0.4 eV, and the π-band maximum shifts in energy by 0.7 eV to higher energies. The experiment was repeated by *Mansour* and coworkers [3.56] in a laboratory electron excitation experiment applied to the first stages of lithium, potassium, and cesium GICs. The spectral results are in good qualitative agreement with the data reported by *Eisberg* et al. [3.55]. *Mansour* et al. [3.56] tried, however, to quantitatively evaluate charge transfer numbers and arrived at a complete charge transfer from the alkali atoms. They assumed rigid-band behavior in their normalization procedure and thus overestimated the intensity of the

Fermi-edge feature. The discrepancy in the interpretation of these results in terms of partial [3.55] versus complete [3.56] charge transfer can be explained by an inappropriate data analysis [3.22].

The synchrotron-excited SXS measurements were extended to aluminum chloride graphite (stage 1) and compared with direct valence-band photoemission data excited with He II radiation [3.57]. In the Fermi-level shift, acceptor GICs exhibit the inverse sign of donor GICs. However, the magnitude was with 0.7 eV exactly the same as for stage-1 potassium GIC. Unlike with the donor GICs, good agreement in the lineshapes of the π and σ bands of the GICs with those of pristine graphite was found, indicating the validity of the rigid-band approximation for even concentrated acceptor GICs. No strong edge feature for the conduction band of the acceptor GICs was found in both SXS and direct photoemission. The lineshapes of the valence bands in both techniques agree surprisingly well, and by direct comparison an unambiguous assignment of the UPS features to the π and σ bands is achieved. A molecular state arising from chlorine $3p$ of the intercalated molecules was identified 3.5 eV below the Fermi edge. This state is a fingerprint for the presence of Al_2Cl_6 dimers, as concluded from comparison with gas phase UPS data of $AlCl_3$.

The issue of charge transfer was also discussed in an experiment carried out with the potassium SXS. In a laboratory X-ray-excited experiment, *Hague* et al. [3.58] analyzed the photon energy distribution from a valence-band transition in the K $1s$ level. After a suitable satellite subtraction arising from the dominant transition of the K $3p$ shallow core level (at ca. 18 eV binding energy), the authors arrived at a valence-band feature for free metallic potassium that agrees well with a theoretical calculation within the ASW frame [3.59]. In this and a further theoretical calculation [3.60], it was pointed out that the conduction band of metallic potassium is not purely $4s$-like, but contains a substantial π character arising from hybridization of K $4s$ with K $4p$ states. Analysis of the same spectrum of stage-1 potassium GIC yielded three distinct features in the conduction-band region at 0.8 eV, 4.3 eV, and 8.8 eV. These features, which are absent in the spectrum of pure potassium metal, coincide well with features found in the SXS experiments of *Eisberg* et al. [3.55] and *Mansour* et al. [3.56]. These results thus very strongly support the incomplete charge transfer from alkali metal to graphite (finite intensity) and the covalent rehybridization model for the mechanism (appearance of "graphite" peaks in the "potassium" spectrum).

The SXS experiments probing the local DOS at the donor sites were continued with stage-1 cesium graphite [3.61]. Again a finite intensity of cesium $6s$ emission was detected. The SXS observations indicate a substantial involvement of π states from both graphite and alkali metal.

In summary, the SXS experiments support the following statements about charge transfer, which were already brought up in the general and theoretical discussion above:

(1) Charge transfer in many acceptor [3.61] and dilute donor GICs can be described within the rigid-band approximation.

(2) The model of intercalation-induced shifting of the Fermi edge through the graphite valence band is qualitatively correct.
(3) In concentrated donor GICs, strong non-rigid-band effects can be observed.
(4) The charge transfer in these GICs is not unity; i.e., the donor atoms are not completely ionized.
(5) The charge transfer involves rehybridization of both the graphite π states and the alkali s/p states and leads to a covalent alkali-graphite bond.

These statements may not be valid for the case of LiC_6, where decisive experiments are still lacking. For all other investigated systems, they are in good agreement with the results from the groundstate spectroscopy of NMR [3.12].

(b) Photoelectron Spectroscopy

The first attempts to apply electron spectroscopy to the problem of charge transfer in GICs were made in the early 1970s. These experiments suffered greatly from their disregard of surface sensitivity. The experiments were done in oil-contaminated vacuum without cleaving the sample in situ. *Bach* and *Thomas* [3.63] give an early valence-band spectrum of the acceptor bromine GIC. The authors claim to have identified the acceptor nature of the GIC by a drastic change in the shape of the overall valence-band spectrum. *Schlögl* and *Boehm* [3.64] investigated a variety of donor and acceptor GICs under ill-defined conditions, and presented the idea of chemical modifications of intercalated acceptor molecules in their interpretation of the spectra. They rejected the prevailing view that donor GICs consist of, e.g., C_8^- anions and K^+ cations but the argument was based on chemical shift arguments that did not fully take into account the effect of Fermi-level shifts.

Joyner and *Vogel* [3.65] investigated acceptor GICs with antimony fluoride. Only one type of pentavalent antimony was detected, despite the conflicting evidence from Mössbauer studies finding clearly two valencies of antimony. This discrepancy may now be understood by taking into account the volatility of the intercalate and the surface sensitivity of the method. The more volatile antimony III species, which does not take part in the charge transfer process [3.4, 6, 8], was probably lost in the vacuum of the instrument.

The side effects of carbon halogenation in antimony fluoride GIC were first studied in detail by *Streifinger* et al. [3.66]. The authors described the extensive surface fluorination in polycrystalline compacts of natural graphites, which spoil the transport properties of such synthetic metals.

(c) Valence Band Spectroscopy

In photoemission experiments the use of well-defined and in situ cleaved samples from HOPG is absolutely necessary for work on donor GICs. These materials can be handled only in UHV, preferably at low temperature. The absence of molecular spectra of the intercalate and related chemical complications (Sect. 3.4) seemed to allow detailed interpretation and comparisons with theoretical models. Reliable valence-band spectra of alkali GICs obtained by UPS were first published by *Oelhafen* et al. [3.67]. An intense conduction-band feature with a pronounced Fermi edge was found. The height of the edge scales with the concentration of the intercalate. The shape of the feature led the authors to conclude that an alkali-like *s* band without hybridization of the graphite states constitutes the conduction band of the GIC.

Oelhafen et al. [3.67] further found a shift of all graphite valence-band features to higher binding energies. The authors attributed shift to a Fermi-level shift (Sect. 3.3) and suggested a method for correcting the binding-energy scale. A characteristic feature of the energy distribution curve found in the spectra of pristine graphite and the GIC is used in this correction method as a marker structure for the Fermi-level shift. The magnitude of the Fermi-level shift was found to depend slightly on the type of intercalated alkali metal and strongly on the concentration of the intercalate.

The interpretation of the conduction-band feature as an alkali-like *s* band was also put forward by the same group to explain a characteristic feature in the Auger KVV transition of donor GICs [3.68]. In this picture the additional peak of twice the width of the UPS conduction band is regarded as a self-convolution of an alkali *s* band. It can only be observed in the carbon Auger transition because of strong transition matrix element effects, allowing electrons from "foreign" atoms to relax the carbon core hole. A ground-state band overlap of the alkali metal with the graphite electronic structure was excluded. This interpretation was supported by theoretical arguments [3.69] that were, however, strongly modified by the same authors in a later publication [3.70].

The whole issue was reanalyzed by *Lagues* et al. [3.71], who experimentally confirmed the results of *Oelhafen* et al. [3.68] and found that the feature is absent in metal chloride acceptor GICs. The explanation given by these authors is the depletion of intercalate from the near-surface region of the GIC for electronic reasons. To the authors mind, however, it was quite likely that electron beam damage of the GIC at room temperature in UHV occurred. Unpublished experiments of the author showed that X-ray excited Auger spectra of acceptor GICs indeed do not contain the extra feature and that the whole spectrum is significantly reduced in intensity, compared with pristine graphite. This observation is in line with the general interpretation of the Auger transition given by *Lagues* et al. [3.71]. Here the extra feature is considered as an integral part of the valence band, which implies hybridization between graphite and the alkali metal. The principal origin of the extra peak is the filling of antibonding states of graphite by alkali *s* electrons. The missing peak in acceptor GICs follows from the depletion of bonding states of graphite. This model fails to quantitatively describe the Auger transition [3.71] for donor GICs, since graphite does not exhibit a sufficient number of empty states in the energy range of discussion. In this calculation no emphasis was given to possible matrix element effects noted in [3.68], which would preclude a direct comparison of transition probabilities in graphite and GICs. Thus the Auger spectra of GICs remain poorly understood.

(d) Angle-Resolved UPS

Another route to check the proposal of alkali-like *s* states as a conduction band is Angle-resolved UPS (ARUPS). This method allows a direct comparison of the experimental DOS with the theoretical calculations presented above. The experimental situation is unfortunately controversial. One group [3.72] reports good agreement between theory and synchrotron experiments at various excitation energies for potassium and cesium graphite for all bands except for the conduction band. The conduction-band features disagree in dispersion as well as in intensity with the theoretical predictions. The band is even not of π type symmetry. The authors offer an interpretation in which this band is ascribed to a surface charge-density wave and not to the electronic structure of the bulk alkali GIC. The charge-density wave is assumed to be caused by a layer of alkali metal floating on top of the GIC. This layer has no structural relationship to the substrate, as was shown by the absence of a LEED pattern of the surfaces under analysis. A possible explanation in which the band arises from a surface state was experimentally excluded.

Another group analyzed the same types of GICs with laboratory ARUPS [3.73] and obtained data compatible with a π-type conduction band arising from graphite states. These authors discuss the new band in pristine graphite [3.34] as the origin of the conduction-band feature.

This experimental situation shows how much the results for the bulk electronic structure depend on experimental uncertainities caused by the undesired surface sensitivity of electron spectroscopy.

(e) Core-Level Spectroscopy

The issue of charge transfer was also investigated using X-ray photoelectron spectroscopy of both core levels and the valence band. In a pioneering study of stage-1 lithium GIC [3.74], complete charge transfer from lithium to graphite was found. This result was reached by comparing the b.e. of the Li 1*s* level for metal and lithium compounds with the b.e. of the GIC after carefully correcting the binding-energy scale for the Fermi-level shift.

DiCenzo [3.75] also used core-level shifts of carbon 1*s* and of intercalate atoms in a study of donor GICs, including ternary potassium mercury compounds. She found complete charge transfer in these compounds. She corrected the Fermi-level shift using optical data, i.e., not spectral features of the surface under investigation. The magnitude of the shift was 10 eV, only about half as large as reported in other studies. This external correction method has the advantage that it requires no assumptions about the rigidity of the bonds upon intercalation. Disadvantages are that it neglects surface effects and possible differences in the interpretation of the origin of band shifts by two different spectroscopic techniques. The conclusions of this paper about the completeness of the charge transfer were supported by the analysis of the XPS valence spectra of potassium graphite, which was also presented in an independent and much quoted publication [3.76] and experimentally confirmed by [3.43]. In this spectrum no peak for the potassium 4*s* conduction band can be observed, although this peak is present in metallic potassium [3.76]. The accuracy to which the absence of the K 4*s* peak can be determined is quite good since the energetically overlapping graphite band exhibits a very low cross section in the XPS experiment. From the absence of this peak it was concluded that the population of the K 4*s* states should be very close to zero, and hence potassium should be fully ionized.

This interpretation did not take into account that the cross section of the K 4*s* states is reduced if they are hybridized with graphite states. Then the K 4*s* states are no longer "atomic" with respect to the transition matrix element: the dependence of the transition probability on excitation energy, which is different for potassium and carbon states, affects the spectral response of the entire hybridized band. As the transition probability at 1254.6 eV is very low for all carbon states, this will reduce the intensity of the peak from the hybrid states. Hybridization further broadens the distribution of states in energy, which results in a "smeared out" peak reducing the sensibility of X-ray valence-band spectroscopy even further. This means that the indisputable fact of the absence of K 4*s* states in the XPS experiment may well arise from a covalent type of interaction between alkali metal and graphite, leading to a rehybridization of the graphite valence states with alkali states. This interpretation is much more in line with other spectroscopic observations and underlines again the inadequate procedure to derive intercalation-induced electronic effects from a linear correlation of spectral parameters in the free and intercalated state.

Ternary donor GICs with potassium–mercury and rubidium–thallium systems were studied by two groups [3.43, 75]. From their results it follows that the addition of the second metal does not increase the amount of charge transferred to graphite but rather decreases the number of conduction electrons in the graphite system relative to the same amount of alkali atoms in binary and ternary GICs. A strong interaction between the two intercalated metals leads to compound formation (Zintl phases). It also explains the sandwich-type structure in which two positively charged alkali atom layers are surrounded by two negatively charged graphene layers (top and bottom) and an additional negative layer of atoms of the second metal (thallium or mercury in the center of the sandwich). This metal layer has no direct electronic contact to the graphene layers [3.43].

There was also an attempt to follow the charge transferred onto graphene layers by carbon 1*s* XPS [3.77]. A series of potassium GICs with varying concentrations of alkali metal was analyzed. The C 1*s* line profile was found to become broader with increasing amounts of alkali metal. This was interpreted as indicating an increasing number of chemically inequivalent carbon atoms located in inequivalent graphene layers, distinguished by their distance to alkali atoms. A localization of the intercalation-induced charge at the neighboring carbon atoms was assumed. No experiments were presented to monitor the influence of phase transitions altering the in-plane structure of the GICs and hence the nearest-neighbor distribution on the C 1*s* spectrum, which in fact is temperature independent (see Sect. 3.3). The magnitude of the shift of the inequivalent carbon atoms quoted in the paper is extremely large compared with the chemical shift of carbon in gas phase ethylene and acetylene of 0.3 eV [3.78].

A much larger shift versus formal charge was obtained by *Siegbahn* and coworkers [3.78] after including molecules with strongly electronegative ligands (oxygen and fluorine) in the correlation. Since alkali metals are certainly not electronegative, this correlation predicting the large chemical shift is here inadequate. This shift corresponds to a charge transfer value of one electron per carbon atom in the model systems, compared with a maximal charge transfer of 0.125 electrons per carbon atom in potassium GICs.

The change in line shape with composition of the GIC can also be interpreted as a change in relaxation of the C 1*s* core hole, which would scale with density of states near the Fermi level. In this view the C 1*s* spectrum consists of only one asymmetric line and not of a superposition of several lines, as assumed by *DiCenzo* and co-workers [3.77]. They used an excitation line profile derived from the spectrum of pristine graphite. This technique is frequently applied in the analysis of alloy spectra and tacitly assumes that final-state distribution and relaxation effects are unchanged in the reference as well as in the compound of interest. For the special case of a GIC with its very large changes in valence-band features, this assumption is at least critical. This view is supported by the observation that in a typical acceptor GIC with very low density of states near the Fermi level, the C 1*s* line is also very narrow but still asymmetric (Sect. 3.3).

In summary, XPS of donor GICs has been used to develop the idea of an ionic electronic structure of these compounds, assuming full charge transfer and a structure of positively charged alkali cations and partly charged negative graphene layers. This idea was used in band-structure calculations assuming empty alkali valence *s* states and partly occupied graphite π states.

Critical reinterpretation of the data in the light of results from other techniques today allows us to discard this oversimplified picture and to suggest an electronic structure with significant covalent interaction between graphite and alkali states, giving rise to an electron density distribution with finite occupation values at the alkali center and at the carbon atoms. GICs may be seen as intermetallic compounds, which are a subgroup of alloys characterized by directed chemical interactions rather than by diffuse mixing of electron gas. In this model a number such as the charge transfer fraction is unnecessary and inadequate to describe the electronic interaction. Chemical modifications such as ternarization can redistribute the charge transferred. The bonding in such a GIC may be described as electrostatic interaction between sandwiches of compounds of intercalate and collectively negatively charged graphene layers, i.e., as a Madelung-energy-driven compound formation. The covalent interaction is reduced in efficiency by electron density removed from the orbitals interacting with graphite and entering into orbitals interacting with metal centers within the intercalate. A model of isolated ions intercalated into graphite, however, is certainly too simple, and determination of the ionicity of a GIC is insufficient as a description of its electronic structure. Note that these statements refer to the GIC analyzed by electron spectroscopy and may not be generalized for all GICs known.

3.3 Principal Results for the Electronic Structure

3.3.1 The Typical Photoelectron Spectrum

This section describes results of core-level spectroscopy carried out with potassium GICs and aluminium chloride GICs. From such experiments information is obtained about chemical shifts of host (carbon 1*s*) and guest species. In addition, we can deduce the chemical composition. i.e., the cleanliness of a freshly cleaved surface and the stage index of the material. In this section the emphasis is on the C 1*s* host spectrum; core-level spectra of the intercalated species will mainly be discussed in Sect. 3.4.

Figure 3.2 shows wide-scan spectra of low- and high-stage potassium GICs taken at room temperature. Compared with spectra of conventional metals these patterns are simpler since the only two prominent absorption edges arise from the host element carbon. The potassium guest atoms exhibit their main emission line (K 2*p*) close to the carbon 1*s* peak (top trace) and thus do not contribute absorption steps of their own.

Some gold from the sample holder and traces of oxygen are detected in addition to the expected elements. The oxygen builds up by diffusion from the

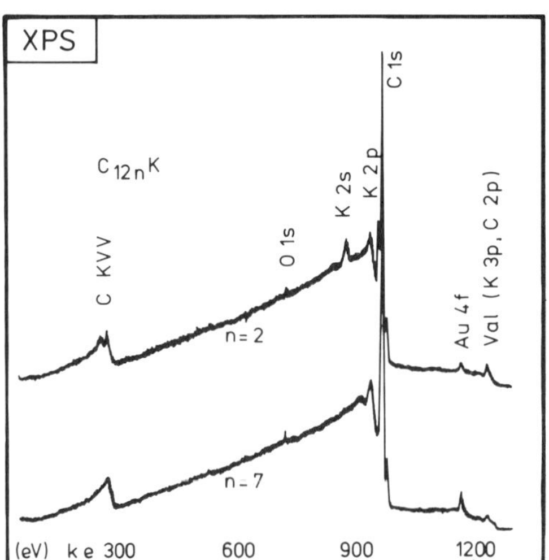

Fig. 3.2. XPS wide scans of potassium graphite of stage 2 and stage 7. The strong modification of the graphite valence band and its reflection in the Auger C KVV transition can be seen in the spectrum of the concentrated GIC

bulk onto the initially completely clean cleaved surface. Cooling to 80 K inhibits this process for several hours. The level of contamination shown in the figure is reached one hour after cleavage. If the sample is exposed to air during transfer, this oxygen level is reached within a few minutes, and the diffusion cannot be stopped by cooling.

The carbon Auger transition is clearly visible in all donor GICs, whereas it is very weak in acceptor GICs of the first to third stages. In the seventh-stage GIC the Auger transition exhibits the same shape as in pristine graphite; the lineshape is modified in the second-stage GIC. This change in the Auger lineshape reflects the changes in the conduction band of the GIC and varies like the height of the Fermi edge with the stage index (see below).

Figure 3.3 shows high-resolution carbon 1s core-level spectra of pristine graphite and first-stage potassium graphite. Referred to a common binding-energy scale (gold 4f) these reveal a significant shift of the line from the GIC relative to that of the host material. In all first-stage heavy alkali-metal GICs, this shift is not constant with time but moves from a typical position indicated in the figure by the dashed line to its displayed position, where it stays constant for a prolonged time. The line profile does not change within this time.

The magnitude of the shift is quite large. Unexpected is the sign of the shift. In a donor GIC charge is transferred from the guest to the graphite host. The increased electron density at the carbon atoms should result in a shift to lower binding energy due to better screening of the core charge. This expected shift is overcompensated for by a shift in the energy scale. The comparison in Fig. 3.3 assumes a common zero point of the energy scale, which is given by the respective Fermi energies of graphite and the GIC. Due to the low density of states in the vicinity of the graphite Fermi energy (Sect. 3.1.1.), the occupation of previously antibonding states by the guest shifts the Fermi edge in the GIC significantly closer to the vacuum level. In the photoemission experiment the Fermi energies of the gold reference, the pristine graphite, and the GIC are pinned to the same value. Its position relative to the vacuum level is given by the

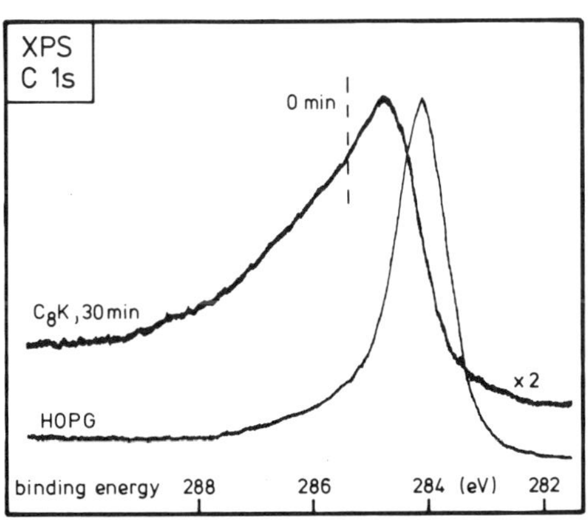

Fig. 3.3. XPS high-resolution scan of the C 1*s* emission from stage-1 potassium graphite and from HOPG as reference. The GIC spectrum is not stable at 300 K but moves to a smaller binding energy without much change in line shape. Only after ca. 60 min measuring time does the spectrum start to split into two components

spectrometer work function. For large shifts of the Fermi energy relative to the spectrometer calibration material, the calibrated energy scale is no longer valid. Independent information is needed to separate the chemical shift from the Fermi-level shift. Their sum determines the apparent shift obtained from experiments, as shown in Fig. 3.3.

The magnitude of the shift, which is about 1.7 ± 0.2 eV for all first-stage alkali-metal GICs and about 1.3 ± 0.2 eV for all second-stage GICs, depends monotonously on the stage index. This is illustrated in Fig. 3.4. The observed behavior of decreasing shift with decreasing concentration of the intercalant is in line with the arguments presented above: less intercalant means fewer electrons transferred into the graphite formerly antibonding states. The reduced filling thus gives rise to a smaller shift of the Fermi level and thus to a smaller apparent core-level shift.

A similar curve was also obtained for rubidium graphite, and for ferric chloride graphite as a typical acceptor system. The curve drawn in Fig. 3.4 serves only to guide the eye. An inflection between the first and second stages might be expected as a consequence of the fundamentally different in-plane structure of the C-8 GICs, compared with all higher-stage C-12 *n* GICs.

The large error bar on the last stage index reflects the difficulty in determining this number from the carbon-to-potassium elemental ratio at the surface. It was generally found with higher-stage samples that the bulk stage index determined by (001) X-ray reflections was lower than the stage index near the surface. This and the observed considerable deviations of the guest-to-host atomic ratio in compounds higher than stage 2 point to undersaturation and an island-like organization of the intercalant. The tendency to increasing non-stoichiometry with the higher-stage index is not only a surface phenomenon: It is also indicated by the shape of the intercalation isotherm and results from the steeper temperature gradient required to obtain these GICs.

The curve of Fig. 3.4 allows one to give an upper estimate on the information depth of XPS in GIC systems. Using the universal escape depth curve, an information depth of 3 ± 2 atomic layers is expected. Successive cleavage of a

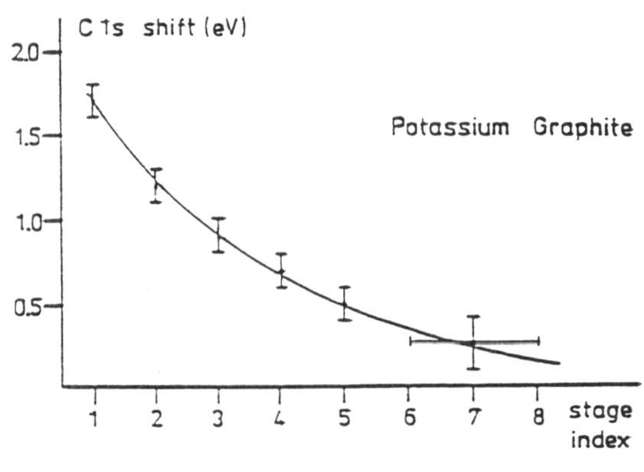

Fig. 3.4. Carbon 1*s* core-level shifts in potassium graphite as a function of the stage index. The line is only to guide the eye

uniform sample of a third-stage compound gave well reproducible carbon-to-potassium ratios. This means that cleavage does not result in a statistical fashion with respect to the type of layer (potassium, next-neighbor and second-neighbor graphene) at which intersection occurs. The fact that for high-stage compounds the potassium intensity almost goes to zero (Fig. 3.2), but not the C 1*s* shift, excludes the potassium layer as the location of cleavage. Thus the potassium signal should be very small for GICs higher than stage 3. The information depth in these compounds seems to be significantly larger than the predicted value.

Except with stage-1 compounds, no time dependence of the nonstoichiometry was found, excluding desorption of potassium from the surface-near regions of the sample as the explanation for the nonstoichiometry.

The instability of stage-1 GICs was reported by several previous groups who also tried to separate a deterioration part of the spectrum from a genuine part of the spectrum. Section 3.3.3 stresses the instability in lineshape; Fig. 3.5 shows the variation in surface composition for a typical sample. Immediately after cleavage there is a period in which the surface composition is close to the expected value with little contamination. This period is temperature dependent and varies from ca. 5 min at 300 K to ca. 60 min at 80 K. Then follows rapid segregation of alkali metal, building up to more than one monolayer. The segregation is coupled with the increase in oxygen contamination. The segregation soon stabilizes, but the level of contamination increases until the alkali metal is oxidized. The whole process can be initiated by chemisorption of oxygen or carbon monoxide at any temperature [3.79]. The instability is probably caused by oxygen-induced segregation of the alkali metal (caused by the residual gas in the form of water, oxygen, or CO). The whole effect is not merely an experimental problem but indicates a substantial difference in the charge transfer properties between first- and higher-stage alkali-metal GICs. In the higher-stage compounds the interaction between guest and host seems to be stronger, so it cannot be perturbed by chemisorption.

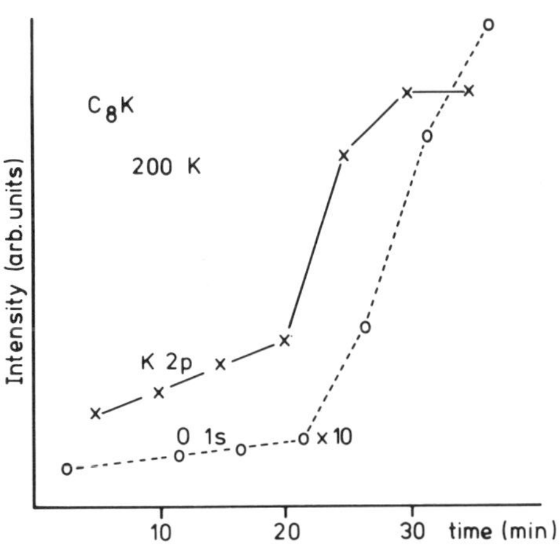

Fig. 3.5. Evolution of surface composition of a stage-1 K-GIC surface at 200 K in UHV. After ca. 20 min potassium begins to segregate and oxidation of the alkali metal starts

Finally, the apparent shift in acceptor compounds should be stressed. Fig 3.6 compares the high-resolution carbon 1 spectra of pristine graphite and graphite intercalated with potassium (donor) and aluminium chloride (acceptor). The position of the donor GIC spectrum has been corrected for the Fermi-level shift according to the procedure described in Sect. 3.3.2. The acceptor GIC spectrum is shown as measured. It is shifted to a lower binding energy, i.e., in the opposite direction to that of the donor spectra. The magnitude of the shift, 1.5 eV, is similar to the first-stage alkali-metal GICs. These observations are in line with the model for the Fermi-level shift. In acceptor GICs charge is removed from the formerly bonding states below the Fermi energy of pristine graphite. The new Fermi-edge is at lower energies with respect to the vacuum level, and thus the energy scale for photoemission is shifted to lower energies.

3.3.2 Fermi-Level Shift – UPS Results

Angle-integrated UPS provides a relatively direct insight into the valence-band structure of a metallic solid. The main difference between true pictures of the DOS and UPS data is the cross-section variations of states of different symmetry with the excitation energy. As discussed in Sect. 3.1.1, these cross-section variations can, on the other hand, be used to derive information about the symmetry of valence-band features if the changes of the UPS patterns are monitored at different excitation energies.

Oelhafen et al. [3.67] summarize the essential information that can be derived from GICs with UPS. In their pioneering work, these authors interpreted the UPS data from heavy alkali-metal GICs similar to the method common for metallic alloys. In the following we discuss their results with emphasis on the Fermi-level shift. Figure 3.7 illustrates He I UPS data measured in the author's laboratory. Valence-band spectra of pristine graphite (top) are compared with those of a typical donor GIC (rubidium). The top two traces were measured with the energy scale of the spectrometer calibrated against the Fermi edge of palladium. The bottom curve was obtained from the same sample after a fresh cleavage and a readjustment that made the peak structures around 4.8 eV coincide for the GIC and pristine graphite.

The most striking difference between the spectra of graphite and the GIC is the new peak at the Fermi edge. The shape of this peak and the bandwidth of ca. 1.0 to 1.5 eV (depending on the element) resemble closely the conduction-band feature of pure alkali metals. The intensity of the new peak decreases within a series of GICs from the same element with increasing stage index. The energy dependence of the cross section of this feature implies at first glance an *s*-type symmetry; i.e., the feature is also present in He II (40.8 eV) spectra and more intense in spectra excited with Ne I (16.8 eV) than in those with He I (21.2 eV) excitation. This and the absence of any feature similar to the observed peak in the DOS of unoccupied states in pristine graphite led *Oelhafen* et al. to conclude that the new feature should be an alkali-like *s* conduction band without participation of graphitic states.

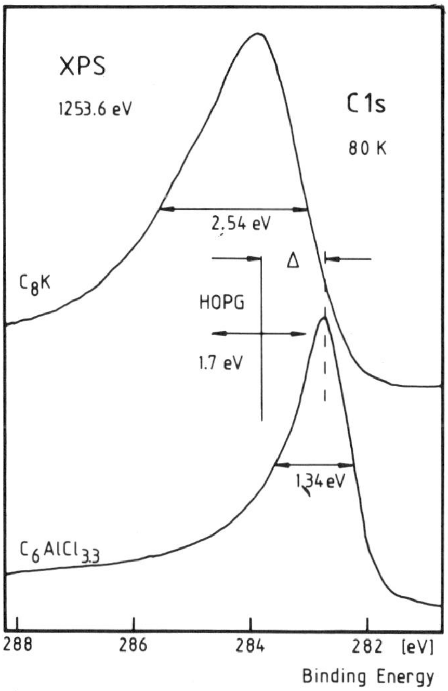

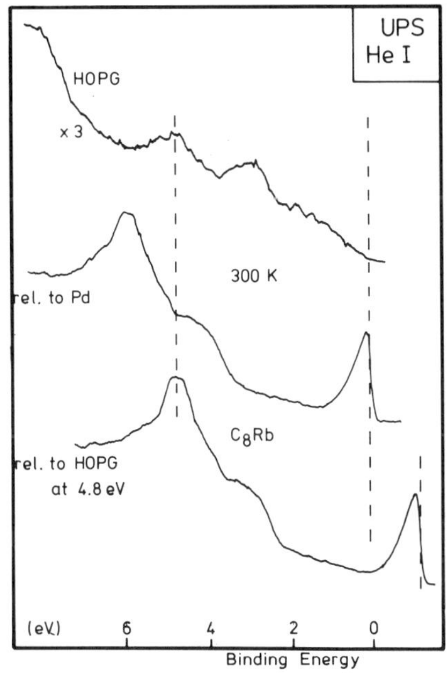

Fig. 3.6. Comparison of the carbon 1*s* line profiles for a donor GIC (stage-1 potassium) and an acceptor GIC (stage-1 $AlCl_3$) at 80 K. The Δ indicates the Fermi-level shift for the acceptor GIC

Fig. 3.7. Valence-band spectra for pristine graphite and stage-1 rubidium GIC. The method of internal correction for the Fermi-level shift is illustrated. The central spectrum is as measured; the bottom spectrum has been shifted to overlap the marker structure (here the top of the graphite π band) with the corresponding peak of the graphite spectrum (dashed line)

The two spectra of the GIC indicate the degree of reproducibility of spectral features that can be reached. The most important source of spectral differences arises from the pronounced angular dependence of all peaks except the additional peak at the Fermi edge. Rotation of the sample by only 5° results in intensity changes of the feature at 4.8 eV of ca. 20%. These findings agree with the band structure, which shows pronounced dispersion of the bands (of π symmetry) close to the Fermi energy. The new conduction band shows little dispersion, as was confirmed by tilting experiments up to 30° with different types of GICs.

For a useful comparison of the GIC spectra with those of pristine graphite, a recalibration of the b.e. scale is essential. This is achieved by determination of the Fermi-level shift by independent methods – e.g., optical spectroscopy ([3.75] and other chapters in this volume) – and subsequent correction of the photoemission data assuming the shift to be constant for a given variety of GICs (e.g., for first-stage potassium graphite).

Another approach is to define a marker structure in the photoelectron spectrum other than the Fermi edge. This marker has to be visible in both the spectrum of pure graphite and in that of the GIC. If we assume that the marker structure is not modified in its energy by intercalation-induced changes of the band structure, we can take the energy difference of the markers in the spectra of the GIC and of the pristine graphite for internal determination of the Fermi-level shift. This is illustrated in Fig. 3.7. The corrected spectrum (bottom) is shifted 1.4 eV closer to the vacuum level, compared with the Fermi energy of pristine graphite.

The advantage of the internal calibration method is its sensitivity to fluctuations in the Fermi-level shift. Unfortunately, this shift is not at all constant for a given type of GIC, but depends critically on minute impurities and defects either from the preparation of the sample or from cleaving the sample. The latter effect on the shift can be reduced by several subsequent cleavage cycles and repeated data acquisition. Compensating for the influence of the preparation, however, was very difficult. In a series of experiments with a second-stage potassium GIC, a clear dependence of the Fermi-level shift on the temperature gradient used for the synthesis was found: the shift varied from 1.1 eV for a 10° gradient to 0.5 eV for an 80° gradient. These data illustrate how a considerable "chemical" shift for, e.g., the carbon 1*s* core level may be erroneously detected if an inappropriate external value for the Fermi-level-shift correction is used. Small differences in sample preparation may well account for such errors.

The correction method relies on the assumption that the marker structures are not influenced in their energies relative to other structures of the DOS, i.e., that the rigid-band approximation holds. Figure 3.7 shows that this holds for the two structures at 2.8 eV and 4.8 eV. The energy separation between these two peaks was also found to be constant for all other heavy alkali-metal GICs.

The valence-band spectrum of graphite exhibits several structures in the energy range up to ca. 10 eV. These structures are largely preserved in all donor GICs. The shape of the peaks, however, is different in the spectra of the GICs compared with the pristine graphite spectrum (Fig. 3.7). These differences can be attributed either to changes in the DOS or to spectroscopic effects such as cross-section variations or final-states effects. The strong dependence of the UPS valence-band spectra of graphite (Sect. 3.1.2) and of GICs (Sect. 3.4) on the excitation energy already implies the relevance of spectroscopic effects. The cross section of all graphite bands and in particular of π states at high excitation energies (XPS) is so low that individual structures can no longer be resolved. It is therefore difficult to quantify with photoemission the relevance of spectroscopic effects on the spectral features (SXS results, Sect. 3.2.2).

It is also suspected that the rigid-band approximation inadequately describes the electronic effects on the DOS of graphite induced by intercalation to low-stage donor compounds [3.55, 57]. Within the UPS data there are several indications of non-rigid-band behavior of these GICs. The changes in shape of several spectral features of the bonding states were already noted above. The antibonding part of the DOS is reflected in a sharp secondary emission peak in pristine graphite (see Sect. 3.1.2). This peak is drastically modified upon intercalation. It moves to lower kinetic energy in accordance with the Fermi-level shift, which can also be calibrated from this lowering. The cross section of the transition is, however, greatly reduced. The separation between the selected marker structure (in the bonding part of the DOS) and the secondary emission peak should be stage independent and remain as large as in pristine graphite. In several GICs, however, it is reduced by ca. 0.5 eV, which is outside the experimental error. The good agreement between the two values for the Fermi-level shift taken from two different marker structures is, however, strong support for the validity of the correction method, i.e., the assumption that the shifts in the angle-integrated UPS data are strongly dominated by the Fermi-level shift.

Reliable determination of the binding-energy scale for all GICs is the prerequisite for any interpretation of intercalation-induced effects on the photoemission spectra. In the discussion so far it has become clear that a direct comparison of positions in the spectra of graphite and GICs is not possible. Any improper correction for the Fermi-level shift obscures real intercalation effects, especially since the Fermi-level shift is large compared with expected "real" shifts. The chosen method of internal calibration is, as discussed above, not completely unambiguous and relies on the experimentally supported assumption that non-rigid-band effects have little influence on the positions of some graphite features in the spectra. To further support this, we compare UPS data with SXS data [3.57]. The spectrum arises from intra-atomic transitions and is thus not affected by the Fermi-level shift.

Figure 3.8 compares the deconvoluted spectra of pristine graphite and first-stage potassium graphite. As in UPS, the most prominent change is the occurrence of a new peak at the high-energy side of the spectrum, i.e., at the low-energy edge of the valence band. This peak is clearly of π symmetry and derives from graphite π states. It also broadens the total width of the valence band. Smaller changes in lineshape deep in the valence band indicate the presence of non-rigid-band effects over the entire width of the valence band. Intercalation shifts the center of gravity energies of the σ band by 0.3 eV and the π band by 0.4 eV to higher photon energy, i.e., to lower binding energy.

Figure 3.9 directly compares the SXS data with UPS data. Good correspondence between the features of the spectra is obtained if the spectra are aligned in the way shown. The different relative intensities illustrate the strong influence of the cross-section modulation on the spectra.

It emerges that the new conduction band in the GIC is of π symmetry and, to a significant extent, of graphitic origin. It is not alkali-*s*-like, as concluded by *Oelhafen* et al. [3.67].

The height of the new feature in SXS falls in the same way with increasing stage index as the Fermi edge in UPS. The two marker structures in Fig. 3.7 arise from the top of the π band and a relative maximum in the σ band, respectively. This accords with the assignments given in the literature. The position and shape of the σ band agree well between SXS and He II spectra, as is expected from the relative cross sections for carbon atomic orbitals.

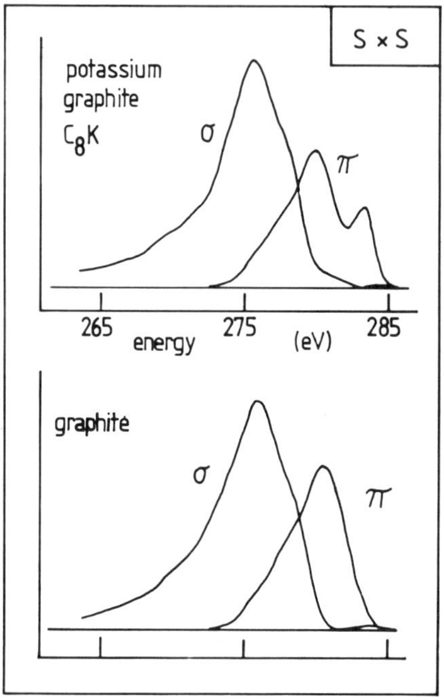

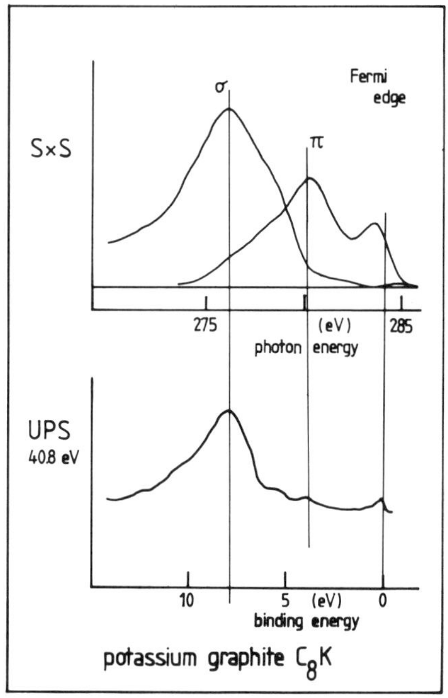

Fig. 3.8. Soft X-ray emission carbon spectra for stage-1 potassium graphite (top) and pristine HOPG (bottom). The σ and π bands have been deconvoluted from two measurements at different orientations. The (0001) surfaces were oriented at 80° and 20° to the incident polarized synchrotron photon beam. Data were taken from in situ cleaved surfaces at 78 K

Fig. 3.9. Comparison of SXS data and direct valence-band spectra (UPS) for stage-1 potassium graphite, showing the assignment of peaks and the cross-section effects in UPS. Note the π symmetry of the conduction-band feature containing the Fermi edge

The Fermi-level shift does not simply arise from the addition of graphite π states and potassium 4*s* states, as emerges from the comparison of the two types of spectra. This was implied, e.g., by the similar dependence of the conduction bandwidth in pure alkali metals and the corresponding GIC on the element number noted by [3.67]. The increase of the total graphitic bandwidth in the GIC (1.0 eV for first-stage potassium graphite) is less than the Fermi-level shift (1.5 eV). This indicates that hybridization between graphite π states and alkali *s* states causes the formation of the conduction band.

Finally, we must show how the intercalation of acceptors influences the UPS spectrum of graphite. Few UPS data on these systems have been published so far [3.26, 80]. This may be due to the instability of all low-stage acceptor GICs in UHV at room temperature. High-stage compounds are more stable, but their UPS spectra are dominated by features arising from irreversible halogenation during intercalation. Nevertheless, an attempt was made to obtain the UPS spectrum of first-stage aluminum chloride graphite. Most acceptor GICs exhibit

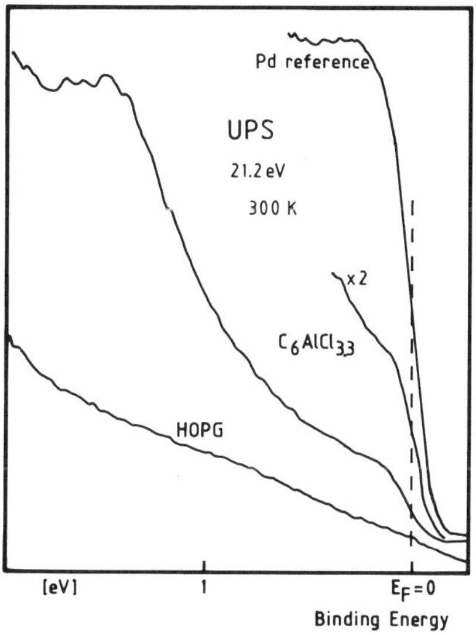

Fig. 3.10. High-resolution UPS data for a stage-1 aluminium chloride GIC in comparison with pristine graphite and palladium metal (intensity reduced by a factor of 22). The transition from a semimetal graphite to a synthetic metal GIC is obvious

a variety of low-temperature phase transformations involving significant changes in the molecular identity of the intercalant, which may influence the spectroscopic results (Sect. 3.4).

Figure 3.10 demonstrates that the removal of valence electrons from graphite leads to a semimetal-metal transition with the formation of a characteristic Fermi edge. This edge is very small, with its height only 0.65% of the Fermi edge of the palladium reference shown. The difference in intensity cannot be correlated directly to the difference in the density of states, because a large variation of the cross section over-emphasizes the DOS in palladium. Even so, the local density of states at the Fermi energy in the synthetic metal is only a small fraction of that in the conventional metal.

The whole He I UPS spectrum of this compound is compared with the spectrum of a typical donor GIC Fig. 3.11. The spectrum of pristine graphite, reduced in intensity, is also shown.

Most striking are the different conduction-band features, if we keep in mind the symmetric shape of the DOS of pristine graphite. The intensity of the two features decreases in a similar manner with increasing photon energy. From these spectra we can conclude that there are large differences in the cross sections caused by different hybridizations of the constituent host and guest atomic orbitals. The intensity differences clearly indicate that the simple addition of states in donor GICs is an inadequate description of intercalation, whereas the removal of graphite electrons from the valence band without the addition of filled acceptor states (hybridization) may be a suitable description for acceptor GICs. The electrons removed from graphite are accommodated in the molecular bonds of the acceptor intercalate with no localization at any certain atom (Sect. 3.4).

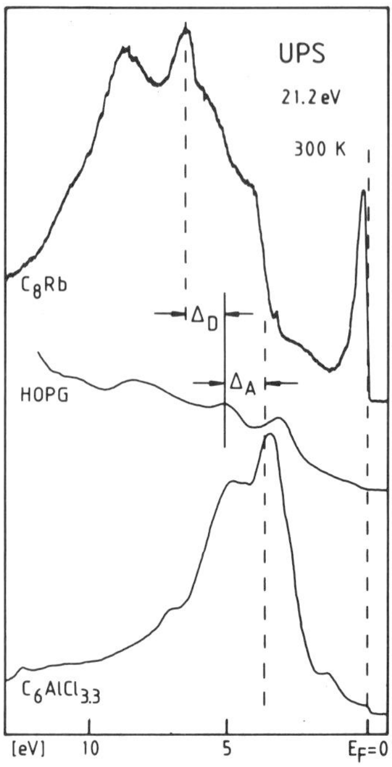

Fig. 3.11. Conduction-band features and Fermi-level shifts for typical donor and acceptor GICs in comparison with pristine graphite. The absolute intensities of the spectra depend on sample quality and photon source conditions, and cannot be compared

The intensity between 3 eV and 6 eV in the aluminum chloride spectrum arises from chlorine 3*p* states and covers most of the graphite spectrum. This becomes apparent from the corresponding He II spectrum (Sect. 3.4.2). On the basis of this spectrum, the structure just below 4 eV was assigned to the marker structure indicated by the solid line in the graphite spectrum (center of Fig. 3.11). With this assignment the Fermi-level shift can be determined as shown in the figure. It moves the apparent energy scale to lower binding energies. The magnitude of this shift is very similar to the shift of corresponding donor GICs of the same stage index.

In summary, the UPS data show that graphite is transformed into a synthetic metal when either donor or acceptor substances are inserted between the graphene layers. Systematic variations of the conduction-band features were found with the nature of the intercalant and with the stage index. In comparison with SXS data and in agreement with their spectral properties, the new features exhibit π symmetry and belong to the graphite valence band. Strong cross-section variations do not allow quantitative determinations of the charge transfer. The UPS data provide a basis for the internal determination of the Fermi-level shift, which is the dominant contribution of the apparent intercalation-induced shift observed in comparison of spectra of pristine graphite and GICs. Various indications for the occurrence of non-rigid-band effects of intercalation have been found.

3.3.3 Lineshape of Core-Level Spectra

The line profile of the carbon 1*s* emission of GICs is important for determining the number of inequivalent carbon atoms in the graphene layer. It has been speculated that the charge transferred from, e.g., alkali might be localized on the nearest-neighbor carbon atoms, giving rise to more or less charged atoms within one graphene layer.

Figure 3.12 shows the experimental situation for donor GICs.

They exhibit much wider and strongly asymmetric lines than those seen in the emission of pristine graphite. The lineshape is almost identical for the two binary donor GICs and slightly less asymmetric for the ternary GICs, which is in line with the reduced charge transfer in this compound [3.43].

The large linewidth, which gradually reduces with increasing stage number [3.77], was taken as indicating the presence of several inequivalent sites in stage-1 compounds and a reduction in the relative amount of these sites with decreasing concentration of intercalate. *DiCenzo* et al. [3.77] decomposed the GIC spectrum in several shifted components of the same lineshape, assuming a line profile of each component similar to that of pristine graphite.

Such an approach assumes that the difference in valence charge of different carbon centers is large enough to produce a chemical shift of up to 1.5 eV. It further assumes that the C 1*s* spectrum is a representation of the electronic ground state and that the spectrum of the synthetic metal GIC is not influenced by relaxation or other final-state effects. The chemical shift sensitivity of carbon in homonuclear bonds can be estimated from the shift difference between acetylene and ethlene in the gas phase, which was found to be 0.3 eV

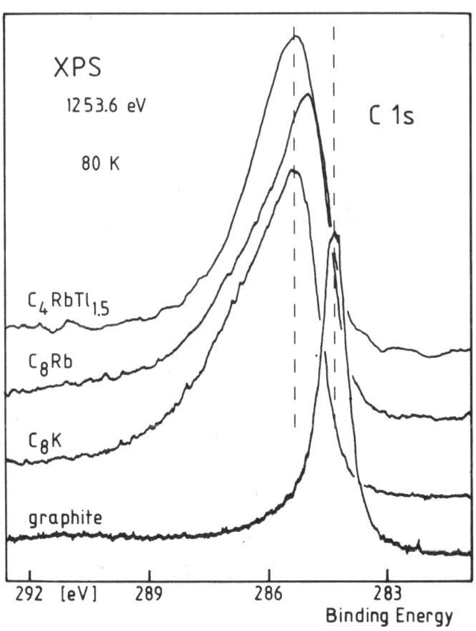

Fig. 3.12. High-resolution carbon 1*s* spectra for various stage-1 donor GICs and HOPG. The GIC spectra are not corrected for the Fermi-level shift

(Sect. 3.2.2c) [3.78]. This value corresponds to localization of a full electronic charge and is significantly smaller than the shifts obtained from the line fits.

The spectroscopic influences on the line profile can be seen from a comparison of the line profiles of acceptor and donor GICs. Figure 3.6 showed that the acceptor GIC exhibited a linewidth of only 1.34 eV compared with the donor GIC of 2.54 eV. Pristine graphite exhibits typical widths between 1.7 eV and 1.2 eV, depending on the defect concentration [3.27, 28]. The very narrow line for the acceptor GIC reflects the loss of π-state density at the Fermi edge. These states are effective in relaxation of the core hole; an increase of their concentration (donor GIC) reduces the lifetime of the core hold and leads to Heisenberg broadening. A reduction of the π DOS (acceptor GIC) increases the lifetime of the core hole and removes the lifetime broadening.

The sensitivity of the C 1*s* spectrum to the charge transfer in the ground state may be rather limited. Ionization of the carbon atoms with the beginning relaxation is such a strong perturbation of the valence electronic system that the intercalation-induced charge transfer may be overlooked. The fact that the changes of interest occur in the low-energy region of the DOS, which does not effectively contribute to electrostatic core-hole screening, accounts for the lack of "chemical shift" in this situation. This insensitivity arises because XPS does not directly probe the electronic groundstate with the spectral parameters, representing a difficult-to-quantify superposition of excited electronic states with the groundstate.

The problem is exemplified in Fig. 3.13, which shows spectra of stage-1 rubidium graphite at low and high temperatures. In analogy to similar observations in the cesium-graphite system [3.81], we assume that the changes in the valence-band spectra and the rubidium spectrum that are fully reversible indicate a phase transition affecting the ordering of the intercalate (rubidium-rubidium interaction). The C 1*s* spectrum is – with a stable position and a change in linewidth of 0.2 eV – relatively, insensitive to the process. The phase transition influences the valence electronic structure, as can be seen in the indirect SES spectrum produced from He I secondary electrons (top left in the figure). A dramatic change in the cross section of the loss structure and a peak splitting for the low-temperature state can be seen. In these spectra the energy separation between the initial and final states is very small. Thus no influences of the reorganization of the photo-ion obscure the differences in the electronic structure of the weakly bound π states. The shallow core level of rubidium 4*p*, which is situated ca. 15 eV below the bottom of the valence band, experiences the phase transition by a change in linewidth and a slight shift of the center of gravity of the emission line (bottom of Fig. 3.13).

The inequivalence of carbon atoms resulting from covalent bonds of, e.g., halogen atoms can easily be detected by a significant "chemical shift". The intercalation of most metal halides into graphite is achieved by the addition of a small amount of free halogen acting as an oxidizing agent. This oxidation of graphite during the intercalation of aluminium chloride produces surface chlorine groups. At these centers the hybridization of graphite is broken and sp^3

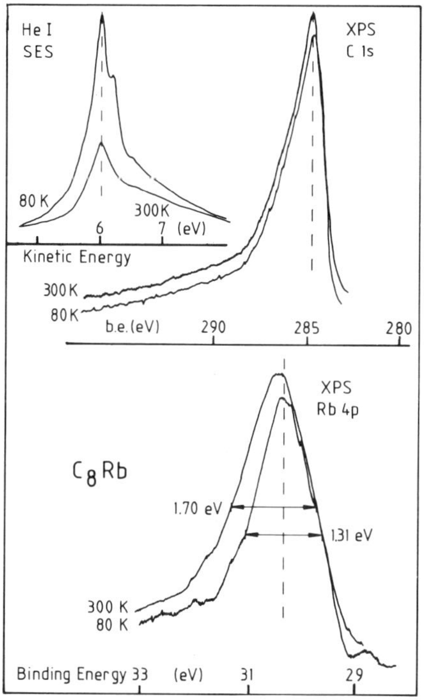

Fig. 3.13. Sensitivity of various photoelectron spectra of stage-1 rubidium GIC to changes in the electronic structure associated with a low-temperature phase transition

centers are formed. The strong electronegative bonding partner chlorine gives rise to substantial deshielding of the carbon atoms, resulting in the large chemical shifts shown in Fig. 3.14. The wide line produced by the overlap of signals with different chemical shifts and the line of the GIC is not a genuine property of the $AlCl_3$ GIC; rather it results from irreversible side reactions. The unperturbed genuine spectrum obtained after cleaving the GIC in situ at low temperature is also shown. The weak structure at ca. 289 eV is not due to C-Cl_3 groups, but arises from a characteristic energy loss of graphite, a plasmon structure, as was checked by energy-loss spectroscopy (not shown here). The occurrence of this structure indicates the close similarity of the overall electronic structure of acceptor GICs and pristine graphite. In concentrated donor GICs the structure is absent. The shift of the main line relative to the position of pristine graphite indicates the integrity of the GIC. Its direction is in line with the Fermi-edge shift model, which predicts a shift of the edge to lower energy for acceptor GICs.

In summary, the line profile of the C 1*s* emission of graphite is strongly affected by intercalation. The assignment of these changes to chemically inequivalent carbon atoms is, however, difficult. Strong spectroscopic effects caused by the modification of the graphite valence electronic structure have been shown to influence the photoemission process, and a direct correlation of the graphite line with the line of the GICs is thus inappropriate. The line profile

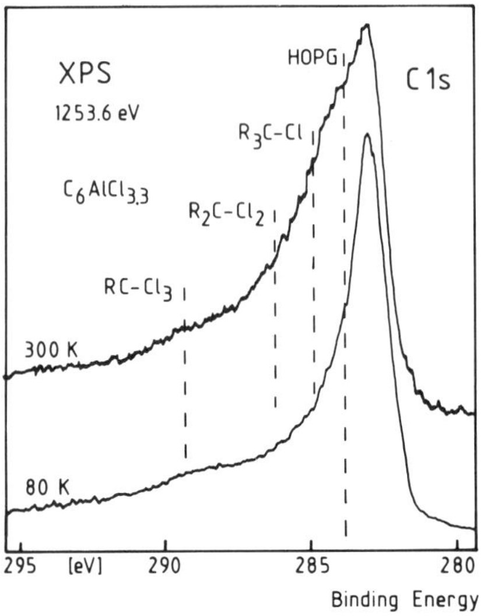

Fig. 3.14. Modification of the carbon 1*s* line profile of the acceptor stage-1 aluminium chloride GIC by irreversible chlorination of the graphite caused by the intercalate at room temperature. The surface chlorination results in a large number of C–Cl groups and small numbers of more highly chlorinated defect centers

does indicate an irreversible carbon oxidation occurring as a side reaction with intercalation of the acceptor species. The non-ground-state character of C 1*s* photoemission does not permit us finally to decide about the existence of inequivalent carbon atoms. The evidence for their existence from XPS seems poor.

3.3.4 Shifts in Core-Level Spectra of GICs

Some aspects of shifts have been discussed already in the preceding sections of this work. This section gives a critical summary. The idea of detecting the charge transfer effects of intercalation directly by chemical shifts between spectra of a GIC and the parent materials was the driving force for many of the XPS studies reviewed here. This idea is based on the assumption that changes in valence DOS of graphite should be reflected in a change of the screening of the carbon nuclear charge – i.e., in donor GICs the C 1*s* line should be at lower b.e. and the emissions of the donor species at higher binding energy; in acceptor GICs the shifts should be reversed. It is further assumed that the spectral properties of graphite (final states, relaxation) remain unchanged upon intercalation. This can be rationalized in terms of the rigid-band approximation and the fact that the total number of modified states is small, much smaller than in conventional alloys.

The first experimental shift data were not in line with these ideas: large shifts in unexpected directions were reported [3.64]. This problem was solved by a correction of the binding-energy scale. The physical reason for this correction relative to the scale used for pure graphite is the very large shift of the Fermi edge relative to the vacuum level caused by the unusually low total density of states. The magnitude depends on the sample and the method of determination, but it is with ca. 1 eV at least as large as the expected chemical intercalation shift. The correction of the b.e. scale described in Figs. 3.6, 11 using the method of internal calibration removes all chemical shifts of the C 1*s* level; the binding energy does not change upon intercalation within the accuracy of the correction procedure (ca. 0.2 eV, caused by non-rigid-band effects of the marker structures).

This somewhat disappointing result does not mean that there is no intercalation-induced charge transfer; rather it indicates that the effect is not detected in the excited state of the photoionized carbon atoms. This means that none of the electrons that contribute significantly to the shielding of the nuclear charge are affected by intercalation, or that intercalation affects only the very weakly bound electrons. This insensitivity of XPS against charge transfer effects is not unusual in covalent solids, where charge transfer occurs as rehybridization of valence electrons [3.82].

Core levels of the intercalate can also be rather insensitive to intercalation effects, as Fig. 3.15 illustrates for potassium graphite. The XPS data of the K 3*p* core level are compared with UPS results for intercalated and deintercalated graphite. Deintercalation was achieved by oxygen-induced segregation of potassium leading to a mixture of potassium oxide and metal at 80 K and

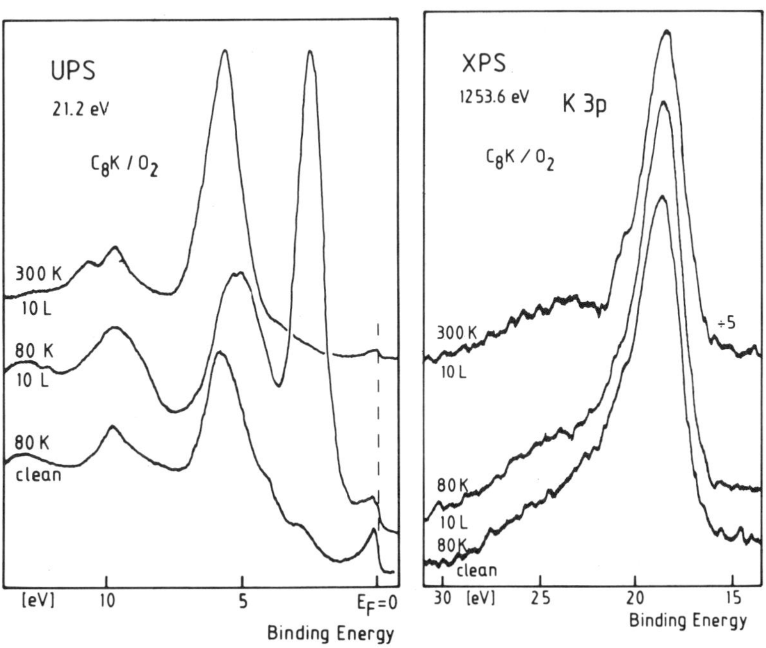

Fig. 3.15. Reaction of stage-1 potassium graphite with oxygen. The potassium core level is insensitive to the valence change, as documented in the UPS data. The asymmetry of the K 3*p* line in the clean state is lost after oxidation, and the intensity at the high b.e. side of the peak is due to the weak O 2*s* peak

potassium peroxide and metal at 300 K. The UPS data indicate the loss of intercalated metal by the decrease in intensity of the Fermi edge after exposure to oxygen. Oxygen $2p$ valence states occur at 2.8 eV for the oxide and at 5.4 eV for the peroxide [3.79]. The corresponding core-level data are completely unaffected by these profound chemical changes. The appearance of the broad line at ca. 22 eV arising from oxygen $2s$ emission and the change in intensity indicate that XPS sees the incorporation of oxygen and that the failure to detect any shift is not due to insufficient oxidation. The position of the K $3p$ line after correction of the binding-energy scale for a constant C $1s$ position with 18.6 eV agrees very well with literature data [3.22] reported to be the same for free and intercalated potassium.

Core-level spectra of intercalated species can be very useful for the analysis of molecular intercalates such as metal halides in acceptor GICs. Figure 3.16 compares the A1 $2s$ peaks of intercalated aluminium chloride and of oxidized aluminium metal. The reference spectrum indicates the chemical shift between Al in the zero-valent and trivalent states. It also gives a typical linewidth for this unsplit core level. Intercalation does not lead to a reduction of the aluminum; i.e., the charge from graphite is not transferred into some or all aluminum ions, as can be seen from the high binding energy of the aluminium chloride in the intercalated state.

The line is, however, much too wide for a single species of aluminum. In Figure 3.17, which compares the core level for the GIC at two temperatures, this is indicated by the dashed line resulting from the spectrum of a thin film of

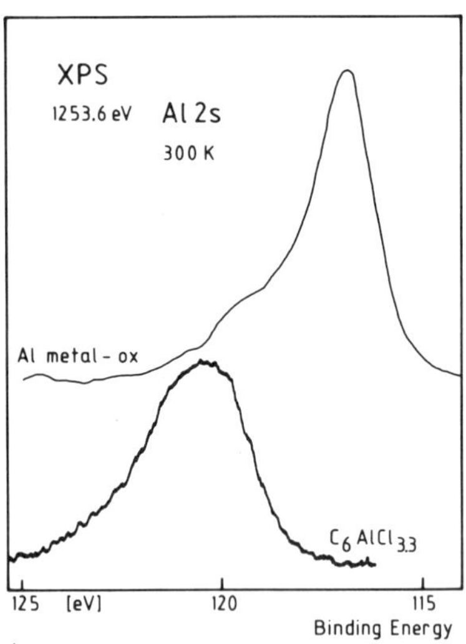

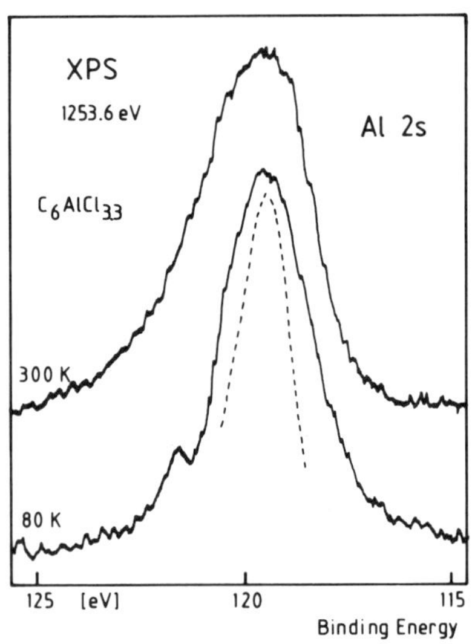

▲

Fig. 3.16. Core levels of intercalated aluminium (in $AlCl_3$) compared with metallic and partly oxidized free aluminium

Fig. 3.17. High-resolution profiles of Al $2s$ core levels of intercalated $AlCl_3$ at high and low temperatures. The dashed line indicates the profile of a single pure aluminium species

aluminum chloride. At low temperature the spectrum splits in two components. This can be explained within the structural model for aluminum chloride GIC derived by *Behrens* et al. [3.37]. The in-plane structure of the GIC consists of islands of Al_2Cl_6 dimers and $AlCl_4^-$ ions. The ratio depends on the island size and hence on the temperature, which controls the agglomeration of the islands. The ionic species is believed to cause the asymmetry and the second peak at higher b.e. in the low-temperature spectrum.

This argumentation formally resembles the attempt to deduce inequivalent carbon atoms from lineshape effects in the C 1*s* spectrum. However, the mentioned spectroscopic effects of final states and Heisenberg broadening do not appear to a measurable extent in the present case of molecular ionic species. According to this interpretation, the intercalation-induced charge transfer from graphite would have reduced some chlorine (added during preparation) to chloride anions. Their presence permits splitting of the Al_2Cl_6 dimers of the gaseous intercalant and formation of the $AlCl_4^-$ units. It would be an over-interpretation, however, to deduce the charge transfer number from a quantitative evaluation of the $AlCl_4^-$ content of the sample. Reduction of chlorine by irreversible carbon chlorination leading to C–Cl_x groups or even to volatile CCl_4 also contributes to the excess chloride ion concentration, and the charge transfer number would thus be too large and depend on details of the sample preparation.

3.4 Photoemission from Acceptor GICs

This section reviews photoemission data of acceptor-halide GICs, in particular, $AlCl_3$, $CuCl_2$, and SbF_5 in their stage-1 structures. The literature is, however, scarce, probably because of experimental difficulties in handling and analyzing the samples in UHV [3.84]. The data used in this section were obtained by the present author and have been published in some conference papers [3.80, 83, 84]. First we review some experimental details typical for most of the GICs discussed here.

3.4.1 Experimental Details

The details of sample preparation and handling are essential in understanding the complexity of chemical effects characteristic of GICs intercalated with reactive molecular species. This chemistry between intercalant species and in the guest-host interaction (irreversible oxidation [3.4, 38]) is by far the dominating effect in the spectra of all acceptor GICs. Its detailed analysis is vital for a correct interpretation of influences of the electronic structure of the synthetic metal on the spectra.

All samples were intercalated from HOPG. $AlCl_3$ was intercalated as stage-1 at 510 K with the graphite held at 500 K in an atmosphere of 50 mbar chlorine

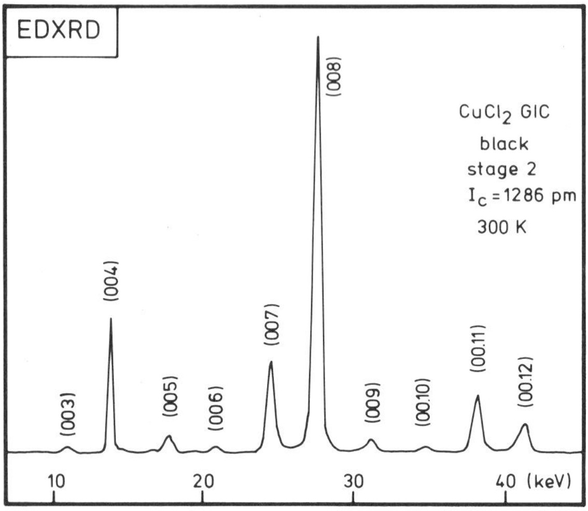

Fig. 3.18. (001) X-ray diffraction pattern of a stage-2 copper chloride GIC. The pattern was obtained by energy dispersive X-ray analysis using white X-rays from a silver target. The sample was kept in a glass ampoule under inert gas

at 300 K. $CuCl_2$ was intercalated to stage-2 with $AlCl_3$ as the complexing agent (50 mg) at 820 K in a temperature difference of 30 K. SbF_5 was intercalated to stage 1 from the liquid held at 305 K. Under these conditions GICs with well-developed staging were obtained in samples of size 9 × 13 × 1 mm. Figure 3.18 gives an example for a 001 diffraction pattern. The intensity distribution of the Bragg peaks indicates stacking disorder, but attempts to improve the ordering led either to partial exfoliation or to unacceptable levels of surface chlorination (see below).

The samples were transferred into the electron spectrometer under argon and cleaned by successive cleavage in situ. This can be done only at 78 K due to the high vapor pressure of the intercalated molecules. Measurements above 200 K led always to surfaces largely depleted in metal chloride. The criteria for a genuine GIC surface were defined as follows: a single line in the C 1*s* spectra to avoid extensive surface halogenation, a Fermi-level shift of larger than 0.5 eV to avoid adsorbed layers of non-intercalated metal halide, and an elemental ratio close to the stoichiometry expected from the XRD examination.

In some preparations it was not possible to fulfill these conditions; i.e., the concentration of intercalate remained always below the expected level, indicating undersaturation of the GIC. This problem arises because of the necessity to cleave off the exterior part of the HOPG sample as much as half its initial thickness. This leaves for examination only the inner most part of the GIC, which is naturally the least completely intercalated section of the sample. Figure 3.19 illustrates the necessity to cleave off a significant fraction of a GIC sample for intercalated $CuCl_2$. The central part of the figure schematically depicts the sequence of materials in the sample as a function of the depth of cleavage. The XPS spectra of characteristic core levels are shown at the sides of the figure. The top 10% of the material was a mixture of carbon and aluminum oxide arising from hydrolysis of the complexing agent $AlCl_3$. This material was

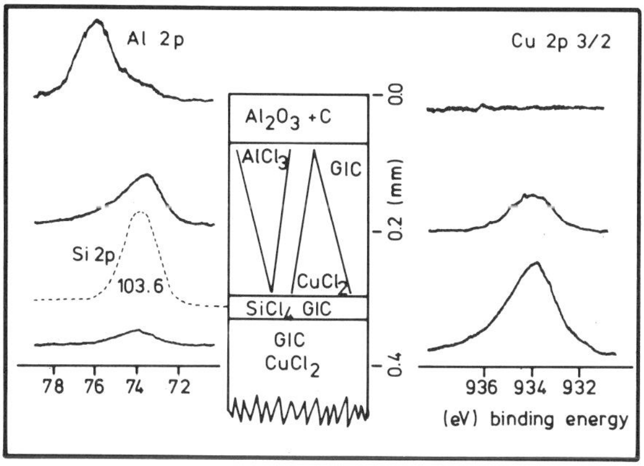

Fig. 3.19. Sequence of layers in the copper chloride GIC sample (total thickness 1.1 mm) as determined by successive cleaving inside the electron spectrometer. The characteristic core-level spectra are correlated with the depth of their occurrence. The arrows in the central sketch indicate the change in relative abundance of the phases. The intercalation compound with $SiCl_4$ was identified by the correct Si:Cl ratio, a Fermi-level shift of 0.7 eV, and the absence of stoichiometric amounts of oxygen

formed as a reaction product of $AlCl_3$ with the quartz glass ampoule during prolonged intercalation time and may have formed a diffusion barrier for complete intercalation of the interior sections. Below this layer, a thick section followed by co-intercalated $AlCl_3$ and $CuCl_2$. During successive cleaving a thin layer of intercalated (Fermi-level shift) $SiCl_4$ was detected. By the time the concentration of metal impurities in the $CuCl_2$ had fallen to below 10 at % in the surface, ca. 0.5 mm of the sample had to be scraped off.

This extreme example shows the difficulty in studying materials intercalated at high temperatures. Irreversible surface halogenation presented the same experimental problem with the other intercalates discussed here.

Finally, note that all acceptor GICs were sensitive to electron beam irradiation as it occurs with electron-induced Auger spectroscopy. This technique always revealed by far too low concentrations of intercalate, suggesting that pure graphite constitutes the top surface after cleaving. This was found to be true for cleaving at 300 K by UPS (valence-band spectrum very similar to that of graphite) but not for cleaving at 78 K (new distinct valence-band spectrum). Under the present experimental conditions (3 kV energy, 30 nA current, bam rastering over 0.5 mm × 0.5 mm) the electron beam caused desorption of a large fraction of the intercalated material within ca. 10 min observation time, as followed from an examination of the area after the Auger experiment by UPS.

3.4.2 Surface Halogenation

GICs with SbF_5 were examined earlier in the literature [3.66] and extensive surface fluorination was detected by XPS. If samples are never exposed to air, the degree of fluorination is much less pronounced and no distinct features for CF_2 and CF_3 groups occur in the C 1*s* spectra. This can be seen in Fig. 3.20, which shows spectra of a HOPG sample cleaved at 300 K, at 80 K, and after warming the cooled sample back to 300 K. Comparison with the spectrum of pristine HOPG reveals the Fermi-level shift ΔE_F. Both 300 K spectra show a shoulder at higher binding energy indicative of carbon fluorine bonds. The low-temperature spectrum is almost free of this feature and exhibits a lineshape similar to that of pristine HOPG (dashed line). The excess width of the line of the GIC is temperature dependent and vanishes at temperatures above 220 K without showing a sudden change, as expected for a phase transition of the in-plane structure.

From the position of the fluorination line we might expect that free graphite occurs as a consequence of deintercalation of the volatile SbF_5. This is the case if the sample is kept overnight in the spectrometer at 300 K. The deintercalation process causes excessive surface fluorination. The extensive surface fluorination upon warming from 80 K to 300 K can be followed clearly by fluorine X-ray excited Auger spectroscopy. An example is given in Fig. 3.21, which shows integrated Auger spectra together with the assignment. The valence transition

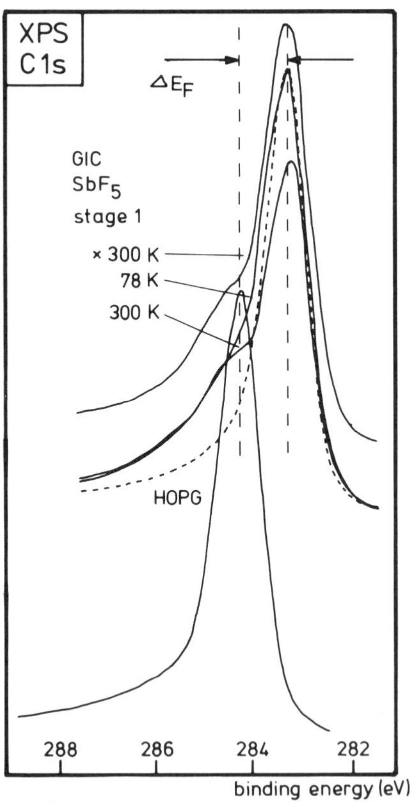

Fig. 3.20. High-resolution scans through the carbon 1*s* emission of stage-1 antimony fluoride GIC and a HOPG sample taken from the same piece as the GIC precursor. The dashed line is the profile of HOPG normalized in intensity into the low-temperature GIC spectrum. The × 300 K run was obtained after warming the cold sample to 300 K inside the spectrometer without cleaving

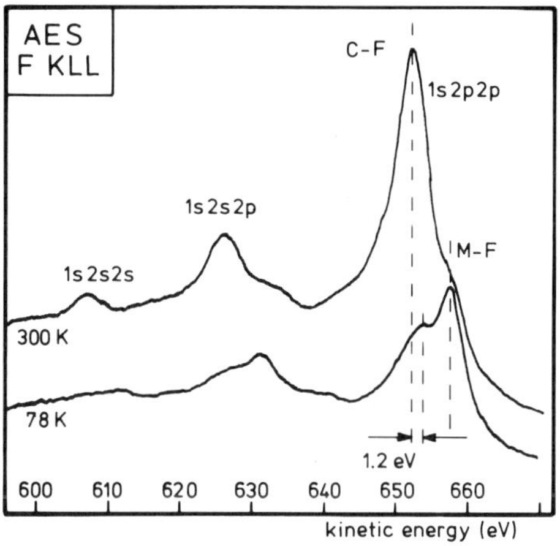

Fig. 3.21. Photon-excited Auger spectra of the fluorine *KLL* transition. The shift of 1.2 eV is the Fermi-level shift, M–F denotes a characteristic position for metal fluorides, and C–F denotes the position of C–F bonds as seen in, e.g., PTFE

exhibits a chemical shift between fluorine bonded to metals and bonded to nonmetals (carbon). The figure shows that warming causes the formation of new carbon fluorine bonds. These bonds are also present to a small extent at low temperatures, permitting the assignment of the excess linewidth of the C 1*s* spectrum to such groups and not to intercalation-induced electronic effects on the host atoms. The fact that the C–F peaks in the Auger spectrum are shifted by the Fermi-level shift of ca. 1.2 eV indicates that the C–F groups at 80 K are part of the electronic structure of the GIC. In other words, they constitute non-insulating groups with sp^3 carbon atoms as they are present in the C–F groups after warming up, now decoupled from the still-existing conduction band of the synthetic metal (compare the Fermi-level shift in the C 1*s* spectrum). This decoupling is further reflected in the intensity change of the spectrum. Comparison with the fluorine 1*s* XPS intensity showed no increase in surface concentration. The increase in the Auger intensity is thus likely to be caused by a cross-section change resulting from the transition of a metallic into a non-metallic valence state.

Surface chlorination was studied in the $AlCl_3$ GIC system [3.28]. Figure 3.22 shows spectra from a warming experiment similar to the one described in the previous section for a low-temperature cleaved sample of $AlCl_3$ GIC, stage-1.

The C 1*s* spectrum of the genuine surface at 80 K is a very narrow line with weak structures arising from residual C–Cl_2 and C–Cl_3 groups [3.28]. Warming causes the formation of a large number of C–Cl groups (double feature) and a slight reduction of the Fermi-level shift (peak at the main line position), as indicated by the difference spectrum in the figure. The material remains at large a GIC characterized by the narrow linewidth of the C 1*s* peak.

The corresponding chlorine XPS data in the figure indicate that the C–Cl groups give rise to a lower b.e. for chlorine than in the metal chloride. The lineshape of the overall spectrum is the result of a superposition of several spectra from chemically inequivalent chlorine species bound to aluminum, in

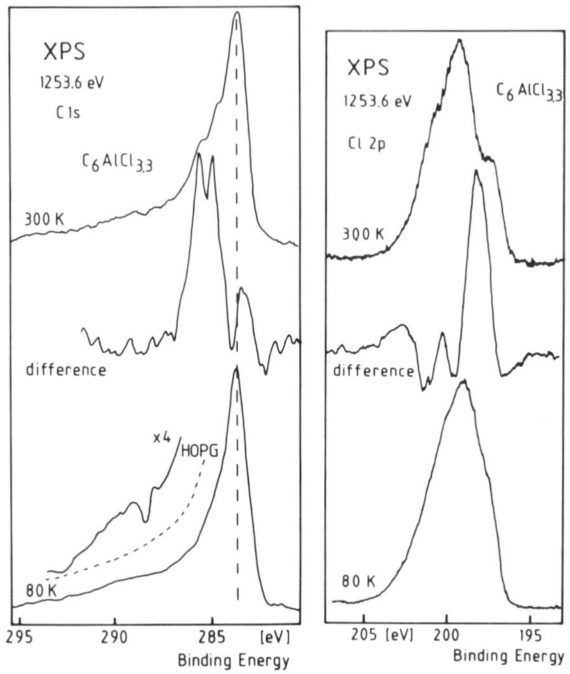

Fig. 3.22. High-resolution scans of the carbon 1*s* and chlorine 2*p* lines of a stage-1 aluminum chloride GIC at 80 K and after warming to 300 K. The difference spectra ($\times 2$) were obtained without normalization. The dashed line indicates the spectrum of a HOPG sample that was briefly Ar etched to remove the surface-plasmon peak

addition to the spectrum of the C–Cl groups. A single species would exhibit a well-resolved spin-orbit doublet structure with a splitting of ca. 1.0 eV. The absence of such a structure points to at least two spectra with a separation of ca. 1.0 eV. These spectra are assigned to terminal and bridging chlorine atoms in the intercalated mixture of Al_2Cl_6 and $AlCl_4^-$.

The situation close to the real surface of the $AlCl_3$ GIC is schematically sketched in Figure 3.23. A stepped graphite surface contains cleavage defects representing sites of carbon chlorination throughout the volume of the GIC [3.62]. The "chemist's magnifying glass" shows the arrangement of C–Cl and C–Cl_2 groups found in the genuine GIC. The intercalate is organized in islands made from a mixture of the two aluminum chloride species sketched in the figure. The islands are nonrigid and can move or even fall apart upon thermal stress, thus providing additional chlorination when the sample is warmed. Little evidence was found for the presence of elemental chlorine encapsulated in the GIC, which may be considered as an alternative chlorine source. Figure 3.24 provides evidence for the thermal sensitivity of the islands. A surface was quickly cleaved at 300 K and cooled within ca. 100 s down to 80 K. A single scan of the

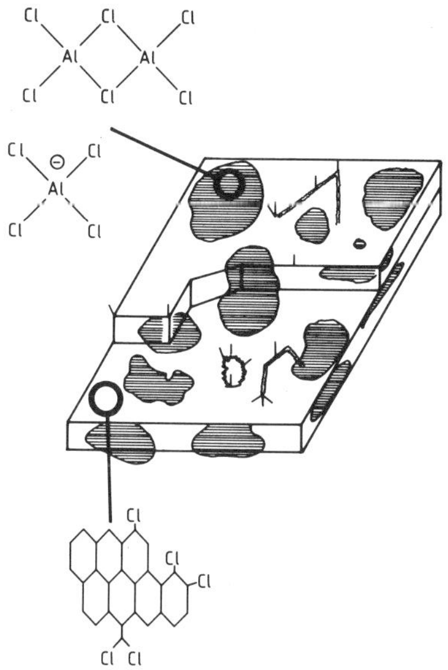

Fig. 3.23. Schematic representation of the surface of an $AlCl_3$ GIC after cleaving in situ. Islands of the composite intercalate (top left) are inserted between defective chlorinated graphene sheets (bottom left). Surface groups with several chlorine atoms are present at low coordinated carbon atoms

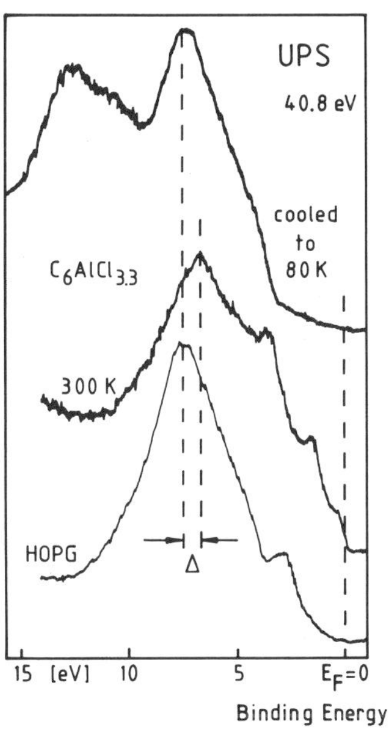

Fig. 3.24. He II valence-band spectra of stage-1 aluminum chloride GIC. The Δ indicates the Fermi-level shift. After cooling, a layer of aluminum chloride has formed on top of the GIC on which the graphite σ peak is superimposed

He II UPS documents the initial state of the GIC, including a Fermi-level shift, a metallic Fermi edge, and a molecular orbital of Al_2Cl_6 at ca. 3.2 eV.

A second single-sweep spectrum was recorded immediately after the sample reached the low temperature. The spectrum had changed completely and consisted now of a truncated graphite peak over a large background of secondary electrons arising from an insulating surface of, e.g., aluminum chloride, which had segregated from the GIC following the thermal stress during rapid cooling.

This state is not stable in UHV. After storage a completely insulating surface results, yielding no more UPS data. Such experiments illustrate the difficulty in deducing valid bulk information from a material with high surface reactivity and a technique probing predominantly the surface region.

3.4.3 Valence-Band Information

The occurrence of a Fermi edge, the most important feature of intercalation-induced valence-bands in acceptor GICs, has already been demonstrated for the $AlCl_3$ GIC. The high-resolution valence-band edges in Fig. 3.25 show a clear transition from the semimetal graphite to the metal $CuCl_2$ GIC. The spectra are dominated by an intensity above ca. 4 eV which is much larger than the intensity of pure graphite. The main peak undergoes a marked reversible change upon temperature change, indicating a phase transition. This transition modifies the height of the Fermi edge, which increases with higher temperatures. The magnitude of the Fermi-level shift indicated by the arrow in the figure does not change with the temperature. The spectra did not require numerical intensity normalization, so we can conclude that the change at the Fermi energy is a real effect and may be due to an increase in intensity of the 4 eV feature extending to the Fermi energy, rather than the reflection of a change in charge transfer from graphite, which should cause an easily detectable change in the Fermi-level shift.

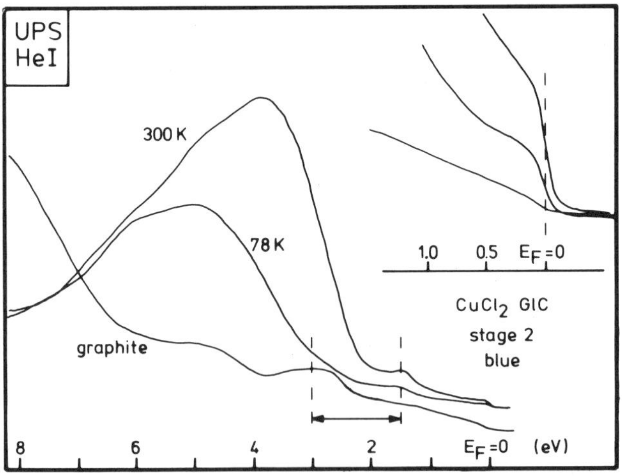

Fig. 3.25. High-resolution valence-band spectra of stage-2 $CuCl_2$ GIC. The arrow indicates the Fermi-level shift; the inset shows the transition from semimetal graphite (bottom) to the metal GIC. The height of the Fermi edge is strongly temperature dependent and larger at higher temperatures

The assignment of the intense valence-band feature to an atomic state is facilitated by comparison of the UPS data with valence-band XPS. Consideration of the spectral cross sections shows that XPS is particularly sensitive to the copper 3*d* states and almost insensitive to carbon states. Figure 3.26 shows experimental verification: spectra of high- and low-stage GICs. The insensitivity of XPS to carbon states is illustrated by the small feature of the graphite σ-band edge in the high-stage GIC with a chemical composition of $C_{85}\ CuCl_{2.05}$. The valence-band maximum at ca. 6 eV is due to copper 3*d* states, which occur at the dashed position in pure $CuCl_2$. The peak is much wider and shifted to higher binding energy than in metallic copper because of the ligand field bonding interaction between copper 3*d* and chlorine 3*p* states. These states do not occur as discernible peaks in the XPS data but contribute as the highest occupied states to the intensity of the spectrum below 5 eV. The UPS structure at ca. 5 eV, which was found to be temperature dependent in the intercalate (Fig. 3.25), characterizes states of the molecular intercalate. The temperature effect points to a rearrangement of copper chloride molecules within a layer of intercalate, rather than to a change of the graphite-$CuCl_2$ relation.

The valence-band XPS data of the two GICs exhibit differences in the position of the Cl 3*s* core level. The spectrum of the high-stage GIC is in all aspects very similar to that of pure $CuCl_2$. The low-stage GIC contains chlorine in a different chemical environment, as indicated by the difference in the distance Cl 3*s* to Cu 3*d*. Differences were also detected in the chlorine 2*p* spectra exhibiting two components at 201.5 and 200.6 eV, besides the C–Cl component at 199.6 eV. At 78 K both metal chloride peaks shift by 2.0 eV to lower binding energy without changes in the lineshapes. Only the position of 200.6 eV with no temperature sensitivity was found in the spectrum of pure $CuCl_2$.

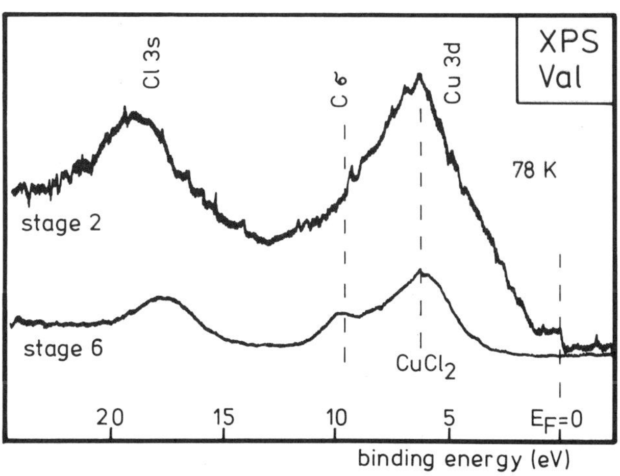

Fig. 3.26. X-ray-excited valence-band spectra of $CuCl_2$ GIC with different concentrations of the intercalate. The shapes of these spectra are dominated by the emissions from the molecular intercalate

All this indicates that in the low-stage GIC the intercalate is present in a form different from the free metal chloride, which seems, however, to constitute the intercalate in the high-stage GIC. The integrated intensity of the Cl 2*p* line without the C–Cl component yields a Cu:Cl ratio of 2.05 for the high-stage compound and 2.6 for the low-stage compound. A chlorocomplex of divalent copper with terminal and bridging chlorine atoms may be a suitable model for the low-stage intercalate.

Any charge transferred from graphite is not localized on the copper ions; i.e., copper is not reduced as follows from the satellite spectrum of the copper 2*p* photoemission. Figure 3.27 shows the data together with the reference spectrum of free $CuCl_2$. There is no chemical shift in the main line transitions after correction for the Fermi-level shift. This is in line with the valence-band data of Fig. 3.26. The main copper spectrum (final state $3d^{10}$L) is associated with an intense satellite arising from the final state $3d^9$. Its existence is unambiguous evidence for divalent copper; chemically reduced monovalent copper exhibits only one final state $3d^{10}$. The distances between the main line and satellites are the same in both spectra, indicating the chlorine environment of the copper ions (copper oxide exhibits satellites at ca. 1 eV further distance). The line shape and the satellite intensity are, however, different for free and intercalated $CuCl_2$ for the low-stage case. This satellite cross-section change indicates a different charge distribution in the Cu–Cl bonds. Although it is tempting to attribute this effect to the presence of additional charge density in the Cu–Cl bond from the graphite, a more realistic interpretation in the light of the effects discussed above

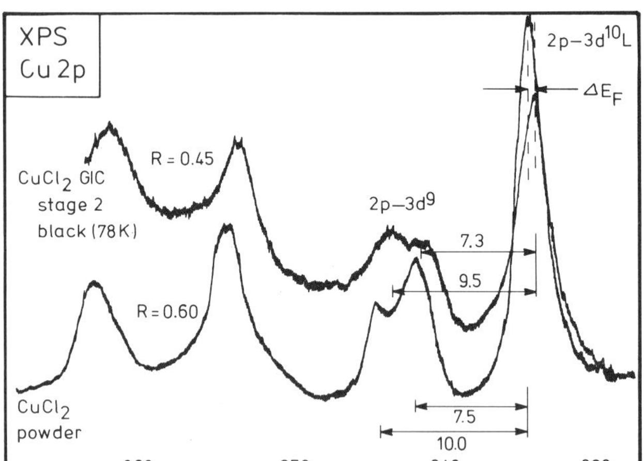

Fig. 3.27. Copper 2*p* core-level spectra of free intercalated $CuCl_2$: R denotes the ratio of areas between the main and satellite transitions. The annotation above the 2*p* 3/2 lines gives the final states of the transitions; the numbers indicate the distances between the main transitions and the satellites in eV

is to assume that in the copper-chlorine complex of the intercalate, bonding is slightly different from the interactions in pure $CuCl_2$.

In summary, photoemission in the $CuCl_2$ GIC system allows us to detect the semimetal–metal transition and a significant Fermi-level shift as intercalation-induced changes of the graphite electronic structure. Evidence was found that the chemical constitution of the intercalate is different in the low-stage GIC and the high-stage GIC, with the latter containing the intercalate in the same form as in free $CuCl_2$. Charge transfer from graphite does not lead to a reduction of the metal cation. The excess charge is stored in the Cu–Cl bonds. The low-stage GIC contains a complex of copper chloride with two inequivalent types of ligands and a more open structure, as in bulk $CuCl_2$.

Figure 3.28 shows the valence-band spectra of stage-1 SbF_5 GIC. Intercalation transforms HOPG into a synthetic metal by removing valence-band electron density from graphite (see shift of the marker structure at ca. 2.8 eV). The resulting Fermi-level shift is also reflected in the secondary electron emission structure seen at 7.2 eV kinetic energy. In the low-stage donor GIC it is shifted to ca. 6.0 eV; in the present acceptor GIC it is shifted closer to the Fermi level and thus to ca. 8.6 eV kinetic energy. The pronounced change in line shape and intensity indicate the presence of significant final-state effects in the spectrum of the GIC, which reduce the accuracy of the correction of the binding-energy scale.

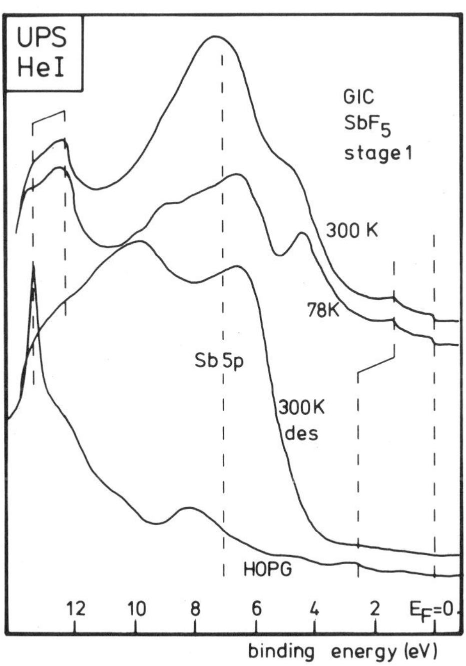

Fig. 3.28. Valence-band spectra of stage-1 antimony fluoride GIC in comparison with the HOPG spectrum. The notation 300 K des indicates the surface after warming from 80 K to 300 K inside the UHV of the spectrometer. The antimony 5*p* valence doublet is located in the free atom at 7.3 eV

Again, the valence-band spectra are dominated by molecular levels of the intercalate arising here from antimony $5p$ states. If the temperature of the surface is raised to 300 K, segregation of antimony fluoride begins without changing the charge transfer (no change in Fermi-level shift, no loss of the Fermi edge). This implies that not all of the intercalated antimony fluoride takes part in the charge transfer process. This is in line with other spectroscopic observations that identify trivalent antimony fluoride as inactive co-intercalate [3.8].

Keeping the sample for more than 15 min at 300 K leads to strong desorption effects and transformation of the synthetic metal surface into an insulating fluorinated carbon with antimony fluorides deposited on it. The corresponding spectrum (center of Fig. 3.28) is essentially a structureless secondary electron emission peak with residual intensity from Sb $5p$ states superimposed. The transformation into a good insulator is clearly seen by the loss of all intensity from the former Fermi edge down to below 4 eV.

Several times in this chapter we have seen how final-state effects in the valence-band spectra interfere with the determination of the Fermi-level shift. A way to overcome this problem is to determine the Fermi-level shift by energy-loss spectroscopy. This works only with high-quality single-crystal surfaces of a GIC because the observation of a surface plasmon as a marker structure is required. The bulk-plasmon features of graphite are too diffuse to allow an accurate determination of an energy shift of ca. 1.5 eV. Figure 3.29 illustrates the experiment. Next to the omitted elastic beam, sharp features at 6.5 eV for pristine graphite and a wide structure at ca. 26 eV result from surface- and bulk-plasmon excitations. The Fermi-level shift also causes a shift of the two structures in the GIC, which can be conveniently determined for the surface plasmon. Its sheer occurrence indicates the formation of a well-ordered graphitic surface after the sample was cleaved at 80 K. The change in shape of the bulk plasmon reflects the presence of the intercalate and the resulting modification of the graphite electronic structure, and can be used to distinguish an intercalate from a chemisorption system.

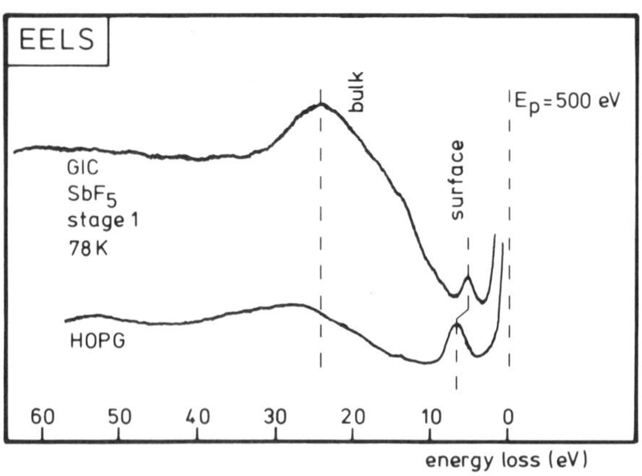

Fig. 3.29. Electron energy-loss spectrum of a mirror-cleaved surface of stage-1 antimony fluoride GIC. The kinetic energy of the primary electrons was 500 eV. The plasmon losses are indicated. The shift of the surface plasmon can be used to independently determine the Fermi-level shift and so to assign the peaks in the UPS valence-band spectrum

Note that the experiment needs low electron energies, very low beam currents, and quick data acquisition. Electron beams destroy the GIC structure, as can be seen from AES by the backshift and intensity loss of the surface plasmon following desorption and fluorination of the irradiated surface.

3.5 Summary and Conclusions

This chapter has described some of the experimental efforts to understand the electronic structure of GICs. Electron spectroscopy and X-ray methods are high-energy spectroscopies that let us study wide regions of the valence band and even deep-lying core levels. The element-specific information was thought to be useful in determining the relocation of electrons as an intercalation-induced effect.

In the course of the investigations several complications arose. Strong final-state effects prevent simple comparison of spectral patterns of "free" and "intercalated" systems. Large shifts of the energy scale and difficulties in finding reliable correction methods preclude the determination of physically accurate binding energies and the identification of true chemical shifts. The surface sensitivity of electron spectroscopy requires careful sample preparation to determine true bulk properties and to prevent such side effects as carbon oxidation, electron-induced segregation, and desorption phenomena. Comparing electron spectra with X-ray spectra, both of which are true bulk methods, we can verify the character of bulk information for the electron spectroscopic results.

A more fundamental problem occurs with molecular and alloy intercalates exhibiting strong interatomic interactions within the intercalate. Often the free compound is not identical in its chemical constitution to the intercalated species. During intercalation, redox reactions and complex formation occur – not as a consequence of the electronic charge exchanged with the host graphite, but solely as a consequence of the thermodynamics of the intercalation process. These difficult-to-quantify processes again preclude the comparison of "free" and "intercalated" states for evaluation of the intercalation-induced charge transfer effects.

Attempts have been made to verify a simple model of charge transfer. Graphite is assumed to exhibit a system of rigid bands overlapping in some spatial regions to form a semimetal. Intercalation of donors leads to filling of conduction bands, resulting in a metallic system; intercalation of acceptors depletes the graphite valence band with the same consequence of an increased metallic character. In these models, the shapes of the graphite bands were assumed to stay rigid and the intercalated species were assumed to interact only by electrostatic forces, with no orbital overlap contributions. The degree of ionicity and the charge transfer number were the desired parameters to describe the electronic structure of GICs.

Looking back, it is easy to see that these parameters are so difficult to define that they can scarcely serve as guidelines, and that electron spectroscopy in all its forms is a technique too difficult to interpret to the desired level of accuracy.

Nevertheless, the scientific controversies about the interpretation of spectra have led to a better understanding of the complex interactions in GICs, and the commonly accepted results of the experimental studies can be summarized as follows.

The crude picture sketched above about the charge transfer models is correct. Intercalation leads in all cases to a metallic Fermi-edge feature. The low total density of states around that feature causes, in the process of charge transfer, large Fermi-level shifts that are chemically equivalent to a significant destabilization of the material. Intercalation of donors leads to significant modifications of the whole graphite π band and also to changes in the DOS of the σ band, i.e., to strong non-rigid-band effects. These effects are by far less pronounced with acceptor intercalates.

No conclusive evidence was found for localization of the charge transfer effects within the graphene layers, i.e., for the inequivalence of carbon atoms within the layer adjacent to the intercalate. GICs are not ionic compounds. All cases investigated show strong covalent interactions with the transferred charge density stored in bonds formed between rehybridized molecular orbitals of graphite and the intercalates. We should not conclude, however, that ionic graphite compounds such as graphite salts with acceptor molecules and solvated intercalated ions do not exist; there have been insufficient spectroscopic studies on such systems up to now. The covalent nature of the interaction in the alkali GIC in particular gives rise to the chemical behavior of these GICs similar to free alkali metals, and to the reactivity of these compounds in ternarization reactions, which are governed by the properties of the intercalated atoms and not by negatively charged aromatic carbon entities. In acceptor GICs the charge transferred from graphite is not located at the central metal ion of the intercalate but is stored in the chemical bonds between the central ion and the electronegative ligands.

The description of the electronic structure of acceptor GICs by a modified graphite structure seems consistent with most of the experimental observations. Donor GICs require, however, a new electronic structure determination with atomic states as ingredients from both carbon and alkali metal. Alkali-metal ions are certainly a very crude approximation to a GIC. Experimentalists and theoretical workers realize this, and full convergence in their views about donor GICs is expected in the near future.

References

3.1 M.S. Dresselhaus, G. Dresselhaus: Adv. Phys. **30**, 139 (1981)

3.2 H. Zabel, S.A. Solin (Eds.): *Graphite Intercalation Compounds I: Structure and Dynamics* Springer Ser Mater. Sci. Vol. 14 (Springer, Berlin, Heidelberg 1990)

3.3 P. Pfluger, H.J. Güntherodt: Festkörperprobleme. **21**, 271 (1981)
3.4 G. Wortmann, W. Krone, G. Kaindl, R. Schlögl: Synth. Met. **23**, 139 (1988)
3.5 D. Bonnin, P. Kaiser: In *Chemical Physics of Intercalation*, ed. by A.P. Legrand, S. Flandrois, NATO ASI Series B. Vol. 172, 319 (1987)
3.6 G. Wortmann: Hyperfine Interact. **27**, 263 (1986)
3.7 H. Touhara, K. Kadono, H. Ionoto, N. Watanabe, A. Tressaud, J. Graunec: Synth. Met. **18**, 549 (1987)
3.8 G. Wortmann, F. Godler, B. Perscheid, G. Kaindl, R. Schlögl: Synth. Met. **26**, 109 (1988)
3.9 S. Tanuma: In Ref. [3.5] p. 165
3.10 A. Koma, K. Miki, H. Suematsu: In *Graphite Intercalation Compounds*, ed. by S. Tanuma, H. Kamimura (World Sientific, Singapore 1985) p. 246
3.11 G. Loupias, J. Chomilier, D. Guerrard: Solid State Commun. **55**, 299 (1985)
3.12 J. Conard: In Ref. [3.5] p. 357
3.13 J.E. Fischer: In *Intercalated Layered Materials* Vol. 6, ed. by F. Levy (Reidel. Dordrecht 1979) p. 481
3.14 E. McRae, J.F. Mareché: J. Mater. Res. **3**, 1 (1988)
3.15 E. McRae, J.F. Mareché, M. Lelaurin: In: Int. Colloq. Layered Materials, ed. by D. Guerrard, P. Lagrange (1988) p. 175
3.16 L. Schade, P.V. Rague-Schleyer: Adv. Organomet. Chem. **27**, 169 (1987)
3.17 T. Takahashi, H. Tokalin, T. Sagawa: Solid State Commun. **52**, 765 (1984)
3.18 N.A.W. Holzwarth, S.G. Louie, S. Rabii: Phys. Rev. B **26**, 5382 (1982)
3.19 F.R. McFeely, S.P. Kowalczyk, L. Ley, R.G. Carwell, R.A. Pollak, D.A. Shirley: Phys. Rev. B **9**, 5268 (1974)
3.20 M. Cardona, L. Ley (Eds.): *Photoemission in Solids I* Topics Appl. Phys. Vol. 26 (Springer, Berlin, Heidelberg 1978) p. 84 ff.
3.21 Ch. Beyreuther, R. Hierl, G. Wiech: Ber. Bunsenges. Phys. Chem. **79**, 1081 (1977)
3.22 A. Simunek, G. Wiech: Solid State Commun. **64**, 1375 (1987)
3.23 G.K. Wertheim, P.M. Th. van Attekum, S. Basu: Solid State Commun. **33**, 127 (1980)
3.24 J.M. Thomas, E.L. Evans, M. Barber, P. Swift: J. Chem. Soc., Faraday II **67**, 1875 (1972)
3.25 P.M. Th van Attekum, G.K. Wertheim: Phys. Rev. Lett. **43**, 1896 (1979)
3.26 Y. Baer: J. Electron Spectrosc. Relat. Phen. **24**, 95 (1981)
3.27 R. Schlögl, H.P. Boehm: Carbon **21**, 345 (1983)
3.28 R. Schlögl: Surf. Sci. **189**, 861 (1987)
3.29 G.S. Painter, D.E. Ellis: Phys. Rev. B. **1**, 4747 (1970)
3.30 J.C. Slonzewski, P.R. Weiss: Phys. Rev. **109**, 272 (1958)
3.31 J.W. McClure: Phys. Rev. **108**, 612 (1960)
3.32 W. Rüdorff, E. Schulze: Z. Anorg. Allg. Chem. **272**, 156 (1954)
3.33 N. Daumas, A. Herold: C.R. Seanc. Acad. Paris C **268**, 373 (1969)
3.34 M. Posternack, A. Baldereschi, A.J. Freeman, E. Wimmer, M. Weinert: Phys. Rev. Lett. **50**, 761 (1983)
3.35 N.A.W. Holzwarth, S.G. Louie, S. Rabii: Phys. Rev. B **28**, 1013 (1983)
3.36 G. Kaindl, J. Feldhaus, U. Ladewig, K.H. Frank: Phys. Rev. Lett. **50**, 123 (1983)
3.37 P. Behrens, U. Wiegand, W. Metz: Ext. Abstracts, Carbon **86**, Baden-Baden 86, 502 (1986); P. Behrens: Dissertation, Hamburg (1988)
3.38 R. Schlögl, W. Jones, H.P. Boehm: Synth. Met. **7**, 131 (1983)
3.39 M.L. Dzurus, G.R. Hennig: J. Am. Chem. Soc. **79**, 1051 (1957)
3.40 A. Herold: Mater. Sci. Eng. **31**, 1 (1979)
3.41 M. El Makrini, P. Lagrange, A. Herold: Carbon **18**, 374 (1980)
3.42 P. Lagrange, A. Bendriss-Rerhrhaye, J.F. Mareché, E. McRae; Synth. Met. **12**, 201 (1985)
3.43 R. Schlögl, V. Geiser, P. Oelhafen, H.J. Güntherodt: Phys. Rev. B **35**, 6414 (1987)
3.44 J. Jegoudez, C. Mazieres, R. Setton: Synth. Met. **7**, 85 (1983)
3.45 R. Setton: Ext. Abstracts Carbon 86, Baden-Baden **86**, 420 (1986)
3.46 R. Schlögl, H.P. Boehm: Carbon **22**, 341, 351 (1984)
3.47 R. Schlögl, B. Wrackmeyer: Polyhedron **4**, 885 (1985)

3.48 H.P. Boehm, R. Schlögl: Carbon **25**, 583 (1987)
3.49 T. Inoshita, K. Nakao, H. Kamimura: J. Phys. Soc. Japan **43**, 1273 (1977)
3.50 T. Ohno, K. Nakao, H. Kamimura: J. Phys. Soc. Japan **47**, 1125 (1979)
3.51 D.P. DiVincenzo, S. Rabii: Phys. Rev. B **25**, 4110 (1981)
3.52 T.C. Tartar, S. Rabii: In *Graphite Intercalation Compounds*, ed. by P.C. Eklund, M.S. Dresselhaus, G. Dresselhaus, MRS Fall Meeting (1984) p. 77
3.53 A. Koma, K. Miki, H. Suematsu, T. Ohno, H. Kamimura: Phys. Rev. B **34**, 2434 (1986)
3.54 S. Mizuno, H. Hiramoto, K. Nakao: Solid State Commun. **63**, 705 (1987)
3.55 R. Eisberg, P. Josuks, G. Wiech, R. Schlögl: Solid State Commun. **60**, 827 (1986)
3.56 A. Mansour, S.E. Schnatterly, J.J. Ritsko: Phys. Rev. Lett **58**, 614 (1987)
3.57 R. Eisberg, G. Wiech, R. Schlögl: Solid State Commun. **65**, 705 (1988)
3.58 C.F. Hague, J.M. Mariot, G. Indlekofer, P. Oelhafen, H.J. Güntherodt: Synth. Met. **23**, 211 (1988)
3.59 A.R. Williams, J. Kübler, C.D. Gelatt Jr.: Phys. Rev. B **19**, 6094 (1979)
3.60 E. Ojala, M. Lähdeniemi: Physica Scripta **25**, 793 (1982)
3.61 C.F. Hague, J.M. Mariot, V. Geiser, P. Oelhafen, H.J. Güntherodt: Synth. Met. **23**, 211 (1988)
3.62 R. Eisberg: Dissertation, University of Hamburg (1987)
3.63 B. Bach, J.M. Thomas: Carbon 72, Int. Carbon Conf., Baden-Baden (1972)
3.64 R. Schlögl, H.P. Boehm: Carbon 80, Int. Carbon Conf., Baden-Baden (1980)
3.65 R.W. Joyner, F.L. Vogel: Synth. Met. **4**, 85 (1981)
3.66 L. Streifinger, H.P. Boehm, R. Schlögl, R. Pentenrieder: Carbon **17**, 195 (1979)
3.67 P. Oelhafen, P. Pfluger, E. Hauser, H.J. Güntherodt: Phys. Rev. Lett. **44**, 197 (1980)
3.68 P. Oelhafen, P. Pfluger, H.J. Güntherodt: Solid State Commun. **32**, 885 (1979)
3.69 J.G. Murday, B.J. Dunlap, F.L. Huston, P. Oelhafen: Phys. Rev. B **24**, 4764 (1981)
3.70 B.J. Dunlap, D.E. Ramarker, J.S. Murday: Phys. Rev. B **25**, 6439 (1982)
3.71 M. Lagues, D. Marchand, C. Fretigny, A.P. Legrand: Solid State Commun. **49**, 739 (1984)
3.72 C. Fetigny, D. Marchand, M. Lagues: Phys. Rev. B **32**, 8462 (1985)
3.73 T. Takahashi, N. Gunasekara, T. Sagawa, H. Suematsu: Solid State Commun. **59**, 105 (1986)
3.74 G.K. Wertheim, P.M. Th. van Attekum, S. Basu: Solid State Commun. **33**, 1127 (1980)
3.75 S.B. DiCenzo: Synth. Met. **12**, 251 (1985)
3.76 M.E. Preil, J.E. Fischer: Phys. Rev. Lett. **52**, 1141 (1984)
3.77 S.B. DiCenzo, S. Basu, G.K. Wertheim, N.E. Buchanan, J.E. Fischer: Phys. Rev. B **25**, 620 (1982)
3.78 K. Siegbahn, C. Nordling, G. Johansson, J. Hedman, P.F. Heden, K. Hamrin, U. Gelius, T. Bergmark, L.O. Werme, Y. Baer: *ESCA Applied To Free Molecules* (North-Holland, Amsterdam 1969)
3.79 R. Schlögl: In *Physics and Chemistry of Alkali Metal Adsorption*, ed. by H.P. Bonzel, A.M. Bradshaw, G. Ertl (Elsevier, Amsterdam 1989) p. 347
3.80 R. Schlögl: Proc. Int. Colloq. Layered Materials, Pont-a-Mousson (1988) p. 237
3.81 H. Estrade-Szwartzkopf, B. Rousseau, M. Malki, P. Lauginie, J. Conard, P. Lagrange, D. Guerard: Synth. Met. **12**, 401 (1985)
3.82 R. Schlögl, W. Bensch: J. Less Common Met. **132**, 155 (1987)
3.83 R. Schlögl, V. Geiser, P. Oelhafen, H.J. Güntherodt: Proc. Int. Carbon Conf. Carbon '86, Baden-Baden 148 (1986)
3.84 D. Marchand, M. Lagues, C. Fretigny: In *Chemical Physics of Intercalation*, ed. by A.P. Legrand, S. Flandrois, NATO ASI Series B, **172**, 331 (1987)

4. Effects of Charge Transfer on the Optical Properties of Graphite Intercalation Compounds

By Peter C. Eklund and Gary L. Doll

With 40 Figures

In this chapter, we review the effects of charge transfer on the optical properties of graphite intercalation compounds (GICs) [4.1]. The principal optical techniques used to study charge transfer and its effects on the electronic and phonon properties of GICs have been reflectance spectroscopy and Raman scattering. Several reviews of Raman scattering [4.2] and the optical properties [4.3, 4] of GICs have appeared in the past few years.

Graphite intercalation compounds are quasi-two-dimensional metals whose highly anisotropic electrical and thermal properties stem from the anisotropic properties of graphite itself. Pristine graphite is a semimetal with small hole and electron pockets that lie along the H-K-H axes in the six corners of a hexagonal Brillouin zone [4.1]. The intercalation of foreign atoms or molecules to form a stage-*n* compound, where *n* refers to the number of carbon (C) layers between periodically placed intercalate (I) layers, is accompanied by a substantial transfer of charge between these two types of layers, leaving the hexagonal intralayer arrangement of the carbon atoms intact. With increasing charge transferred to the C layers, the small electron and hole pockets found in pristine graphite evolve into larger cylinders that eventually exhibit trigonal warping. The cylindrical surfaces of a low charge transfer stage-2 GIC are shown schematically in Fig. 4.1.

Of course, charge neutrality demands that an equal and opposite net charge resides in the C and I layers. For stage $n > 2$ compounds, two types of C layers are possible: C layers that lie adjacent to an intercalant layer, termed bounding carbon layers (C_b), and C layers that lie between other C layers, termed interior carbon layers (C_i). Experiment and theory have shown that most of the net layer charge in the C layers resides in the C_b layers. This uneven *c*-axis distribution of layer charge removes degeneracies in the graphitic intralayer phonons, and new "interior" and "bounding" layer mode frequencies are observed [4.1, 2]. Furthermore, the uneven net layer charge distribution results in *n* conduction and *n* valence graphitic π bands in a stage *n* compound [4.1–4]. Optical data can be used with other experimental results – (e.g., Shubnikov de Haas (SdH), nuclear magnetic resonance (NMR), Pauli spin susceptibility) – to determine parameter values for electron energy band models. Once the parameter values are known, the net layer charge on the C_i and C_b layers can then be evaluated from the band models.

If electrons are "donated" by the I layers to the C layers, the Fermi level, E_F, is raised in the carbon p_z (or π) bands, and the GIC is termed a "donor-type"

Springer Series in Materials Science, vol. 18
H. Zabel · S.A. Solin (eds.): Graphite Intercalation Compounds II

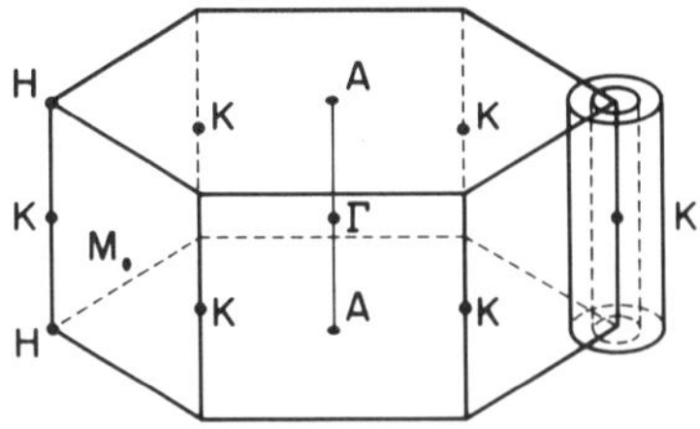

Fig. 4.1. Graphitic Brillouin zone: concentric cylinders represent the Fermi surface of a stage-2 GIC at low charge transfer

compound. If, on the other hand, electrons are transferred from the C layers and "accepted" by the I layers, an "acceptor"-type GIC is formed. X-ray and neutron scattering experiments have shown directly that increasing charge transfer is accompanied by a measurable expansion (donor-type GICs) or contraction (acceptor-type GICs) in the in-plane C–C bond length, as well as a contraction in the thickness of the three-layer C_b–I–C_b sandwich.

In Sect. 4.1 we discuss experimental considerations pertinent to Raman scattering and reflectance spectroscopy in GICs, and describe the safe handling of air-sensitive samples for optical study. In Sect. 4.2 we indicate the relationship between the optical reflectance, the dielectric constant, and the electron energy band structure in GICs. We discuss the process for the Kramers-Kronig determination of the GIC (basal plane) dielectric function using "density-scaled graphite". In Sect. 4.3 we review results of optical studies on donor and acceptor (binary and ternary) GICs that have led to values for the charge transferred to the carbon layers, or the effects of charge transfer on the optical-phonon frequencies. Finally, in Sect. 4.4 we present a summary of the optical determination of charge transfer, and offer directions for future research.

4.1 Experimental Considerations

4.1.1 Optical Measurement of Air-Sensitive Compounds

Most GICs are moderately to very air sensitive, although there are a few noteworthy exceptions (i.e., graphite-$SbCl_5$ [4.5], graphite-$CsBi_x$ [4.6], and graphite-F [4.7]). Therefore, for most GICs, considerable care is required in handling samples to prevent changes in composition before the actual measurements. This is particularly true when carrying out short-probe-depth experiments. Almost all the optical data presented here were taken on the *c* face of the GIC, where a typical skin depth for the incident radiation is ~ 1000 Å. Changes in composition at this depth can occur, in some cases, in a matter of seconds under ambient conditions. In this section we discuss briefly methods for handling samples before optical measurement. Various methods to prepare GICs have been reviewed by *Hérold* [4.8a] and *Eklund* [4.8b].

Often the best way to obtain the optical spectrum of an air-sensitive GIC is to do so without removing the sample from the intercalation chamber (or reactor). If this approach is chosen, care must be taken during synthesis not to precipitate, sublime, or condense reactant onto the sample surface. This can occur, for example, in a two-zone reaction [4.8] when the zone temperatures of the sample (T_s) and the reactant (T_r) are such that $T_s \leqslant T_r$. Caution must also be taken to ensure that the optical window is not clouded by the intercalation reaction.

Quartz or Pyrex windows – the latter being sufficient for Raman scattering, and the former for reflectance studies over the range 0.5–6 eV – can be fused onto glass reactors. Figure 4.2 shows a convenient optical cell used for both reflectance and Raman scattering. It employs quartz or Pyrex rectangular cross-section glass tubing near the sample (7 × 3 mm inside dimensions, 1/2 mm wall). The sample shown in the figure is a $\sim 6 \times 8 \times 0.5$ mm^3 slab of synthetic, highly oriented pyrolitic graphite (HOPG; Union Carbide). The slab is cut to fit in the diagonal of the tubing as shown, and stainless steel clips can also be used to keep the sample from shifting. A constriction is shown at "C" about 4–5 cm from the sample. It can be sealed off with a torch, thereby separating a small, sealed cell for experiments. If low-temperature measurements are necessary, He gas can be diffused through the walls of this cell (after sealing it off) to promote heat transfer between the sample and cell walls.

If it is necessary to transfer a GIC from the reactor to a different cell for optical study, this should be done under vacuum or in a very dry, inert-atmosphere glove box. Even in a fairly dry ($\sim$ 5–10 ppm O_2/H_2O) He-atmosphere glove box, a gold-colored stage-1 alkali-metal GIC turns blue in several seconds, indicating the formation of stage-2 or higher compounds in the optical skin depth. These GICs are perhaps best transferred under vacuum. We have sometimes used a small vacuum chamber, equipped with a hammer to break open glass reaction ampoules, and transferred very sensitive samples under

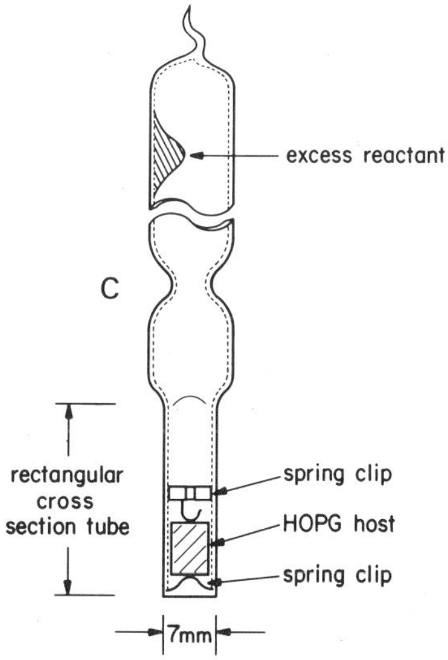

Fig. 4.2. Glass cell suitable for optical studies of air-sensitive GICs

vacuum to a second glass ampoule that could then be sealed with a torch and used as an optical cell.

However, not all GICs need to be handled this carefully. Stage-2 alkali-metal GICs, for example, have been found to transfer quite well in a dry (< 5 ppm O_2/H_2O) He glove box. For mid-infrared optical measurements (2–15 microns) requiring salt or sapphire windows, we have transferred GIC samples in a He glove box to a small, stainless-steel optical cell equipped with a 3/4 l/s ion pump and valve. The cell is removed and evacuated on a diffusion pump station. The ion pump is then started and maintains a vacuum of 10^{-5}–10^{-6} Torr.

Usually it is important to determine the stage index of the sample under optical study. Traditionally, the stage index is determined from (001) X-ray diffraction. Mo K_α radiation easily penetrates 0.5 mm thick glass ampoule walls. The X-ray probe depth is $\sim$ 20–50 microns, which is much larger than the optical skin depth. Several samples are usually studied to ensure that the optical spectrum is properly identified with the correct stage index.

Raman scattering experiments probe approximately the same depth as reflectance measurements. Thus the same care used in handling samples for reflectance studies is needed in Raman scattering experiments. However, Raman scattering studies benefit from the fact that the Raman-active carbon intralayer phonons are stage-specific [4.2]. Raman scattering has, therefore, become a commonly used tool for determining the stage index of samples studied in reflectance. The technique is particularly useful in identifying low-stage-index $n \leqslant 3$ compounds, where the stage dependence of the graphitic intralayer $\sim 1600\ \mathrm{cm}^{-1}$ phonons is the strongest [4.1.2].

4.1.2 Optical Reflectance Spectroscopy

Reflectance spectra with the radiation incident at near-normal incidence to the c axis have been used with electron energy band models to determine the in-plane dielectric function, the charge transfer between C and I layers, and the distribution of net c-axis charge among the C_i and C_b layers. Most optical studies have been made on samples prepared from HOPG [4.9] – a polycrystalline synthetic graphite – because large samples with reasonably smooth surfaces can be obtained (typical dimensions $5 \times 5 \times 1\ \mathrm{mm}^3$). The c axes of the $\sim 1\ \mu$m diameter crystallites in HOPG are aligned to better than 1 degree, so it is a suitable host for basal (a-b) plane (or c-face) optical studies of GICs. HOPG can be readily cleaved parallel to the basal plane using transparent tape. With some difficulty, a faces can be prepared [4.10] by careful polishing, followed by ion sputtering of the polished surface to remove surface damage. The HOPG host may then be reacted to study the c-axis dielectric constant. Only a few such studies have been carried out to date. Thin single crystals of graphite are also available, but they are small (typically $\sim 1 \times 1\ \mathrm{mm}^2$), and difficult to hold in place in the optical cell.

Care must be taken to collect most, if not all, of the light reflected from the GIC, if quantitative reflectance results are required. Generally speaking, the effect of surface roughness (incurred during GIC synthesis) on the optical spectrum can be accounted for, if necessary, by multiplying R by a wavelength independent scale factor $\beta \sim 1$. This procedure is possible because the surface roughness usually has a scale that is large compared with the wavelength of the incident light. However, the reflectance spectrum (i.e., $R \to \beta R$) should be scaled carefully and, in connection with the Drude analysis of the reflectance spectrum, an additional fitting parameter β is introduced to the three (ε_ω, ω_p, τ; Sect. 4.2.1) normally used in this analysis. Other forms of graphite, such as Grafoil (Union Carbide) exhibit a surface roughness on a shorter length scale, which introduces a wavelength-dependent correction factor to R (Sect. 4.3.1b).

4.1.3 Raman Spectroscopy

For the most part, *c*-face Raman scattering studies of GICs have been carried out to obtain information about the intercalate-perturbed modes involving intralayer displacement of the carbon atoms [4.1.2]. Only in a few cases have intercalant modes been observed, and probably their detection resulted from a resonantly enhanced scattering cross section. The difficulty in seeing the potentially very interesting intercalate modes stems from both the 1000 Å skin depth that reduces the size of the scattering volume, and the fact that the intercalated atoms or molecules are diluted in concentration by the carbon layers. A few Raman scattering studies have been reported with the laser radiation incident on an *a* face of the GIC, where the *a* face was prepared by tearing open HOPG under tension applied parallel to the basal plane [4.11]. The *a* face affords the possibility of studying interlayer modes, when the radiation is polarized parallel to the *c* axis.

Raman scattering on opaque materials in general, and on GICs in particular, can best be carried out in the Brewster-angle backscattering geometry [4.2]. For *c*-face studies, the incident beam strikes the *c* face of the GIC at an angle $\sim 45°$ with respect to the *c* axis, with the electric field polarized in the plane of incidence. This angle is close enough to the pseudo Brewster angle to approximately minimize the reflection coefficient for the incident radiation. The Raman scattered light is collected at normal incidence to the *c* face.

GICs have been found to decompose or change stage during Raman scattering experiments because of laser heating. Low powers (~ 100 mW or less) and a cylindrical lens are therefore used to reduce the power density on the sample.

4.2 Deducing Charge Transfer from Optical Studies

4.2.1 The Relationship Between the Optical Reflectance, Dielectric Function, and Electronic Band Structure of GICs

Most reflectance studies on GICs have been carried out at near-normal incidence to the *c* face. The quantity measured is the intensity ratio $R = |E_r/E_i|^2$, where E_i and E_r are the incident and reflected electric fields perpendicular to the

c axis. Then $R(\omega)$ is governed by the basal (i.e., a-b) plane dielectric function $\varepsilon_\perp$, which is related to the reflectivity r by the usual expression [Ref. 4.12, p. 232]

$$r = \frac{E_r}{E_i} = \varrho e^{i\theta} = \frac{1 - \sqrt{\varepsilon_\perp}}{1 + \sqrt{\varepsilon_\perp}}, \tag{4.1}$$

where ϱ and θ are the reflectivity amplitude and phase shift, respectively.

In a solid, the dielectric function $\varepsilon(\omega) = \varepsilon_1(\omega) + i\varepsilon_2(\omega)$ can be represented as a sum of three terms:

$$\varepsilon(\omega) = \varepsilon_{free}(\omega) + \varepsilon_{inter}(\omega) + \varepsilon_{phonon}(\omega), \tag{4.2}$$

where $\varepsilon_{free}(\omega)$, $\varepsilon_{inter}(\omega)$, and $\varepsilon_{phonon}(\omega)$ refer to contributions from free carriers, interband transitions, and infrared-active phonons. The real and imaginary parts of $\varepsilon(\omega)$ are associated with dispersion and absorption, respectively. In the Drude approximation, the free-carrier term is written as [Ref. 4.12, p. 53].

$$\varepsilon_{free}(\omega) = \varepsilon_\infty - \frac{\omega_p^2}{\omega(\omega + i/\tau)}, \tag{4.3}$$

where ε_∞, ω_p, and τ are, respectively, the core dielectric constant, plasma frequency, and carrier lifetime. The interband contribution from vertical (i.e., k-conserving transitions is given by [4.12, p. 114]

$$\varepsilon_{inter}(\omega) = \frac{e^2h^2}{m^2\omega^2} \int_s \frac{|\boldsymbol{n} \cdot \boldsymbol{M}_{ss'}(\boldsymbol{k})|}{|\nabla_{\boldsymbol{k}}[E_{s'}(\boldsymbol{k}) - E_s(\boldsymbol{k})]|} dS\,. \tag{4.4}$$

The integration is over the k-space surface $S(\boldsymbol{n} \perp S)$ defined by $\{E_{s'}(\boldsymbol{k}) - E_s(\boldsymbol{k})\} = \hbar\omega$, where ω is the photon frequency.

In a stage-n GIC, the low-energy ($\omega < 3$ eV) interband transitions between valence (v or π) and conduction (c or π^*) bands are of two principal types: $v \to v$ and $v \to c$. These bands are centered at the K point and are shown schematically in Fig. 4.3a, b for stage-1 and stage-2 donor compounds; k represents the in-plane wave vector measured relative to the K point (Fig. 4.1) in a graphitic

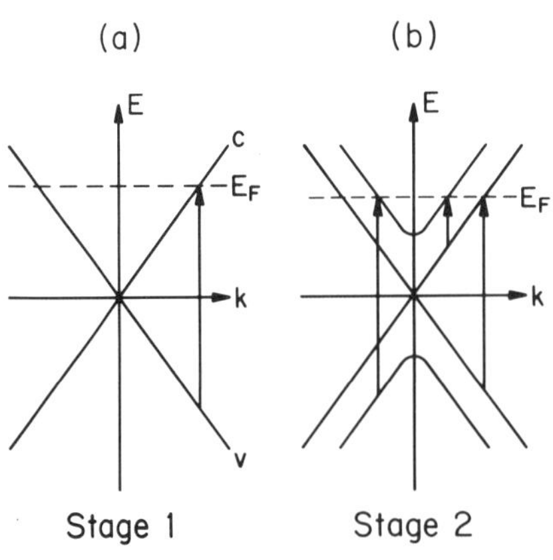

Fig. 4.3. Stage-1(**a**) and stage-2(**b**) energy bands near the K point in a graphitic Brillouin zone. The radial wave vector k is measured from the K point and refers to the momentum in the basal plane

Brillouin zone. The vertical arrows in the figure indicate the strongest interband transitions, as first found by *Blinowski* et al. [4.13]. Structure in the optical spectrum of stage $n > 1$ GICs associated with $v \to v$ transitions (acceptor-type) or $c \to c$ transitions (donor-type) have been found to be reasonably sharp, indicating that these energy bands are nearly parallel over a considerable volume of k space. In the case illustrated in the figure, $v \to c$ absorption exhibits a threshold energy $E_T \sim 2E_F$; i.e., the strongest transitions occur between bands of near mirror symmetry [4.13]. Figure 4.1 shows a schematic two-dimensional Fermi surface for the stage-2 π-band model of Fig. 4.3b. A pair of concentric cylinders is located at each K point, with the cylinder axes parallel to the c axis. For large enough charge transfer, the cylinders will eventually exhibit trigonal warping [4.4]. The π- or π^*-band contribution to $\varepsilon_{\text{free}}(\omega)$ comes from the response of graphitic carriers in these cylinders – which are two-dimensional holes (electrons) in acceptor-type (donor-type) compounds. However, other carrier pockets may also contribute to the free-carrier response. For example, in stage-1 binary donor GICs (Sect. 4.3.1a), a three-dimensional piece of the Fermi surface located at the Γ point has also been proposed to contribute to $\varepsilon_{\text{free}}(\omega)$.

The principal contribution to $\varepsilon_{\text{phonon}}(\omega)$ in GICs is from zone-center transverse-optical (TO) phonons whose contribution can be represented by a sum of Lorentz oscillators [4.14]:

$$\varepsilon_{\text{phonon}}(\omega) = \sum_{j=1}^{N} \frac{\Omega_j^2}{(\omega_{\text{TO},j}^2 - \omega^2) - \mathrm{i}\gamma_j\omega} , \tag{4.5}$$

where $\omega_{\text{TO},j}$, γ_j, and Ω_j are the frequency, damping, and strength of the jth TO ($q = 0$) phonon.

To determine the values of the parameters included in the model dielectric function (4.2–5), we can either fit a reflectance calculated from a model dielectric function to the R data or fit the model dielectric function itself directly to $\varepsilon_1, \varepsilon_2$ data obtained from a Kramers-Kronig (KK) analysis of the R data. In a KK analysis, the reflectivity phase shift θ is calculated, and (4.1) is inverted to obtain the experimental values for $\varepsilon_1, \varepsilon_2$. The KK phase integral is given by the usual expression [4.15]

$$\theta = -\frac{\omega}{\pi} \mathscr{P} \int_0^\infty \frac{\ln[R(\omega')/R(\omega)]}{\omega'^2 - \omega^2} d\omega' , \tag{4.6}$$

where $\mathscr{P}$ denotes the principal value. Although the $R(\omega)$ data is obtained in a limited frequency range, the phase θ may still be calculated approximately by extrapolating the reflectance to higher and lower frequencies. Consistent with the metallic character of GICs, $R(\omega)$ is extended to lower frequencies according to the Drude model (4.3). For frequencies $\omega > 8$ eV, an approximate reflectance for the GIC can be generated according to the model dielectric function given by *Eklund* et al. [4.4]:

$$\varepsilon_{\text{DSG}}(\omega) = 1 + \frac{3.35n}{l_c(\text{Å})} [\varepsilon_{\text{G}}(\omega) - 1] , \tag{4.7}$$

where ε_G is the experimental, in-plane dielectric function of pristine graphite obtained by *Taft* and *Phillip* [4.16], n is the stage index, I_c is the spacing between successive intercalate layers, and 3.35 Å is the distance between C layers in graphite. The physical basis for this approximation to the high-energy dielectric function of a GIC, referred to as "density-scaled graphite" (DSG) [4.4], is summarized by three key points:

(1) The GIC is mostly graphitic carbon.

(2) For $\omega > 8$ eV, $\varepsilon_G(\omega)$ is dominated by the contribution from σ electrons.

(3) The character of the σ band is determined primarily by very strong C–C intralayer interactions (i.e., σ bonds) that are largely unaffected by intercalation (i.e., the hexagonal C network is preserved).

Thus the high frequency $\varepsilon(\omega)$ for the GIC should be similar to $\varepsilon_G(\omega)$, but scaled according to the number of carbon layers per unit distance along the c axis of the GIC.

For comparison, Fig. 4.4 shows $\varepsilon_2(\omega)$ data for LiC_6, KC_8, and $KHgC_4$ (solid lines) calculated according to the DSG model (dotted line) by *Fischer* et al. [4.17]. Above 8 eV, the agreement between the DSG appproximation and the data is good. Note, however, that the peak at ~ 5 eV in these donor compounds, associated with transitions between π bands near the M point in a graphitic Brillouin zone, is not well described by the DSG model. This disagreement near 5 eV indicates intercalate-induced changes in π-band dispersion near the M point – i.e., π-band hybridization with intercalate states – and/or an intercalate layer contribution to $\varepsilon_{\text{inter}}$ in this frequency range.

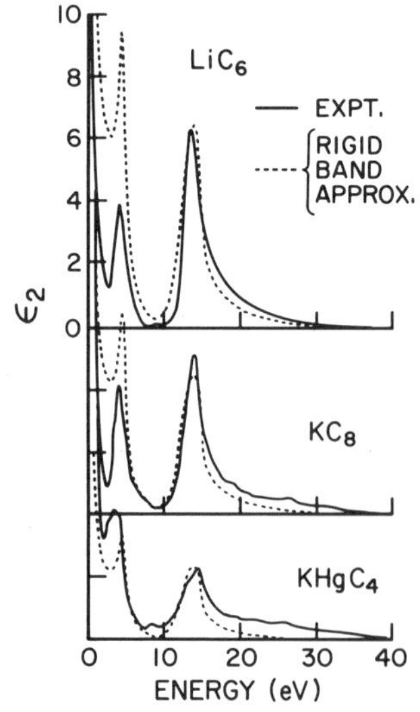

Fig. 4.4. Kramers-Kronig derived values of $\varepsilon_2(\omega)$ for LiC_6, KC_8, and $KHgC_4$ (solid line). The dashed lines represent values calculated according to (4.7) [4.17]

4.2.2 Electronic Band Structure Models

In chap. 2 *Holzwarth* reviews in detail electron energy band calculations in GICs. In the following three sections, we briefly discuss three models that have been used to analyze quantitatively charge transfer and the dielectric response of GICs.

(a) The *K*-Point Tight-Binding Model of Blinowski and Rigaux

Blinowski et al. [4.13] developed an early tight-binding band model (referred to here as the BR model) for stage-1 and stage-2 GICs that has been used extensively for the analysis of electrical transport, magnetic susceptibility, NMR, and optical data. Using proper values for the model band parameters (γ_0 and γ_1), good agreement with optical data can be obtained, provided the charge transfer is not too high. In the BR model, an n-stage compound is viewed as a collection of uncoupled subsystems consisting of n graphite layers bounded by intercalate (I) layers, where the I layers are viewed as a structureless, uniformly charged sheet. The BR model does not consider C-layer interaction across an I layer, and thus it is two dimensional. Considering the lowest order intralayer (γ_0) and interlayer (γ_1) C–C transfer integrals, *Blinowski* et al. [4.13] obtained analytic expressions for the stage-1 and stage-2 π-band dispersion near the K point (Figs. 3a, b) in a graphitic Brillouin zone. Because of the relatively simple analytic form obtained for the band dispersion, which stems from ignoring all higher-order C–C transfer integrals, useful analytic expressions can be derived for many physical quantities. In the BR model, the carbon π-band dispersion is given in terms of the nearest-neighbor C–C intralayer (γ_0) and interlayer (γ_1) transfer integrals by [4.13]

$$\text{Stage 1:}\quad E(k) = \pm\tfrac{3}{2}\gamma_0 bk\,, \tag{4.8a}$$

$$\text{Stage 2:}\quad E(k) = \pm\tfrac{1}{2}[\gamma_1 \pm (\gamma_1^2 + 9\gamma_0^2 b^2 k^2)^{1/2}]\,, \tag{4.8b}$$

where the wave vector k is measured relative to the K point, and $b = 1.42$ Å is the intralayer C–C distance. The BR model neglects longer range C–C transfer integrals. Most optical studies of GICs seem to indicate that the BR model parameters should be $\gamma_1 \approx 0.38$ eV and $\gamma_0 \approx 2.8 - 3.0$ eV. We discuss the experimental determination of these values later.

Several other important expressions useful to our purposes here follow from the BR model. For example, charge transfer f_C and the position of the Fermi level E_F are related by

$$\text{Stages 1 and 2:}\quad E_F = \gamma_0\sqrt{\pi f_C\sqrt{3}}\,, \tag{4.9}$$

where f_C is the net charge per C atom in units of electrons (e) per C atom. Note that the stage-2 relation between f_C and E_F given in the original work [4.13] is incorrect. The "bare" or "unscreened" plasma frequency ω_p in the Drude expression (4.3) is related to the free-carrier concentration N by

$$\omega_p^2 = 4\pi Ne^2/m^*\,, \tag{4.10}$$

where m^* refers to the optical mass, and represents the average band mass – i.e., $\hbar^2(\partial^2 E/\partial k^2)^{-1}$ averaged over occupied states. The bare plasma frequency ω_p should not be confused with the screened plasma frequency ω_p^*. In the Drude model, $\omega_p^* = \omega_p/\sqrt{\varepsilon_\infty}$. The screened plasma frequency is usually identified with the lowest energy zero crossing for $\varepsilon_1(\omega)$. Note that ω_p^* is close to the position of the "Drude edge" in $R(\omega)$, where by "Drude edge" we mean the portion of the reflectance curve associated with the rapid decline in reflectance from high values in the infrared region to quite low values (in some cases as low as 1–2%). Calculation of the plasma frequency using the BR-model band dispersion equations (4.8a, b) results in the following dependence on the Fermi energy [4.13]:

$$\text{Stage 1:} \quad \omega_p^2 = \frac{4e^2}{\hbar^2 l_c} E_F , \tag{4.11a}$$

$$\text{Stage 2:} \quad \omega_p^2 = \frac{8e^2 E_F}{\hbar^2 l_c} \left(\frac{E_F^2 - \gamma_1^2/2}{E_F^2 - \gamma_1^2/4} \right) , \tag{4.11b}$$

where l_c refers to the distance between intercalate layers, and differs in meaning from the same quantity used by *Blinowski* et al. [4.13]. A typical value for the stage-1 graphitic carrier optical mass m^* can be obtained with (4.9, 10, 11a). Using the values $E_F = 0.8$ eV, $\gamma_0 = 3$ eV, and $l_c = 10$Å, we obtain $m^*/m_0 = 0.2$, where m_0 is the free-electron mass. This light carrier mass is responsible for the high in-plane electrical conductivity observed in GICs, despite the fact that the carrier density N is at least a factor of 10 lower than in Cu.

(b) LCAO Model of Holzwarth

Holzwarth [4.18] has considered a π-band model for a stage-n GIC using the method of linear combination of atomic orbitals (LCAO) as developed by *Johnson* and *Dresselhaus* for graphite [4.19]. The assumptions of the *Holzwarth* model (or "H model") regarding the carbon bands in the GIC are similar to those of the BR model discussed above; i.e., the π bands are only weakly perturbed by the insertion of the I layers, and the C layers on either side of an I layer are not coupled (i.e., the model is two dimensional). For stage $n > 2$, the H model does not consider the effects of the I-layer charge on the net C-layer charge in the C_i and C_b layers. Later theoretical work (Chap. 2) has shown these c-axis electrostatic effects to be important: they give rise to charge in the graphitic layers residing primarily in the C_b layers. For stage $n \leqslant 2$, the H and BR models should provide a reasonable description of the carbon π-band contribution to the electronic properties of GICs; the former being more appropriate for higher charge transfer compounds. However, *Holzwarth* [4.18] gives the numerical results for stage-1 compounds only.

The quantitative differences between these two models stem from the neglect in the BR model of more distant intralayer C–C interactions. For increasing Fermi wave vector k_F (measured relative to the K point), the higher-order interactions considered in the H model (and neglected in the BR model) eventually introduce trigonal warping of the cylindrical Fermi surfaces, and destroy the exact mirror symmetry between the valence (π) and conduction (π^*) bands. *Zhang* et al. [4.20, 21] have considered the differences in the H and BR models that affect the determination of f_c from the π-electron plasma frequency $\omega_{p,\pi}$. Figures 4.5, 6 summarize the H and BR model results for stage 1 GICs. The

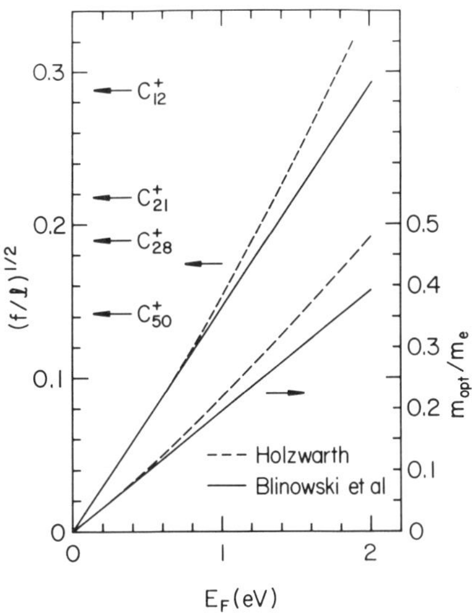

Fig. 4.5. Comparison of results for the stage-1 Holzwarth and Blinowski-Rigaux models for acceptor-type GICs. The quantity (f/l) is equivalent to f_C in the text ($\gamma_0 = 2.9$ eV) [4.20]

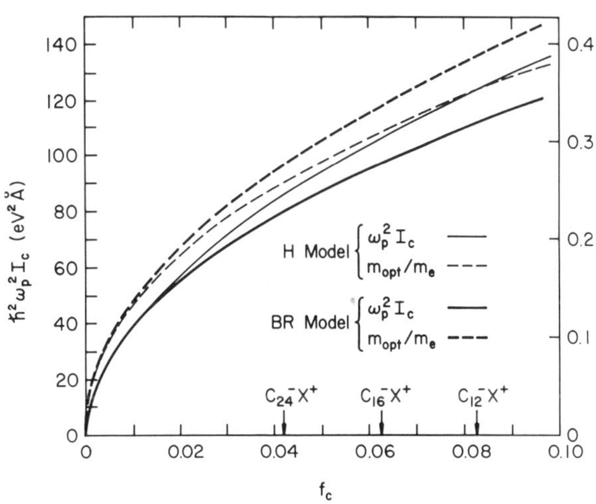

Fig. 4.6. Comparison of the stage-1 Holzwarth (H) and Blinowski-Rigaux (BR) model results for donor-type GICs ($\gamma_0 = 2.9$ eV) [4.20]. The quantity I_c is equivalent to I_c in (4.11a, b).

notation C_p^+ (C_p^-) in the figures indicates a charge transfer f_C equivalent to one hole (one electron) per p C atoms, i.e., $f_C = 1/p$. The convergence of the results for the two models at low charge transfer was accomplished [4.20] by setting $\gamma_0 = 2.93$ eV in the BR model.

(c) Tight-Binding Model of Saito and Kamimura

The two-dimensional tight-binding band model of *Saito* and *Kamimura* [4.22] (SK) was also calculated according to the two basic assumptions listed in the previous section for the H and BR models. However, for the stage $n > 2$ compounds, *Saito* and *Kamimura* treated the c-axis charge distribution in a self-consistent way, as explained below. Figure 4.7 shows the C-layer structure of a stage-3 compound, which illustrates the SK band model parameters. The γ_j's are tight-binding transfer integrals between neighboring C atoms in the same and adjacent layers, and δ and δ_j are linear combinations of self-energy integrals from different carbon sites. To determine values for these band parameters for stage $n = 2$–6 KC_{12n}, *Saito* and *Kamimura* carried out a least-squares fit of their tight-binding bands to the self-consistent band structures of *Ohno* and *Kammiura* [4.23], where the value $f_C = 1/12n$ was assumed (i.e., the K 4s band is empty).

To achieve good agreement between their optical results and the SK band model, *Yang* and *Eklund* [4.24] had to adjust two SK parameters. First, to fit the position of observed $c \rightarrow c$ interband structure in the dielectric function at 0.6 eV, they lowered the value of the SK electrostatic parameter δ from 0.46 to 0.3 eV. Second, they found that a k-dependent form for γ_0 was necessary to fit a broad peak observed in $\varepsilon(\omega)$ near 5 eV. This peak is the analog of the 4.6 eV peak in pristine graphite, identified by *Johnson* and *Dresselhaus* [4.19] with transitions between π and π^* bands at the M point. As *Johnson* and *Dresselhaus* pointed out [4.19], for pristine graphite the π–π^* band splitting is approximately $2\gamma_0$, and the second-neighbor C–C interactions (also neglected in the SK model)

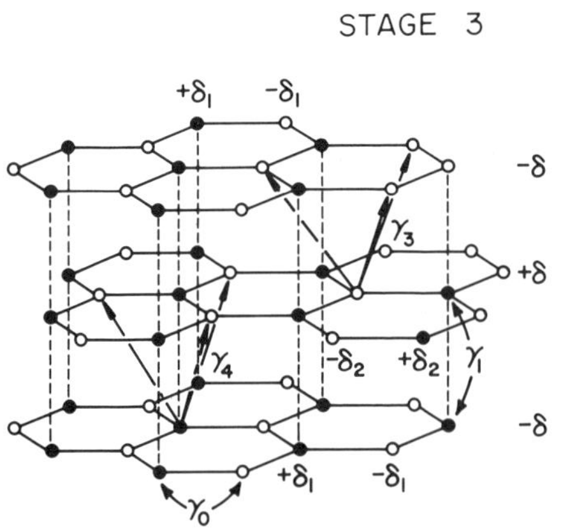

Fig. 4.7. Graphitic carbon structure of a stage-3 compound indicating the tight-binding band parameters for stage $n \geqslant 3$ GICs [4.22]

make significant contributions to the π-band structure at points far removed from the K-point, leading to "effective" values of $\gamma_0 = 3.2$ eV (K point) and 2.3 eV (M point). Consistent with this observation, *Yang* and *Eklund* [4.24] introduced the following k-dependent form for γ_0:

$$\gamma_0 - \frac{3.0\,\text{eV}}{1 + \chi(\kappa/\kappa_0)}\,, \tag{4.12}$$

where κ is the length of the wave vector measured relative to the K point, and κ_0 is the wave vector joining K and Γ. A *stage-independent* constant χ in (4.12) was chosen so that $\gamma_0 = 3.0$ eV at the K point and $2\gamma_0 = 4.6$ eV at the M point. Figure 4.8a–c shows the bands resulting from the calculation of *Yang* and *Eklund* [4.24] for KC_{12n} $n = 2, 3, 4$. The vertical arrows in the figure indicate the allowed interband transitions, and the dashed horizontal line indicates the position of E_F consistent with complete charge transfer of one K(4s) electron to

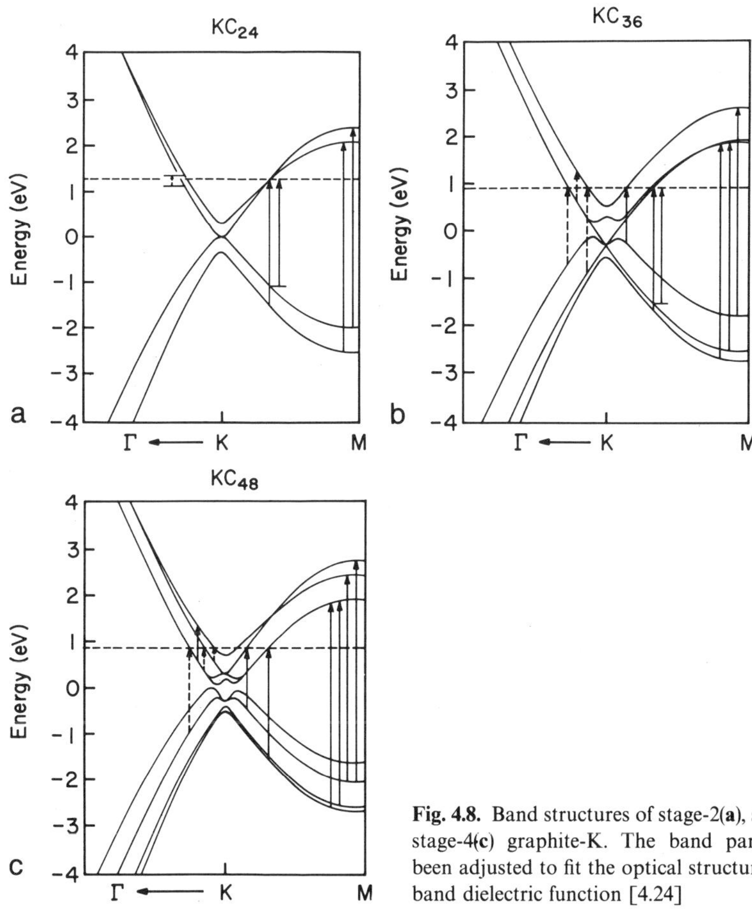

Fig. 4.8. Band structures of stage-2(**a**), stage-3(**b**), and stage-4(**c**) graphite-K. The band parameters have been adjusted to fit the optical structure in the interband dielectric function [4.24]

the C layers. *Yang* and *Eklund* [4.24] have calculated the interband dielectric function for the π bands shown in the figure, including the $\boldsymbol{A} \cdot \boldsymbol{p}$ matrix elements. Their analysis, which leads to values for the net charge distributed to the C_i and C_b layers, is discussed in Sect. 4.3.1a

4.2.3 Charge Transfer and Graphitic Intralayer Phonon Frequencies

In previous contributions to these volumes, *Zabel* [4.25] and *Solin* [4.2] reviewed the lattice dynamics of GICs emphasizing, respectively, the experimental contributions from neutron scattering and optical spectroscopy. Readers are referred to these and previous reviews [4.1, 2]. Here we give a brief introduction to infrared- and Raman-active graphitic intralayer phonons that can be observed optically from experiments on the c face, and discuss the theoretical progress in understanding the effects of charge transfer on the respective mode frequencies.

A group-theoretical analysis of the hexagonal structure of pristine graphite (D^4_{6h} space group) predicts 12 zone-center (i.e., $q = 0$) phonons that transform according to the irreducible representations of the D^4_{6h} space group [4.26]. Three of these modes are acoustic ($A_{2u} + E_{1u}$), three are infrared-active ($A_{2u} + E_{1u}$), four are Raman-active ($2E_{2g}$), and two are optically silent ($2B_{1g}$). The (E_{1u}, E_{2g}) and (A_{2u}, B_{1g}) symmetry modes involve, respectively, intralayer and interlayer displacements of the carbon atoms.

The frequencies, and sometimes the line shapes, of optically active, graphitic vibrations in GICs are strongly influenced by intercalation through the phenomenon of staging and charge transfer. For stage $n \geqslant 3$, the high-frequency Raman-active E_{2g} singlet observed in graphite at $1582\,\mathrm{cm}^{-1}$ is split into a doublet. The splitting is a consequence of the different net charge per carbon atom on the interior (C_i) and bounding (C_b) layers. The lower (E_{2gi}) and higher (E_{2gb}) frequency components of the doublet have been identified [4.1.2] with intralayer C-atom motion in the C_i and C_b layers, respectively. The frequencies of the E_{2gi} modes have been found to be typically within ~ 5–$10\,\mathrm{cm}^{-1}$ of the E_{2g} mode in pristine graphite, whereas the high-frequency E_{2gb} component has been found to exhibit a frequency ~ 15–$50\,\mathrm{cm}^{-1}$ higher than the E_{2g} mode. Breit-Wigner lineshapes are sometimes observed, indicating a coupling of the intralayer modes to a Raman-active continuum.

Figure 4.9a shows the strong stage dependence for the $\sim 1600\,\mathrm{cm}^{-1}$ E_{2gi} and E_{2gb} modes in the acceptor-type GIC graphite-$FeCl_3$ reported by *Underhill* et al. [4.27]. It shows that the E_{2g} mode in pristine graphite evolves from a singlet (stage $n = \infty$), to a doublet for stage $11 \leqslant n \leqslant 3$, and finally a higher frequency singlet is obtained when interior carbon layers are no longer present (i.e., $n \leqslant 2$). Figure 4.9b shows the IR spectra of *Underhill* et al. [4.27] that show the stage dependence of the E_{1u} modes for graphite-$FeCl_3$ [stage $n = 2, 4, 6, 11$, and ∞ (HOPG)]. The E_{1u} mode lineshape must be analyzed using (4.5) to determine the mode parameters. Note that for stage $n > 2$, independent oscillator contributions to the overall lineshape are not well resolved, indicating that

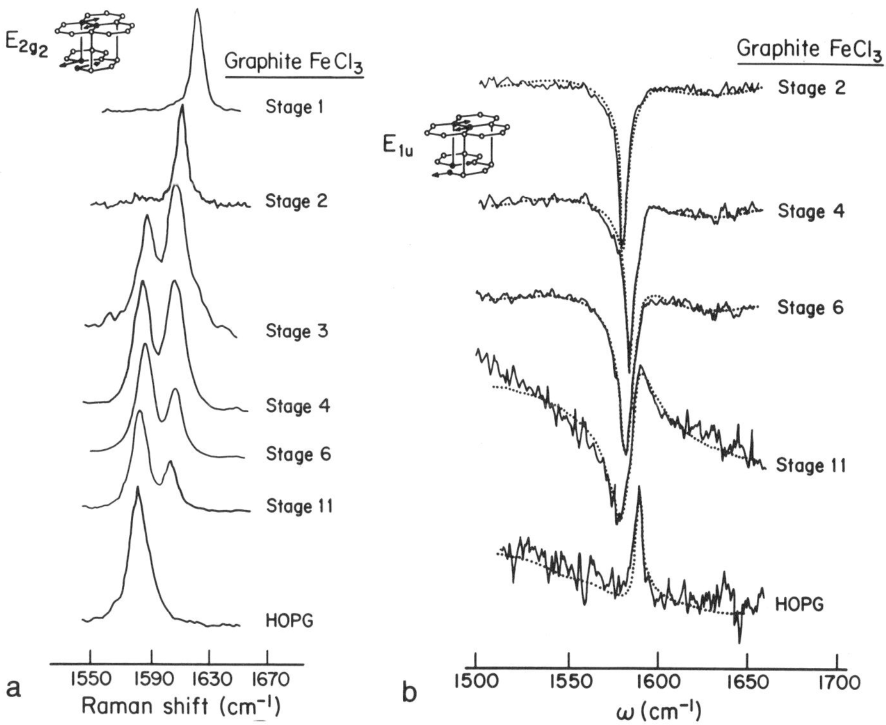

Fig. 4.9. High-frequency, Raman-active (**a**) and infrared-active (**b**) graphitic intralayer modes in graphite $FeCl_3$ [4.27]

it is more difficult to use IR-mode spectra than Raman-mode spectra to determine the stage index of a GIC.

Pietronnero and *Strassler* [4.28] first appreciated the connection between charge transfer and the frequency of the zone-center graphitic modes in GICs. By applying a model used successively to describe several properties of carbon molecules, they derived the stage dependence of the E_{2gi} component for donor- and acceptor-type GICs, obtaining numerical results for the derivative of the mode frequency with respect to reciprocal stage index $(1/n)$. They obtained $[\delta(\Delta\omega/\omega)/\delta(1/n)] = \pm 0{\cdot}007$ for acceptor-type $(+)$ and donor-type $(-)$ GICs, consistent with the trends in the data. Somewhat later, *Chan* et al. [4.29] carried out self-consistent electronic band structure calculations on a simulated stage-1 structure containing point ions centered above the hexagons in an AA stacked graphitic structure. By varying the charge transfer between the ions and C layers, they obtained the dependence of the intralayer C–C bond length d_{CC} on f_C. Furthermore, by introducing two interlayer coupling parameters, they carried out a frozen phonon calculation to obtain the stage dependence of the E_{1u} and E_{2g} zone-center mode frequencies, which was found to be in general agreement with the trends in the experimental data for acceptor-type and donor-type GICs. Unlike the changes in d_{CC}, which in their calculations are completely determined by f_C, the in-plane phonon frequencies were found to be sensitive to both f_C and the interlayer coupling. For comparison Figs. 4.10a, b and 4.11a, b

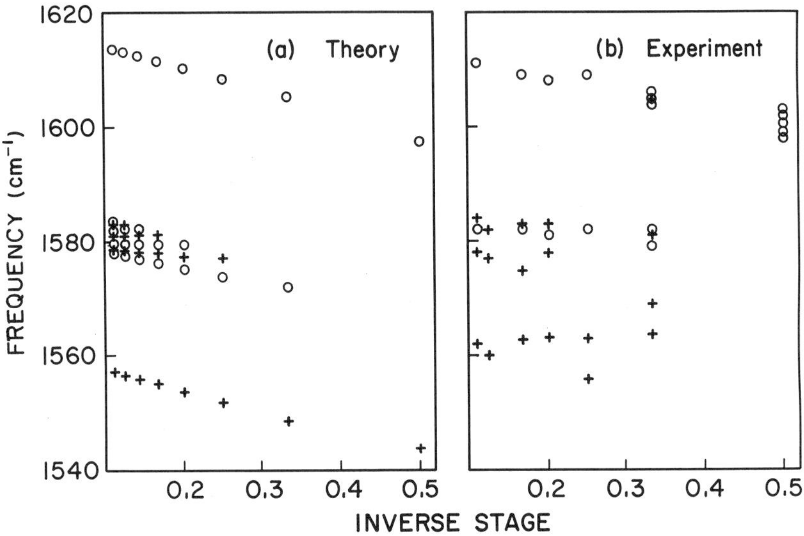

Fig. 4.10. Calculated (**a**) and experimental (**b**) stage dependence of the high-frequency graphitic intralayer mode frequencies for donor-type GICs [4.29]

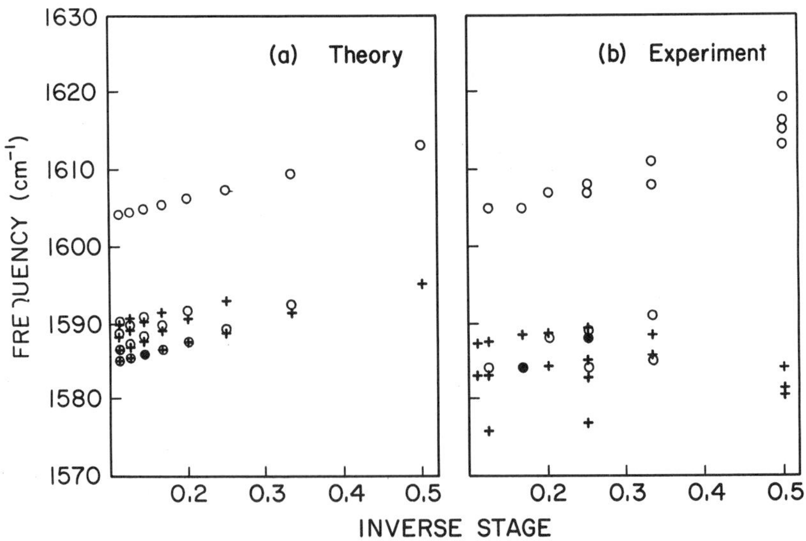

Fig. 4.11. Calculated (**a**) and experimental (**b**) stage dependence of the high-frequency graphitic intralayer mode frequencies for acceptor-type GICs [4.29]

give the model and experimental results, respectively, for donor-type and acceptor-type GICs. Their calculations agree better with the Raman-active mode data than with IR-active mode data. However, the calculations reproduce the salient features (mode stiffening for acceptors and mode softening for donors).

4.3 Experimental Results and Discussion

In this section we review experimental optical studies that have yielded values for the charge transfer and net layer charge distribution in GICs. Other reviews of the optical properties of GICs have appeared recently [4.3, 4]. We endeavor here to consider in some depth the results of a large (but not exhaustive) set of binary and ternary, donor and acceptor GICs. We hope that the progress made in the last ten years in understanding the optical properties of GICs and the effects of charge transfer is apparent.

4.3.1 Donor-Type GICs

(a) Potassium GICs

Binary alkali-metal GICs prepared by the reaction of alkali metals M = K, Rb, or Cs exhibit at stages $n = 1$ and $n > 1$ nominal stoichiometries MC_8 and MC_{12n}, respectively. For $n \geqslant 2$ M GICs, it appears that the charge transfer to the C layers is very nearly 1 electron per M atom [4.1, 2]. However, charge transfer in the stage-1 MC_8 compounds has been somewhat of a controversial issue [4.30–32]. Most experimental effort on the optical properties of binary alkali-metal GICs has focused on K-graphite (stages 1–4), including a pressure-dependent study [4.33] and an in situ electrochemical study [4.34]. Therefore, we have chosen the K-graphite system as the representative binary donor GIC in this review.

KC_8. Over the last 12 years, considerable controversy has developed over the electrical properties of stage-1 MC_8 (M = K, Rb Cs) compounds [4.30–32]. The debate centers around two principal results obtained from two competing band-structure calculations:

(1) the degree of charge transfer $f = 8f_C$ between the K(4s) band and the carbon π band

(2) the s character of a three-dimensional (3D) carrier pocket in the center of the Brillouin zone

Both competing band models place two-dimensional (2D) graphitic π^* carrier pockets at the Brillouin-zone corners. The volume of the occupied π^* states, of course, determines the value of f; i.e., $f = 1$ means the K 4*s* states are empty. The central three-dimensional pocket has been identified with hybridized K 4*s* and graphitic "interlayer" states [4.31, 35, 36], weakly hybridized graphitic states [4.37, 38], or graphitic 3D "interlayer" states [4.32]. *Posternak* et al. [4.39] first proposed the existence of 3D "interlayer" bands in graphitic thin-film systems (e.g., C_6–Li–C_6, C_6–C_6, C_8–K–C_8). The term "interlayer" is derived from the fact that the associated charge density is found between the C and I (or C)layers. Figures 4.12, 13 show for comparison the Fermi surfaces of the two competing band models of *DiVicenzo* and *Rabii* [4.37] and *Ohno* et al. [4.36]. Note that while both calculations result in 2D carriers in the Brillouin-zone corners, only the calculation of *Ohno* et al. [4.36] exhibits a significant population of (zone-center) 3D carriers.

Figure 4.14 shows the *c*-face reflectances $R(\omega)$ for KC_8 of *Yang* [4.40] (dotted line) and *Fischer* [4.17] (crossed line), which are in good agreement. The inset to the figure, displays the results to *Yang* [4.40] for $\varepsilon_2(\omega)$ plotted as $(\omega\varepsilon_2)^{-1}$ vs. ω^2. In this plot, Drude-like (4.3), free-carrier behavior is linear, and the slope and intercept of the linear portion of the curve determine values for the "effective" plasma frequencies $\omega_p = 4.6$ eV and $\omega_p\,\tau = 14.5$. We use the term "effective" when more than one set of carriers may contribute to the free-carrier dielectric response. For energies $\omega > 2.9$ eV, the calculated $(\omega\varepsilon_2)^{-1}$ values

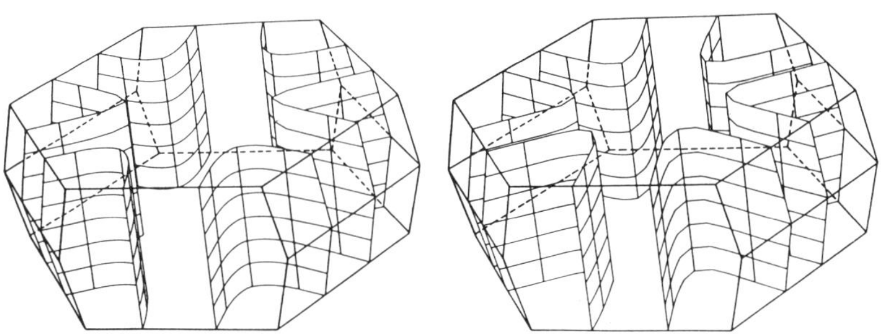

Fig. 4.12. Fermi surface of KC_8 according to the calculations of *DiVicenzo* and *Rabii* [4.37]

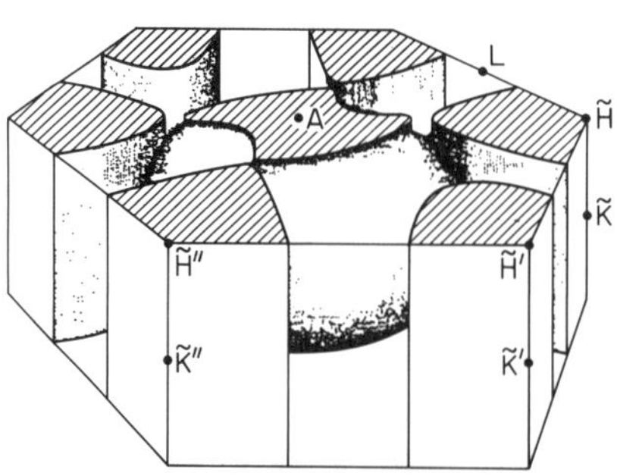

Fig. 4.13. Fermi surface of KC_8 according to the calculations of *Ohno* et al. [4.36]

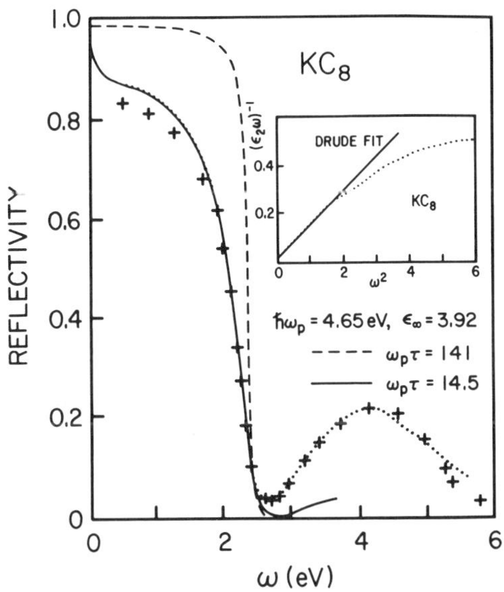

Fig. 4.14. Reflectance ($E \perp c$) of KC_8 in the energy range 0.5–6 eV: dotted line [4.40], crosses [4.17]. The solid [4.40] and dashed [4.17] curves are calculated according to the parameters given in the respective references, and their values are indicated in the figure. The inset is a plot of $1/\omega\varepsilon_2$ versus ω^2; Drude behavior is linear in this plot [4.40]

deviate from the data because of the onset of interband absorption. The solid and dashed lines in Fig. 4.14 represent reflectances calculated according to the Drude model using the same plasma frequency ($\omega_p = 4.6$ eV) and core dielectric constant ($\varepsilon_\infty = 3.9$). However, the dashed curve, which is too steep in the vicinity of the Drude edge, was calculated according to $\tau = 141/\omega_p$ [4.17], indicating this value of τ is too large.

Fischer and coworkers were the first to consider a two-carrier contribution to the plasma frequency of MC_8 compounds [4.41, 42]. They pointed out that the in-plane *dc* conductivity obtained from transport measurements ($\sigma_{dc} \sim 1 \times 10^5$) was significantly greater than the value obtained from a single-carrier Drude analysis, i.e., $\sigma_{opt} = (\omega_{p,\perp}^2 \tau/4\pi) \ll \sigma_{dc}$. They proposed both 2D π^* electrons and a 3D set of electrons as contributors to the effective plasma frequency [4.41, 42].

$$\omega_{p,\perp}^2 = \omega_{p,\pi}^2 + \omega_{p,3D}^2 , \tag{4.13a}$$

$$\omega_{p,\parallel}^2 = \omega_{p,3D}^2 , \tag{4.13b}$$

where $\omega_{p,\perp}$ and $\omega_{p,\parallel}$ refer to the basal plane and c-axis effective plasma frequencies, and $\omega_{p,\pi}$ and $\omega_{p,3D} = 4\pi e^2(N/m^*)_{3D}$ are the plasma frequencies of the 2D π^* and 3D Γ-point carriers, respectively. Using this two-carrier model, *Doll* et al. [4.43] estimated the charge transfer in KC_8 from their experimental value for $\omega_{p,\perp} = 4.5$ eV by assuming values for the 3D carrier mass. Assuming $m^*/m_0 = 2$, they found $f = 8f_C = 0.45 \pm 0.15$; in their work [4.43] the 2D π^* electron contribution was calculated using *Holzwarth*'s LCAO model (Sect. 4.2.2b).

Alstrom [4.32] also analyzed charge transfer in the MC_8 compounds using a two-carrier model in which the π^* electrons were described in the BR model approximation, and the 3D carriers were assumed to have an isotropic effective mass. *Alstrom* [4.32] identified these 3D carriers with the "interlayer" states proposed by *Posternak* et al. [4.39]. Fitting his model parameters to specific heat, Fermi-level shift, de Haas–van Alphen (dHvA) data, and the optical results of *Guerard* et al. [4.41], *Alstrom* [4.32] obtained $(m^*/m_0)_{3D} \sim 1.8$, $f = 8f_C = 0.59$, and $\gamma_0 = 2.29$ eV. Note that his value of γ_0 is much lower than most values typically used in analyzing optical results in GICs. For example, we list values of γ_0 used in the analysis of optical data and supported by a second experiment (e.g., dHvA, SdH, electrochemistry): graphite-$SbCl_5$, $\gamma_0 = 2.9$ eV – stages 1 and 2 (Sect. 4.3.2b); graphite-H_2SO_4, $\gamma_0 = 2.62 - 2.93$ – stages 1 and 2 (Sect. 4.3.2a); $KH_{0.8}C_{4n}$, $\gamma_0 = 2.9$ eV – stages 1 and 2 (Sect. 4.3.1b); and KC_{24}, $\gamma_0 = 2.9$ eV yields $f = 1.01$ (Sect. 4.3.1b). The lower value for $\gamma_0 = 2.29$ eV obtained by *Alstrom* [4.32] from his careful analysis of KC_8 data may indicate that significant hybridization between π, π^* and K(4*s*) states has occurred, and/or that the charge transfer into the π^* bands is too high to use the BR model approximation.

Zanini and *Fischer* [4.42] carried out difficult polarized reflectance experiments on the *a* face of KC_8 and CsC_8; they reported $\omega_{p,\parallel} \sim 2$eV for both compounds. Figure 4.15 shows their spectra from the *a* and *c* faces of these compounds. Using their value for $\omega_{p,\parallel}$, the plasma frequency is relatively isotropic, i.e., $\omega_{p,\perp}/\omega_{p,\parallel} = 4.5\ \text{eV}/2.0\ \text{eV} \sim 2.3$. This value favors significant population of 3D states at the zone center. *DiVicenzo* and *Rabii* [4.37] calculated $\omega_{p,\perp} = 4.65$ eV in good agreement with the experimental value 4.5 eV [4.17, 40]. However, they calculate $\omega_{p,\perp}/\omega_{p,\parallel} \sim 15$, much larger than the value obtained from experiment.

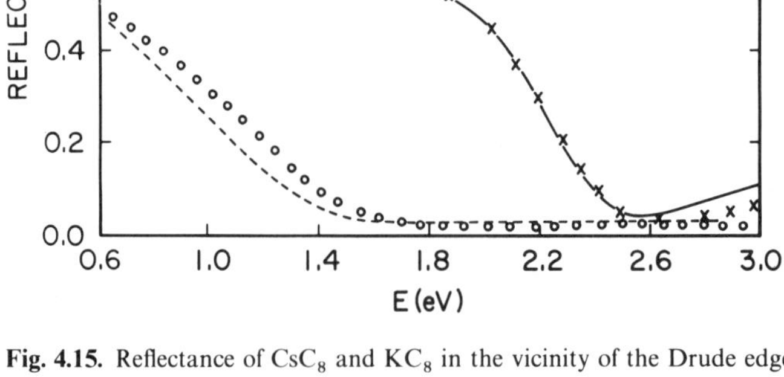

Fig. 4.15. Reflectance of CsC_8 and KC_8 in the vicinity of the Drude edge for $E \perp c$ and $E \parallel c$ [4.10]

In his analysis of KC_8 data, *Alstrom* [4.32] found a set of Drude parameters consistent with both optical and transport results (parallel and perpendicular to the c axis). In the Drude two-carrier model, the zero-frequency optical conductivities parallel and perpendicular to the c axis are given by

$$\sigma_{\text{opt},\parallel} = \omega_{\text{p},3\text{D}}^2 \tau_{3\text{D}}/4\pi \; , \tag{4.14a}$$

$$\sigma_{\text{opt},\perp} = (\omega_{\text{p},3\text{D}}^2 \tau_{3\text{D}} + \omega_{\text{p},\pi}^2 \tau_\pi)/4\pi \; . \tag{4.14b}$$

Setting $\sigma_{\text{opt},\perp} = \sigma_a$ and $\sigma_{\text{opt},\parallel} = \sigma_c$, where σ_a and σ_c are the values obtained from dc transport data, *Alstrom* [4.32] obtained the following values for the Drude parameters: $\omega_{\text{p},3\text{D}}\tau_{3\text{D}} \sim 12.5$, $\omega_{\text{p},3\text{D}} = 1.7$ eV, $\omega_{\text{p},\pi}\tau_\pi = 187$, and $\omega_{\text{p},\pi} = 3.9$ eV.

We conclude that the two-carrier model analysis of the optical data in KC_8 indicates incomplete charge transfer $f = 8f_c \sim 0.5$–0.6, which favors the band structure proposed by *Ohno* et al. [4.36] ($f \sim 0.6$), rather than that proposed by *Rabii* and coworkers [4.37, 38] ($f \sim 1$). Several articles address in detail the controversy about the electronic band structure of KC_8 [4.17, 30, 31, 44].

We turn next to the results of diamond anvil studies of KC_8 (with excess K in the cell) carried out by *Sonnenschien* et al. [4.33]. Using pressures up to 370 they observed sizable shifts in the position of the Drude edge, which they attributed to potassium uptake by the sample. At $p \sim 4$, they observed a discontinuous blue shift of the Drude edge, followed by a weaker, continuous blue shift for increasing pressure up to 70 kbar. Figure 4.16 shows their data for the pressure dependence of ω_p^*, i.e., the zero crossing of $\varepsilon_1(\omega)$ and the approximate position of the Drude edge. The inset to the figure shows the actual shift of the Drude edge

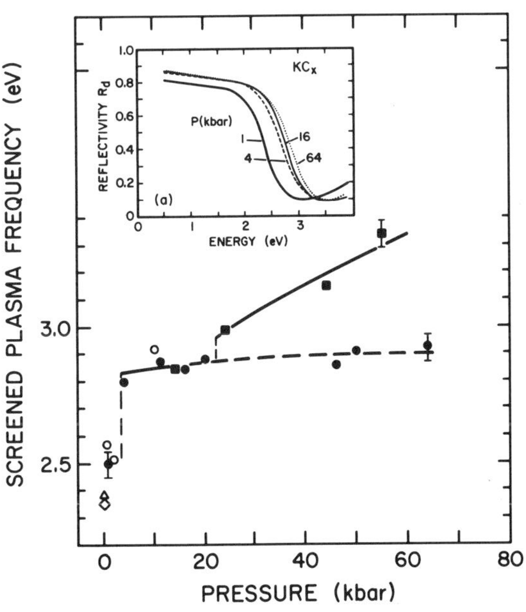

Fig. 4.16. Screened plasma frequency ω_p^* of KC_x as a function of pressure. The inset shows the actual shift of the Drude edge with pressure [4.33]

with pressure. *Sonnenschien* et al. [4.33] found the position of the blue-shifted Drude edge of 4 kbar to be almost identical to the position of the Drude edge in LiC_6 ($f \sim 1$ for LiC_6) [4.17], leading them to interpret the shift as a pressure-driven intralayer densification of potassium resulting in the formation of KC_6, with a charge transfer $f = 6f_C = 1$. If we accept the ambient pressure value $f = 8f_C \sim 0.5$–0.6 discussed above for KC_8, then the pressure and/or the higher K density must raise the K 4*s* band minimum above E_F, leading to $f = 1$ in KC_6. Above 20 kbar, heat-treated samples were found to exhibit a further blue shift in ω_p^* to a maximum value $\omega_p^* = 3.3$ eV (upper branch of ω_p^* versus pressure data in Fig. 8.16), which they similarly identified with further densification leading to a $KC_{4.4}$ compound. Experiments on samples cycled between 1 and 370 showed that the optical effects were reversible.

KC_{12n}. Using a small ion-pumped optical cell with a KBr window, *Yang* and *Eklund* [4.24] (YE) measured the *c*-face reflectance of stage $n = 2$–4 KC_{12n} over the range 0.2–6 eV. Figure 4.17a shows their spectra (dots), together with a "modified" Drude fit (solid line) to the data. Here *R* was calculated according to a Drude-Lorentz form for the dielectric function given by

$$\varepsilon(\omega) = \varepsilon_\infty - \frac{\omega_p^2}{\omega(\omega + i/\tau)} - \sum_j \frac{f_j}{\omega^2 - \omega_j^2 + i\omega\Gamma_j}, \tag{4.15}$$

where the sum is over Lorentz oscillators whose frequencies ω_j correspond to interband transition energies between approximately parallel conduction bands (i.e., $c \to c$ transitions). Kramers-Kronig analyses using density-scaled graphite (Sect. 4.2.1) for the high-energy data extension were carried out to determine the dielectric function.

Yang and *Eklund* [4.24] analyzed the interband contribution to $\varepsilon_2(\omega)$ via the tight-binding π-band model of *Saito* and *Kamimura* (SK) [4.22] (Sect. 4.2.2c). A charge transfer $f = 12nf_C = 1$ was assumed. Figure 8.17b shows their experimental results (crosses) for stages 2–4 interband contributions to $\varepsilon_2(\omega)$ determined by $\varepsilon_{2,\text{inter}} = \varepsilon_2 - \varepsilon_{2,\text{Drude}}$, where the Drude contribution had been determined from fits to both *R* and $1/\omega\varepsilon_2$. The dashed and solid curves in the figure represent, respectively, the results calculated by *Yang* and *Eklund* [4.24] using the SK band parameters (dashed line) and the YE band parameters (solid line). The interband matrix elements – M_{ss} in (4.4) – were calculated according to a method developed for pristine graphite by *Johnson* and *Dresselhaus* [4.19]. As the figure shows, the energy-band parameter adjustments made by *Yang* and *Eklund* [4.24] to the SK band model significantly improve the fit. Overall, the fits using the YE parameters are fairly good, except for the stage-2 and stage-4 compounds at the lowest energy. Below 0.2 eV, both the YE and the SK band parameters predict a sharp peak in $\varepsilon_{2,\text{inter}}$ near 0.15 eV, identified [4.24] with the lowest-energy $c \to c$ transitions. However, the high-energy tails of these low-energy peaks were not observed.

For stage $n > 2$, the distribution of net layer charge between interior (C_i) and bounding (C_b) layers is sensitive to the δ parameter in the SK model. According

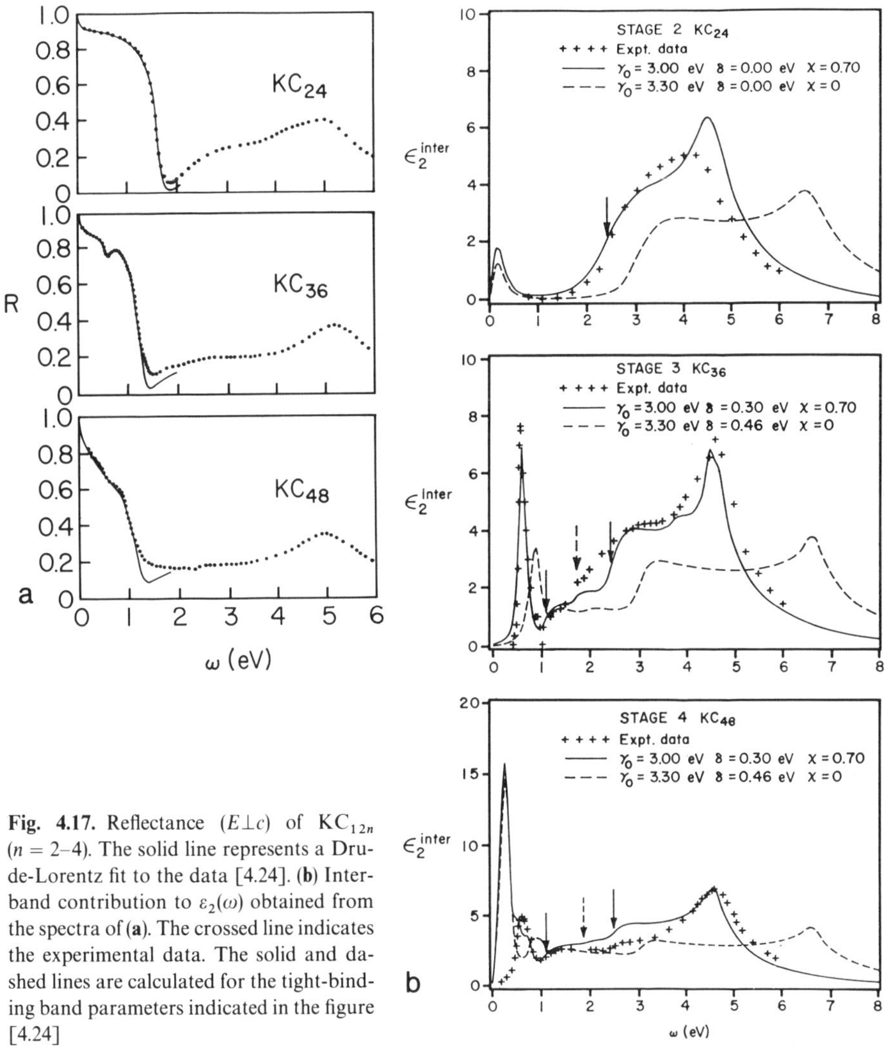

Fig. 4.17. Reflectance ($E \perp c$) of KC_{12n} ($n = 2$–4). The solid line represents a Drude-Lorentz fit to the data [4.24]. (**b**) Interband contribution to $\varepsilon_2(\omega)$ obtained from the spectra of (**a**). The crossed line indicates the experimental data. The solid and dashed lines are calculated for the tight-binding band parameters indicated in the figure [4.24]

to the results of *Yang* and *Eklund* [4.24], strong $c \to c$ transitions should occur between parallel bounding-layer and interior-layer π^* bands separated in energy by 2δ. These transitions were identified [4.24] with the ~ 0.60 eV peak in the stage-3 and stage-4 data. Thus, the 0.6 eV peak is a direct experimental measure of the electrostatic energy difference between the bounding and interior carbon layers in these compounds. *Yang* and *Eklund* [4.24] compared values for the fractional layer charge in the interior and bounding layers of KC_{12n} obtained from the SK band structure using both the SK and YE band parameters. These values can also be compared with those obtained by *Kume* et al. [4.45] by applying the Koringa relation to the spin-lattice relaxation times they obtained from their ^{13}C NMR experiments. For example, in stage-3 KC_{36}, the NMR and optical results are in good agreement, with $\sim 90\%$ of the K(4*s*) electrons transferred to the bounding C_b layers, and 10% to the interior C_i layers. In the SK band structure, the interior layer charge for stage-3 is noticeably lower (1.3%). Unfortunately, there are no NMR results for stage-4. However, the stage-4 optical results (84%, bounding layer; 16%, interior layer) and the stage-6 NMR results (82%, bounding layer; 18%, interior layer) are in good agreement, whereas the values for the interior layer charge in the stage-4 and stage-6 SK band structure are lower (i.e., 3.0%).

Comparing the band-structure electronic density of states at the Fermi energy $D(E_F)$ with that obtained from specific-heat measurements [4.46, 47] $D^*(E_F)$, *Yang* and *Eklund* [4.24] determined values for the electron-phonon coupling constant λ via the relation $D^*(E_F) = D(E_F)(1 + \lambda)$. They found $\lambda = 0.3$, 0.4, and 0.5 for stages 2, 3, and 4, respectively. These values may be compared with the value $\lambda = 0.3$ obtained by *Alexander* et al. [4.48] for the stage-1 MC_8 (M = K, Rb, CC) from the transition temperature T_c to the superconducting state.

Nogami et al. [4.34] performed in situ optical studies on thin films of K-graphite in an electrochemical cell. Poly(*p*-phenylene vinylene) films (5 microns thick) were first graphitized at 2900 °C for 1 hour in high-purity argon. The cell electrolyte was prepared by dissolving $KAsF_6$ in dimethyl sulfoxide (DMSO). A Pt wire was used as the counterelectrode. The films turned green at a cell potential of − 3.5 volts. Response times as short as 50 ms were reported for their "color switching" cell.

The Drude edge was observed at 0.8 eV (midpoint) and the Drude minimum occurred at $E_m = \sim 0.9$ eV, compared with the value $E_m = 2.6$ eV found for KC_8 [4.17, 40, 43]. *Nogami* et al. [4.34] analyzed their reflectance data using a modified Drude model, in which the interband transitions were modeled as a complex dielectric constant, similar to that used in early studies of the optical properties of GICs by *Zanini* and *Fischer* [4.42]. From their Drude analysis, they obtained a "bare" plasma frequency for their electrochemically prepared K-graphite of $\omega_p = 2.6$ eV, almost 2 eV lower than obtained [4.17, 40, 43] for chemically prepared KC_8. As a result, *Nogami* et al. concluded [4.34] that only carbon π electrons were generated in eletrochemically prepared KC_x. However, this conclusion appears unfounded, since they made no attempt to measure the K/C stoichiometry or to determine if DMSO solvent molecules may have been co-intercalated. The co-intercalation of NH_3 or tetrahydrafuran (THF) into the K layers of ternary solvent K GICs has been shown to lower the plasma frequency (Sect. 4.3.1c).

(b) Potassium-Hydrogen GICs

Shortly after the report of the synthesis of the alkali-metal graphite compounds by *Nixon* and *Parry* [4.49], hydrogen and hydrogenic gases were found by *Colins* and *Herold* [4.50] to react chemically (i.e., chemisorption) and by

Watanabe et al. [4.51] to react physically (i.e., physisorption) with the binary alkali-metal host.

For the $T \sim 300$ K reaction of H_2 with KC_8, *Enoki* et al. [4.52] determined that the hydrogen molecules dissociated, becoming stabilized as atomic hydrogen. From X-ray studies, *Conard* et al. [4.53] found that the potassium ions were rearranged in the I layer to form a three-layer sandwhich (K-H-K), similar to that found in the ternary $KHgC_{4n}$ (Sect 4.3.1d). As the hydrogen concentration (x) was increased in KH_xC_8, *Enoki* et al. [4.52] found that the material evolved from a stage-1 phase ($x \leqslant 0.10$), to a mixture of stages 1 and 2 ($0.10 = x < 0.67$), and then to a homogeneous (hydrogen-saturated) stage-2 material ($x = 0.67$). In an interesting observation, *Kaneiwa* et al. [4.54] found that as x was changed from 0 to ~ 0.19, the superconducting transition temperature of KH_xC_8 first increased and then disappeared entirely. *Enoki* et al. [4.55] found the introduction of hydrogen to C_8K to reduce the in-plane electrical conductivity by about 50% as x increased from 0 to 0.67, and attributed the decrease to a reduction in the number of carriers in the carbon π bands. Hydrogen was therefore proposed to act as an acceptor ion (H^-), causing the back-donation of electrons from the carbon to the intercalate layers [4.55].

Watanabe et al. [4.56] first proposed that at low temperatures ($T = 77$ K) hydrogenic *molecules* physisorb into the alkali-metal layers of stage $n \geqslant 2$ alkali-metal GICs (MC_{12n}). For the potassium compounds, the low-temperature adsorption of hydrogen is accompanied by a 0.3 Å c-axis lattice expansion [4.56, 57]. The mechanism for adsorption has been recently addressed by *Lee* et al. [4.58]. Despite the earlier proposals of hydrogen physisorption in KC_{24} by *Watanabe* et al. [4.51, 56], *Kondow* et al. [4.59–61] reported the results of several experiments that they interpreted to indicate that 40%–60% of the electrons originally donated by the K 4s states to the C layers were back-donated to the intercalate layers. This large amount of electron back-donation was disputed by *Terai* et al. [4.62], who claimed that there was little, if any, back-donation of electrons during hydrogen uptake.

Hydrogen Chemisorption in KC_8. *Doll* et al. [4.43, 63] carried out optical studies of electron back-donation in KH_xC_8. Since the $T = 300$ K reaction rate of hydrogen with KC_8 is prohibitively slow in well-ordered graphite such as HOPG (which is a desirable host material for optical measurements), *Doll* and *Eklund* [4.63] first examined the hydrogen chemisorption in Grafoil-based KC_8, where the absorption kinetics are comparable with graphite powder. However, the surface of Grafoil is not optically smooth. To account for the loss of specularly reflected light from Grafoil-based samples, they determined a wavelength-dependent correction factor for Grafoil and assumed that this correction factor was not appreciably altered by intercalation. To test the validity of this approach to Grafoil-based GICs, they first examined the optical relectance of KC_8 Grafoil. Figure 4.18 shows their data, uncorrected for diffuse scattering loss (dotted line), together with the corrected data (crosses). For comparison, the dashed line is the spectrum of KC_8 HOPG, and the solid line is a Drude fit to

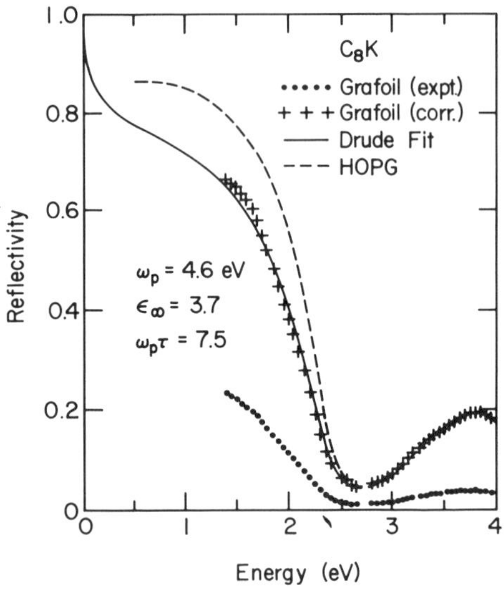

Fig. 4.18. Reflectance ($E \perp c$) of KC_8 Grafoil (dotted line) and KC_8 HOPG (dashed line). The crosses represent the spectrum of KC_8 Grafoil corrected for diffuse scattering loss. The solid curve is a Drude fit to the corrected spectrum for the parameters indicated in the figure [4.63]

the corrected spectrum of KC_8 Grafoil. The figure shows the Drude parameters obtained from the fit, and ω_p and ε_∞ are found to be in excellent agreement with the parameters previously reported for KC_8 HOPG [4.17, 43]. *Doll* and *Eklund* further observed that the free-carrier lifetime in KC_8 Grafoil is $\sim 1/2$ that of the KC_8 HOPG, which they attributed to a larger impurity scattering in the Grafoil-based material [4.63].

Having shown that a surface roughness correction can be successfully applied to Grafoil-based GICs, *Doll* and *Eklund* [4.63] obtained the corrected reflectance spectrum of the stage-2 (hydrogen-saturated) compound $KH_{0.67}C_8$ Grafoil in the same way (Fig. 4.19). The dots show the uncorrected data, and the crosses represent the spectrum corrected for surface roughness. The solid line is a Drude fit to the corrected spectrum; the figure lists the parameters. Converting the plasma frequency to f_C in the framework of the BR model [4.13] (Sect. 4.2.2a), *Doll* and *Eklund* obtained $f_C \sim 0.040 \pm 0.002$. They assumed that the K 4s band is empty and that the contribution to the effective plasma frequency from H 1s carriers is negligible in comparison with that from the light and more numerous carbon π^* electrons. *Doll* and *Eklund* [4.63] calculated the population F_H in the hydrogen 1s band using $F_H = 1 + (1 - 8f_C)/0.67$, where F_H is therefore in units of electrons per hydrogen atom. They obtained $F_H = 2.01 \pm 0.03$, indicating that the hydrogen band is full or nearly full; i.e., the minimum value of $F_H = 1.98$, or 0.02 holes per H atom. On the basis of later, more sensitive, proton NMR measurements by *Miyajima* et al. [4.64] and ultraviolet photoelectron spectroscopy (UPS) by *Yamamoto* et al. [4.65], a small number of holes were indeed found in the hydrogen band.

Guerard et al. [4.66] later discovered that stage $n = 1$ and $n = 2$ potassium-hydride GICs with stoichiometry $KH_{0.8}C_{4n}$ could be obtained by reacting KH

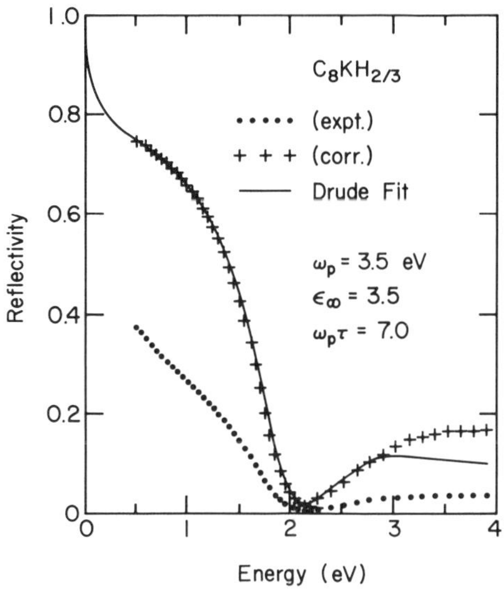

Fig. 4.19. Reflectance of $KH_{0.67}C_8$ Grafoil. The spectrum corrected for diffuse scattering loss is indicated by the crossed line. The solid curve is a Drude fit to the corrected spectrum for the parameters indicated in the figure [4.63]

salt with well-ordered graphite (e.g., HOPG). Charge transfer studies of these compounds were also carried out by *Doll* et al. [4.43] and compared with their binary GIC counterparts KC_8 and KC_{24}. Kramers-Kronig analyses to determine $\varepsilon(\omega)$ were carried out using density-scaled graphite for the high-energy data extension (Sect. 4.2.1). Figures 4.20, 21 show the results of *Doll* et al. [4.43] for $\varepsilon_2(\omega)$. The arrows in the figures show the midpoints of the respective $\pi \rightarrow \pi^*$ thresholds, identified as E_T. *Shung* [4.67] has shown theoretically that the $\pi \rightarrow \pi^*$ interband threshold is broadened about $\hbar\omega \sim 2E_F$ due to electron scattering. The Drude behavior of the free-carrier contribution to $\varepsilon_2(\omega)$ in KC_8, KC_{24}, $KH_{0.8}C_4$, and $KH_{0.8}C_8$ is evident in the data of *Doll* et al. [4.43] (Fig. 4.22), where $1/\omega\varepsilon_2$ is plotted against ω^2. The slope and intercept of the low-energy linear portion of the data were used to deduce values for the Drude parameters (ω_p, τ).

Using the reported stoichiometry KH_xC_{4n} (stage $n = 1, 2$) and charge neutrality, *Doll* et al. [4.43] expressed the relation between the fractional band occupations as

$$(F_K - 1) + x(F_H - 1) + 4nf_C = 0 \ , \tag{4.16}$$

where F_K, F_H, and f_C represent the number of electrons in the respective K 4*s*, H 1*s*, and C π^* bands. Assuming the H 1*s* band is essentially full (i.e., $F_H = 2$) and $x = 0.8$, (4.16) reduces to $f_C = (0.2 - F_K)/4n$. Using this result, *Doll* et al. [4.43] then interpreted the values for the effective plasma frequency in terms of a "two-carrier" model (Sect. 4.3.1a, Eq. 4.13a), with the π^* electrons described according to either the BR model [4.13] (stage 1 and 2) or the H model [4.18] (stage-1 only) (Sects. 4.2.2a, b). The second carrier type was assigned a range of effective mass $m = 1 - 2\,m_0$. Their minimum experimental value for ω_p (i.e.,

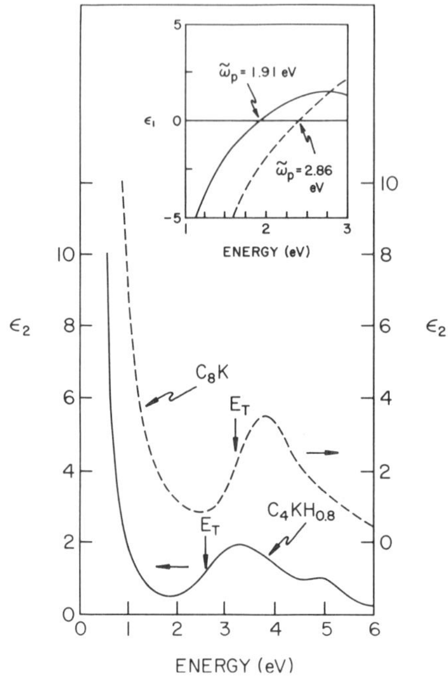

Fig. 4.20. Plot of Kramers-Kronig derived values for $\varepsilon_2(\omega)$ for KC_8 and $KH_{0.8}C_4$. The inset shows the zero-crossing of $\varepsilon_1(\omega)$. The arrows labeled E_T indicate the mid points of the respective inter-π-band threshold [4.63]

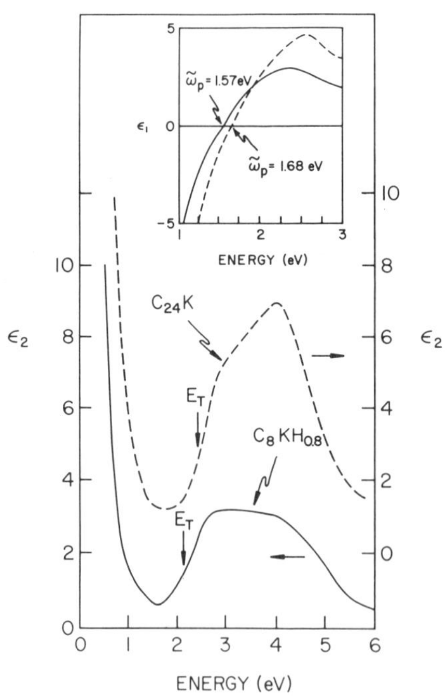

Fig. 4.21. Plot of Kramers-Kronig derived values for $\varepsilon_2(\omega)$ for KC_{24} and $KH_{0.8}C_8$. The inset shows the zero-crossing of $\varepsilon_1(\omega)$. The arrows labeled E_T indicate the midpoints of the respective inter-π-band threshold [4.63]

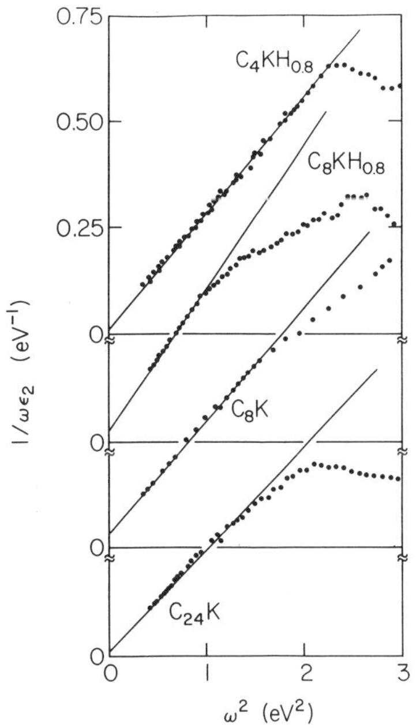

Fig. 4.22. Plot of $1/\omega\varepsilon_2$ versus ω^2 for KC_8, KC_{24}, $KH_{0.8}C_8$, and $KH_{0.8}C_4$. Drude behavior is linear in this plot, and the slope and intercept determine ω_p and τ [4.43]

$\omega_p - \Delta\omega_p$, where ω_p is the value obtained from the Drude analysis, and $\Delta\omega_p = 0.1$ is the experimental uncertainty) was found consistent with $F_K \sim 0.02$ electrons per K atom in the K(4*s*) band (where $\gamma_0 = 3.0$, $m = 1$). Of course, if the K(4*s*) band is empty, and the second carrier type was assigned instead to holes in an H(1*s*) band, their result would indicate 0.025 holes per H atom. The H and BR models ($\gamma_0 = 2.9$–3.0) were found to yield essentially the same values for f_C and E_F for $KH_{0.8}C_4$: $f_C = 0.05$ and $E_F = 1.5$ eV.

For $KH_{0.8}C_4$, the values obtained for E_F from the position ($\sim 2E_F$) of the interband threshold were found to be in good agreement with those obtained from ω_p. The H model and BR model gave slightly different results, and the difference was attributed to the asymmetry of the π and π^* bands obtained only in the H model. In the BR model, the π (valence) and π^* (conduction) bands exhibit mirror symmetry near the K point, and the dominant transitions between these mirror image bands yield an inter-π-band threshold with midpoint $E_T = 2E_F$. In the H model, the small asymmetry in the π and π^* bands near the K point leads to a midpoint $E_T \sim 2E_F - \Delta E$, where the correction factor ΔE increases slowly with increasing $E_F(\Delta E \sim 0.1$ eV at $E_F = 1.4$ eV).

Note that in the course of studying the charge transfer in the potassium-hydride GICs, *Doll* et al. [4.43] also carried out a BR-model analysis of the reflectance of stage-2 KC_{24} using $\gamma_0 = 2.9$ eV and $\gamma_1 = 0.375$ eV. From ω_p they obtained the value $f = 1.0 = 24f_C$ for the charge transfer from the K(4*s*) band to the π^* band (i.e., $f_K = 0$). Using the BR model to convert the experimental value of ω_p into E_F, they obtained $2E_F = 2.84$ eV, in excellent agreement with the midpoint ($E_T = 2.80$ eV) of the interband threshold (Fig. 4.21).

The values for f_C and E_F obtained optically by *Doll* et al. [4.43] for the $KH_{0.8}C_{4n}$ compounds can be compared with those obtained from Shubnikov-de Haas (SdH) experiments by *Enoki* et al. [4.68]. Whereas good agreement was obtained between optical and SdH results for the stage-2 hydride (SdH [4.68]: $f_C = 0.033$, $E_F = 1.15$ eV; Opt [4.43]: $f_C = 0.026$, $E_F = 1.13$ eV), a large disagreement arose for the stage-1 compound (SdH [4.68]: $f_C = 0.020$, $E_F = 0.65$ eV; Opt [4.43]: $f_C = 0.05$, $E_F = 1.62$ eV). However, high-frequency SdH oscillations are experimentally difficult to observe. If the observed stage-1 SdH frequencies of *Enoki* et al. [4.68] are reindexed assuming that the highest frequency oscillation was not observed, then the stage-1 values for f_C and E_F obtained from the SdH and optical studies can be brought into good agreement.

Hydrogen physisorption in KC_{24}. *Doll* and *Eklund* [4.69] carried out an optical study of the low-temperature physisorption of H_2 into stage-2 KC_{24} to investigate the controversy regarding the back-donation of electrons from the C layers to the $K(H_2)_x$ layers as a result of the hydrogen uptake. Figure 4.23 shows the reflectance spectra reported by *Doll* and *Eklund* [4.69] for $K(H_2)_x C_{22.6}$ before ($x = 0$) and after ($x = 1.6$) the $T = 77$ K physisorption of H_2. The C/K stoichiometry was determined by the change in weight of the HOPG-based sample, and the saturation hydrogen concentration at $p = 1000$ Torr and $T = 77$ K was measured to be $x = 1.6$. The inset shows the energy region near the reflectance minimum (E_m) on a magnified scale. Visual inspection of the sample through the optical dewar window revealed that, during production of the hydrogen-saturated compound, the sample surface had transformed from the specular, steel-blue color of $KC_{22.6}$ to a dull gray surface. *Doll* and *Eklund* concluded that

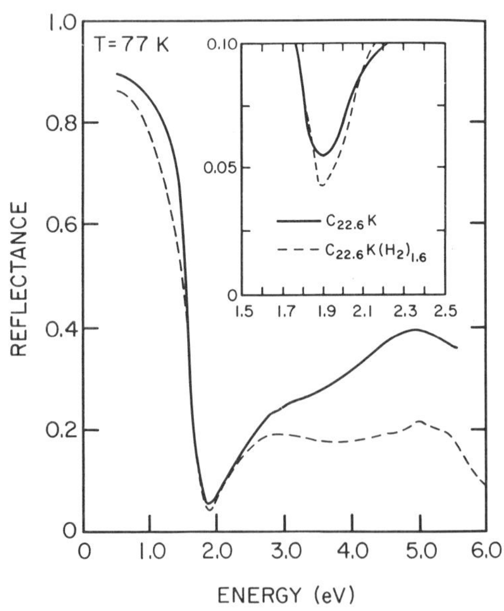

Fig. 4.23. The reflectance ($E \perp c$) spectra of $KC_{22.6}$ (solid line) and $K(H_2)_{1.6}C_{22.6}$ (dashed line). The latter compound is formed by the $T = 77$ K adsorption of H_2. The inset shows the energy region near the reflectance minimum (E_m) [4.43]

this transformation was, in part, due to a surface roughness induced by the intercalation of hydrogen. Unlike the studies in Grafoil-based $KH_{0.67}C_8$, a wavelength-dependent correction factor was not obtained. However, this factor should be a slowly varying function of energy. Therefore, it should not seriously distort a sharp Drude edge or induce an artificial shift in the position of the following reflectance minimum (E_m).

To determine the degree of the electron back-donation $f_B = (1 - 22.6 f_C)$ under these experimental circumstances, *Doll* and *Eklund* used the stage-2 BR-model dielectric function [4.13] to calculate the f_B dependence of the reflectance minimum E_m. The parameter f_B represents the fraction of the K 4*s* electron originally donated to the C layers that is *returned* to the I layers by virtue of the H_2 uptake. Values for E_m were calculated using $\tau = 2.2 \times 10^{-15}$ s, $T_{eff} - 700$ K, i.e., parameters that fit the spectrum of the binary host $KC_{22.2}$ before the physisorption of hydrogen. For small f_B, they obtained $f_B = (1.21\ eV^{-1})\ \Delta E_m$, where $\Delta E_m = E_{m0} - E_m$, and $E_{m0} = 1.90$ eV is the calculated position of the Drude minimum for $f_B = 0$. Experimentally, they found $E_m = 1.88 \pm 0.01$ eV for $C_{22.6}K(H_2)_{1.6}$, with the uncertainty arising from the point spacing of the $R(\omega)$ data. From the shift $\Delta E_m = 0.02$ eV, they concluded that $f_B = 0.01 \pm 0.01$, or an upper bound for $f_B = 0.02$ *e*/K atom or 0.0065 *e*/H atom. This small value for f_B is consistent with the results of *Terai* et al. [4.62] and *Sano* and *Inokuchi* [4.70], and disagrees sharply with the much larger value $f_B \sim 0.40$ *e*/K atom reported by *Kondow* and coworkers [4.59–61]. According to the calculations of *Doll* and *Eklund* [4.69], the value $f_B \sim 0.40$ would have caused a redshift $\Delta E_m \sim 0.5$ eV, much greater than the small shift observed.

(c) Potassium-NH_3 and Potassium-THF GICs

Ternary GICs containing both alkali-metal atoms and solvent molecules (e.g., ammonia NH_3, benzene C_6H_6, and tetrahydrofuran (THF) C_2H_8O) in the intercalate (I) layers can be prepared by a two-step process [4.8, 71, 72]. The first step is the synthesis of an alkali-metal GIC, and the second step is the reaction of this binary GIC with the particular solvent vapors. The intercalation of the large solvent molecules increases considerably the intercalate layer thickness and induces a measurable back-transfer of electrons to the intercalate layers as observed in NMR, X-ray, and optical studies (to be discussed below). The back-transfer is observed optically as a redshift of the Drude edge, consistent with a decrease of the electron population in the carbon π^* bands.

$K(NH_3)_xC_{12n}$. Stage $n = 1$ and $n = 2$ GICs with stoichiometry $K(NH_3)_xC_{12n}$ $(0 \leq x \leq 4.4)$ can be prepared, respectively, by the exposure of stage-2 KC_{24} [4.72] and stage-3 KC_{36} [4.73] to the vapors of ammonia (NH_3). The initial binary GIC has an empty K(4*s*) band [4.1, 2], and the subsequent intercalation of NH_3 into KC_{24} results in a back-transfer of electrons from the carbon to the $K(NH_3)_x$ layers, as first reported in X-ray studies by *York* and *Solin* [4.72] At $x \sim 1.4$ a stage $2 \rightarrow 1$ transition occurs, and further NH_3 can be accommodated in the stage-1 compound, up to a maximum value $x = 4.4$ [4.72]. *Huang* et al.

[4.74] first reported the results of in situ optical reflectance studies and electrical transport measurements of the c-axis (ϱ_c) and a-axis (ϱ_a) resistivity experiments in $K(NH_3)_xC_{24}$. At $x \sim 4$, they observed a sudden small drop in ϱ_a, which they identified with a Mott transition in the $K(NH_3)_x$ layers.

According to their model [4.74], as x increases from 1.5 to ~ 4 in the stage-1 compound, electrons continue to be transferred back from the C layers to localized states in the $K(NH_3)_x$ layers. At $x \sim 4$, the density of these electrons is sufficient for the Mott transition to occur and a metallic band is formed. This band then conducts in parallel with the carbon π^* band. Nonmetal-metal transitions had been observed previously in 3D alkali-metal-NH_3 liquids [4.75] and identified with Mott transitions.

Figure 4.24 shows the reflectance spectra reported by *Zhang* et al. [4.76] for the $x = 0$, 1.49, 4.11, 4.38 $K(NH_3)_xC_{24}$ compounds; the spectra are the same as those reported previously by *Huang* et al. [4.74]. The solid and dashed lines represent, respectively, the data and Drude fits to the data. Values for the plasma frequency were obtained from the Drude fits to the $R(\omega)$ data. The experimental values of ω_p obtained by *Zhang* et al. [4.76] are $\omega_p = 4.12$ eV ($x = 0$), 2.04 eV ($x = 1.49$), 3.33 eV ($x = 4.11$), and 3.27 eV ($x = 4.38$). The value of ω_p obtained for the stage-2 KC_{24} host is 0·08 eV, slightly lower than the value $\omega_p = 4.2 \pm 0.1$ eV reported by *Doll* et al. [4.43]. Note that the plasma frequency of the $x = 1.49$ compound has no simple interpretation, since this is a "residue" compound consisting of KC_8 and stage-1 and stage-2 $K(NH_3)_xC_{24}$ [4.72]. Using the H and BR models and assuming an empty K 4s band, *Zhang* et al. [4.76] converted their experimental plasma frequencies for the stage-1 $K(NH_3)$ GICs into values

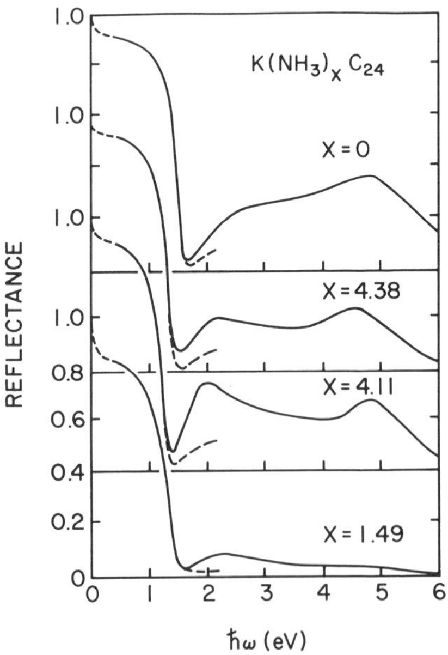

Fig. 4.24. Reflectance ($E \perp c$) of $K(NH_3)_xC_{24}$. The dashed lines represent Drude fits to the data [4.76]

for f_C. This conversion assumed that the contribution to ω_p from the intercalate layer is small enough to be neglected. The degree of back-donation is perhaps easiest to appreciate in terms of $f_K = (24 + \delta) f_C$, i.e., the charge per K atom in the carbon π^* band, where δ is the deviation from the nominal K/C stoichiometry determined by weight uptake measurements. *Zhang* et al. [4.76] obtained the following results: $f_K^{(BR)} = 0.80$ and $f_K^{(H)} = 0.74$ for $x = 4.38$ compound, and $f_K^{(BR)} = 0.85$ and $f_K^{(H)} = 0.78$ for the $x = 4.11$ compound. In other words, about $\sim 20\%$ of the electrons originally obtained from the K layers were back-donated to the intercalate layers during the formation of $x > 4$ compounds. The values obtained from the H model are about 0.07 lower than those obtained from the BR model (using $\gamma_0 = 2.9$ eV). The optical value $f_K \sim 0.80 \pm 0.07$ reported by *Zhang* et al. [4.76] for the $x > 4$ compounds can be compared with other experimental results: $f_K = 0.84$ (NMR) by *Tsang* et al. [4.77], $f_K = 0.88$ (diffuse X-ray scattering) by Quian et al. [4.78], and $f_K = 0.95$ (001 X-ray diffraction studies of c-axis contraction) by *York* and *Solin* [4.72].

Zhang et al. [4.76] carried out Kramers-Kronig analyses using density-scaled graphite for the high-energy data extension (Sect. 4.2.1). Figure 4.25 shows their results for $\varepsilon_2(\omega)$ for the $x = 0$, 1.49, 4.11, and 4.38 compounds. The $\varepsilon_2(\omega)$ data falls as $\sim \omega_p^2/\omega^3$ in the low-energy region, typical of free-carrier absorption, followed first by the threshold for interband absorption, which should contain the $\pi \to \pi^*$ interband threshold at $h\omega \sim 2E_F$. The broad peak at ~ 4.5 eV is the analog of the M-point peak in graphite at ~ 5 eV, and the strength is reduced in the GIC because of the reduced c-axis density of the carbon layers. In the $x = 1.49$ compound, the M-point peak is further reduced due to surface roughness arising during the removal of the NH_3 from the sample.

The optical evidence cited by *Huang* et al. [4.74] and *Zhang* et al. [4.76] for the transfer of electrons to *localized* states in the $K(NH_3)_x$ layers is the presence in $\varepsilon_2(\omega)$ (Fig. 4.25) of a narrow absorption band at 1.8 eV. This peak is most prominent in the $x = 4.11$ data, where a narrow (~ 0.4 eV) peak (dashed line, modeled as a Lorentzian lineshape) is superposed on the $\pi \to \pi^*$ threshold (dash-dotted line). In 3D $M(NH_3)_x$ liquids, a band at 0.8eV was observed previously (with the same width) and identified with the $1s \to 2p$ transition of an electron localized in a "hydrogenic" potential [4.75]. The electron was said to be "solvated" by a cage of NH_3 molecules [4.75]. However, the oscillator strength of the 1.8 eV band found by *Zhang* et al. [4.76] is ~ 4, much larger than found in the 3D metal–NH_3 liquids. They attributed this enhancement of the oscillator strength to a coupling of the "solvated" electrons in the intercalate layers to π^* electrons on adjacent C layers. The oscillator strength of the 1.8 eV band was found to be lower in the $x = 4.38$ compound than in the $x = 4.1$ compound, consistent with the formation of a conduction band from the solvated electron states.

Zhang et al. [4.73] performed a similar in situ optical reflectance study of the ammonia-saturated *stage-2* $K(NH_3)_{4.5}C_{36}$ to see if solvated electrons could also be detected in a *stage-2* $K(NH_3)_x$ GIC. The $K(NH_3)_{4.5}C_{36}$ compound was

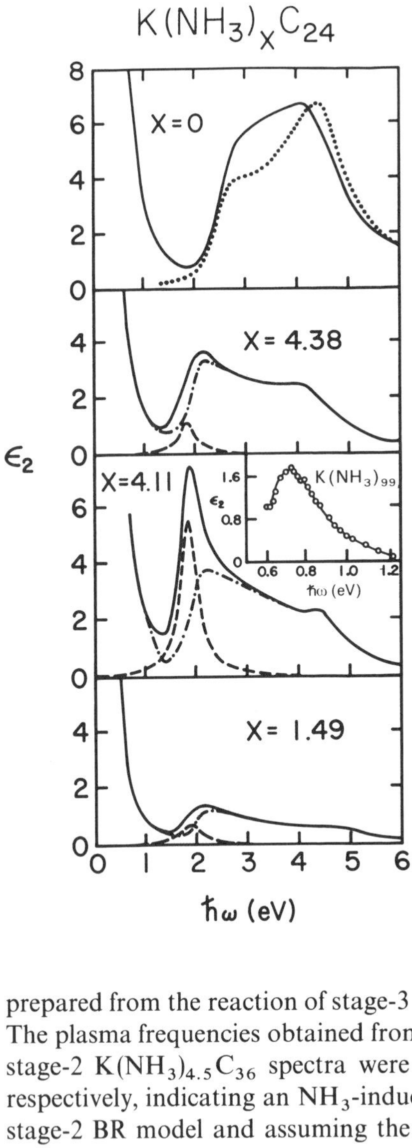

Fig. 4.25. Kramers-Kronig derived $\varepsilon_2(\omega)$ for the $K(NH_3)_xC_{24}$ spectra in Fig. 4.24. The data are shown as the solid lines. The dashed lines represent a Lorentz oscillator, identified with absorption by "solvated" electrons in the $K(NH_3)$ intercalate layers. The dash-dotted line corresponds to the inter-π-band contribution and the data. The dotted line in the upper panel is the calculated inter-π-band contribution [4.24]. The inset to the $x = 4.11$ panel is the absorption band associated with "solvated" electrons in 3D liquid $K(NH_3)_x$ [4.76]

prepared from the reaction of stage-3 KC_{36} with NH_3 at a pressure of ~ 9 atm. The plasma frequencies obtained from a Drude fit to the stage-3 KC_{36} host and stage-2 $K(NH_3)_{4.5}C_{36}$ spectra were found to be $\omega_p = 3.40$ eV and 3.12 eV, respectively, indicating an NH_3-induced redshift $\Delta\omega_p = -0.28$ deV. Using the stage-2 BR model and assuming the K(4*s*) band is empty, *Zhang* et al. [4.73] found that $f_K = (36 + \delta) f_C = 0.70 \pm 0.07$, which is somewhat lower than the value $f_K = 0.8$ reported [4.74, 76] for stage-1 $K(NH_3)_{4.1}C_{24}$ and $K(NH_3)_{4.4}C_{24}$ compounds.

Zhang et al. [4.73] also detected a narrow absorption band in the $\varepsilon_2(\omega)$ data for stage-2 $K(NH_3)_{4.5}C_{36}$, which, as in their work in the stage-1 compounds, they correlated with the presence of solvated electrons. Figure 4.26 shows their $\varepsilon_2(\omega)$ data for $K(NH_3)_{4.5}C_{36}$ together with that of the $x = 0$ compound (KC_{36}). For the $x = 4.5$ compound, the figure shows several curves. The solid and dashed curves represent, respectively, the data and the calculated (BR model)

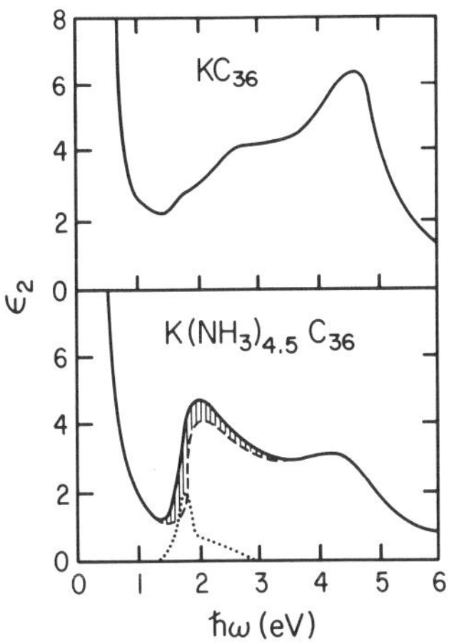

Fig. 4.26. Kramers-Kronig derived $\varepsilon_2(\omega)$ for stage-2 $K(NH_3)_{4.5}C_{36}$ and KC_{36}. The data are shown as the solid lines. The dashed line in the lower panel is the graphitic inter-π-band contribution calculated according to the BR model, and the dotted curve represents the difference between the data and the inter-π-band contribution. The dotted curve peaks at 1.7 eV and is identified with "solvated electrons" in the $K(NH_3)$ layers [4.73]

π, π^* contribution, while the dotted curve represents the difference between these two curves. The dotted curve therefore represents the non-π, π^* contribution. As the figure shows, the non-π, π^* contribution exhibits a narrow, asymmetric peak at 1.7 eV, similar to the peak reported [4.71, 73] at 1.8 eV in the stage-1 $K(NH_3)$ – GICs, suggesting the existence of solvated electrons in the intercalate layers of the stage-2 $K(NH_3)_xC_{36}$ compound.

$K(THF)_xC_{24}$, $x=1, 2$. *Zhang* and *Eklund* [4.79] also carried out in situ optical studies of stage-1 $K(THF)_xC_{24}$, $x = 1,2$, to investigate the differences in the $K(THF)$ and $K(NH_3)$GICs. Previous studies in the 3D alkali-metal-solvent (MS) liquids have shown NH_3 to be a rather unique solvent [4.80]. The much smaller size and higher dielectric constant of NH_3 (compared with other polar solvents) has been suggested as the origin of the significant differences observed in the physical properties of the 3D MS liquids [4.80]. For example, K readily dissolves in NH_3 to form K^+ ions and solvated electrons, but K does not dissolve in THF. However, *Dye* and coworkers [4.81, 82], used "crown" ethers or "cryptates" as complexing agents to stabilize K^+ in liquid THF and thereby produced solvated electrons, detecting them optically via a narrow absorption band at 0.58 eV.

Stage-1 K(THF) GICs with the nominal stoichiometry $K(THF)_xC_{24}$ ($x = 1$–3) can be obtained by the reaction of stage-2 KC_{24} with vapors of THF [4.83]. *Zhang* and *Eklund* [4.79] were able to prepare single-stage HOPG-based compounds with $x = 1$ and $x = 2$ for optical study. Figure 4.27 shows their reflectance spectra for $K(THF)_xC_{24}$ ($x = 0, 1, 2$) as solid lines. The dashed lines represent fits to the data calculated according to the stage-1 BR dielectric constant model [4.13] for $\gamma_0 = 2.9$ eV and $\gamma_1 = 0.375$ eV (Sect. 4.2.2a). From the value of the adjustable parameter E_F, *Zhang* and *Eklund* [4.79] determined values for f_K (the charge per K atom in the carbon π^* band) to be $f_K = 0.53 \pm 0.07$ ($x = 1$) and $f_K = 0.49 \pm 0.07$ ($x = 2$). The small difference between the

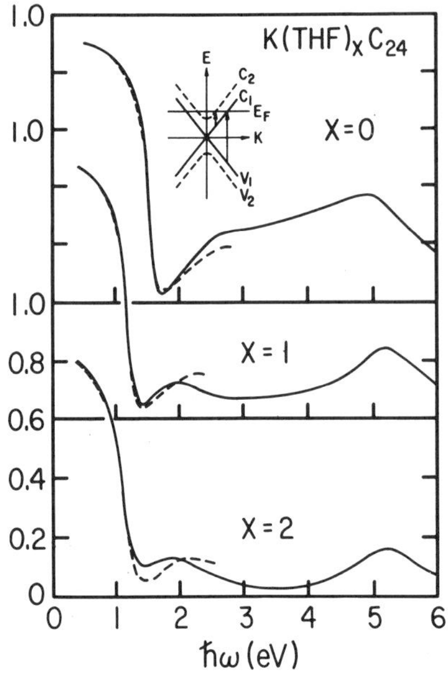

Fig. 4.27. Reflectance ($E \perp c$) for $K(THF)_xC_{24}$. The dashed lines represent BR-model fits to the data [4.79]

values of f_K for the $x = 1$ and $x = 2$ compounds was therefore found to be insignificant. However, the backdonation of electrons $f_B = (1 - f_K)$ to the intercalate layers induced by solvent molecule uptake is larger in the K(THF) GICs than in the $K(NH_3)$ GICs. *Zhang* and *Eklund* [4.79] concluded that this increase in f_B indicated that different chemical mechanisms were responsible for the back-donation in these two GIC systems, and argued that the larger back-donation of electrons to the K(THF) layers was consistent with the insolubility of the K^+ ion in 3D THF liquids.

The values of f_K obtained by *Zhang* and *Eklund* [4.79] for $K(THF)_xC_{24}$ are somewhat lower than values obtained from ^{13}C NMR experiments. Using the NMR chemical shift data of *Quinton* et al. [4.84] and the empirical relation determined by *Tsang* et al. [4.77] that links chemical shift and charge transfer in donor GICs, *Zhang* and *Eklund* [4.79] determined a value for $f_K = 0.81$, similar to the optical values obtained [4.74, 76] for f_K in the stage-1 $K(NH_3)$ GICs. Using their chemical shift data and the BR model for $D(E_F)$, the density of π^* states at the Fermi energy, *Quinton* et al. [4.84] found f_K (ternary) $= 0.72 \pm 0.05$, in better agreement with the optical value.

In contrast to their studies of stage-1 and stage-2 $K(NH_3)C_{24}$ compounds, but consistent with the insolubility of K in THF 3D liquids, *Zhang* and *Eklund* [4.79] found no evidence for the presence of solvated electrons in the $K(THF)_x$ layers; i.e., no narrow absorption band in $\varepsilon_2(\omega)$ was detected.

(d) Cesium-Bismuth and Potassium-Mercury GICs

The $Cs(Bi)_x$GICs and (KHg) GICs have both been reported to superconduct at temperatures $T > 1$ K. However, several groups have since failed to confirm the report of superconductivity in the $Cs(Bi)_x$GICs. Both GIC systems exhibit an intercalate layer structure composed of a three-layer sandwich M-X-M, where the two outer alkali-metal (M) layers are next to the bounding carbon layers. However, the central X = Bi [4.85] or Hg [4.86] layer may be slightly split. Optical reflectance studies of stage-1 and stage-2 $CsBi_x$GICs [4.87] and KHg GICs [4.17, 87–90] have been carried out.

$CsBi_x$-GICs were reported by *Lagrange* et al. [4.91] to superconduct at the following critical temperatures (T_c): stage-1 $CsBi_{0.5}C_4$ (α) ($T_c = 4.0$ K), stage-1 $CsBi_{1.0}C_4$ (β) ($T_c = 2.4$ K). However, several groups have failed to observe superconductivity in $CsBi_x$ GICs. [4.88, 92, 93] Even minority phase superconductivity (< 2% volume superconducting) was not detected [4.87] for temperatures $T > 1.5$ K and pressures $p < 7$ kbar, in several stage-1 α- or β-phase samples, although the I_c values agreed to $\pm$ 0.02 Å with those reported by *Lagrange* et al. [4.91]. In the KHg GICs, stage-2 $KHgC_8$ has been reported to superconduct at $T_c \sim 2.0$ K, 1.2 K higher than stage-1 $KHgC_4$, where $T_c \sim 0.8$ K [4.94–96].

Fig. 4.28 shows the optical reflectance data (dotted line) reported by *Yang* et al. [4.86] for the stage-1(α), stage-1(β), and stage-2(β) $CsBi_x$GICs, together with stage-1 and stage-2 (KHg) GICs. The dot-dashed and solid lines in the figure represent, respectively, the high-energy data extension (density-scaled graphite) and the Drude model fit to the data. As can be seen, the Drude fit near the reflectance minimum is well below the data, indicating the presence of low-energy interband transitions near the screened plasma frequency. Above 0.5 eV, the data shown for $KHgC_{4n}$ are in good agreement with the data of *Preil* et al. [4.88, 89] and *Heinz* et al. [4.90]. Below 0.5 eV, *Yang* et al. [4.87] did not

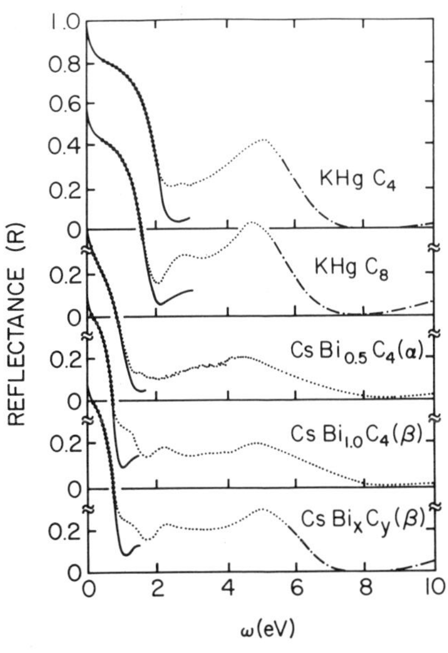

Fig. 4.28. Reflectance ($E \perp c$) for stage-n $KHgC_{4n}$, stage-1 $CsBi_xC_4$ (α and β phases), and stage-2 $CsBi_xC_y$ (β). The solid lines represent Drude fits to the data, and the dash-dotted lines are the data extensions based on density-scaled graphite [4.87]

observe the weak periodic structure reported previously by *Heinz* et al. [4.90]. This periodic structure is likely due to a surface condensate.

Figures 4.29, 30 show the corresponding $\varepsilon_2(\omega)$ data [4.87] obtained by a Kramers-Kronig analysis using density-scaled graphite for the high-energy data extension (Sect. 4.2.1). The $\varepsilon_2(\omega)$ data (dots) are shown together with $\varepsilon_{2,\mathrm{Drude}}$ (solid line) and $\varepsilon_{2,\mathrm{inter}} = \varepsilon_2 - \varepsilon_{2,\mathrm{Drude}}$ (dashed line). *Yang* et al. [4.87] identified the onset of interband absorption in all five compounds with interband transitions involving intercalate states. As they pointed out, the plasma frequencies obtained from Drude analyses of the (KHg) and $CsBi_x$GIC data are too high for these onsets to be identified with $\pi \to \pi^*$ transitions. As Fig. 4.29 shows, the onsets for interband absorption in all three $CsBi_x$GICs occur near $E_{T,i} = 0.5$ eV (the subscript "i" signifies "intercalate"). For the (KHg) GICs, the onsets are somewhat higher, at $E_{T,i} = 1.3$ eV. This agrees well with the optical results of

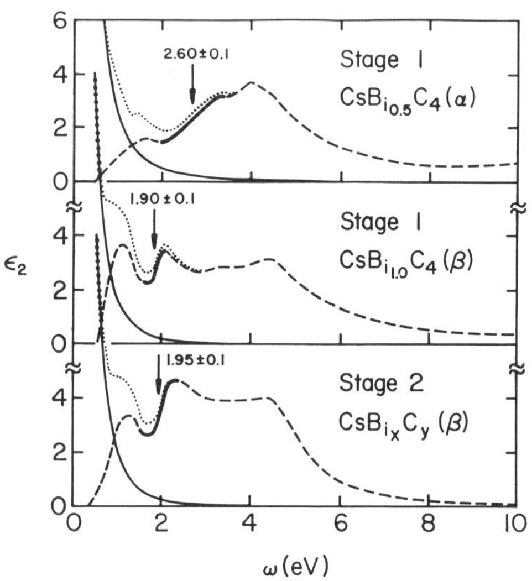

Fig. 4.29. Kramers-Kronig derived $\varepsilon_2(\omega)$ for the CsBi-GIC spectra shown in Fig. 4.28. The dots are the data, the solid line represents the Drude contribution, and the dashed line represents the interband contribution. The inter-π-band thresholds are highlighted with the bold lines, and the arrows indicate their mid points, which are identified with the inter-π-band threshold energy E_T [4.87]

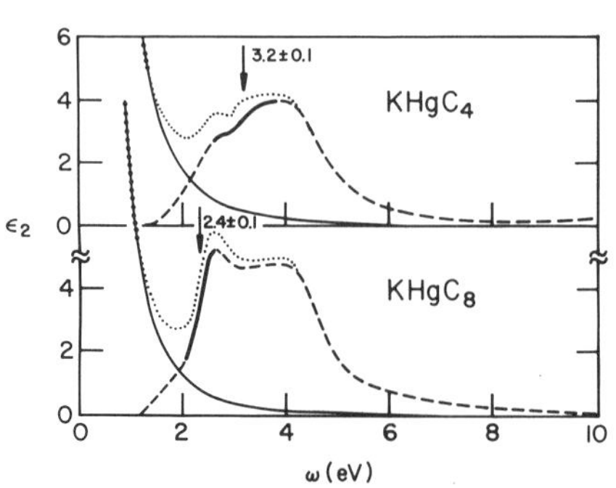

Fig. 4.30. Kramers-Kronig derived $\varepsilon_2(\omega)$ for the KHg-GIC spectra shown in Fig. 4.28. The dots are the data, the solid line represents the Drude contribution, and the dashed line represents the interband contribution. The inter-π-band thresholds are highlighted with the bold lines, and the arrows indicate their mid points, which are identified with the inter-π-band threshold energy E_T [4.87]

Preil and *Fischer* [4.89] and *Fischer* et al. [4.17]. The arrows in the figure indicate the midpoint of interband thresholds (highlighted by the bold lines) identified by *Yang* et al. [4.87] with $\pi \to \pi^*$ interband thresholds. As mentioned earlier, *Shung* [4.67] has shown that these thresholds are broadened about the midpoint $E_T \sim 2E_F$. *Yang* et al. [4.87] interpreted their experimental values for the in-plane plasma frequency ω_p in terms of a two-carrier model, similar to that discussed in Sect. 3.1a for KC_8. Free-carrier contributions to the in-plane plasma frequency were considered from both the carbon π^* ($\omega_{p,\pi}$) and intercalate ($\omega_{p,i}$) bands:

$$\omega_p^2 = \omega_{p,\pi}^2 + \omega_{p,i}^2 = 4\pi e^2[(f_C/\Omega_C m_{opt,\pi}) + (f_i/\Omega_i m_{opt,i})] , \tag{4.17}$$

where Ω_c and Ω_i refer to the volume per C atom and intercalate unit (i.e., KHg), respectively. The other symbols have their usual meaning (e.g., f_i is the itinerant charge per intercalate unit). For the carbon π^* electrons, *Yang* et al. [4.87] used the BR model to obtain values for $\omega_{p,\pi}$ ($\gamma_0 = 2.9$ eV and $\gamma_1 = 0.375$ eV) from the experimental values for E_F obtained from the midpoints of the inter-π-band thresholds. Then using (4.17) they calculated $\omega_{p,i}$. Finally, by assuming a value for f_i, they calculated $m_{opt,i}$, or vice versa. A significant difference between $CsBi_x$ and KHg GICs emerged from their free-carrier analysis. The intercalate bands in the KHg GICs were found to exhibit a much larger contribution to the free-carrier response than in the $CsBi_x$GICs.

To interpret all their results, *Yang* et al. [4.87] proposed the schematic density of states (DOS) models for the $CsBi_x$ GICs and KHg GICs shown in Figs. 4.31, 32. The interband energies $E_{T,i}$ and $E_{T,\pi}$ are indicated by the arrows in the figures. In the case of the $CsBi_x$ GICs (Fig. 4.31), the Fermi energy E_F is seen to cut through the top of the $CsBi_x$ valence band, as well as the carbon π^* band. For comparison, the inset shows the DOS for $CsBi_{0.33}$ determined from

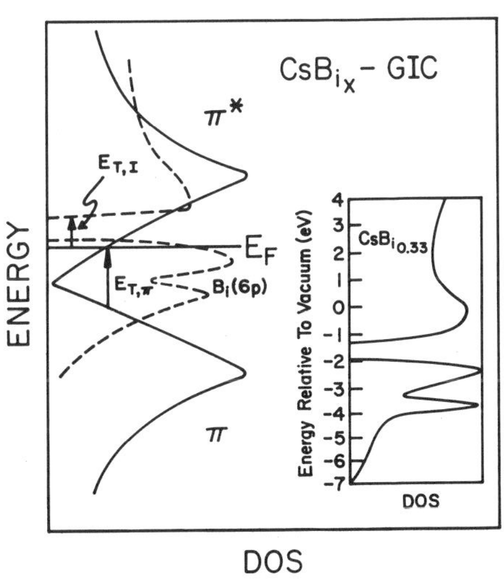

Fig. 4.31. Schematic electronic density of states for CsBi GICs. The inset is from UPS studies of the alloy $CsBi_{0.33}$ [4.87]

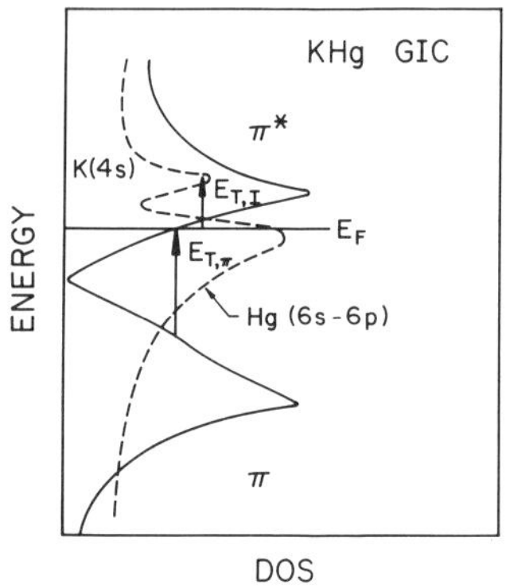

Fig. 4.32. Schematic electronic density of states for KHg GICs [4.87]

photoemission studies by *Spicer* and coworkers [4.97]. The valence bands in $CsBi_{0.33}$ were identified [4.97] as spin-orbit split Bi(6*p*) bands, separated 0.7 eV from a higher lying conduction band.

For the KHg GICs (Fig. 4.32), the schematic DOS model of *Yang* et al. [4.87] has the Fermi energy also cutting both the π^* band and an intercalate band that they have identified as a hybridized Hg(6*s*, 6*p*)–K(4*s*) band. Previous core-level and valence-band photoemission studies of stage-1 and, stage-2 KHg GICs by *DiCenzo* et al. [4.98] reported a greater occupancy of states with K 4*s* character in $KHgC_{4n}$ than in the KC_8 compounds. They also found that the stage-2 $KHgC_8$ compound exhibited a higher occupancy of states with Hg 6*s* character than did stage-1, suggesting that the Hg 6*s* states play an important role in the superconductivity of these compounds.

By assuming that the KHg layer provides 2 (3) electrons per KHg formula unit to fill both the π^* and intercalate conduction band(s), *Yang* et al. [4.87] determined values via (4.17) for the occupation f_i of the intercalate band(s): $f_i = 1.78$ (2.78) for stage-1 and $f_i = 1.75$ (2.75) for stage-2. Therefore, the principal difference between the KHg and $CsBi_x$ GICs obtained from their optical study is that the former compounds have a large fraction of their conduction electrons remaining in the intercalate layers, whereas the latter compounds do not (i.e., $f_i \sim 0.0$–0.15 for the $CsBi_x$ GICs and $f_i \sim 2$–3 for the KHG GICs).

4.3.2 Acceptor-Type GICs

In all cases known to date, the optical properties of acceptor-type GICs are dominated in the 0.5–6 eV range by the dielectric response of the carbon π, π^* bands. The intermolecular coupling does not appear to lead to free carriers in the intercalate layers, and the contribution to the dielectric response from intramolecular or charge transfer absorption has not, in most cases, been found to be significant. The optical determination of charge transfer in acceptor GICs then reduces to measuring the in-plane plasma frequency $\omega_p = \omega_{p,\pi}$ and using an appropriate π-band model to convert $\omega_{p,\pi}$ into f_C. Since most acceptor GICs have moderately small charge transfer, the rigid-band model of Blinowski-Rigaux [4.13] (and the associated dielectric functions) can be used to determine f_C and E_F in stage-1 and stage-2 compounds. Consistent with this picture, acceptor-type GICs are found to exhibit a much higher electrical and thermal anisotropy than donor-type GICs.

Below we review optical studies of acceptor-type GICs that have resulted in the determination of the charge transfer or the effects of charge transfer on the graphitic intralayer phonons. Some of the results have been summarized recently by *Rigaux* [4.3]. All the optical data discussed below were collected from the *c* face (*ab* plane) of HOPG host material.

(a) Sulfuric Acid GICs

Perhaps the most important optical studies of charge transfer in acceptor-type GICs have been carried out on H_2SO_4 GICs. These compounds can be prepared electrochemically, and the charge flowing in the external circuit can be used to determine the hole population in the carbon π band(s). A low current (10–50 microamps) is passed between a graphite (HOPG) electrode (+) and Pt counter electrode (−) in a cell containing H_2SO_4 to drive the reaction slowly forward. As time progresses, the graphite electrode evolves through a succession of well-staged GICs (. . . $4 \to 3 \to 2 \to 1$). Electrochemical techniques used to prepare other GICs (including donor-type GICs) have recently been reviewed by *Eklund* [4.99].

The electrochemical reaction of graphite with concentrated H_2SO_4 has been proposed by *Aronson* et al. [4.100] and *Bessenhard* et al. [4.101] to lead to the stoichiometry $C_p^+ HSO_4^- (H_2SO_4)_x$ for $p > 21$. A rather unique property of these GICs is the existence of a series of stage-1 ($21 < p < 28$) and stage-2 ($48 < p < 60$) compounds with continuously variable charge transfer. In these ranges of p, charge transfer has been identified [4.101] with the electrochemical conversion of neutral H_2SO_4 into ionized HSO_4^-, i.e., only protons need to be removed from the I layers to increase f_C. According to the electrochemical model for these GICs [4.100, 101], each HSO_4^- molecule in the intercalate layer is linked to both one hole in the carbon layers and to one electron flowing in the external electrochemical circuit. Therefore, f_C can be determined from the net charge Q passed in the electrochemical circuit; i.e., $f_C = 1/p = m_C Q/eM$, where e is the electronic charge, M is the mass of the graphite host, and m_C is the mass of a carbon atom.

The first in situ optical study of $C_p^+ HSO_4^- (H_2SO_4)_x$ was performed by *St. Jean* et al. [4.102]. The sample was only partly immersed in the acid, so that the light was incident on the sample above the acid level. Later, Raman scattering experiments by *Eklund* et al. [4.103] showed that at certain times during the electrochemical intercalation process, even the stage indexes of the sample above and below the acid level were different. However, during the variable p and fixed stage-1 and stage-2 intervals, the Raman studies [4.103] showed that sample surfaces above and below the acid level exhibited the same spectra. This observation is undoubtedly connected with the high mobility of protons in the intercalate layers. To characterize the stage index of their samples in the optical skin depth, *St. Jean* et al. [4.102] monitored optical structure in the infrared range, which they identified with inter-π-band ($v \to v$) band transitions. Their assignment of some of the observed infrared structure has been questioned [4.104] and will be discussed below. From their interpretation of the infrared structure, *St. Jean* et al. [4.102] reported that uniform stage-2 and stage-1 phases extended over the charge transfer interval corresponding to $C_{58}^+ \to C_{48}$ (stage-2) and $C_{28}^+ \to C_{21}^+$ (stage-1), in good agreement with the earlier results of *Bessenhard* et al. [4.101]. Recording several reflectance spectra during these variable charge transfer intervals, they observed a continual blueshift in the Drude edge with decreasing *p*. *St. Jean* et al. [4.102] fit their spectra to the stage-1 and stage-2 dielectric function models of *Blinowski* et al. [4.13] to determine values for E_F as a function of *p*. These data were used to deduce the first experimental values for the intralayer transfer integral (γ_0). From the BR model (4.9).

$$p = 1/f_C = \gamma_0 \pi \sqrt{3}/E_F^2 \,. \tag{4.18}$$

This behavior was indeed observed in both the stage-1 and stage-2 graphite-H_2SO_4 compounds by *St. Jean* et al. [4.102], and is shown in Fig. 4.33, where the quantities pE_F^2 (crosses) and pE_F (dots) are plotted against $p (p = x$ in their figure). Values for $\gamma_0 = 2.63 \pm 0.03$ eV (stage 1) and $\gamma_0 = 2.81 \pm 0.04$ eV (stage 2) were determined from the data. The stage-1 value appears to be somewhat lower than the typical range of values 2.8–3.0 used in a variety of later optical studies.

As part of a series of studies on the graphite-H_2SO_4 system that included neutron diffraction [4.105, 106] and Raman scattering [4.107, 108], *Zhang* et al. [4.104] also performed in situ optical reflectance measurements on the stage-1

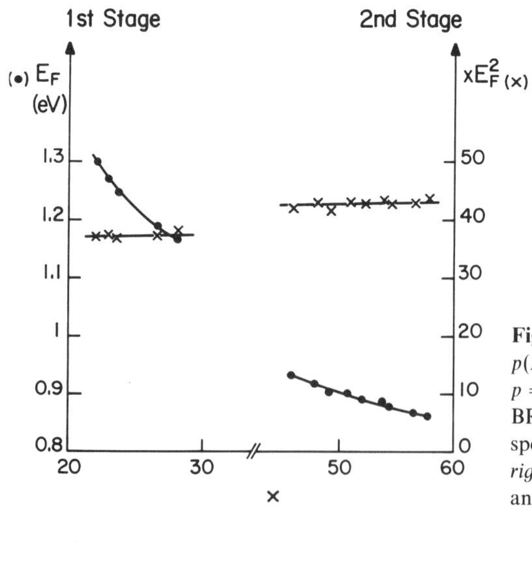

Fig. 4.33. Plot of E_F (dots, left scale) and $p(E_F)^2$ (crosses, right scale) against *p*, where $p = x = 1/f_C$. The data were obtained from BR-model fits to the reflectance ($E \perp c$) spectra for $C_p^+(HSO_4)^- \cdot H_2SO_4$. In the *rigid* π-band model, $p(E_F)^2$ is a constant and is proportional to γ_0 [4.102]

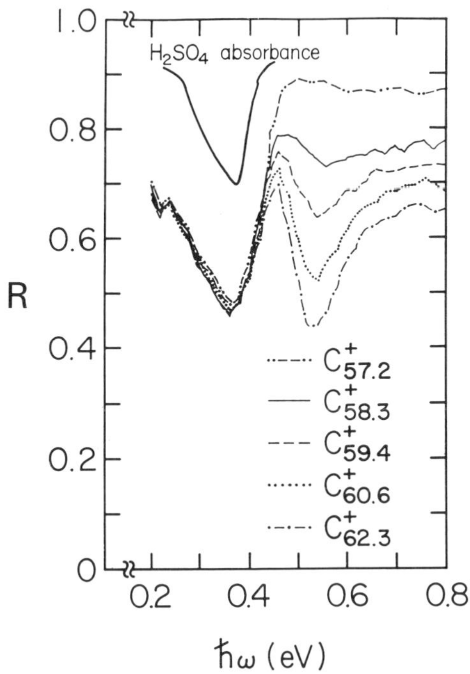

Fig. 4.34. The low-energy reflectance of graphite H_2SO_4 for values of the electrochemical charge near the stage 3 to 2 transition. The feature at 0.57 eV in the GIC spectra is identified with $v \rightarrow v$ interband absorption in the stage-3 GIC. The upper trace is the characteristic absorbance of liquid H_2SO_4. The 0.38 eV feature is identified with absorption in H_2SO_4, which masks the weaker stage-2 $v \rightarrow v$ absorption at 0.375 eV [4.76]

and stage-2 $C_p^+HSO_4^-(H_2SO_4)_x$ compounds in the energy range 0.2–6.0 eV. Figure 4.34 shows the reflectance spectra (reported by *Zhang* et al. [4.104] in the low-energy region ($E < 0.8$ eV) for several spectra near the stage $3 \rightarrow 2$ transition at $p = 60$. A prominent feature in the spectra is at 0.38 eV (which remains unchanged throughout the range of p studied). Another at 0.57 eV disappears near C_{60}^+ compound, in agreement with the previous work of *St. Jean* et al. [4.102]. The top trace in the figure is the absorbance spectrum obtained by *Giguere* and *Savoie* [4.109] of liquid H_2SO_4, which also exhibits structure at 0.38 eV. The 0.57 eV feature, identified previously by *St. Jean* et al. [4.102] with $v \rightarrow v$ interband absorption in the stage-3 compound, is similar in strength and position to the $v \rightarrow v$ structure observed, for example, in graphite-$SbCl_5$ by *Hoffman* et al. [4.110] and in graphite-$FeCl_3$ by *Smith* and *Eklund* [4.111]. In Fig. 4.34, the disappearance of the 0.57 eV feature in the data of *Zhang* et al. [4.104] at $p = 60$ agrees with the previous reflectance studies of *St. Jean* et al. [4.102] and accords with Raman scattering studies [4.103] of the evolution of the graphitic intralayer modes in this GIC system. However, the much stronger 0.38 eV structure in the figure, identified previously by *St. Jean* et al. [4.102] with stage-2 $v \rightarrow v$ interband transitions and reported by them to disappear at the end of the stage-2 interval, was found by *Zhang* et al. [4.104] to remain in the stage-1 spectra, growing in strength. *Zhang* et al. [4.104] were then forced to conclude that this structure should be identified with intramolecular optical absorption in H_2SO_4, either within the intercalate layers or in a thin film on the sample surface. The dip in $R(\omega)$ in the stage-2 spectra of *St. Jean* et al. [4.102] is $\sim 40\%$, almost a factor of ten larger than the 3–5% dip in the reflectance observed at 0.375 eV by *Hoffman* et al. [4.110] in stage-2 graphite-$SbCl_5$ and by *Smith* and *Eklund* [4.111] in stage-2 graphite-$FeCl_3$.

Zhang et al. [4.104] carried out Drude analyses of their data for the low- and high-charge transfer forms of stage-1 and stage-2 graphite-H_2SO_4 (i.e., C_{28}^+, C_{21}^+, C_{60}^+, and C_{48}^+). They obtained $\omega_p = 3.31, 3.06, 3.10$, and 2.90 eV, respectively, for the $p = 21$, 28, 48, and 60 compounds. As *St. Jean* et al. [4.102] had done previously, *Zhang* et al. [4.104] used the BR model to convert the experimental plasma frequencies into values for γ_0. However, in contrast to the results of *St. Jean* et al. [4.102], they found γ_0 to be nearly stage-independent, reporting an average value for $\gamma_0 = 2.9$ eV.

For comparison with these values obtained for γ_0, Shubnikov-de Haas (SdH) experiments by *Zaleski* et al. [4.112, 113] on stage-1 and stage-2 graphite-$SbCl_5$ found orbit areas in good agreement with the charge transfer values obtained optically by *Hoffman* et al. [4.110] using $\gamma_0 = 2.9$ eV. Furthermore, the H model and the BR model can be brought into good agreement with each other (at low charge transfer), if the value $\gamma_0 = 2.93$ eV is used in the BR model. Also note that a BR-model analysis of the optical reflectance of KC_{24} is consistent with complete charge transfer of the K 4*s* electron to the carbon layers – if the value $\gamma_0 = 2.9$ eV is used in the data analysis [4.57].

Eklund et al. [4.103] examined the charge transfer dependence of the Raman-active E_{2g} graphitic intralayer phonons in the graphite-H_2SO_4 system. Figure 4.35 shows their results for the charge transfer dependence of the mode fre-

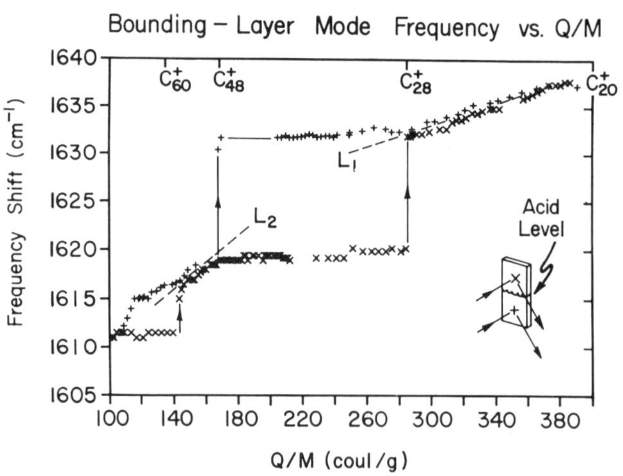

Fig. 4.35. Raman-active E_{2g} graphitic intralayer phonon frequency against Q/M for the stage-1 and stage-2 graphite-H_2SO_4 compounds ($Q/M \propto f_C$). The (×) and (+) symbols represent data taken above and below the acid level, respectively. The lines L1 and L2 determine the charge transfer dependence of the bounding layer mode frequency for the stage-1 and stage-2 compounds [4.105]

quency. The experiments were carried out in situ in the electrochemical cell with the HOPG-based samples half submerged in H_2SO_4, with the upper half in N_2 gas. The data in the figure indicated by (+) and (×) were taken, respectively, with the laser radiation incident on the acid/sample and the N_2 gas/sample interface. As mentioned earlier, the data coincide in the regions of constant stage index and variable charge transfer ($60 < p < 48$, stage-2, and $28 < p < 21$, stage-1). In these fixed-stage regions, the phonon frequency exhibits a linear dependence on $Q/M \propto f_C$. From the slopes of the lines L1 and L2 in the figure, *Eklund* et al. [4.103] obtained $d\omega_{2g}/df_C = 460 \pm 30\ \mathrm{cm}^{-1}$ (stage 1) and $d\omega_{2g}/df_C = 1050 \pm 120\ \mathrm{cm}^{-1}$ (stage 2). Thus the E_{2g} mode frequency dependence on charge transfer is given by

$$\text{Stage 1:} \quad \omega_{2g}(\mathrm{cm}^{-1}) = \omega_{2g0} + (460 \pm 30) f_C\,, \tag{4.19a}$$

$$\text{Stage 2:} \quad \omega_{2g}(\mathrm{cm}^{-1}) = \omega_{2g0} + (1050 \pm 120) f_C\,, \tag{4.19b}$$

where $\omega_0 = 1615$ and $1635\ \mathrm{cm}^{-1}$, for stages 2 and 1, respectively. Note that these values for ω_{2g0} are strictly applicable only for stage-1 and stage-2 graphite H_2SO_4. Other acceptor-type GICs might exhibit a stronger (or weaker) c-axis coupling between the C and I layers, which might influence ω_{2g0}. However, the values they obtained for the derivatives ($d\omega_{2g}/df_C$) in H_2SO_4 might be expected to be more generally applicable to other acceptor-type GICs. In this regard, *Ohana* et al. [4.114] have used the derivative values to interpret the effects of charge transfer in metal-fluoride GICs. We discuss this further below.

Finally, we mention briefly the results of in situ studies by *Rao* et al. [4.115] of the infrared-active (E_{1u}) $\sim 1600\ \mathrm{cm}^{-1}$ modes in graphite-H_2SO_4 during electrochemical intercalation ($\infty < p < 24.3$). Although they did not carry out a detailed line-shape analysis for these intralayer graphitic modes, they obtained a few interesting results. First, they noted a clear *upshift* in the E_{1u} mode frequency as the stage index decreased from stage $4 \to 3$, followed by a clear *downshift* in the E_{1u} mode frequency at the stage $3 \to 2$ transition. The stage $3 \to 2$ *downshift* contrasts with the *upshift* reported previously by *Gualberto* et al. [4.116] in graphite-$AlCl_3$, but is consistent with the results of *Eklund* et al. [4.117] in graphite-$SbCl_5$ and *Underhill* et al. [4.27] in graphite $FeCl_3$. *Rao* et al. [4.115] suggested that these results, taken collectively, indicate that both intralayer and interlayer force constants have a significant effect on the frequencies of the infrared- and Raman-active graphitic modes. This point of view is further strengthened by the observation [4.115] that the stage-2 E_{1u}-mode frequency remained constant during the stage-2 $p = 28 \to 21$ interval, whereas the E_{2g} mode frequency upshifted by $5\ \mathrm{cm}^{-1}$ [4.103, 105]. *Rao* et al. [4.115] therefore concluded that a theory for the charge transfer dependence of the $\sim 1600\ \mathrm{cm}^{-1}$ mode frequencies must also include c-axis force constants, i.e., the change in the graphitic intralayer mode frequency is not just a simple consequence of charge-transfer-induced expansion or contraction of the in-plane C–C bond length. This conclusion agrees with the theoretical results of *Chan* et al. [4.29].

(b) Metal-Chloride GICs

$SbCl_5$ and $FeCl_3$ GICs. Because of its stability in air [4.5], graphite-$SbCl_5$ is one of the most heavily studied acceptor-type GICs. The compounds can be prepared by the reaction of the Lewis acid $SbCl_5$ with graphite. By analogy with the disproportionation chemistry suggested for graphite-AsF_5 by *Bartlett* et al. [4.118], the intercalation chemistry of graphite-$SbCl_5$ has been described in terms of the disproportionation of $SbCl_5$ into $SbCl_3$ and $SbCl_6^-$, although various concentrations of reaction products or adducts (e.g., $SbCl_5$, $SbCl_6^-$, $SbCl_3$, $SbCl_4^-$, and $SbCl_5 . SbCl_6^-$) [4.119–122] have been proposed to reside in the intercalate layers, depending on the details of the sample preparation.

High-resolution, scanning transmission electron microscopy studies of stage-4 graphite-$SbCl_5$ by *Hwang* et al. [4.123] found clear evidence for segregated phases of differing Sb/Cl stoichiometry in the intercalate layers. Islands with average stoichiometry $SbCl_{3.2}$ accounted for 11% of the area and were found amidst a sea of $\sqrt{7} + \sqrt{7}\,(R^\circ 19.1)$ material with an average stoichiometry $SbCl_{7.2}$. The islands were observed to be 500–1000 Å in diameter and did not exhibit crystalline order. *Hwang* et al. [4.123] suggested that the various species of $SbCl_x (x = 3, 4, 5, 6)$ could be distributed in the I layer to obtain an equal charge per unit area in both the sea and the island regions.

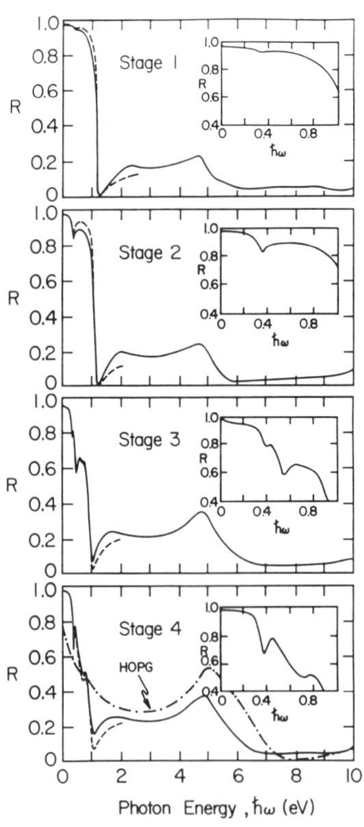

Fig. 4.36. Reflectance ($E \perp c$) for stages 1–4 graphite $SbCl_5$. The dashed lines represent Drude fits to the data. The inset to each figure magnifies the region associated with $v \rightarrow v$ interband absorption. The spectrum of HOPG is also shown in the bottom panel [4.110]

The optical properties of stages 1–4 graphite-$SbCl_5$ were first studied in detail by *Hoffman* et al. [4.110], who measured the optical reflectance over the energy range 0.08–10 eV. Figure 4.36 shows the spectra of stages 1–4 graphite-$SbCl_5$ from their study. The dashed lines represent fits to the respective spectra using a Drude–Lorentz model (4.15), with the Lorentz oscillators representing low energy ($v \to v$) absorption between nearly parallel valence (π) bands. The spectra in the figure agree well with those reported earlier by *Eklund* et al. [4.124] in the energy range 0.2–2.5 eV. Each spectrum exhibits features characteristic of acceptor-type GICs, including the low-energy ($E < 0.6$ eV) structure associated with $v \to v$ inter-π-band absorption (except stage 1), a sharp Drude edge located near ~ 1 eV, and the analog of the pristine graphite M-point peak at ~ 5 eV. The weak feature at 0.38 eV in the stage-1 spectrum was attributed by *Hoffman* et al. [4.110] to a small inclusion of stage-2 material in the optical skin depth.

A Kramers-Kronig analysis of the reflectivity using density-scaled graphite (Sect. 4.2.1) was used to obtain ε_1 and ε_2. The reflectance and $(\omega\varepsilon_2)^{-1}$ were both fit to the Drude-Lorentz model to determine the Drude parameters. Figure 4.37 shows the results obtained by *Hoffman* et al. [4.110] for the $v \to v$ interband contribution to the dielectric function $\varepsilon_{2,\text{bound}} = \varepsilon_{2,\text{inter}} = \varepsilon_2(\omega) - \varepsilon_{2,\text{Drude}}$. The solid lines represent the oscillator fit to the data, and the center frequency of

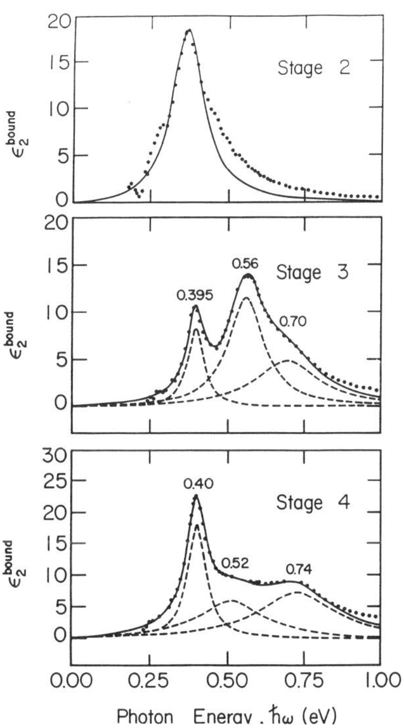

Fig. 4.37. $v \to v$ inter-π-band absorption for stages 2–4 graphite $SbCl_5$. The data (dots) were obtained from a Kramers-Kronig analysis of the spectra in Fig. 4.36. The solid lines represent Lorentz oscillator fits to the $v \to v$ structure [4.110]

each oscillator determines the splitting between various valence (π) bands. Without making contact with an energy-band model, such as was done by *Yang* and *Eklund* [4.24] in the KC_{12n} compounds, information regarding the net layer charge distribution in stage-3 and stage-4 graphite-$SbCl_5$ cannot be obtained. However, from the inter-π-band structure at ~ 0.6 eV in both these GIC systems, the electrostatic band parameter 2δ(SK model, Sect. 4.2.2c), which controls the distribution of net layer charge between the interior and bounding layers, appears to have nearly the same value in graphite-$SbCl_5$ and graphite-K.

The systematics of the free-carrier contribution to σ_{opt} in *high-stage* acceptor-type GICs were examined in graphite-$SbCl_5$ and graphite-$FeCl_3$ by *Hoffman* et al. [4.110]. They proposed an independent layer model in which all the charge from the intercalate layer was considered to be transferred to the bounding carbon layers. The interior layers remained graphitic (i.e., they exhibited a plasma frequency $\omega_p = 0.44$ eV equal to that of pristine graphite [4.16]). In their model, the effective plasma frequency ω_p is due to a sum of contributions from the bounding (C_b) and interior (C_i) carbon layers, leading to the expression [4.110]

$$\omega_p^2 = \frac{(n-2)\Omega_i^2 + 2\Omega_b^2}{D/d_C + n}, \tag{4.20}$$

where $I_c = D + nd_C$ is the distance between intercalate layers, d_C and D are the thickness of a carbon and an intercalate layer, and Ω_i and Ω_b are, respectively, the plasma frequency of the C_i and C_b layers. Figure 4.38 shows the fit of the independent layer model of *Hoffman* et al. [4.110] to the data for graphite-$FeCl_3$ [4.111] and graphite-$SbCl_5$. The data are plotted as ω_p^2 vs reciprocal stage index ($1/n$), and the solid lines in the figure were calculated according to (4.20) using stage-independent parameters: $\Omega_i = 0.44$ eV [4.16], $D(FeCl_3) = 6.09$ Å, and $D(SbCl_5) = 6.02$ Å. The values of $\Omega_b(FeCl_3)$ and $\Omega_b(SbCl_5)$ were chosen to fit

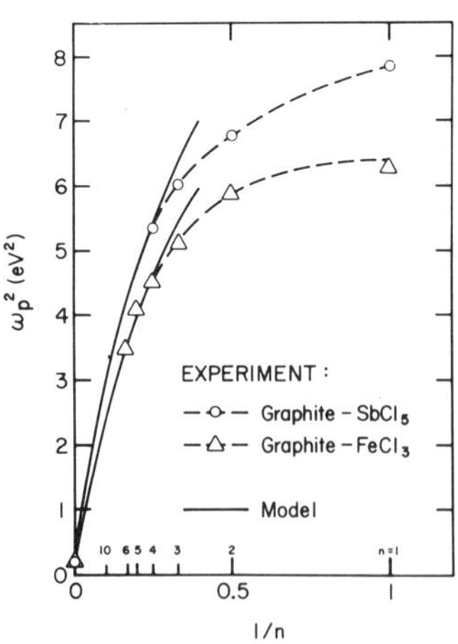

Fig. 4.38. Plot of ω_p^2 against the reciprocal stage index ($1/n$) for graphite $FeCl_3$ and graphite $SbCl_5$. The solid lines are calculated from the independent layer model described in the text. The dashed lines are guides for the eye [4.110]

the $n = 4$ data. As the figure shows, the model performs reasonably well for the intended high stage index region (i.e., stage $n \leqslant 4$), but underestimates the plasma frequency for lower stage index compounds.

Hoffman et al. [4.110] used the BR model to convert the experimental values of the plasma frequency into values for f_C. For the stage-1 $SbCl_5$, they determined $f_C = 0.033 \pm 0.005$ using $\gamma_0 = 3.0 \pm 0.1$ eV holes per C atom. From the reported range of stoichiometry $C_{12-14}SbCl_5$ [4.125, 126], this value of f_C translates into 0.43 ± 0.1 holes per intercalated $SbCl_5$. However, if complete disproportionation occurs ($3SbCl_5 \rightarrow 2SbCl_6^- + SbCl_3$ [4.118]), 2/3 of a π-band hole per intercalated $SbCl_5$ is anticipated. Thus *Hoffman* et al. [4.110] deduced that the disproportionation is 71% complete in the stage-1 compound. Results of Shubnikov-de Haas experiments on stage-1 graphite-$SbCl_5$ reported by *Zaleski* et al. [4.112], which yield the charge transfer directly from the orbit area, found 0.37 holes per intercalated $SbCl_5$, in good agreement with the optical results of *Hoffman* et al. [4.110]. For the stage-2 compound, using a BR-model analysis of their data ($\gamma_0 = 3.0 \pm 0.1$ eV and $\gamma_1 = 0.375$ eV), they obtained $f_C = 0.013 \pm 0.003$, or 0.34 ± 0.1 holes per intercalated $SbCl_5$ – i.e., $\sim 50\%$ disproportionation, where we have corrected their reported value for f_C using (4.9). These stage-2 values for f_C also agree well with the SdH results of *Zaleski* et al. [4.113].

(c) Fluorine and Metal-Fluoride GICs

Fluorine and metal-fluoride GICs have received a great deal of attention recently because the conductivities of some of these compounds rival that of copper. The relationship between the conductivity and the charge transfer in these compounds is therefore of special significance. In this section, we examine the results of several studies that have attempted to deduce optically the charge transfer in fluorine or metal fluoride GICs.

Graphite-AsF_5. Graphite-AsF_5 is the most conductive of the GIC synthetic metals with the electrical conductivity of the stage-2 material comparable to that of copper [4.127]. The compound also exhibits one of the largest electrical anisotropies reported for GICs [4.128], i.e., $\sigma_a/\sigma_c > 10^6$. *Fachini* et al. [4.129] interpreted their NMR data to indicate that the charge transfer in grahite-AsF_5 is a function of temperature, but this result conflicts with optical data presented below.

St. Jean et al. [4.130] preformed optical studies of f_C in the stage-1 and stage-2 $C_{8n}AsF_5$ compounds over the energy range 0.5–2.2 eV, and at temperatures $T = 300$ K, 77 K, and 10 K. The positions of the Drude minimum in the stage-1 and stage-2 compounds (at $T = 300$ K) agree well with those reported previously by *Hanlon* et al. [4.131]. Furthermore, *St. Jean* et al [4.130] found the position of the Drude edge to be *temperature-independent*, direct evidence that f_C does not exhibit significant temperature dependence, and in sharp contrast to the NMR study [4.129] of *Fachini* et al. *St. Jean* et al. [4.130] fit their stage-1 and stage-2 reflectance data at $T = 300$ K using the dielectric function model of

Blinowski et al. [4.13]. They obtained $E_F = 1.28$ eV, $\tau = 5 \times 10^{-14}$s for stage 1, and $E_F = 1.02$ eV, $\tau = 5 \times 10^{-14}$s for stage 2. Using the values of $\gamma_0 = 2.6$ eV (stage 1) and $\gamma_0 = 2.8$ eV (stage 2) obtained from their study of graphite-H_2SO_4 [4.102], they calculated that $f_C = 0.41$ and 0.53 for stages 1 and 2, respectively, agree with the observation σ_a (stage 1) < σ_a (stage 2). Furthermore, they found their f_C value for stage 2 to agree with the results of *Weinberger* et al. [4.132], who interpreted their Pauli spin susceptability data within the framework of the BR model. Results for f_C from SdH studies on stage-1 graphite-AsF_5, also interpreted within the BR model [4.13], are in good agreement: *Markiewicz* et al. [4.133] ($f_C = 0.41$) and *Fischer* [4.128] ($f_C = 0.42$).

Graphite-MF_6(M = Os, Mo). The nature of the interaction between the graphite host and the intercalant species has been found to vary with the oxidizing strength of the intercalant species. For instance, weakly oxidizing species display an electrostatic interaction with the C layers in the GIC, leaving itinerant holes in the carbon π bands. *Henning* [4.134] has pointed out that strongly oxidizing intercalant species have a tendancy to bond covalently with the carbon layers, creating a localized charge at the site of the covalent bond.

Two stage-1 metal-fluoride GICs that have similar stoichiometries, but stem from intercalants with much different oxidizing strengths, are stage-1 $C_x(Os)F_6$ and stage-1 $C_x(Mo)F_6$, where $8 < x < 11$. The charge transfer in these compounds was examined by *Ohana* et al. [4.114] and *Vaknin* et al. [4.135]. They carried out a series of experiments sensitive to the free holes in the C layers (optical reflectance) [4.114] as well as the localized charge in the intercalate layers (electron-spin resonance and magnetic susceptibility) [4.135].

The reflectance spectra of C_xMoF_6 and C_xOsF_6 were collected by *Ohana* et al. [4.114] at room temperature over the energy range 0.8–3.0 eV. The data were fit with the BR model with $\gamma_0 = 3.0 \pm 0.3$. For C_xOsF_6, they found $f_C = 0.035 \pm 0.003$ with $\tau = 1.8\,\mathrm{eV}^{-1}$ (7.44×10^{-15} s), while the C_xMoF_6, $f_C = 0.029 \pm 0.003$ and $\tau = 4\,\mathrm{eV}^{-1}$ (1.65×10^{-14} s). *Vaknin* et al. [4.135], using ESR and magnetic-susceptibility measurements to determine the charge on the molecular species in the intercalate layer, reported that in stage-1 C_xOsF_6 there is a charge of nearly one electron residing on each intercalate molecule (i.e., $f_M = 1$). On the other hand, in stage-1 C_xMoF_6 they found $f_M = 0.2$ electrons per molecule, much lower than found for the C_xOsF_6 compound. These values of $f_M = (1 - xf_C)$ can be compared with the optical values of *Ohana* et al. [4.114], who assumed that $x = 9$: $f_M = 0.261$ (C_9MoF_6) and 0.315 (C_9OsF_6). Thus, for C_xMoF_6, the optical, ESR, and magnetic-susceptibility results are in apparent agreement for f_M. However, there is noteworthy disagreement between the f_M values obtained by these experimental techniques for C_xOsF_6 (i.e., $\Delta f_M \sim 0.7$). This was attributed [4.114] to the formation of covalent bonds, so that the electrons transferred to the I layers did not leave itinerant holes behind in the C layers, but rather electrons localized in covalent bonds. While this explanation seems plausible, the size of the discrepancy is quite large, suggesting that C_xOsF_6 might exhibit significant covalent bonding between the C and I layers – so that the BR model might not be a good description under these circumstances.

Note that in both C_xMoF_6 and C_xOsF_6, the free-carrier lifetimes τ observed by *Ohana* et al. [4.114] are quite short for GICs (i.e., $\tau = 1.8 - 4\,eV^{-1}$), consistent with the formation of a significant density of covalent C–I bonds. These bonds, of course, introduce in-plane disorder in the C layers, which is known [4.2] to "turn on" the disorder-induced Raman scattering feature at $1360\,cm^{-1}$, as observed by *Ohana* et al. [4.114] for the C_xOsF_6 compound.

Ohana et al. [4.114] also made a determination for f_C via the frequency of the Raman-active E_{2g} phonon frequency. Using the result expressed in eq. 17a, obtained by *Eklund* et al. [4.105] for stage-1 graphite-H_2SO_4, they obtained a value for the difference Δf_C in charge transfer between C_xOsF_6 and C_xMoF_6 in agreement with their optical results. However, this approach assumes that ω_{2g0} is the same for both compounds, an assumption that should be questioned because the two materials appear to have quite different *c*-axis coupling between the C and I layers.

Fluorine and BF_3 GICs. The onset of covalent bond formation in carbon-fluoride compounds was studied optically by *Ohana* et al. [4.7] by examining the optical reflectance spectra of a series of compounds with increasing fluorine concentration. A series of air-stable compounds C_xF with $3.3 < x < 9.0$ were produced by the reaction of graphite (HOPG) with dilute HF + F_2 gas (1:1). *Ohana* et al. reported that stage-2 C_xF compounds with variable charge transfer can be obtained over a wide range of x, i.e., $4 < x < 8$. Thus this situation is similar to the stage-2 graphite-H_2SO_4 compounds, where a particularly simple electrochemical model allows the direct determination of the charge transfer from the charge passed in the reaction (Sect. 4.3.2a). The mixtures stage-1/stage-2 and stage-2/stage-3 were obtained, respectively, for $x < 4$ and for $x > 8$.

Ohana et al. [4.7] recorded optical reflectance spectra for the C_xF compounds over the energy range 0.8–3.2 eV, and carried out Drude analyses to determine the plasma frequency as a function of fluorine concentration ($1/x$). In the single-phase, stage-2 region, the data were analyzed using the BR model ($\gamma_0 = 3.0$ eV and $\gamma_1 = 0.375$ eV) to obtain values for E_F and f_C. The plasma frequency in the stage-2 region exhibited a maximum value for $1/x \sim 0.16$ (Fig. 4.39) near the middle of the concentration range. The Fermi energy E_F also exhibited unusual behavior in this range of concentration, first falling slightly and then rising with increasing ($1/x$).

This unusual behavior was fit [4.7] to a two-acceptor-state model. For $x > 6$ *Ohana* et al. proposed the existence of one acceptor level with energy $E_1 = -1.034$ eV. For $x < 6$ they proposed the existence of two acceptor levels, one at the previous energy, and a second at $E_2 = -0.84$ eV. The model appears to fit the $1/x$ dependence of f_C reasonably well (Fig. 4.40). Of course, the proper application of the BR model to these particular stage-2 compounds presumes that a significant number of covalent bonds have not been formed between the

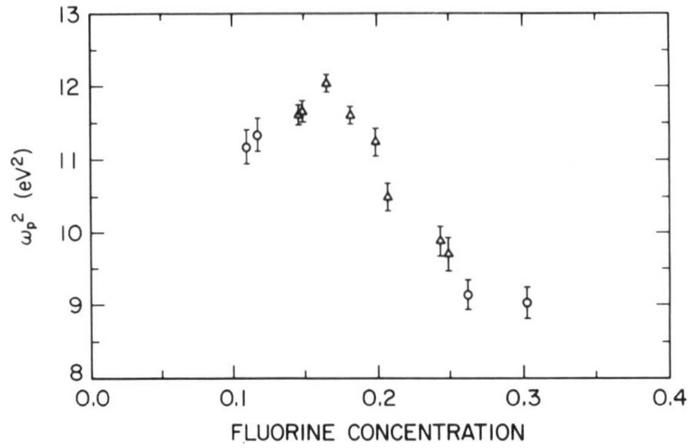

Fig. 4.39. Plot of ω_p^2 against fluorine concentration ($1/x$) in the stage-2 CF_x compounds. The unusual behavior was interpreted in terms of an x-dependent, two-level acceptor stage model [4.7]

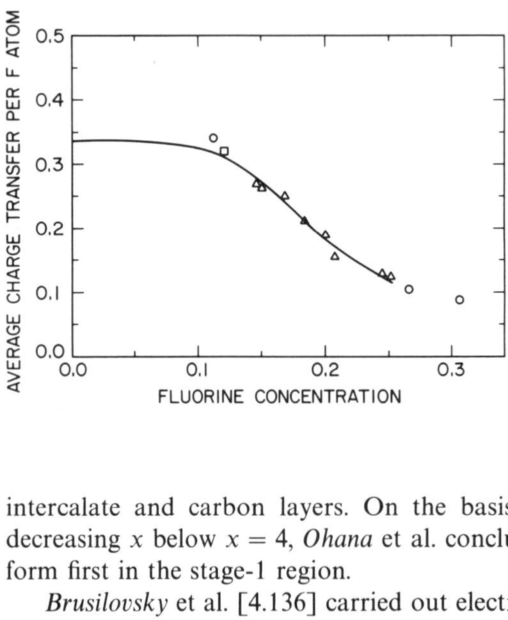

Fig. 4.40. Plot of the charge transfer f_F against ($1/x$) for stage-2 CF_x, where f_F is the charge transferred to the π band per F atom. The solid line was calculated according to a two-level acceptor state model [4.7]

intercalate and carbon layers. On the basis of a sudden decline in σ_a for decreasing x below $x = 4$, *Ohana* et al. conclude that covalent bonds begin to form first in the stage-1 region.

Brusilovsky et al. [4.136] carried out electrical conductivity (σ_a and σ_c) and optical reflectance studies of the charge transfer in $C_{7-9}BF_x$ (stage 1), $C_{12}BF_x$ (stage 2), and $C_{24}BF_x$ (stage 3), where the B/F ratio is unknown, but presumed to be close to 4. The materials were prepared by the reaction of graphite (HOPG) with a 1:1 mixture of BF_3 and F_2 gas. The reflectance spectra of stage-1 and stage-2 samples were measured over the energy range 0.8–3.2 eV, and the data for the stage-1 and stage-2 compounds were fit to the BR-model dielectric functions to determine values for f_C using $\gamma_0 = 2.9$ eV and $\gamma_1 = 0.38$ eV. From the fits of the BR-model dielectric functions to their data, they obtained the following results: $f_C = 0.032 \pm 0.002$ for $C_{7.3}BF_x$ (stage 1), $f_C = 0.028 \pm 0.220$ for

C_9BF_x (stage 1), and $f_C = 0.027 \pm 0.002$ for $C_{14}BF_x$ (stage 2). Assuming that the only intercalate species is BF_4^-, the charge per intercalate molecule was then determined to be $f_M = 0.23$, 0.25, and 0.39 for the respective compounds. However, later NMR studies by *Davydov* and *Selig* [4.137] reported that this assumption was incorrect. The reported value for the room-temperature in-plane resistivity of the stage-3 compound is $\varrho_a = 4\ \mu\Omega\,\text{cm}$; this suggests that no significant covalent bonding is taking place. However, for stage 1, $\varrho_a = 100\ \mu\Omega\,\text{cm}$, which is quite high, suggesting the formation of a considerable number of covalent bonds.

4.4 Summary and Conclusions

The results reviewed in this chapter indicate clearly that optical methods can be used to study charge transfer in acceptor- and donor-type GICs. We have reviewed both Raman scattering and optical reflectance studies of charge transfer.

In the case of Raman scattering, it has been demonstrated clearly in graphite H_2SO_4 that the frequency of the $\sim 1600\ \text{cm}^{-1}$ graphitic intralayer mode is sensitive to the net charge in the bounding C layers. This result is perhaps expected, since it is known that the in-plane C–C bond length undergoes contraction (expansion) with the removal (addition) of charge from (to) the graphitic carbon layers. However, the mode frequency is also sensitive to the nature of the *c*-axis coupling between the layers of the GIC, and this interlayer coupling is expected to vary from one compound to another.

For graphite-H_2SO_4, the $\sim 1600\ \text{cm}^{-1}$ bounding layer mode upshifts linearly with charge transfer for both stage-1 and stage-2 compounds. However, the results for the rate of change of the mode frequency with increasing charge transfer are stage-dependent – again suggesting that *c*-axis coupling is important. Similar conclusions about charge transfer and mode frequency can be made from the limited data on the infrared-active graphitic intralayer modes. These observations also accord with results from lattice dynamics calculations discussed here.

Of course, it can be argued that a phenomenological correlation between the Raman-active mode frequency $\omega(E_{2gb})$ and charge transfer f_C should nevertheless be attempted. We have considered this correlation – for seven stage-1 acceptor-type GICs: graphite-AsF_5, -$SbCl_5$,-$FeCl_3$,-OsF_6,-MoF_6,-H_2SO_4 ($p = 21, 28$). All the f_C values came from the in-plane plasma frequency using a BR-model analysis with the same value for γ_0. No correlation could be discerned. However, this may in part be due to the small size of the shift in mode frequency, coupled with the experimental uncertainty in the charge transfer $\Delta f_C \sim 10\%$. However, the lack of a correlation may simply indicate the importance of the interlayer coupling on the mode frequency. At the time of writing, it still appears that more theoretical and experimental work is needed before the graphitic intralayer mode frequencies can be used to obtain reliable values for the charge transfer.

On the other hand, it does seem that reasonable success has been achieved in using optical reflectance and the associated dielectric function to obtain values for the charge transfer. For stage 1, both the H and BR models have been shown to provide a means to obtain f_C from the plasma frequency and the midpoint E_T of the inter-π-band absorption threshold. For stage-2 compounds, the BR model has also been used extensively to obtain f_C from ω_p and E_T.

There still appears to be some uncertainty ($\sim$ 10%) about the appropriate value on the principal tight-binding parameter γ_0 to be used in the BR model, which should be addressed in future work. Most recent work in both donor- and acceptor-type GICs has used values for γ_0 in the range $2.9 < \gamma_0 < 3.0$. For γ_0 in this range, the H model (stage 1) will give answers for f_C similar to the BR model, as long as the charge transfer is not too high. As we have discussed (Sect. 4.3.2a), values for γ_0 can be obtained directly from the experimental plasma frequency in graphite-H_2SO_4, since the charge transfer is known from the electrochemical charge. However, two experimental groups have obtained somewhat different values for γ_0 in the stage-1 compounds – i.e., $\gamma_0 = 2.6$ and 2.9 eV, or a 10% discrepancy. For the case of stage-2 graphite-H_2SO_4, the disagreement in acceptable – i.e., $\gamma_0 = 2.8$ and 2.9 eV, or a 3% discrepancy. The universality of γ_0 should be investigated in several simple donor and acceptor GICs by performing *on the same sample* optical reflectance and at least one other experiment sensitive to charge transfer (i.e., Shubnikov–de Haas, de Haas–van Alphen).

A few optical studies (e.g., of KC_{12n}) have demonstrated that optical structure in the interband contribution to the dielectric function can be used to determine important higher-order band-structure parameters. Further work of this nature might be carried out to determine indirectly, for example, the distribution of net layer charge along the c-axis. Since some of the most conductive GICs are stage $n \geqslant 3$ acceptor compounds, it is both technologically and fundamentally important to develop quantitative band models for these GICs.

Finally, it has been demonstrated that a two-carrier model can be used successfully in understanding the free-carrier optical response in donor GICs. The inter-π-band absorption thresholds seems to provide, via the linking of E_T with f_C, a self-consistent check of the π-electron contribution to the plasma frequency. Further donor GIC optical studies should also strive to collect data from the a face to learn more about the anisotropy of the Fermi surface.

Acknowledgements. It is a pleasure to acknowledge the support of our GIC research at the University of Kentucky by the United States Department of Energy under grant #DE-FGO5-84 ER45151. One of us (P.C.E.) would like to express his sincere appreciation for the many contributions from former students and postdoctoral researchers which have led to his better appreciation of the fundamental issues in the optical properties of GICs. Explicit thanks are therefore due to J.M. Zhang, M.H. Yang, C.H. Olk, D.S. Smith, D.M. Hoffman, and V.R.K. Murthy.

References

4.1 For general recent reviews of the properties of GICs published prior to this two volume series *see* M.S. Dresselhaus, G. Dresselhaus: Adv. Phys. **30**, 139 (1981); S.A. Solin: Adv. Chem. Phys. **49**, 445 (1982)

4.2 S.A. Solin: In *Graphite Intercalation Compounds I*, Topics Curr. Phys., ed. by H. Zabel, S.A. Solin, (Springer, Berlin, Heidelberg 1989); P.C. Eklund: In *Intercalation in Layered Materials*, ed. by M.S. Dresselhaus (Plenum, New York 1986) p. 323; M.S. Dresselhaus, G. Dresselhaus: in *Light Scattering in Solids III*, ed. by M. Cardona, G. Guntherodt (Springer, Berlin, Heidelberg 1982) Chap. 2

4.3 C. Rigaux: In *Intercalation in Layered Materials*, ed. by M.S. Dresselhaus (Plenum, New York 1987) p. 235

4.4 P.C. Eklund, M.H. Yang, G.L. Doll: In *Intercalation in Layered Materials*, ed. by M.S. Dresselhaus (Plenum, New York 1987) p. 257

4.5 V.R.K. Murthy, D.S. Smith, P.C. Eklund: Mater. Sci. Eng. **45**, 77 (1980)

4.6 M.H. Yang, P.A. Charron, R.E. Heinz, P.C. Eklund: Phys. Rev. B **37**, 1711 (1987)

4.7 I. Ohana, I. Palchan, Y. Yacoby, D. Davidov: Phys. Rev. B **38**, 12627 (1988)

4.8 A. Herold: In *Physics and Chemistry of Materials with Layered Structures*, Vol. 6., ed. by F. Levy (Reidel, Dordrecht 1979) p. 323; P.C. Eklund: In *Intercalation in Layered Materials*, ed. by M.S. Dresselhaus (Plenum, New York 1986) p. 163

4.9 A.W. Moore: In *Physics and Chemistry of Carbon*, Vol. 11, ed. by P.L. Walker, P.S. Thrower (Dekker, New York 1973) p. 69

4.10 M. Zanini, J.E. Fisher: Mater. Sci. Eng. **31**, 169 (1977)

4.11 I. Ohana, Y. Yacoby: Phys. Rev. Lett. **57**, 2572 (1986)

4.12 F. Wooten: *Optical Properties of Solids* (Academic, New York 1972)

4.13 J. Blinowski, N.H. Hau, C. Rigaux, J.P. Viren, R. LeToullec, G. Furdin, A. Herold, J. Melin: J. de Phys. **41**, 47 (1980)

4.14 F. Gervais: In *Infrared and Millimeter Waves*, ed. by K.J. Button (Academic, New York 1983) Vol. 8, p. 306

4.15 see p. 244 of Ref. [4.12]

4.16 E.A. Taft, H.R. Phillip: Phys. Rev. **138**, A 197 (1965)

4.17 J.E. Fischer, J.M. Bloch, C.C. Shieh, M.E. Preil, K. Jelley: Phys. Rev. B **31**, 4773 (1985)

4.18 N.A.W. Holzwarth: Phys. Rev B **21**, 3665 (1980)

4.19 L. Johnson, G. Dresselhaus: Phys. Rev. B **7**, 2275 (1973)

4.20 J.M. Zhang, D.M. Hoffman, P.C. Eklund: Phys. Rev. B **34**, 4316 (1986)

4.21 J.M. Zhang, P.C. Eklund, Y.B. Fan, S.A. Solin: Phys. Rev. B **37**, 6226 (1988)

4.22 R. Saito, H. Kamimura: Synth. Met. **12**, 295 (1985); R. Saito: Ph.D. thesis, University of Tokyo (1984)

4.23 T. Ohno, H. Kamimura: J. Phys. Soc. Jpn. **52**, 223 (1983)

4.24 M.H. Yang, P.C. Eklund: Phys. Rev. B **38**, 3505 (1988)

4.25 H. Zabel: In *Graphite Intercalation Intercalaton Compounds I*, ed. by H. Zabel, S.A. Solin, Topics Curr. Phys. (Springer, Berlin, Heidleberg 1989)

4.26 L.J. Brillson, E. Burstein, A.A. Maradudin, T. Stark: In *Physics of Semimetals and Narrow gap Semiconductors*, ed. by D.L. Carter, R.T. Bates (Pergamon, Oxford 1971) p. 187

4.27 C. Underhill, S.Y. Leung, G. Dresselhaus, M.S. Dresselhaus: Solid State Commun: **29**, 769 (1979)

4.28 L. Pietronero, S. Strassler: In *Physics of Intercalation Compounds*, ed. by L. Pietronero, E. Tossati (1981) p. 23.

4.29 C.T. Chan, K.M. Ho, W.A. Kamitakahara: Phys. Rev. B **36**, 3499 (1987)

4.30 J.E. Fischer: In *Graphite Intercalation Compounds- Proc. of Symp. I*, ed. by P.C. Eklund, M.S. Dresselhaus, G. Dresselhaus (Mater. Res. Soc., Pittsburgh 1984) p. 33

4.31 H. Kamimura: In *Graphite Intercalation Compounds- Proc. of Symp. I*, ed. by P.C. Eklund, M.S. Dresselhaus, G. Dresselhaus (Mater Res. Soc., Pittsburgh 1984) p. 36; H. Kamimura: Annales de Phys. **11**, 39 (1986)

4.32 P. Alstrom: Synth. Met. **15**, 311 (1986)

4.33 R. Sonnenschien, M. Hanfland, K. Syassen: Phys. Rev. B **38**, 3152 (1988)

4.34 K. Nogami, H. Ueno, K. Yoshino: J. Appl. Phys. **64**, 6460 (1988)

4.35 T. Inoshito, K. Nakao, H. Kamimura: J. Phys. Soc. Jpn. **43**, 1237 (1977)

4.36 T. Ohno, K. Nakao, H. Kamimura: J. Phys. Soc. Jpn. **47**, 1125 (1979)

4.37 D.D. DiVicenzo, S. Rabii: Phys. Rev B **25**, 4110 (1982)

4.38 T.C. Tatar: Ph.D. thesis, University of Pennsylvania (1985); R.C. Tatar, S. Rabii: in *Graphite Intercalation Compounds- Proc. of Symp. I*, ed. by P.C. Eklund, M.S. Dresselhaus, G. Dresselhaus (Mater. Res. Soc., Pittsburgh 1984) p. 71

4.39 M. Posternak, A. Baldereschi, A.J. Freeman, E. Wimmer, M. Weinert: Phys. Rev. Lett. **50**, 761 (1983)

4.40 M.H. Yang: Ph.D. Thesis, University of Kentucky (1988)

4.41 D. Guerard, G.M.T. Foley, M. Zanini, J.E. Fischer: Nuovo Cimento **38B**, 410 (1977)

4.42 M. Zanini, J.E. Fischer: Mater Sci. Eng. **31**, 169 (1977)

4.43 G.L. Doll, M.H. Yang, P.C. Eklund: Phys. Rev. B **18**, 9790 (1987)

4.44 A. Koma, K. Miki, H. Suematsu, T. Ohno, H. Kamimura: Phys. Rev. B **34** 2434 (1986)

4.45 K. Kume, K. Nomura, Y. Hioryama: Synth. Met. **12**, 307 (1985)

4.46 U. Mizutani, T. Kondow, M. Suganuma: Phys. Rev. B **17**, 3165 (1978)

4.47 M. Suganuma, T. Kondow, U. Mizutani: Phys. Rev. B **23**, 706 (1981)

4.48 M.G. Alexander, D.P. Goshorn, D. Guerard, P. Lagrange, M. EI Makrini, D.G. Onn: Solid State Commun. **38**, 103 (1981)

4.49 D.E. Nixon, G.S. Parry: J. Phys. D 201 (1968)

4.50 M. Colin, A. Herold: C.R. Acad. C 269 (1969)

4.51 K. Watanabe, T. Kondow, M. Soma, T. Onishi, K. Tamura: Nature **233**, 160 (1971)

4.52 T. Enoki, H. Inokuchi, M. Sano: Chem. Phys. Lett. **86**, 285 (1982)

4.53 J. Conard, H. Estrade-Szwarckopf, P. Laugins, M. El Makrini, P. Lagrange, D. Guerard: Physica B **105**, 290 (1981)

4.54 S. Kaneiwa, M. Kobayashi, I. Tsuijikawa: J. Phys. Soc. Jpn. **51**, 2375 (1982)

4.55 T. Enoki, M. Sano, H. Inokuchi: Synth. Met. **12**, 207 (1985)

4.56 K. Watanabe, T. Kondow, M. Soma, T. Onishi, K. Tamaru: Proc. R. Soc. London A **333**, 51 (1973)

4.57 G.L. Doll, P.C. Eklund, G. Senatore: In *Intercalation in Layered Materials*, ed. by M.S. Dresselhaus (Plenum, New York 1986) p. 309

4.58 S. Lee, H. Miyazaki, S.D. Mahanti, S.A. Solin: Phys. Rev. Lett. **62**, 3066 (1989)

4.59 T. Kondow, K. Ando, Y. Tomono: In *Physics of Intercalation Compounds*, ed. by L. Pietronero, E. Tosatti (1981) p. 315

4.60 T. Kondow, U. Mizutani: Synth. Met. **6**, 141 (1983)

4.61 T. Kondow, M. Sagawa, T. Takeyama, Y. Tomono, K. Ando, U. Mizutani Synth. Met. **12**, 213 (1985)

4.62 T. Terai, Y. Nonaka, M. Ohira, Y. Takahashi: Synth. Met. **12**, 219 (1985)

4.63 G.L. Doll, P.C. Eklund: J. Mater. Res. **2**, 638 (1987)

4.64 S. Miyajima, T. Chiba, T. Enoki, H. Inokuchi, M. Sano: Phys. Rev. B **37**, 3246 (1988)

4.65 H. Yamamoto, K. Seki, T. Enoki, H. Inokuchi: Solid State Commun. **69**, 425 (1989)

4.66 D. Guerard, C. Takoudjou, F. Rousseaux: Synth. Met. **7**, 43 (1983)

4.67 K.W.-K Shung: Phys. Rev. B **34**, 1264 (1986)

4.68 T. Enoki, N.C. Yeh, S.T. Chen, M.S. Dresselhaus: Phys. Rev. B **33**, 1292 (1986)

4.69 G.L. Doll, P.C. Eklund: Phys. Rev. B **36**, 9191 (1987)

4.70 M. Sano, H. Inokuchi: Chem. Lett., • 4405 (1979)

4.71 F. Beguin, R. Setton, L. Fachini, A.P. Legrend, G. Merle, C. Mai: Synth. Met. **2**, 161 (1980) and references therein

4.72 B.R. York, S.A. Solin: Phys. Rev. B **31** 8206 (1985)

4.73 J.M. Zhang, P.C. Eklund, Y.B. Fan S.A. Solin: Phys. Rev. B. **38**, 10878 (1988)
4.74 H.H. Huang, Y.B. Fan, S.A. Solin, J.M. Zhang, P.C. Eklund, J. Heremans, G.G. Tibbetts: Solid state Commun. **64**, 443 (1987)
4.75 For a general review, see J.C. Thompson: *Electrons in Liquid Ammonia* (Clarendon, Oxford 1976)
4.76 J.M. Zhang, P.C. Eklund, Y.B. Fan, S.A. Solin: Phys. Rev. B **37**, 6226 (1988)
4.77 T. Tsang, R.M. Fronco, H.A. Resing, X.W. Qian, S.A. Solin: Solid State Commun. **62**, 117 (1987)
4.78 X.W. Qian, D.R. Stump, S.A. Solin: Phys. Rev. B **33**, 5756 (1986)
4.79 J.M. Zhang, P.C. Eklund: Synth. Metals **26**, 357 (1988)
4.80 For a general review, see *Electrons in Fluids*, ed. by J. Jortner, N.R. Kestner (Springer, New York 1972)
4.81 M.T. Lok, F.J. Tehan, J.L. Dye: J. Phys. Chem. **76**, 2975 (1972)
4.82 J.L. Dye: In *Electrons in Fluids*, ed. by J. Jortner, N.R. Kestner (Springer, New York 1972) p. 77
4.83 F. Beguin, R. Setton, A. Hamwi, P. Touzain: Mater Sci. Eng. **40**, 167 (1979)
4.84 M.F. Quinton, A.P. Legrand, L. Facchini, F. Beguin: Synth. Met. **14**, 179 (1988)
4.85 M.H. Yang, P.C. Eklund: J. Mater. Res. **1**, 827 (1987)
4.86 M.H. Yang, W.A. Kamitakahara, P.C. Eklund: In *Graphite Intercalation Compounds- Proc. of Symp. I*, ed. by P.C. Eklund, M.S. Dresselhaus, G. Dresse lhaus (Mater. Res. Soc., Pittsburgh 1984) p. 125
4.87 M.H. Yang, P.A. Charron, R.E. Heinz, P.C. Eklund: Phys. Rev. B **37**, 1711 (1988)
4.88 M.E. Preil, J.E. Fischer, P. Lagrange: Solid State Commun. **44**, 357 (1982)
4.89 M.E. Preil, J.E. Fischer: Synth. Metals **8**, 149 (1983)
4.90 R.E. Heinz, P.C. Eklund: Mat. Res. Soc. Symp. Proc. **20** (1983) p. 81
4.91 P. Lagrange, A. Bendriss-Rerhrhaye, J.F. Mareche, E. McRae: Synth. Met. **12**, 201 (1985)
4.92 A. Chaiken: Ph.D. Thesis, Massachusetts Institute of Technology (1986); A. Chaiken, N.C. Yeh, G. Dresselhaus, P. Tedrow: In *Graphite Intercalation Compounds- Proc. of Symp. K*, ed. by M.S. Dresselhaus, G. Dresselhaus, S.A. Solin (Mater. Res. Soc., Pittsburgh 1986) p. 158
4.93 I. Stang, K. Lüders, Z. Geiser, H.J. Güntherodt: Syn. Met. **23** (1988) p. 371
4.94 M.G. Alexander, D. Goshorn, D. Guerard, P. Lagrange, M. El Makrini, D. Onn: Solid State Commun. **38**, 103 (1981)
4.95 Y. Iye, S. Tanuma: Phys. Rev. B **25**, 4583 (1982)
4.96 L.E. DeLong, P.C. Eklund: Synth. Met. **5**, 291 (1983)
4.97 W.E. Spicer: Phys. Rev. Lett. **11**, 243 (1963); F. Wotten, J.P. Hernandez, W.E. Spicer: Appl. Phys. **44**, 1112 (1973)
4.98 S.B. DiCenzo, P.A. Rosenthal, H.J. Kim, J.E. Fischer: Phys. Rev. B **34**, 3620 (1986)
4.99 P.C. Eklund: In *Intercalation in Layered Materials*, ed. by M.S. Dresselhaus (Plenum, New York 1986) p. 337
4.100 S. Aronson, S. Lemont, J. Weiner: Inorg. Chem. **10**, 1296 (1971)
4.101 J.O. Bessenhard, E. Wudy, H. Moewald, J.J. Nickl, W. Biberacher, W. Foag: Synth. Met. **7**, 185 (1983)
4.102 M. St. Jean, M. Menant, N. Hy Hau, C. Rigaux, A. Metrot: Synth. Met. **8**, 189 (1983)
4.103 P.C. Eklund, C.H. Olk, F.J. Holler, J.G. Spolar, E.T. Arakawa: J. Mater . Res. **1**, 361 (1986)
4.104 J.M. Zhang, D.M. Hoffman, P.C. Eklund: Phys. Rev. B **34**, 4316 (1986)
4.105 P.C. Eklund, E.T. Arakawa, J.L. Zarestky, W.A. Kamitakahara, G.D. Mahan: Synth. Met. **12**, 97 (1985)
4.106 W. Kamitakahara, J.L. Zarestky, P.C. Eklund: Synth. Met. **12**, 301 (1985)
4.107 P.C. Eklund, G.D. Mahan, J.G. Spolar, E.T. Arakawa, J.M. Zhang, D.M.Hoffman: Solid. State Commun. **57**, 567 (1986)
4.108 P.C. Eklund, G.D. Mahan, J.G. Spolar, J.M. Zhang, E.T. Arakawa, D.M. Hoffman: Phys. Rev. B **37**, 691 (1988)
4.109 P.A. Giguere, R. Savoie: Can. J. Chem. **38**, 2467 (1960)
4.110 D.M. Hoffman, R.E. Heinz, G.L. Doll, P.C. Eklund: Phys. Rev. B **32**, 1278 (1985)

4.111 D.S. Smith, P.C. Eklund: In *Intercalated Graphite*, Vol. 20 of Symposia Proceedings, ed. by M.S. Dresselhaus, G. Dresselhaus, J.E. Fischer and M.J. Moran (Mater. Res. Soc., Boston 1983) p. 99
4.112 H. Zaleski, P.K. Ummat, W.R. Datars: J. Phys. C **17**, 3167 (1986)
4.113 H. Zaleski, P.K. Ummat, W.R. Datars: Phys. Rev. B **35**, 2958 (1987)
4.114 I. Ohana, D. Vaknin, H. Selig, Y. Yacoby, D. Davidov: Phys. Rev. B **35**, 4522 (1987)
4.115 A.M. Rao, J.M. Zhang, G.W. Lehman, P.C. Eklund: In *Graphite Intercalation Compounds-Proc. of Symp. K*, ed. by M.S. Dresselhaus, G. Dresselhaus, S.A. Solin (Mater. Res. Soc., Pittsburgh 1986) p. 158
4.116 G.M. Gualberto, C. Underhill, S. Y. Leung, G. Dresselhaus: Phys. Rev B **21**, 862 (1980)
4.117 P.C. Eklund, D.S. Smith, V.R.K. Murthy, S.Y. Leung: Synth. Met. 2,99 (1980)
4.118 N. Bartlett, R.N. Biagioni, B.W. McQuillan, A.S. Robertson, A.C. Thompson: J. Chem. Soc., Chem. Commun. **200** (1978)
4.119 L.B. Ebert, D.R. Mills, J.C. Scanlon: Mater. Res. Bull. **18**, 1505, (1983)
4.120 P. Boolchand, W.J. Bresser, D. McDaniel, K. Sisson, V. Yeh, P.C. Eklund: Solid State Commun. **40**, 1049 (1981)
4.121 D. McDaniel, P. Boolchand, W.J. Bresser, P.C. Eklund: In *Intercalated Graphite*, ed. by M.S. Dresselhaus, G. Dresselhaus, J.E. Fisher, M.J. Moran (Elsevier, New York 1983) p. 377
4.122 M.J. Moran, G.R. Miller, R.A. DeMarco, H.A. Resing: J. Phys. Chem. **88**, 1473 (1984)
4.123 D.M. Hwang, X.W. Qian, S.A. Solin: Phys. Rev. Lett. **53**, 1473 (1984)
4.124 P.C. Eklund, D.S. Smith, V.R.K. Murthy: Synth. Met **3**, 111 (1981)
4.125 J. Melin, A. Herold: C. R. Acad. Sci. **269**, 877 (1969)
4.126 M.H. Boca, M.L. Saylors, D.S. Smith, P.C. Eklund: Synth. Met. **6**, 39 (1983)
4.127 G.M.T. Foley, C. Zeller, E.R. Falardeau, F. Vogel: Solid State Commun. **24**, 371 (1977)
4.128 J.E. Fischer: In *Physics and chemistry of Materials with Layered Structures*, ed. by F. Levy, Vol. 6 (Reidel, Dordrecht 1981) p. 481
4.129 L. Fachini, J. Bouat, H. Sfihi, A.P. Legrand, G. Furdin, J. Melin, R. Vangelisti: Synth. Met. **5**, 11 (1982)
4.130 M. St. Jean, N.H. Hau, C. Rigaux, G. Furdin: Solid State Commun. **46**, 55 (1983)
4.131 L.R. Hanlon, E.R. Falardeau, J.E. Fischer: Solid State Commun. 24, 377 (1977)
4.132 B.R. Weinberger, J. Kaufer, A.J. Heeger, J.E. Fischer, M. Moran, N.A. Holzwarth: Phys. Rev. Lett. **41**, 1417 (1978)
4.133 R.M. Markiewicz, H.R. Hart, J.L.V. Interrante, J.S. Kasper: Synth. Met. **2**, 331 (1980)
4.134 G.R. Henning: Prog. Inorg. Chem. **1**, 125 (1959)
4.135 D. Vaknin, D. Davidov, H. Selig, Y. Yeshurun: J. Chem. Phys. **83**, 3859 (1985); D. Vaknin, D. Davidov, H. Selig, V. Zevin, I. Felner, Y. Yeshurun: Phys. Rev. B **31**, 3212 (1985)
4.136 D. Brusilovsky, H. Selig, D. Vaknin, I Ohana, D. Davidov: Synth. Met. **23**, 377 (1988)
4.137 D. Davydov, H. Selig: In *Intercalation in Layered Materials*, ed. by M.S. Dresselhaus (Plenum, New York 1986) p. 433

5. Superconductivity of Graphite Intercalation Compounds

By Sei-ichi Tanuma

With 25 Figures

The superconductivity of graphite intercalation compounds was first reported by *Hanney* et al. [5.1], who observed the superconduction of the first-stage compounds of K, Rb, and Cs GICs, after an initial unsuccessful search by *Henning* and *Meyer* [5.2]. The highest transition temperature T_c among the reported specimens was 0.55 K for the excess-intercalated GIC with potassium. Because neither graphite nor alkali metal is a superconducting material, the report by *Hanney* et al. attracted much interest, although subsequent investigations such as the paper by *Poitrenaud* [5.3], who searched in the temperature range down to 300 mK, failed to reproduce their work.

The superconductivity of C_8K was reconfirmed by *Kobayashi* and *Tsujikawa* [5.4] for powder samples and by *Koike* et al. [5.5] for samples using PG, but the measured T_c was as low as 80 mK and 139 mK, respectively. The highest T_c value for the first-stage C–K compound that was reported later was 198 mK, being much different from the value of the first finding by *Hanney* et al. The graphite intercalation compounds of amalgams, thallides, and bismuthides of alkali metals were revealed to have much higher T_c values – 0.7 K to 4 K – as described later.

5.1 Superconductivity of C_8M (M = K, Rb, Cs)

Koike et al. [5.5] extended the low-temperature search region down to 50 mK by the use of a ^{3}He-^{4}He dilution refrigerator. Figure 5.1 shows the main part of their cryostat for measuring the magnetic susceptibility with a dilution refrigerator. The sample was wrapped in a copper coil made of fine parallel insulated copper wires. The coil was connected thermally to the mixing chamber. Apieson N grease was added between the sample and the coil to ensure good thermal contact. The temperature was measured by a CMN (cerium magnesium nitrate) magnetic thermometer and a carbon resistor. The ac magnetic susceptibility χ was measured with a small ac field (350 Hz, 0.2 Oe). The magnetic field in any direction with respect to the crystal c axis was applied by means of two superconducting magnets, a vertical coil, and a horizontal Helmholtz-type coil, by applying an appropriate ratio of the currents to these coils.

The superconducting transition was observed on a sample of C_8K derived from HOPG supplied by Union Carbide. Figure 5.2 shows the susceptibility

Springer Series in Materials Science, vol. 18
H. Zabel · S.A. Solin (eds.): Graphite Intercalation Compounds II

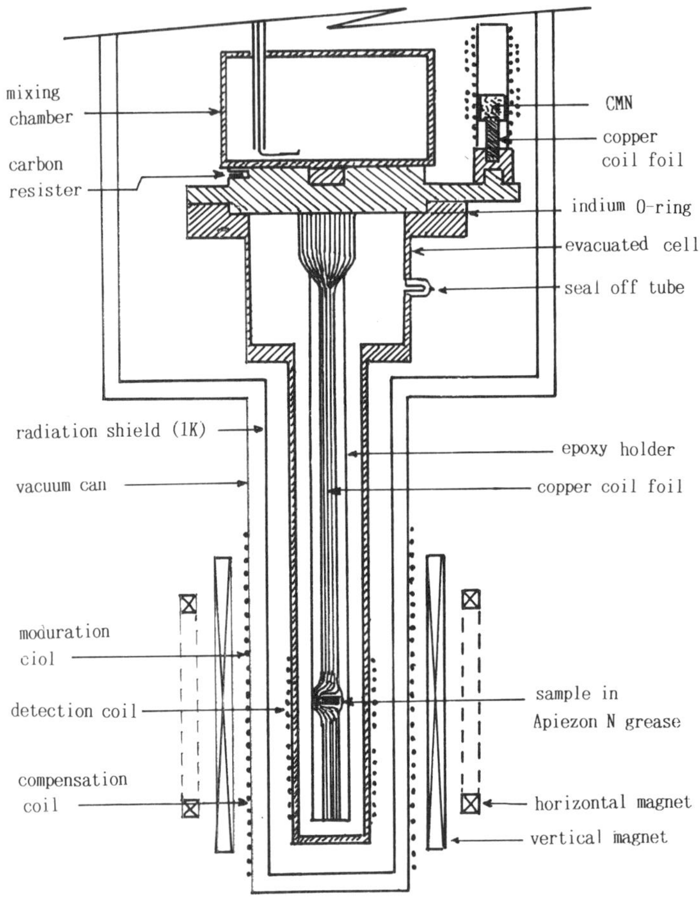

Fig. 5.1. Mixing chamber and sample holder in the ^3He–^4He dilution refrigerator [5.5]. The GIC sample is wrapped in a thin insulated copper foil of which the top is soldered to the bottom of the mixing chamber

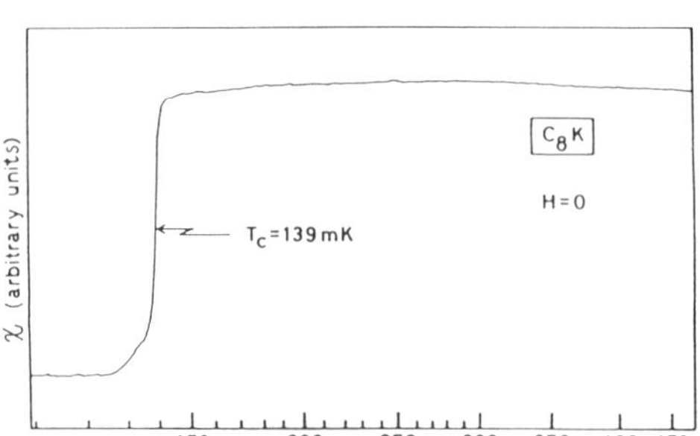

Fig. 5.2. Temperature dependence of the ac magnetic susceptibility showing the superconducting transition of C_8K [5.5]

versus temperature curve with the T_c value of 139 mK. The stepwise change of χ is on the order of $|-1/4\pi|$, i.e., perfect diamagnetism. *Koike* et al. reported the character of the specimen as an anisotropic type-II superconductor. They explained the overall behavior using the anisotropic effective-mass model of the three-dimensional superconductor.

In successive work, *Koike* et al. [5.6, 7] revealed the remarkable fact that this material is probably the first superconducting layered material that provides either type-I or type-II characteristics, depending on the magnetic field direction with respect to the crystal axis. Figure 5.3 shows the susceptibility under a dc magnetic field parallel (Fig. 5.3a) and perpendicular (Fig. 5.3b) to the c axis. There is a distinct difference between the traces. That is, χ measured with the magnetic field parallel to the c axis shows the type-I characteristic accompanying a supercooling effect, while χ with the field perpendicular to the c axis

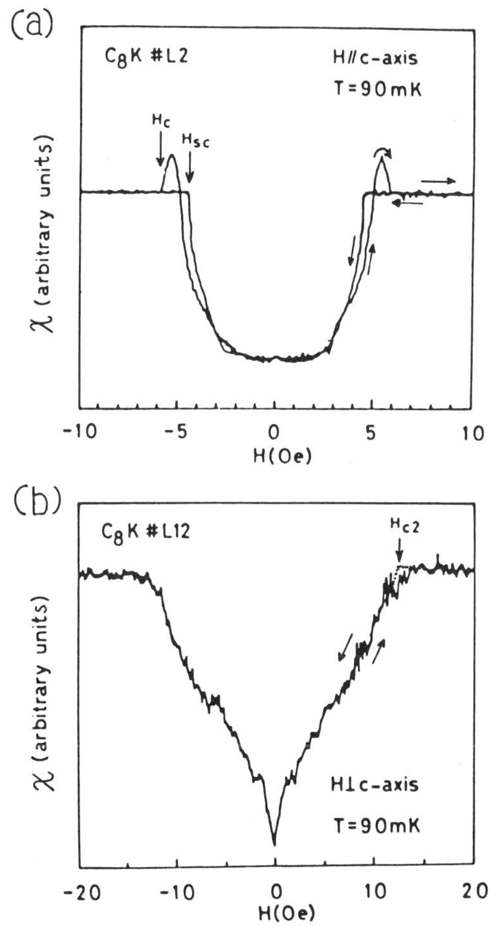

Fig. 5.3. Observation of the superconducting transition of C_8K by the ac susceptibility under a dc magnetic field H: (**a**) $H \parallel c$ (**b**) $H \perp c$ [5.6]

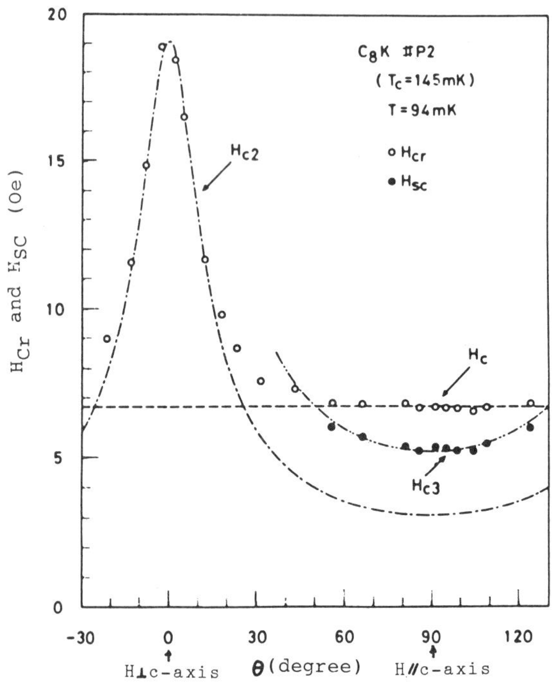

Fig. 5.4. Magnetic field angle dependence of the critical fields. H_c and H_{sc} correspond to Fig. 5.3. H_{sc} (the supercooling field) is regarded as H_3 (the surface nucleation field); $H_{c2\parallel}$ is determined by $H_{c3}/1.695$ [5.6]

shows the type-II characteristic with no hysteresis. The thermodynamic critical field H_c and the supercooling field H_{sc} were determined as indicated in Fig. 5.3a. The peak appearing just under H_c in the increasing field seems to be the differential paramagnetic effect. The upper critical field H_{c2} in the type-II case was determined as the field strength where the extrapolated χ value reached the normal state value as indicated by the dotted line in Fig. 5.3b. The value of H_{c2} for $H \perp c$ axis was a few times larger than H_c for $H \parallel c$ axis, showing a remarkable anisotropy. Figure 5.4 shows the angular dependences of H_{cr} (H_c or H_{c2}) and H_{sc}. When θ is defined as the angle between the applied dc magnetic field and the layer plane, C_8K behaves as a type-II superconductor for $0° \leqq |\theta| < 25°$, and a type-I superconductor for $25 \leqq |\theta| < 90°$.

According to the BCS theory of superconductivity, T_c is described as follows, where θ_D is the Debye temperature, $N(0)$ is the electronic density of states, and V is the pairing potential arising from the electron-phonon interaction:

$$T_c \simeq 0.85\,\theta_D \mathrm{e}^{-1/N(0)V}. \tag{5.1}$$

Putting in the values of $\theta_D = 234.8$ K determined by *Mizutani* et al. [5.8] and $T_c = 130$ mK, $N(0)V$ is obtained as 0.14. This value is as small as the value 0.12 for a weak coupling superconductor Ir of which T_c is 112.5 mK. It is much smaller than the values of Hg and Pb – 0.35 and 0.39, respectively – which are known as strongly coupled superconductors. Thus C_8K should be regarded as a weakly coupled superconductor. Table 5.1 lists the characteristic values.

Table 5.1. Fundamental parameters characterizing the superconductivity of C_8K.

T_c	[mK]	128–198
θ_D	[K]	234.8
$N(0)V$		0.14
$H_c(0)$	[Oe]	10–13
$\kappa_{\parallel}(T_c)$		0.22
$\kappa_{\perp}(T_c)$		0.78
$\lambda_{\parallel}(0)$	[Å]	8100
$\lambda_{\perp}(0)$	[Å]	2300
$\xi_{\parallel}(0)$	[Å]	2900
$\xi_{\perp}(0)$	[Å]	10000

κ: Landau-Ginzburg parameter, λ: penetration depth,
ξ: Ginzburg-Landau coherence length
Subscripts $\parallel$ and $\perp$ mean the directions parallel and perpendicular to the c axis, respectively

Kats [5.9] predicted such a superconductivity in which the type changes from I to II with the change of field direction in relation to the major crystal axis. The compound C_8K appears to be the first example of this type of material. To explain the anisotropy of the critical field, the effective-mass models by *Kats* [5.9] and *Morris* et al. [5.10] were tested. The model gives the upper critical-field anisotropy, as follows:

$$\frac{H_{c2}(\theta)}{H_{c2\parallel}} = (\sin^2\theta + \varepsilon^2\cos^2\theta)^{-1/2} . \tag{5.2}$$

Here the constant ε^2 is the ratio of effective masses $m_{\perp}$ and $m_{\parallel}$, namely,

$$\varepsilon^2 = \frac{m_{\perp}}{m_{\parallel}} = \frac{H^2_{c2\perp}}{H^2_{c2\parallel}} = \frac{\xi^2_{\parallel}}{\xi^2_{\perp}} . \tag{5.3}$$

In the above formula, H_{c2} is not determined directly because the type-I behavior is realized for the field parallel to the c axis. Rather H_{c2} is obtained from the surface nucleation field using the equation $H_3(T) = 1.695\,H_{c2}(T)$, which was obtained by *Saint-James* and *de Gennes* [5.11]. H_{c3} could be replaced by the experimental value of H_{sc} in Figs. 5.3, 4. Thus the parameter values used for these calculations are 0.026, 39, 6.2, 0.32, and 2.0 for ε^2, $m_{\perp}/M_{\perp}$, $\xi_{\perp}/\xi_{\parallel}$, $\kappa_{\parallel}$, and $\kappa_{\perp}$, respectively.

Figure 5.4 shows that type-II superconductivity ($H_{c2} < H_2 < H_{c3}$) is obtained in the range $0° \leqslant |\theta| < 25°$, while the sample is type I ($H_{c2} < H_c$) in the range $25° \leqslant |\theta| < 90°$. On the basis of hysteresis, the angular range of type I is divided in two. Except for the nonhysteresis range $25° < |\theta| \leqslant 50°$ the experimental points agree with the calculated lines, and the effective-mass model seems to provide a valid interpretation of the superconductivity of C_8K.

Kobayashi and *Tsujikawa* [5.4] first reconfirmed the superconductivity of C_8K with graphite powder using the refrigeration method of CrK-alum adiabatic demagnetization. They measured the superconducting transition by measuring the magnetic susceptibility. Next they observed [5.12] the superconductivity using Grafoil made by Union Carbide. The transition temperature was 120 mK for the compound whose composition was $C_{8.8}K$ as determined by weight measurement. The susceptibility χ gradually decreased toward the perfect diamagnetic value from about 130 to 100 mK. (T_c was determined as the temperature for the half value of perfect diamagnetism.) This was discussed in terms of fluctuation in the T_c region.

Subsequently, *Kobayashi* and *Tsujikawa* [5.13] investigated the variation of the transition temperature due to the composition x of the compound in C_xK using HOPG (grade ZYH from Union Carbide). They determined the composition x by weight uptake. Figure 5.5 shows for $C_{8.1}K$ an example of the same kind of ac measurement as that for the HOPG-based GIC mentioned above. The real and imaginary components of the susceptibility, χ' and χ'' respectively, are plotted versus temperature near the superconducting transition. The extrapolated value of T_c at zero ac field was 152 mK. They found that T_c was fixed at 152 mK for x in the range 8–9.4 and slightly decreased to 147 mK for $x = 9.4$. But no superconducting transition was observed down to 60 mK for x of 16.7 and 21.6. Figure 5.6 shows this behavior.

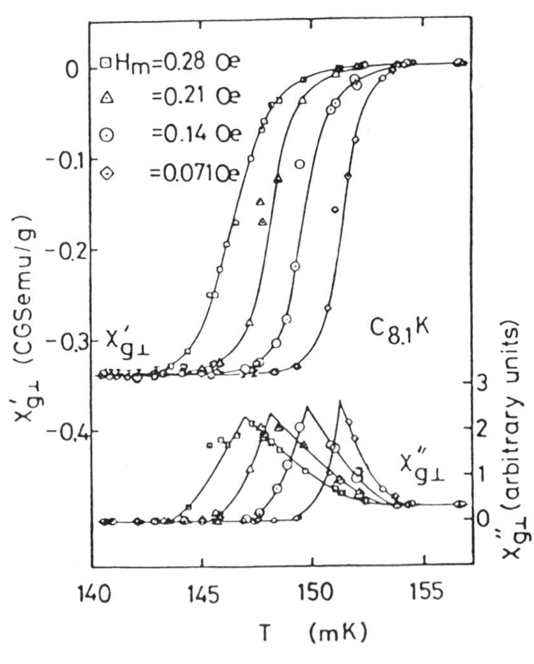

Fig. 5.5. Plot of ac susceptibility of HOPG-based $C_{8.1}K$ for the field orientation perpendicular to the c plane [5.13]

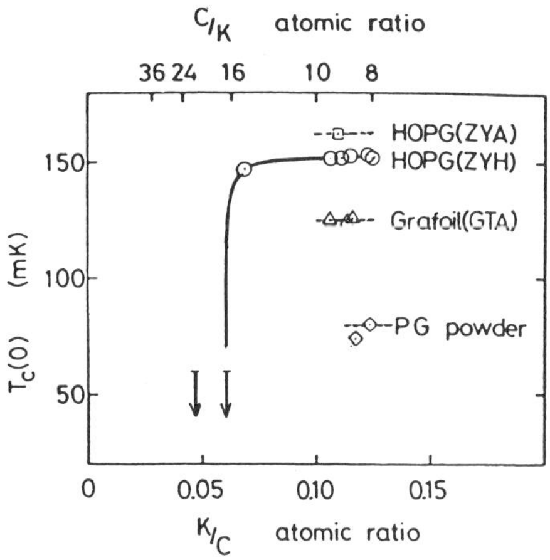

Fig. 5.6. Potassium concentration dependence of superconducting transition temperature T_c on C_xK GICs [5.13]. The HOPG-based compounds of $x = 16.7$, 21.6 are not superconducting above 60 mK

For the x range of the constant transition temperature, *Kobayashi* and *Tsujikawa* [5.13] found the interesting result that the anisotropy of the superconductivity decreases with a decrease of potassium content. The strongest indication of this was the appearance of supercooling and a differential paramagnetic effect in the ac susceptibility versus the magnetic field. This was very similar to the behavior shown in Fig. 5.3 for the field orientations of H being not only parallel but also perpendicular to the c axis in the case of $C_{9.4}K$. For $C_{8.1}K$, such behavior for type I was not observed for the field perpendicular to the c axis, as was the case with the sample prepared by *Koike* et al. Thus $C_{9.4}K$ could be a type-I superconductor over the entire field direction.

Another related property is the temperature dependences of H_c and H_{c2}. Figure 5.7 plots the fields $H_c^{\parallel}$, $H_{c2}^{\parallel}$, $H_c^{\perp}$, and $H_{c2}^{\perp}$ against $T_c/T(0)$ for $C_{8.0}K$, $C_{8.1}K$, and $C_{9.4}K$. The $H_{c2}^{\perp}$ and $H_{c2}^{\perp}$ curves for $C_{8.0}K$ have positive curvatures, those for $C_{8.1}K$ are almost linear, and those for $C_{9.4}K$ have negative curvatures. The positive curvature is known in superconducting layered materials such as 2H-$NbSe_2$ and alkali and alkali-earth intercalates of MoS_2, which are very anisotropic. The negative curvature is common for three-dimensional materials. Later, in relation to the band-structure model, I discuss the anisotropy of C_8K and the related materials.

Except for *Hanney* et al., who reported T_c for Rb GICs in the range 20–135 mK without mentioning detailed superconducting properties, no one else has detected superconductivity in this material. Rather recently, *Kobayashi* et al. [5.14] found a definite superconducting transition in a C_8Rb sample by an ac magnetic-susceptibility measurement. They used a ^3He–^4He dilution refrigerator for which the lowest temperature was 10 mK. The earth's field was shielded to a strength below 1 m Oe. The T_c value as the temperature corresponding to half of perfect diamagnetism was measured as a function of the ac measuring field amplitude, and extrapolated to a zero measuring field to determine an accurate value of T_c. The value of T_c was 25.2 mK. The magnetization curve at

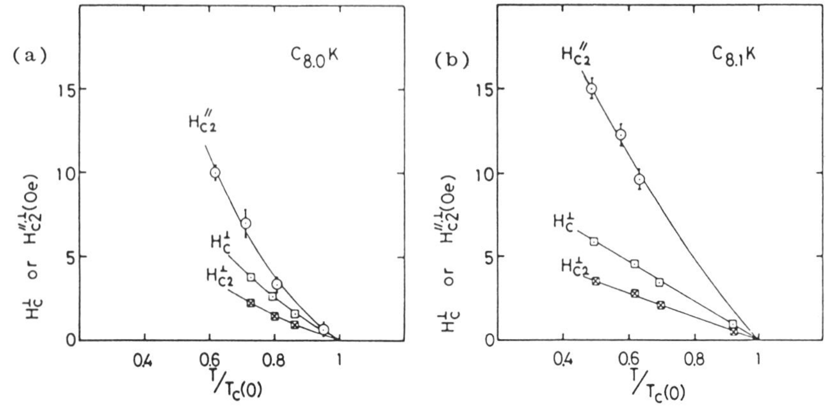

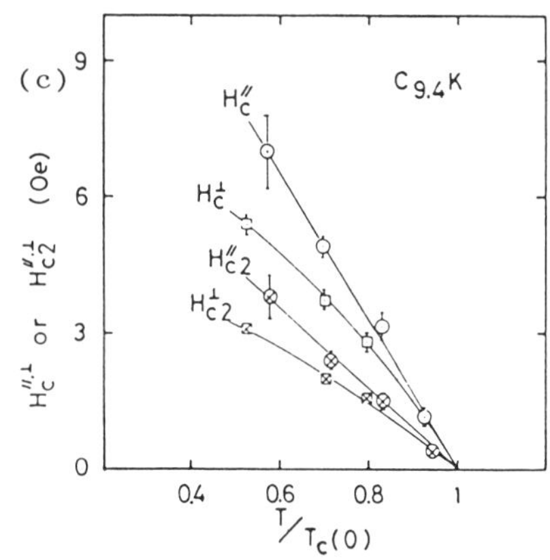

Fig. 5.7. Temperature dependence of the critical magnetic fields on C_xK GICs [5.13]: (**a**) $C_{8.0}K$, positive curvature; (**b**) $C_{8.1}K$, an almost linear relationship; (**c**) negative curvature

12.5 mK for the field parallel to the *c* axis was similar to that in Fig. 5.3a showing a hysteresis and a differential paramagnetic effect, so the material was recognized as being type I in this configuration.

No lower-stage GIC of sodium had been known, except in cases of poorly oriented pyrolytic graphite, in which sodium atoms could be adsorbed in mismatched vacant positions.

Belash et al. [5.15] synthesized C_2Na and C_3Na using high pressure. They placed the weighed amounts of HOPG and sodium for the stoichiometric compositions of C_2Na or C_3Na into a self-sealed copper tube in an inert atmosphere. Such compositions were chosen because the formulas C_2M and C_3M were theoretically predicted by *Guerard* et al. [5.16] as the limiting saturation of alkali-metal atoms, M. First the tubes were heat treated at 450°–700°C with a quasi-hydrostatic pressure of 45 kbar for 10 min. Then the pressure was reduced to atmospheric pressure and the temperature was kept at the room temperature for 10 min, after which the tube was immersed in liquid nitrogen. The samples were taken out in the liquid from the copper tube and transferred into Teflon tubes. The type was treated for 24 hours at 300 K with different pressures of 16 and 37 kbar. The samples were again cooled and kept in liquid nitrogen under pressure until the measurement.

The magnetic susceptibility χ of the high-pressure phase C_2Na ($p = 37$ kbar) was measured down to 0.3 K in a ^3He cryostat. The χ versus T curve showed a rather sharp jump attributed to the occurrence of superconductivity; T_c determined from the midpoint of the susceptibility jump was 5 K. The values of $H^{||}_{c2}$ were 490 and 1150 Oe respectively. The sample of C_3Na (37 kbar) revealed a superconducting transition in the temperature range 3.8–2.3 K. These measurements were done immediately after the synthesis. After keeping the sample of C_3Na in liquid nitrogen for six days, the superconducting transition was shifted by 1.2 K to a lower temperature; the same amount of lowering of T_c occurred with the annealing of the sample at 100 K for only 25 min. After 10 min annealing of the sample at room temperature, no superconducting transition was found down to 0.3 K, although some indication of Na intercalation remained in the X-ray diffraction measurement.

Belash et al. [5.17, 18] synthesized and examined by similar methods, different kinds of alkali-metal GICs with high-pressure phases, such as C_2Li, C_3Li, C_3K, C_6K, and C_8K. Most samples showed superconductivity by the ac susceptibility measurement. The abrupt changes of χ with temperature variation, which were recognized as the superconducting transitions, are illustrated in Fig. 5.8a for the samples C_6K, C_4K, C_3K, and C_2Li. Figure 5.8b illustrates the dependence of T_c on the alkali-metal content. The T_c values are 1.5 K, 3 K, and 1.9 K for C_6K, C_3K, and C_2Li, respectively. The sample of C_3Li did not change its susceptibility value down to 0.35 K. In Figure 5.8a, three curves corresponding to the potassium content of C_4K show a stepwise change, which may suggest that C_4K samples have nonuniform composition, but are a mixture of compounds of other stoichiometries.

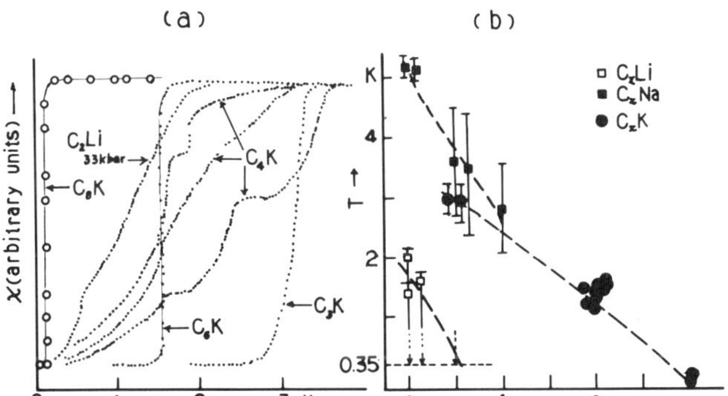

Fig. 5.8. (**a**) Dependence of magnetic susceptibility of C_8K, C_6K, C_4K, and C_3K on temperature. (**b**) Dependence of the superconducting transition temperature on the alkali-metal content of the graphene layer [5.17, 18]

Table 5.2. Superconducting parameters of GICs with potassium

GIC	T_c [K]	$-H_{c2}/dT$ [Oe/K]	$dH_{c2}^{\perp}/dT$ [Oe/K]	$\xi^{\parallel}(0)$ [Å]	$\xi^{\perp}(0)$ [Å]	ε^{-1}
C_3K	3	2045	2273	209	232	1.11
C_6K	1.5	219	437	500	1000	2
C_8K	0.13	–	–	2420[a]	6150[a]	2.5[a]
C_8K (13 kbar)	1.5	175	950	205	1125	5.43
C_8K (21 kbar)	1.5	120	580	280	1350	4.83
C_8K (30 kbar)	1.4	150	855	218	1253	5.75
C_8K (37 kbar)	1.13	125	1222	155	1528	9.8

[a] at $T = 0.067$ K

Table 5.2 gives the values of the superconducting parameters of various GICs with potassium as deduced and tabulated by *Belash* et al. [5.18].

5.2 Superconductivity of Binary Intercalants

5.2.1 Superconductivity of Potassium Hydride GICs, $C_{4n}KH$

The first-stage potassium GIC is known to chemisorb hydrogen, resulting in a second-stage hydride compound with the chemical formula of $C_8KH_{2/3}$ [5.19]. *Kaneiwa* et al. [5.20] investigated the effect of hydrogenation on superconductivity. The problem is interesting by analogy with the addition of hydrogen to, e.g., nonsuperconducting palladium metal, which gives rise to superconductivity with a T_c of 4.6 K, while hydrogenation of V_3Ga, which has a T_c of 15.3 K, decreases the value to 8.1 K. In a preceding report, *Kaneiwa* et al. did not find any large change of susceptibility down to 52 mK in $C_8HK_{2/3}$, which was made from grade GTA graphite film (Grafoil) from Union Carbide.

Kaneiwa et al. prepared a partially hydrogenated $C_8KH_{0.19}$ sample from HOPG-derived C_8K (grade ZYH from Union Carbide) by chemisorption of hydrogen at room temperature under a constant pressure. It took a hundred days to reach the above-mentioned concentration of hydrogen, due to the low degree of crystal imperfections. This sample showed superconductivity in an ac susceptibility measurement. Figure 5.9 shows the real and imaginary parts of susceptibility χ' and χ'' as functions of temperature. This should be compared with the same kind of measurement on HOPG-derived $C_{8.1}K$, which is shown in Fig. 5.5 using data from the same research group (*Kobayashi* and *Tsujikawa* [5.13]). The T_c was determined as 195 mK in the manner described in the preceding section. This value is higher than the T_c value of C_8K.

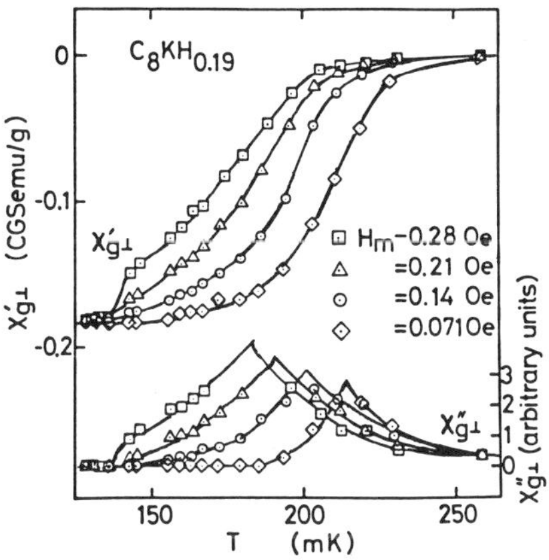

Fig. 5.9. Temperature dependences of ac magnetic susceptibility $\chi_{g\perp}$ and $\chi_{g\parallel}$ [5.20]

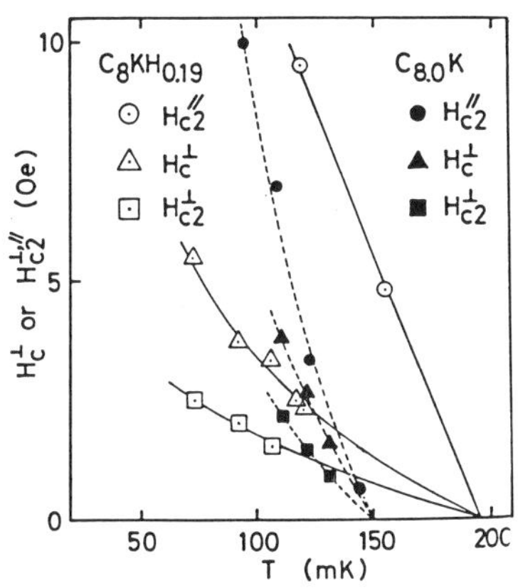

Fig. 5.10. Temperature dependences of the magnetic critical fields $H_{c2}^{\parallel}$, $H_c^{\perp}$, and $H_{c2}^{\perp}$ on $C_8KH_{0.19}$ and $C_{8.0}K$ under the ac magnetic field of 0.14 Oe [5.20]

Figures 5.5, 9 show distinctly that the temperature transition range is several mK for $C_{8.1}K$, while it is several tens of mK for $C_8KH_{0.19}$. This may be due partly to the enhancement of the superconducting fluctuation by hydrogenation, but mainly it is due to the inhomogeneity of the hydrogen concentration in the hydrogenated sample.

The superconducting behavior of the sample was proved to be similar to C_8K: the type-I character appeared in the field configuration parallel to the c axis and the type-II character appeared in the perpendicular field [5.20].

Figure 5.10 shows the temperature dependences of the critical fields for $C_8HK_{0.19}$ and $C_{8.0}K$. If the anisotropic Ginzburg-Landau (GL) theory is applicable, an effective-mass ratio $m_{\parallel}/m_{\perp} = 150$ is obtained at T_c, where $m_{\parallel}$ and $m_{\perp}$ are the effective masses of the Cooper pairs for the magnetic field parallel and perpendicular to the c axis, respectively. This value is one order of magnitude larger than that of C_8K.

Suzuki and *Tsujikawa* [5.21] prepared samples of potassium hydride GICs by the method of direct interaction of potassium hydride and HOPG – a process developed by *Guerard* et al. (1983) [5.22] – and observed the superconductivity. These compounds contain more hydrogen than the above-described chemisorbed compounds. Furthermore, they have a peculiar intercalant structure: the hydrogen layer is sandwiched between two potassium layers in a manner similar to KHg GICs, which will be mentioned later.

The first- to third-stage compounds were obtained by reacting HOPG and KH powder, which were sealed in Pyrex glass tubes, for 7 to 20 days at the appropriate temperatures. The first-stage C_4KH_x (violet) was obtained for the temperature range 450°–290°C, the second-stage C_8KH_x (blue) was obtained for 280°–220°C, and the third-stage $C_{12}KH_x$ (dark blue) was obtained for temperatures lower than 200°C. The (001) X-ray diffraction spectra showed that C_4KH_x had a *c* axis repeat distance of 8.43–8.53 Å mixed with 9.13 Å of a minority phase. The majority and minority phases possibly had in-plane potassium structures of $(2 \times 2)R0°$ and $(\sqrt{3} \times \sqrt{3})R30°$, respectively.

Superconductivity was examined for the stage-1 and stage-2 compounds by ac susceptibility measurements down to 70 mK. The stage-1 compound showed no indication of superconductivity. As Fig. 5.11 shows, the stage-2 compound C_8KH_x indicated a large diamagnetic signal at a temperature around 360 mK, although the value of diamagnetism was about 60% of the perfect diamagnetic value for the measuring ac field of 0.013 Oe [5.21].

It is supposed that the hydrogenation suppresses the superconductivity of the first-stage potassium compound but enhances that of the second stage. This may result from the extraction of electrons from the Fermi surface of potassium graphite to hydrogen atoms in the case of stage-1 and from the increased packing of potassium atoms on the graphene plane compared with $C_{24}K$ in stage 2 [5.21]

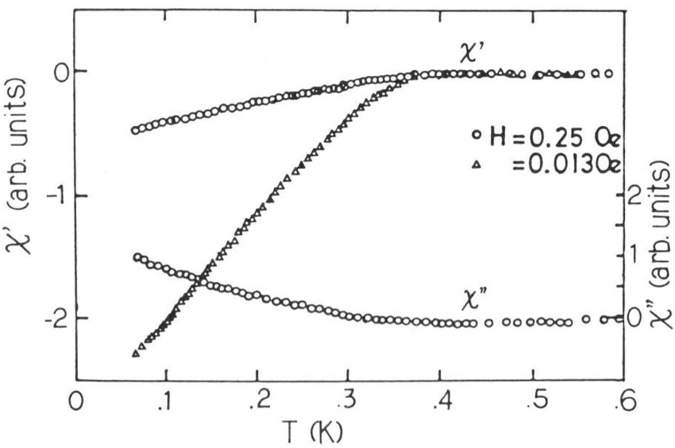

Fig. 5.11. Temperature dependence of ac susceptibility for C_8KH_x; *H* denotes the measuring magnetic field [5.21]

5.2.2 Superconductivity of GICs of Alkali-Metal Amalgams, $C_{4n}MHg$

The intercalation of alkali-metal amalgam into graphite was first performed by *Lagrange* et al. [5.23] and *El Makrini* et al. [5.24]. They obtained C_8KHg and C_8RbHg as second-stage compounds by heating graphite with MHg_2 (M is K or Rb) in a vacuum-sealed Pyrex glass tube in a furnace at 300°C for a week. Beforehand, MHg_2 was prepared by heating alkali metal and mercury to amalgamate in a vacuum sealed glass tube at about 400°C for one day. Thus the resultant C_8KHg was found to be a superconductor at 1.93 K, according to the specific-heat measurements reported by *Alexander* et al. [5.25], whose results are shown in Fig. 5.12. *Koike* and *Tanuma* [5.26] prepared C_8KHg by the same method using HOPG and confirmed the superconductivity of this material by an ac susceptibility measurement. Two samples showed transition temperatures of 1.90 K and 1.70 K and showed a very anisotropic critical magnetic field H_{c2}; the anisotropy was well explained by the three-dimensional effective-mass model resulting in a mass ratio $m_{\|}/m_{\perp} = 90.9$. *Pendrys* et al. [5.27] also observed the superconductivity of C_8KHg and further discovered C_8RbHg to be superconductive at 1.44 K. *Alexander* et al. [5.28] also found superconductivity in C_8RbHg at about the same T_c.

The most densely intercalated alkali-metal amalgam GICs are first-stage C_4KHg and C_4RbHg. *Tanuma* et al. [5.29] and *Iye* and *Tanuma* [5.30, 31] discovered the superconductivity of C_4KHg and C_4RbHg with transition temperatures of 0.73 K and 0.99 K, respectively. They made a detailed study of the first- and second-stage compounds with emphasis on the stage dependence of the anisotropy of the critical magnetic field in order to distinguish the dimensionality of the superconducting features.

The alkali-metal amalgam GIC has a stoichiometry of $C_{4n}MHg$ ($n = 1, 2$, M: K, Rb). The intercalates consist of two layers of alkali-metal and a mercury layer sitting between the alkali-metal layers, as depicted in Figs. 5.13a, b. The in-plane configuration of the alkali-metal atoms is presumably (2×2) R0°, i.e., the same as C_8M. Synthesis of the second-stage compound was accomplished by heat treating C_8M with mercury vapor at about 100°C. The color changed from the

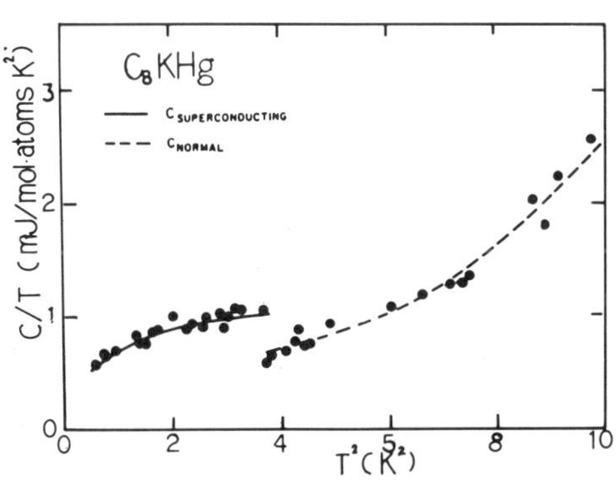

Fig. 5.12. Temperature dependence of the low-temperature specific heat C_p of C_8KHg; C_p/T is plotted versus T^2. The jump corresponds to T_c of 1.93 K ($T^2 = 3.72$ K^2) [5.28]

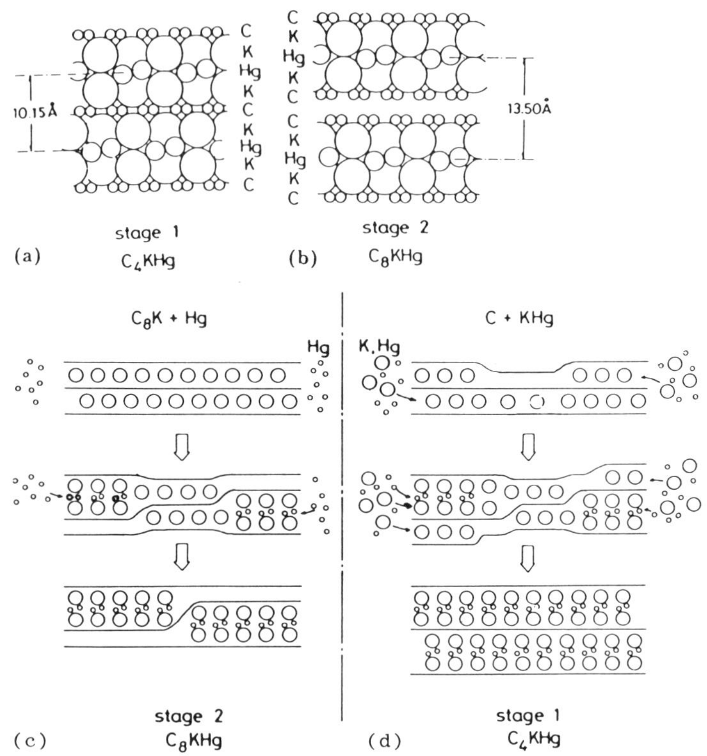

Fig. 5.13. Layer stacking order (**a**, **b**) and reaction process (**c**, **d**) of first-stage C_4KHg and second-stage C_8KHg [5.31]

brass yellow of C_8M to blue by mercury intercalation. Figure 5.13c illustrates the reaction. The first-stage compound was made by reaction of graphite with the amalgam MHg (one-to-one ratio) at about 220°C. The binary compound C_8M seemed to form first, and then mercury was intercalated under the presence of both alkali-metal vapor and mercury vapor, as illustrated in Fig. 5.13d. The color of this compound was copper pink.

The superconductivity of these GICs was investigated with an ac susceptibility measurement. For all the specimens and any field orientation, the recorder traces of the superconducting/normal transition showed no indication of the supercooling that appeared for C_8K in Fig. 5.3. That is, alkali-metal amalgam GICs are type-II superconductors for all field orientations. The data of the angular dependence of the critical field ($H_{c2}(\theta)$) proved to fit to the three-dimensional anisotropic effective-mass model, i.e., (5.2), as exemplified in Fig. 5.14.

The temperature dependence of the coherence length is given by

$$\xi(T) = \xi(0)(1 - T/T_c)^{-1/2} \tag{5.4}$$

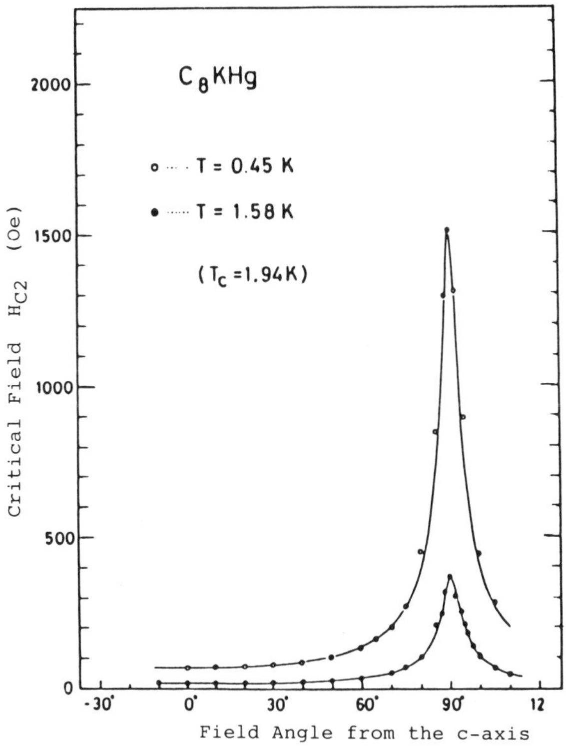

Fig. 5.14. Angular dependence of the upper critical field H_{c2} of C_8KHg [5.31]

in the GL region. The following expressions are obtained by combining (5.2) and (5.4) for the temperature dependence of the parallel and perpendicular critical fields:

$$H_{c2}^{\parallel}(T) = \frac{\phi_0}{2\pi\,\xi_\perp^2(0)}\left(1 - \frac{T}{T_c}\right) = -\,T_c\left(\frac{dH_{c2}^{\parallel}}{dT}\right)_{T_c}\left(1 - \frac{T}{T_c}\right), \tag{5.5}$$

$$H_{c2}^{\perp}(T) = \frac{\phi_0}{2\pi\,\xi_\perp(0)\,\xi_\parallel(0)}\left(1 - \frac{T}{T_c}\right)$$

$$= -\,T_c\left(\frac{dH_{c2}^{\parallel}}{dT}\right)_{T_c}\left(1 - \frac{T}{T_c}\right). \tag{5.6}$$

Figure 5.15 shows the temperature dependence of $H_{c2}^{\parallel}$ and $H_{c2}^{\perp}$. The linear relationship predicted by (5.5, 6) is obeyed, except in the lower temperature range. At low temperatures, an upward deviation is observed. This behavior, the so-called positive curvature, is also seen in C_8K, as shown in Fig. 5.7, and is a fairly common characteristic of layered superconductors.

Iye and *Tanuma* [5.31] investigated the stage dependence of the anisotropy in the coherence length of $C_{4n}M$Hg, where n is the stage number. The coherence

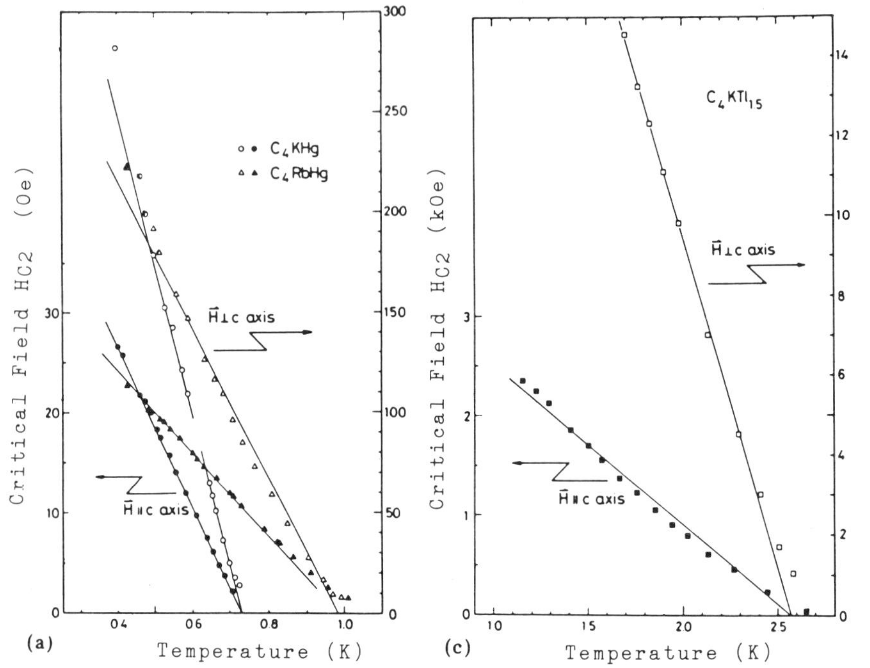

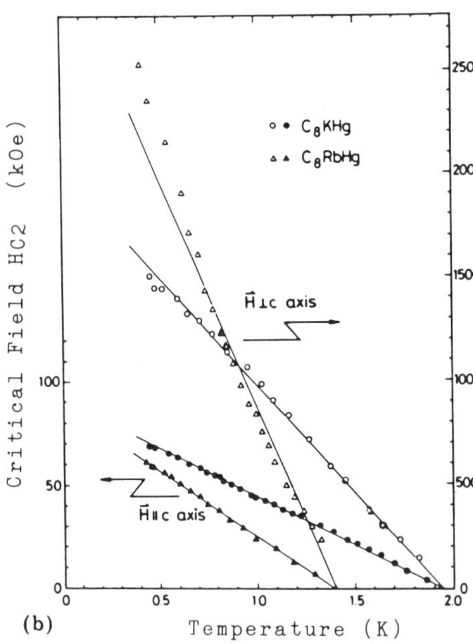

Fig. 5.15. Temperature dependence of the critical fields of (**a**) C_4KHg and C_4RbHg, (**b**) C_8KHg and C_8RbHg, and (**c**) $C_4KTl_{1.5}$ [5.31]

lengths were evaluated by (5.5, 6). Figure 5.16 summarizes the results. Denoted by stage zero in the figure are the respective pristine alkali-metal amalgams, KHg and RbHg. The basal plane coherence length ξ_a is about 2000 Å and only weakly stage-dependent. By contrast, the c-axis coherence length ξ_c rapidly decreases with stage. This clearly indicates an enhancement of the two-dimensionality of the superconductivity with an increase of carbon layers that separate the adjacent alkali-metal layers. It is thus presumed that the major origin of the superconducting electrons is not the graphite-oriented π-like carriers, but the electrons of alkali-metal amalgam.

Figure 5.16 also indicates that even for stage-2, the c-axis coherence length is still much longer than the c-axis repeat distance. This fact justifies the application of the effective-mass model to the explanation of the superconductive anisotropy of these materials, and indicates that the high anisotropy is essentially three dimensional rather than two dimensional. *Iye* and *Tanuma* conjectured that this kind of compound of stage-3 or higher could be a two dimensional superconductor, if it was synthesized and became superconductive, because the c-axis coherence length is possibly shorter than the c-axis repeat distance. So there should be an interesting crossover from three to two dimensions.

Timp et al. [5.32] studied the effect of in-plane density of KHg intercalant on the superconducting transition temperature. The first-stage KHg GIC samples were prepared under temperature differences ΔT of the two-zone method of $\Delta T = 0°C$ (predominantly $(2\times 2)R0°$), $\Delta T = 5°C$ (predominantly $(\sqrt{3}\times\sqrt{3})R30°$), and $\Delta T = 10°C$ (mostly in-plane disordered). The superconducting transition temperatures T_c were respectively 1.56 K, 1.6 K, and 1.09 K. The second-stage compound with predominantly $(2\times 2)R0°$ ordering

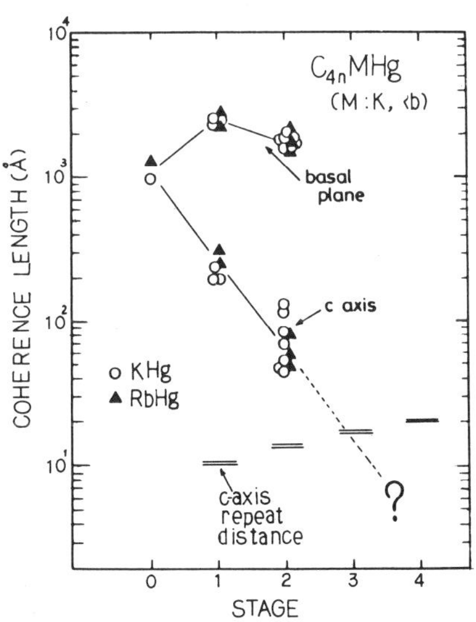

Fig. 5.16. Stage dependence of the coherence length of $C_{4n}M$Hg (M:K, Rb) [5.31]. Stage 0 denotes the pristine intercalant materials. A pair of horizontal bars represents the c axis repeat distance of each stage; the upper bar corresponds to C_{4n}RbHg and the lower to C_{4n}KHg

exhibited a T_c of 1.9 K in agreement with the former reports. *Timp* et al. also observed a superconducting transition in the third-stage samples at 1.94 K. It is interesting that T_c values are higher for higher stage compounds. As a possible explanation for the observed stage dependence of T_c, they took into account an increase of Hg/K ratio with stage index.

It is worth noting that *Timp* et al. [5.32] investigated the Shubnikov–de Haas oscillations on the stage-2 and stage-3 samples. For the stage-2 compound, they detected six oscillation frequencies ranging from 1460 tesla to 55 tesla for $H \parallel c$. From the angular dependences of these frequencies, all except for the lowest one showed three-dimensional Fermi surfaces.

Roth et al. [5.33] also studied the effect of the sample preparation temperature in the isothermal reaction on the superconductivity. The first-stage KHg GICs were synthesized by reacting HOPG with K_xHg_y alloys having the ratios x/y of 50/50 to 56/44 at about 200 K.

They obtained two types of stage-1 compounds; one had a c-axis repeat distance I_c of 10.22 Å corresponding to a (2×2) in-plane structure, and the second one showed an admixture of a second repeat distance of 10.83 Å, probably corresponding to $(\sqrt{3} \times \sqrt{3})R30°$ or $(\sqrt{3} \times 2)(30°, 0°)$ in-plane structure. For the single-phase sample, the highest T_c was 1.53 K with a very narrow transition of 0.03 K. The second type of sample with the two-phase admixture had a T_c of 1.48 K and 1.51 K with a broad transition width of about 0.2 K. For a sample reacted at the higher temperature of 260°C, T_c was 1.32 K with a broad transition width of 0.24 K.

After comparing various data, *Roth* et al. [5.33] suggested that the spread of T_c from 0.8 to 1.5 K is not due to a variation in mercury content but rather due to the varying degree of order of the in-plane structure.

Chaiken [5.34], *Chaiken* et al. [5.35], and *Dresselhaus* et al. [5.36] investigated the superconductivity of the first-stage compound C_4KHg and its hydrogenated form. They also identified two kinds of structural difference, as *Timp* et al. [5.32] and *Roth* et al. [5.33] had reported earlier. One is the pink-colored so-called α-phase compound having an I_c of 10.83 Å; the other is the gold- or copper-colored β phase having an I_c of 10.84 Å. The pure α-phase specimens were routinely produced, while the β phase was always accompanied by a mixing of α phase. The temperature difference 5°–10°C enhanced the amount of β phase. The T_c value ranged from 0.7 to 1.5 K, and lower values corresponded to the β-phase sample, while higher values were seen for the α phase.

For lower T_c samples (e.g., 0.95 K and 0.73 K [5.29]), the upper critical field H_{c2} was well described by (5.2), which is based on the effective-mass model. In contrast, the $H_{c2}(\theta)$ values of higher T_c (1.5 K) samples showed a deviation from (5.2). To understand this, note that the type-II superconductivity appears only for magnetic field angles perpendicular to the c axis, while type-I behavior appears for the rest of the field directions. This feature was also found in C_8K by *Koike* et al. [5.6].

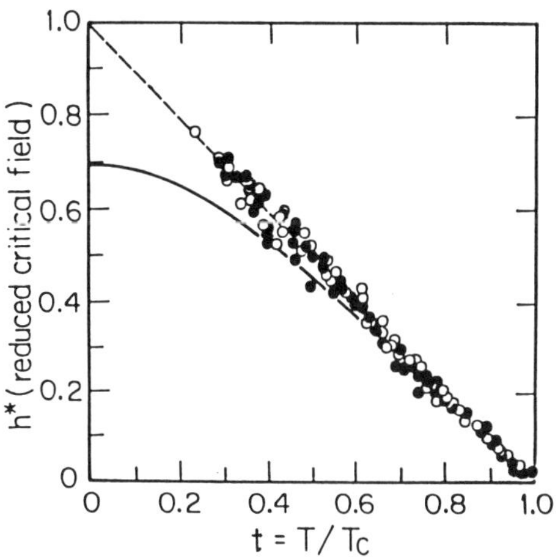

Fig. 5.17. Linear temperature dependence of H_{c2} for five C_4KHg specimens with T_c between 1.5 K and 0.7 K [5.36]. The open circles and full circles correspond respectively to the magnetic fields parallel and perpendicular to the c axis

Considerable attention was paid to the anomalous temperature dependence of H_{c2}. In field directions both parallel and perpendicular to the c axis, H_{c2} showed a linear relationship to temperature below $t = 0.3$. This is shown in Fig. 5.17, where $t \equiv T/T_c$ and $h^* = H_{c2} \Big/ \left(T_c \dfrac{dH_{c2}}{dT} \right)$. Five specimens with $T_c = 1.5$ K to 0.7 K, including magnetic field orientations both parallel and perpendicular to the c axis, gave a single straight line. As Fig. 5.15 showed, many kinds of superconducting GICs manifest this linear relationship in an extended temperature range. In many layered superconductors, a superlinear relationship or the so-called positive curvature is recognized, and this feature is understood as a dimensionality crossover from three dimensions (higher temperature) to two dimensions (lower temperature). The precise linear relationship in these GIC superconductors needs more theoretical investigation.

5.2.3 Superconductivity of Alkali-Metal Thallide and Bismuthide GICs

Analogous ternary compounds were synthesized by *El Makrini* et al. [5.37] using thallium in place of mercury, that is, $C_{4n}MTl_{1.5}$, ($n = 1, 2$, M = K, Rb). The stage-1 compound was prepared by reaction of HOPG and $KTl_{1.5}$ alloy at 320°C and showed a silver luster. The c axis repeat distance was relatively large and equal to 12.08 Å for the first-stage compound. In the case of $C_4RbTl_{1.5}$, there were two varieties, α and β, of the repeat distances, i.e., 12.65 Å and 13.40 Å, respectively. Superconductivity was found in the first-stage $C_4KTl_{1.5}$ and the second-stage $C_8KTl_{1.5}$ by *Vogel* et al. [5.38] with a T_c of 2.7 K and 1.3 K, respectively. *Iye* and *Tanuma* [5.31] obtained T_c for $C_4KTl_{1.5}$ 2.56 K.

Lagrange et al. [5.39] and *McRae* et al. [5.40] synthesized alkali-metal bismuthide GICs and observed superconductivity in some of those compounds. The chemical formula is $C_{4n}MBi_x$, in which M = K, Rb, or Cs and $x \sim 0.6$. The same method of reaction as above was adopted. The c axis repeat distance had

Table 5.3. Structural parameters of Hg, Tl, and Bi GICs ($C_{4n}MT_x$)

T	M	x	α		β	
			n	d_i[Å]	n	d_i[Å]
Hg	K	1	1–2	10.20 ± 0.04	–	–
	Rb	1	1–2	10.76	–	–
Tl	K	1.5	1–2	12.08	–	–
	Rb	1.5	1	12.65	1–2	13.40
Bi	K	0.6	2–5	$9.94 \pm .02$	2	10.86
	Rb	0.6	1–4	$10.10 \pm .02$	4, 7	10.52
	Cs	0.55	1	$10.61 \pm .02$	1–2	11.48

two values as with the thallium GICs. Table 5.3 shows the structural parameters of Hg, Tl, and Bi GICs for comparison.

Diffraction studies indicated that bismuth atoms were located between two alkali-metal planes, but the structure was more complicated than that of Hg GICs. The planar unit cell was much larger and the diffraction patterns were different between K, Rb, and Cs bismuthide GICs.

The in-plane resistivity ϱ_a of KBi and RbBi GICs was metallic, and the values at room temperature were not very dependent on the stage number n; e.g., the values of ϱ_a of RbBi GICs were distributed from 6.9 to 10.7 μΩ cm. The c axis resistivity ϱ_c had a negative temperature coefficient for high-stage compounds, while it had a positive one for the first- and second-stage compounds over the temperature range lower than about 200 K.

The superconductivity was found in both α and β structures by the resistance measurements along c axis as well as in the planar direction. Figure 5.18 shows the observed behaviors. Figures 5.18a and 5.18b are the resistance versus temperature characteristics of CsBi and RbBi GICs, respectively. Table 5.4 gives the superconducting transition temperature T_c, defined as the temperature that gives the half value of the normal resistance.

The steepest transition appears on the first-stage CsBi (α) as Curve α of Fig. 5.18a. Notice that the samples having mixed phases $\alpha + \beta$ show no step-

Table 5.4. Superconducting transition temperatures (in Kelvin) of alkali-metal bismuthide GICs

	Cs-Bi		Rb-Bi
	$n = 1$	$n = 2$	$n = 2$
α	4.05		1.25–1.5
β	2.3	2.7	

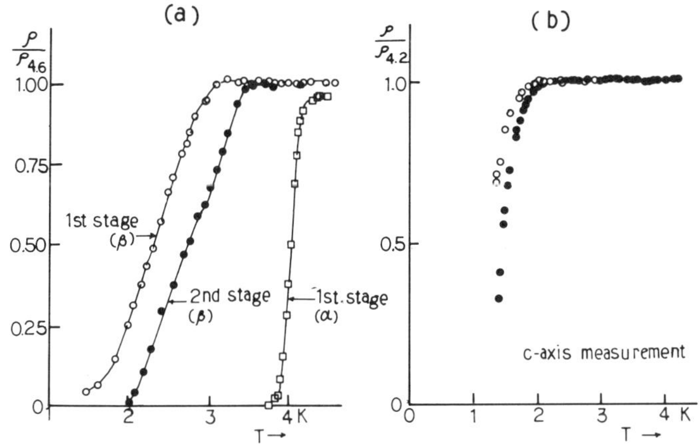

Fig. 5.18. (**a**) Superconducting transition of CsBi GICs observed by resistivity. (**b**) Rapid fall of resistivity of a second-stage RbBi GIC at about 2 K by lowering the temperature. Empty circles: first run; full circles: second run after the configuration change [5.40]

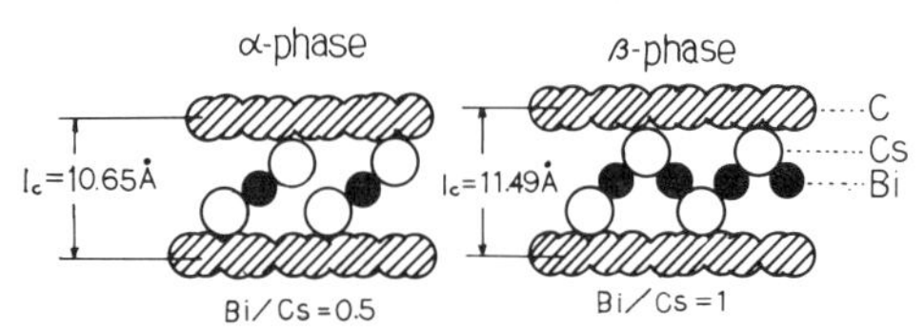

Fig. 5.19. Scale drawing of layer structures of $CsBi_x$ GICs [5.41]

wise transition, but single and slow transitions, such as Curves β. For the β phase, T_c is higher for the second stage than for the first stage. These two facts were similar to those reported by *Tanuma* et al. [5.29] for RbHg GICs. Some doubt remains about the bulk superconductivity of RbBi GICs because the resistance did not take the zero value above the lowest measured temperature of 1 K, as shown in Fig. 5.18b.

The *c* axis structures of $CsBi_x$ GICs were investigated by *Yang* and *Eklund* [5.41] by X-ray (00*l*) scans and analyzed by the use of a three-layer sandwich structure. For the first-stage compounds, two distinct Bi/Cs ratios could explain the observed diffraction data, i.e., Bi/Cs ~ 0.5 for α phase and Bi/Cs ~ 1.0 for β phase. Figure 5.19 shows the scale illustration of the obtained *c* axis structures.

Stang and *Lüders* [5.42] investigated the superconductivity of the first- to third-stage CsBi GICs. They prepared the samples by dipping an HOPG plate into a liquid alloy of $Cs_{1-x}Bi_x$, where x was selected between 0.3 and 0.6. The reaction temperature was varied between 450°C and 650°C and the reaction time between 0.5 and 92 hours. With increasing Bi content, such GICs were obtained as the stage 1 (α) and ($\alpha + \beta$), stages 1 + 2 ($\alpha + \beta$), stage 2 (β), and stage 3 (β). Superconductivity was measured by a mutual inductance method using a ^{4}He cryostat (20–1.2 K) and a ^{3}He/^{4}He dilution refrigerator. For the samples of stage 1 (α), no superconducting transition was detected, while some of the samples of stage 2 and 3 (β) showed T_c values of 4.71 K and 4.73 K but with the volume fractions of 1–5%, and stage 1 ($\alpha + \beta$) 4.69 K with volume fractions of 0.1%. As the bulk $CsBi_2$ alloy is known to have its superconducting transition at 4.15 K, the faintness of the volume superconductivity of these GICs led Stang and Lüders to infer some inclusion effect of bulk $CsBi_2$ alloys into the sample.

5.2.4 Pressure Dependence of the Anisotropy of Superconductivity

The effect of pressure on the superconductivity of KHg, RbHg, and KT l GICs were studied in detail by *Iye* and *Tanuma* [5.31, 43] and *Iye* [5.44] with a view to

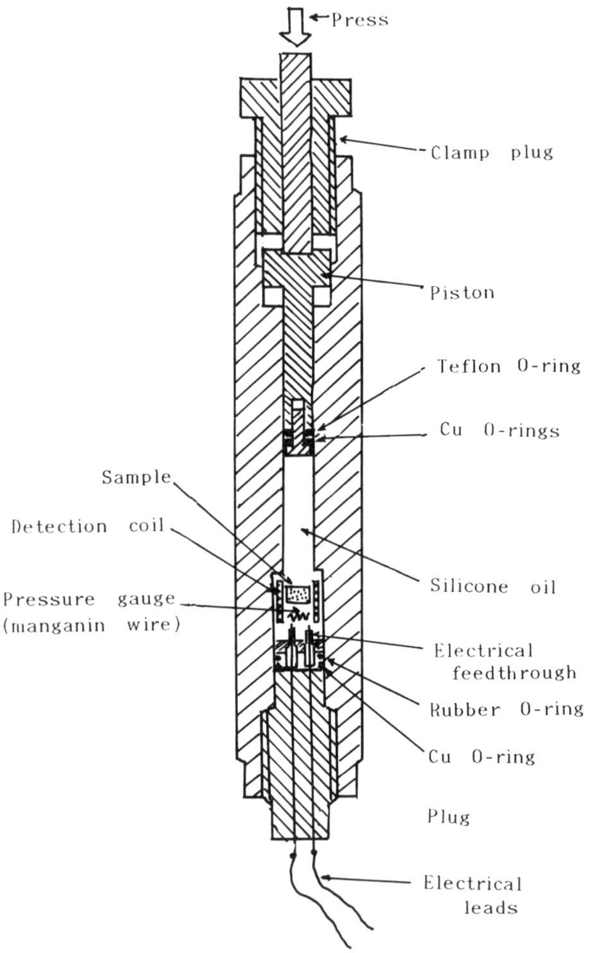

Fig. 5.20. Copper-beryllium clamp cell for the low-temperature and high-pressure experiment with GICs [5.43]

changing the anisotropy of superconductivity as well as the transition temperature. Pressure experiments were carried out using the so-called clamped cell method. Figure 5.20 illustrates the cell. This piston and cylinder type cell was made of Cu-Be hardened alloys, and the sample space was filled with silicone oil, which works as the force-transmitting medium. High pressure was attained by driving the piston with a hydraulic press and subsequently clamping it tightly at room temperature with a screwed plug. Because of thermal contraction and solidification of the oil, the pressure within the clamped cell is reduced upon cooling down to liquid-helium temperature. To calibrate the pressure at lower temperatures, a manganin wire resistance gauge was used. The superconducting transition was detected by a magnetic induction method using a modulation frequency of 27 Hz and a modulation field of 0.2 Oe.

Figure 5.21 shows the pressure dependence of the relationships of the critical field versus temperature for C_8KHg. From these data, the coherence lengths and the transition temperatures were deduced for various pressure values. Figure 5.22 illustrates the results for $C_{4n}M$Hg and $C_4KTl_{1.5}$. The basal plane coherence lengths of $C_{4n}M$Hg are pressure independent, while the c-axis coherence length shows a fairly large dependence on pressure. For example, in the case of C_8RbHg (stage 2), ξ_c is nearly doubled at 5 kbar compared with the case of normal pressure. This clearly shows that the application of pressure suppresses the two-dimensional anisotropy of the superconductivity by virtue of the preferential compression in the c-axis direction. In the case of C_4KHg (stage 1),

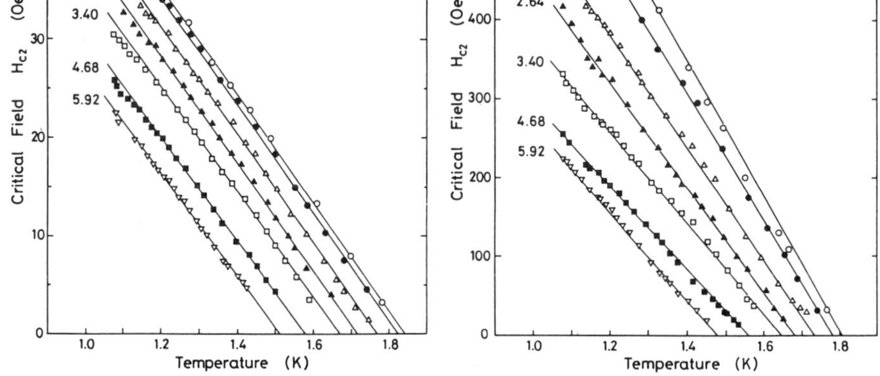

Fig. 5.21. Temperature dependence of the critical field H_{c2} of C_8KHg under various pressure values [5.43]: **(a)** $H \parallel c$, **(b)** $H \perp c$

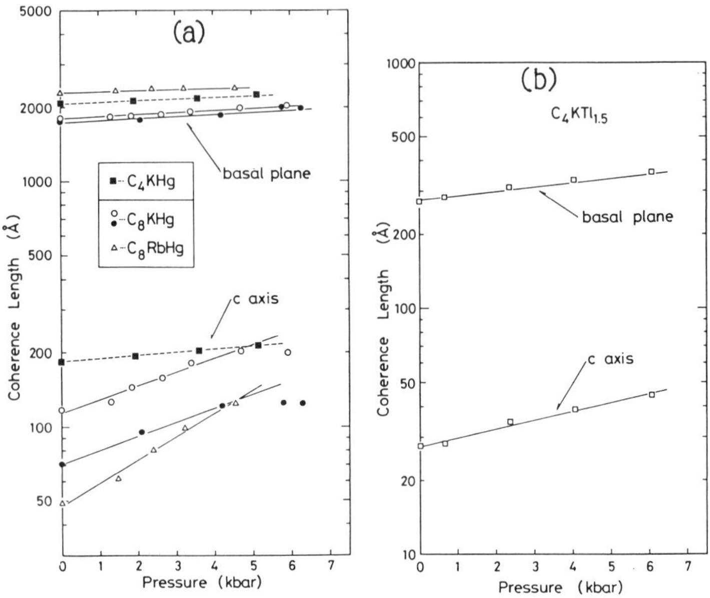

Fig. 5.22. Pressure dependence of the coherence lengths: (**a**) C_4KHg, C_8KHg, and C_8RbHg; (**b**) $C_4KTl_{1.5}$ [5.31]

the increase of ξ_c by pressure is much smaller than the case of stage 2. Therefore, it is inferred that the large suppression of anisotropy of superconductivity is due to the compression of *c-c* interlayer spacing. Figure 5.22b shows the result for $C_4KTl_{1.5}$ (stage 1), where the pressure dependence is moderate both for ξ_a and ξ_c. This is because the coherence length is much shorter than that of *M*Hg GICs, corresponding to the higher T_c.

Figure 5.23 shows the change of T_c due to the application of pressure: T_c is lowered by pressure. For C_4KHg, C_8KHg, C_8RbHg, and $C_4KTl_{1.5}$, the linear reduction of T_c occurs at the following rates: -3.4×10^{-5}, -6.5×10^{-5}, -5.0×10^{-5}, and -6.4×10^{-5} K/bar, respectively.

Prediction of the pressure effect on T_c is not simple, because T_c is determined by a delicate interplay of electronic and phononic properties of the material which are influenced by pressure in different ways. The transition temperature is basically given by the BCS expression (5.1), that is, $T_c \sim (\hbar\omega)\exp(-1/g)$ and $g = N(0)V = N(0)[I^2]/(M[\omega^2])$, where ω is the phonon frequency, $N(0)$ is the density of states at the Fermi level, M is the ion mass, and $[I^2]$ is the matrix element of the electron–phonon interaction. The mean square of phonon frequency, $[\omega^2]$, is increased by pressure, which preferentially affects the exponential factor to lower the transition temperature. A change in g is also brought about through the changes of $N(0)$ and $[I^2]$ by pressure. This change depends on the band structure of the material. So the observed descending trend of T_c by

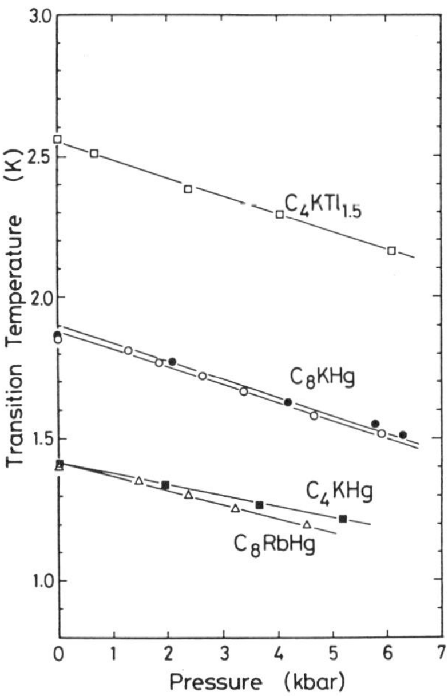

Fig. 5.23. Pressure dependence of the transition temperatures of C_4KHg, C_8KHg, C_8RbHg, and $C_4KTl_{1.5}$ [5.31]

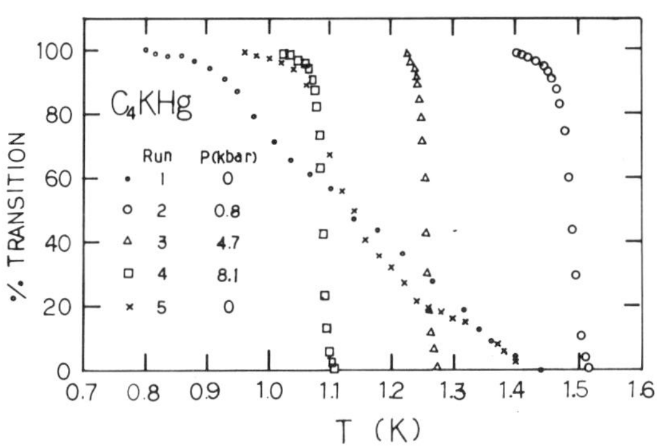

Fig. 5.24. Superconducting volume change of C_4KHg measured by an inductive method under various pressures [5.45]

pressure is not easily understood, although the phonon hardening by the application of pressure should play an important role.

Delong et al. [5.45] investigated the effect of pressure on T_c for C_4KHg, C_8KHg, $C_4KTl_{1.5}$, and C_8K. The decrease rate of T_c with pressure in C_8KHg was -6.5×10^{-5} K/bar in agreement with the data of *Iye* and *Tanuma* [5.31]. $C_{24}K$ was found to be nonsuperconducting down to 0.07 K for $p < 10$ kbar.

Figure 5.24 shows the superconducting volume transition of C_4KHg measured by an inductive method. An initial normal pressure measurement indicated a very broad transition extending from 1.42 K to 0.85 K. Even with a modest pressure of 0.8 kbar, the transition was very much sharpened with the T_c value of

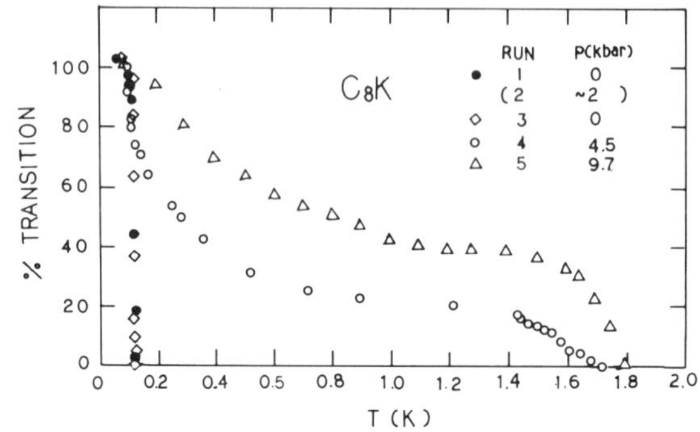

Fig. 5.25. Superconducting volume change of C_8K under various pressures [5.45]

1.486 K. Further increase in pressure reduced T_c at a rate of -5×10^{-5} K/bar. Upon the full release of pressure, the sample showed again a broad transition. For $C_4KTl_{1.5}$, the pressure dependence of T_c of -4.5×10^{-5} K/bar was obtained, a value somewhat smaller than that obtained by *Iye* and *Tanuma.*

On the other hand, the superconducting transition of C_8K is sharp at normal pressure but very much broadened by applying as small a pressure as 2 kbar, as shown in Fig. 25. Further increase of pressure to 9.7 kbar resulted in roughly a "two-step" transition consisting of an "upper T_c" of 1.7 K and a broad "lower T_c" of 0.5 ± 0.4 K. This behavior indicates the presence of two crystalline phases. Another sample showed a narrow single transition at 1.8 K with the transition width of 50–80 mK. The high-pressure phase showed an increase of T_c with pressure at the rate of $+1.5 \times 10^{-5}$ K/bar.

5.3 Theoretical Aspects of the Origin of Superconductivity in GICs

The occurrence of superconductivity in C_8M (M: K and Rb) has been an object of theoretical interest because neither graphite nor the alkali metal is a superconducting material. Typical calculations have been done on C_8K, because the band calculations were most readily carried out for this material.

The first band calculation was done by *Inoshita* et al. [5.46] and *Ohno* et al. [5.47]. Their important findings were that there are two distinct types of Fermi surface. One is predominantly spherical and is located in the center of the Brillouin zone, which comes from the three-dimensional (3D) potassium 4*s* band, and the others are cylindrical surfaces in the peripheral region along the H-K-H axes, which originate from the two-dimensional (2D) graphite π band. Which kind of carriers, 3D or 2D, are mainly responsible for the superconductivity was the important and interesting problem. *Inoshita* and *Kamimura* [5.48] calculated the electron-phonon coupling constant between 2D electrons and carbon and alkali-metal phonons, ignoring 3D electrons. They obtained a very large coupling constant with mainly carbon atoms, because they included contributions from both inter- and intrapocket electron-phonon interactions corresponding to two inequivalent pockets of 2D electron Fermi surfaces. The dimensionless coupling constant λ was 0.31 even for the second-stage potassium GIC in which no 3D Fermi surface around the Γ point exists. This λ value may yield the transition temperature of 0.23 K if the intervalley electron–electron interaction is not taken in account. However, superconductivity is not found for $C_{24}K$ down to the 10 mK region. This means that the interpocket Coulomb interaction should not be negligibly small because the electron concentration is much larger than in the case of semiconductors, and the interpocket interaction makes the repulsive force larger.

A speculation such as the above was formulated by *Takada* [5.49] with regard to the superconductivity of C_8K, assuming that the 3D electrons effectively interact to form Cooper pairs, because the interpocket Coulomb scattering of the 2D electrons is large enough to cancel a large interpocket contribution of these electrons. Alkali-metal layers were assumed to transfer a fraction f of their valence charges to the graphite layers, so some alkali atoms M lose a valence electron and become M^+ ions. From the band calculation by *Ohno* et al. [5.47], the fraction f is 0.5–0.6 for C_8M, so a fraction $(1-f)$ of electrons form the 3D Fermi surface. The rest reside in the graphite layers and are responsible for the 2D Fermi surface. Thus each carbon atom has an average charge of $-f/8$. In this situation, the 3D electrons should feel the large electric field of the polarization wave associated with both partially ionized $8C^-$ and M^+ ions. Namely, as the longitudinal phonons are considered, there exist two kinds of oscillations. One is the optical mode of the in-phase motion of carbon atoms and out-of-phase motion of M^+ ions referred to carbon ions, and the other is the acoustic mode of the in-phase motions between carbon ions and M^+ ions. The latter also produce an electric polarization if the charge transfer rate f is less than 1.

Takada showed that the 3D electrons in C_8M interact strongly with both longitudinal optical and acoustic phonons, and this yields the strong phonon-mediated attractive force between 3D electrons. Further, the repulsive interaction is much weaker than in alkali metals because the concentration of 3D electrons is considerably smaller in C_8M than in alkali metals. Therefore, the net attractive electron–electron interaction between 3D electrons gives rise to the superconductivity. *Takada* calculated the gap equation within the weak coupling approximation and obtained a value of T_c of about 100 mK by using the proper material constants for C_8K.

Shimizu and *Kamimura* [5.50, 51] used another theoretical approach for calculating the superconducting nature of C_8M. They adopted the idea of Coulomb pseudopotential μ introduced by *Morel* and *Anderson* [5.52], taking into account the contribution of 3D electrons, and calculated the dimensionless coupling constant λ of the phonon-mediated interaction between 3D electrons. The value of λ for C_8K was obtained as 0.31; this agrees well with the experimental value of 0.31 obtained from the experimental T_c value of 150 mK by using McMillan's formula.

As for the anisotropy of the superconductivity, λ consists of $\lambda_K^{\perp}$ and $\lambda_L^{||}$, which are the terms corresponding to the vibration mode of K layers perpendicular to the c axis and to the longitudinal optical and acoustic modes along the c axis. The difference of $\lambda^{\perp}$ and $\lambda^{||}$ gives a certain measure of anisotropy. As $\lambda^{||}$ is not large enough compared with $\lambda^{\perp}$ ($\lambda^{||} \sim 2\lambda^{\perp}$) to explain the measured anisotropy, it is strongly suggested that besides the 3D electrons the anisotropic 2D electrons also contribute to the superconductivity of C_8K. This is due to the interplay of both kinds of electrons through the exchange interactions between them.

Shimizu and *Kamimura* [5.50, 51] extended their theoretical consideration to explain the anisotropy of the critical magnetic field. The relevant effective-mass ratio, $m_{||}/m_{\perp}$, should be about 25 according to the experimental data, but such a large value cannot be ascribed to the anisotropic coupling between the 3D electrons. The band mixing between 3D and 2D electrons only increases the mass ratio slightly. The Fermi surface of C_8K calculated by *Inoshita* et al. [5.46] consists of 2D surfaces along H-K-H edges and a 3D surface centered at the Γ point which overlap with each other, making multiply connected shapes. In such a Fermi surface, they showed that band mixing occurs to give a strong anisotropy. So far, many discussions and experiments have been presented about the Fermi surface of this material in connection with the f value; some have claimed the existence of a 3D Fermi surface while others have denied it.

Recently, *Mizuno* et al. [5.53] performed a new band calculation on C_8K taking into account two kinds of nonequivalent carbon atoms in a graphite layer, which are referred to the position of potassium atoms. They found that there exist 3D Fermi surfaces centered at the Γ point, but the character of the electron wavefunction of that surface is not K 4s but mainly graphite π^*. The shapes of their 3D Fermi surface in the center of the Brillouin zone are convex and concave cylinders that are not connected to the 2D Fermi surfaces. Thus, more detailed experimental investigations of the Fermi surface and theoretical studies of the observed strong anisotropic superconductivity are required.

Note Added in Proof

Superconductivity of Solid State Derivates of Fullerite C_{60}

Many kinds of carbon clusters C_n are known to exist for n values ranging from several to a few hundred. Recently, *Kroto* et al. [5.54] revealed by means of a time-of-flight mass spectrometry that C_{60} and C_{70} clusters are especially stable among other clusters made of 40 or more atoms which were collected after rapid cooling of a high temperature carbon vapor. The vapor was generated by laser evaporation of graphite rods.

The proposed structure of C_{60} is just the same as the patch pattern of a soccer ball, i.e., a truncated icosahedral structure. The structure of C_{70} is also proposed to be ball-like (somewhat like a rugby football). The name "fullerene" has been given to these ball-like carbon clusters. On the spherical or spheroidal fullerene structure, every carbon atom has three neighbors, so that like the carbon atoms in the graphite plane it satisfies the stable sp^2-valence. The outer and inner surfaces of the balls should be covered with π-electron sheets.

It was further proven by *Krätschmer* et al. [5.55] that the soccer-ball-like C_{60} molecules form a solid crystalline phase, "fullerite", with a disordered hexagonal close packed structure. The lattice constants are $a = 10.02$ Å and $c = 16.36$ Å. Comparing to the calculated diameter of the C_{60} fullerene, 7.1 Å, the nearest neighbor distance a is 2.92 Å longer. This difference is close to the van der Waals diameter of carbon 3.3–3.4 Å; therefore the C_{60} fullerite seems to be approximately a van der Waals type molecular crystal. The c axis stacking is ABAB . . . for the hcp structure and ABCABC . . . for the fcc structure. The above authors also reported X-ray diffraction results on carefully grown large-sized and well-faced crystals showing a fuzzy spot pattern, which indicates the existence of stacking disorder between AB and ABC sequences as well as a disordered arrangement of the molecules in the planes normal to the c axis.

Ajie et al. [5.56] found a method of producing high yields of C_{60} and C_{70} by making carbon soot by Joule-heating of a graphite rod with a high current in 0.3 bar He atmosphere. The soot-like material collected on a surrounding glass wall was extracted with boiling benzene or toluene to give a brown-red solution consisting of a mixture of C_{60} and C_{70}. The separation of C_{60} and C_{70} was made by column chromatography on alumina with hexanes giving 99.85% C_{60} and $> 99\%$ C_{70}. *Hauffer* et al. [5.57] developed a simple method which yields a production rate of 10 g/h. The method uses a sharpened graphite rod kept in contact with a graphite disk by gentle spring pressure. The surrounding wall was a water-cooled copper shroud forming a collector. The atmosphere was 100 torr He gas. Vaporization of graphite was achieved by a high current. The best situation obtained by controlling the spring pressure was called "contact arc", in which the electric power was so concentrated on the contact point that a high enough temperature for making C_{60} and C_{70} clusters could be attained. The extraction was made by processing the collected material with boiling toluene until a dark red-brown liquid was obtained. Upon vaporizing the solvent, solid state fullerite was obtained comprising 10% of the original soot.

Since these molecular crystals of C_{60} and C_{70} represent the third form of crystalline carbon, various physical and chemical properties are now being investigated. Similar to the intercalation of graphite, metallic species have been reacted with these materials. A striking event is the discovery by *Hebard* et al. [5.58] of high T_c superconductivity in $C_{60}K_x (x \simeq 3)$. A thin film showed a zero resistance transition. Bulk samples showed a well-defined Meissner effect and magnetic-field-dependent microwave absorption. The doped film was prepared in a glass tube in such a way that the reaction with potassium was performed at an elevated temperature in He gas atmosphere until the resistivity had fallen to

$5 \times 10^{-3}\,\Omega$ cm. The resistivity increased by a factor of two by cooling the sample to 20 K. Below 16 K, the resistance started to decrease; zero resistance was obtained below 5 K. The 10–90% transition width was 4.6 K. A bulk sample of nominal composition $C_{60}K_3$ was prepared in a fused silica tube. The reaction with potassium was done at 200 K for 36 h. The temperature dependence of the dc magnetization of this sample was measured by a SQUID magnetometer. After zero-field cooling (ZFC) to 2 K, a weak magnetic field of 50 Oe was applied. On warming, this field was excluded by the sample up to 18 K, showing the superconductivity with a T_c of 18 K. Upon cooling (FC) with a 50 Oe external field, a well-defined Meissner effect (flux expulsion) developed below 18 K. This transition temperature is the highest of all superconducting molecular crystals investigated so far.

The location of the potassium atoms is not inside the ball but on interstitial sites of the hcp (fcc) molecular crystal. *Zhou* et al. [5.59] investigated the K-doped and Cs-doped C_{60} crystal structure. The pristine C_{60} crystal showed the fcc structure (different from the report of reference [5.55]; and the fcc structure was later confirmed), while $C_{60}M_x$ (M:K, Cs) exhibited a phase transition from fcc to bcc with a saturation composition close to $x = 6$. The closest distance of C_{60} molecules in the undoped fcc phase was 10.02 Å, whereas the corresponding values for $C_{60}K_6$ and $C_{60}Cs_6$ were 9.86 Å and 10.21 Å, respectively.

The experimental work on normal transport properties and superconductivity of fullerite-based derivates is rapidly expanding. In this group of materials, the highest T_c fullerite so far announced is $C_{60}Cs_xRb_y$. *Tanigaki* et al. [5.60] observed the superconductivity on this newly synthesized material. Magnetization curves for ZFC and FC described above [5.59] have been reported, and the T_c was determined to be as high as 33 K for the concentration $x = 2$ and $y = 1$.

Tanigaki et al. [5.60] discussed the high T_c nature of this material as follows. According to the theoretical calculation by *Saito* and *Oshiyama* [5.61], the Fermi level of $C_{60}K_x$ is located at the maximum position of the density of states curve with the highest T_c at a composition of $x = 3$. This result suggest that a BCS type mechanism might operate in alkali-metal-doped fullerides. The increase in T_c with increasing ionic radius may be explained by the expected concomitant increase in the lattice constant, thereby reducing the bandwidth and increasing the density of states at the Fermi level. The tendency should raise the value of T_c of $C_{60}M_3$ in the sequence M = K, Rb and Cs. Actually, $C_{60}Rb_3$ exhibits a T_c of 29 K, but $C_{60}Cs_3$ failed to show superconductivity. The size of the interstitial sites in the C_{60} crystal is such that three Cs atoms cannot be accommodated without severely disrupting the close packed fullerites. Therefore, the value of T_c should be enhanced by the partial substitution of Rb atoms by Cs atoms as far as the close packing is maintained.

The latest news before finishing this note is that the value of T_c is further increased up to 42.5 K in the fulleride $C_{60}Rb_{3-x}Tl_x$ as reported by *Iqubal* and *Raughman* [5.62].

References

5.1 N.B. Hanney, T.H. Geballe, B.T. Matthias, K. Andres, P. Schmidt, D. MacNair: Phys. Rev. Lett. **14**, 225 (1965)
5.2 G.R. Henning, L. Meyer: Phys. Rev. **87**, 439 (1952)
5.3 J. Poitrenaud: Rev. de Phys. Appl. (Suppl. J. de Phys.) **5**, 275 (1970)
5.4 M. Kobayashi, I. Tsujikawa: Annual Meeting of Phys. Soc. Jpn. (1977)
5.5 Y. Koike, H. Suematsu, K. Higuchi, S. Tanuma: Sol. St. Commun. **27**, 623 (1978)
5.6 Y. Koike, S. Tanuma, H. Suematsu, K. Higuchi: J. Phys. Chem. Sol. **41**, 1111 (1980)
5.7 Y. Koike, H. Suematsu, K. Higuchi, S. Tanuma: Physica **99B**, 503 (1980)
5.8 U. Mizutani, T. Kondow, T.B. Massalski: Phys. Rev. B **17**, 3165 (1978)
5.9 E.I. Kats: Sov. Phys. -JETP **29**, 897 (1969)
5.10 R.C. Morris, R.V. Coleman, R. Bhandari: Phy. Rev. B **5**, 895 (1972)
5.11 D. Saint-James, P.G. de Gennes: Phys. Lett. 7, 306 (1963)
5.12 M. Kobayashi, I. Tsujikawa: J. Phys. Soc. Jpn. **46**, 1945 (1979)
5.13 M. Kobayashi, I. Tsujikawa: J. Phys. Soc. Jpn. **50**, 3245 (1981); Physica **105B**, 439 (1981)
5.14 M. Kobayashi, T. Enoki, H. Inokuchi, M. Sano, A. Sumiyama, Oda, H. Nagano: J. Phys. Soc. Jpn. **54**, 2359 (1985); Synth. Met. **12**, 341 (1985)
5.15 I.T. Belash, A.D. Bronnikov, O.V. Zarikov, A.V. Pal'nichenko: Solid State Commun. **64**, 1445 (1987)
5.16 D. Guerard, P. Lagrange, M. El Makrini, A. Herold: Synth. Met. **3**, 15 (1981)
5.17 I.T. Belash, A.D. Bronnikov, O.V. Zhrikov, A.V. Pal'nichenko: Solid State Commun. **69**, 921 (1989)
5.18 I.T. Belash, O.V. Zarikov, A.V. Pal'nichenko: Synth. Met. **34**, 415 (1989)
5.19 P. Lagrange, A. Metrot, A. Herold: CR Acad. Sci. C **278**, 701 (1974)
5.20 S. Kaneiwa, M. Kobayashi, I. Tsujikawa: J. Phys. Soc. Jpn. **51**, 2375 (1982)
5.21 K. Suzuki, I. Tsujikawa: Synth. Met. **12**, 389 (1985)
5.22 D. Guerard, C. Takoujou, F. Rousseaux: Synth. Met. **7**, 43 (1983)
5.23 P. Lagrange, M. El Makrini, D. Guerard, A. Herold: Physica B **99**, 473 (1980); Synth. Met. **2**, 191 (1980)
5.24 M. El Makrini, P. Lagrange, D. Guerard, A. Herold: Carbon **18**, 211 (1980)
5.25 M.G. Alexander, D.P. Goshorn, D. Guerard, P. Lagrange, M. El Makrini, D.G. Onn: Synth. Met. **2**, 203 (1980)
5.26 Y. Koike, S. Tanuma: J. Phys. Soc. Jpn. **51**, 1964 (1981)
5.27 L.A. Pendrys, R. Wachnik, F.L. Vogel, P. Lagrange, G.F. Furdin, M. El Makrini, A. Herold: Solid State Commun. **38**, 677 (1981)
5.28 M.G. Alexander, D.P. Goshorn, D. Guerard, P. Lagrange, M. El Makrini, D.G. Onn: Solid State Commun. **38**, 103 (1981)
5.29 S. Tanuma, Y. Iye, Y. Koike: *Physics of Intercalation Compounds* ed. by L. Pietronero, E. Tossati, Springer Ser. Solid-State Sci., Vol. 38 (Springer, Berlin, Heidelberg 1981) p. 298
5.30 Y. Iye, S. Tanuma: Phys. Rev. B **25**, 4583 (1982)
5.31 Y. Iye, S. Tanuma: Synth. Met. **5**, 257 (1983)
5.32 G. Timp, B.S. Elman, M.S. Dresselhaus, P. Tedrow: Proc. Mat. Res. Soc. Symp. **20**, 201 (1982)
5.33 G. Roth, N.C. Yeh, A. Chaiken, G. Dresselhaus, P. Tedrow: Extend. Abst. Proc. Symp. I, Fall Meeting Mater. Res. Soc. (1984) p. 149
5.34 A. Chaiken: The Superconducting Properties of Ternary Graphite Intercalation Compounds. Ph.D Thesis, MIT (1988)
5.35 A. Chaiken, M.S. Dresselhaus, T.P. Orlando, G. Dresselhaus, P.M. Tedrow, D.A. Neumann, W.A. Kamitakahara: Private communication
5.36 M.S. Dresselhaus, A. Chaiken, G. Dresselhaus: Synth. Met. **34**, 449 (1989)
5.37 M. El Makrini, P. Lagrange, A. Herold: Carbon **18**, 375 (1980)

5.38 F.L. Vogel, R. Wachnick, L.A. Pendrys: *Physics of Intercalation Compounds*, ed. by L. Pietronero, E. Tossati, Springer Ser. Solid-State Sci. Vol. 38 (Springer, Berlin, Heidelberg 1981) p. 288
5.39 P. Lagrange, A. Bendriss-Rehehayer, J.F. Mareche, E. McRae: Synth. Met. **12**, 201 (1985)
5.40 E. McRae, J.F. Mareche, A. Bendriss-Rerhhayer, P. Lagrange, M. Leleurain: Ann de Phys. Colloq. No. 2 (Suppl.) **11**, 13 (1986)
5.41 M.H. Yang, P.C. Eklund: Extend. Abst. Proc. Symp. K, Fall Meeting Mat. Res. Soc. (1986) p. 161
5.42 I. Stang, K. Lüders: Abst. 4th Int. Symp. GICs, Jerusalem (1987) p. 9
5.43 Y. Iye, Tanuma: Solid State Commun. **44**, 1 (1982)
5.44 Y. Iye: Mater. Res. Soc. Symp. Proc. Vol. **20** (Elsevier, Amsterdam 1983) p. 185
5.45 L.E. Delong, V. Yeh, V. Tondiglia, P.C. Eklund: Mater. Res. Soc. Symp. Proc. Vol. **20** (Elsevier, Amsterdam 1983) p. 195
5.46 T. Inoshita, K. Nakao, H. Kamimura: J. Phys. Soc. Jpn. **43**, 1237 (1977)
5.47 T. Ohno, K. Nakao, H. Kamimura: J. Phys. Soc. Jpn. **47**, 1125 (1979)
5.48 T. Inoshita, H. Kamimura: Synth. Met. **3**, 223 (1981)
5.49 Y. Takada: J. Phys. Soc. Jpn. **51**, 63 (1982)
5.50 A. Shimizu, H. Kamimura: Synth. Met. **5**, 301 (1983)
5.51 A. Shimizu, H. Kamimura: *Graphite Intercalation Compounds, Progress of Research in Japan*, ed. by S. Tanuma, H. Kamimura (World Scientific, Singapore 1985) p. 157
5.52 P. Morel, P.W. Anderson: Phys. Rev. **125**, 1263 (1962)
5.53 S. Mizuno, H. Hiramoto, K. Nakao: J. Phys. Soc. Jpn. **56**, 4466 (1987)
5.54 H.W. Kroto, J.R. Heath, S.C. O'Brien, R.F. Curl, R.E. Smalley: Nature **318**, 162 (1985)
5.55 K. Krätschmer, Lowell D. Lamb, K. Fostiropolos, Donald R. Huffman: Nature **347**, 354 (1990)
5.56 Henry Ajie, Marcos M. Alvarez, Samir J. Anz, Rainer D. Beck, Francois Diederich, K. Fostiropoulous, Donald R. Hoffman, Wolfgang Krätschmer, Yves Rubin, Kenneth E. Scriver, Dilip Sensharma, Robert L. Whetten: J. Phys. Chem. **94**, 8630 (1990)
5.57 R.E. Haufler, J. Conceicao, L.P.F. Chibante, Y. Chai, N.E. Birne, S. Flanagan, M.M. Haley, S.C. O'Brien, C. Pan, Z. Xiao, W.E. Billups, M.A. Cuifolini, R.H. Hauge, J.L. Margrave, L.J. Wilson, R.F. Curl, R.E. Smallay: J. Phys. Chem. **94**, 8634 (1990)
5.58 A.F. Hebard, M.J. Rosseinsky, R.C. Haddon, D.W. Murphy, S.H. Glarum, T.T.M. Palstra, A.P. Ramirez, A.R. Kortan: Nature **350**, 600 (1991)
5.59 Otto Zhou, John E. Fischer, Nicole Coustel, Stefan Kycia, Quin Zhu, Andrew R. McGhie, William J. Romanow, John P. McCauley Jr., Amos B. Smith III, David A. Cox: Nature **351**, 462 (1991)
5.60 K. Tanigaki, T.W. Ebbesen, S. Saito, J. Mizuki, J.S. Tsai, Y. Kubo, S. Kuroshima: Nature **351**, 467 (1991)
5.61 S. Saito, A. Oshiyama: Phys. Rev. Lett. **66**, 2637 (1991)
5.62 Z. Iqubal, R. Baughman: Workshop on Fullerites and Solid State Derivates sponsored by DuPont and Xerox, Pennsylvania. August 1–2 (1991)

6. Transport Properties of Metal Chloride Acceptor Graphite Intercalation Compounds

By Jean-Paul Issi

With 19 Figures

Since the beginning of the last decade, a large body of experimental data has been gathered on the temperature variation of the electrical resistivity of graphite intercalation compounds (GICs) and extensive measurements have been performed on their thermal conductivity and thermoelectric power. A decade ago, only scattered data existed about the in-plane and c axis resistivities [6.1–3].

Indeed, most of the early experimental data relating to the electrical resistivity of GICs were aimed at establishing the conductivity level of a given compound – i.e., classifying it electronically – rather than at investigating the microscopic mechanisms responsible for transport. Thus, only the most prominent features of these compounds were pointed out, mainly at room temperature: their high anisotropy and their high in-plane electrical conductivity and its stage dependence.

Nowadays, numerous sets of data are available on the temperature variation of the electrical resistivities of some highly oriented pyrolytic graphite (HOPG)-based compounds taken over a wide temperature range from liquid helium up to room temperature in-plane and in the c axis direction. Also, while no data on the thermal conductivity and only a few on the thermoelectric power existed in the early 1980s, now these properties have been extensively measured and the thermal conductivity data are well explained using standard models for conduction [6.2].

Along these lines, recently the properties of intercalated fibers of various origins have been studied extensively. While data on intercalated fibers were almost nonexistent a few years ago, a large number of transport measurements have now been published and the large variety of host materials available has also allowed new physics to be tested on these fiber compounds. We discuss here the new information relevant to transport revealed by these studies of fiber-based intercalation compounds.

Much work on electrical transport has been reported for fluorides, halogen and acid compounds, and more recently on bi-intercalation compounds comprising these intercalates. However, in the following we concentrate only on the most recent data obtained on metal-chloride acceptor GICs and on their interpretation.

In a previous pedagogical review [6.2], we introduced the main physical ingredients necessary for understanding conduction phenomena in acceptor GICs. We thus refer to this paper wherever needed.

Springer Series in Materials Science, vol. 18
H. Zabel · S.A. Solin (eds.): Graphite Intercalation Compounds II

For the discussion of electronic conduction, we consider the charge carriers, electrons for donor compounds or holes for acceptor compounds, which may either carry the electrical current (electrical conductivity) or the heat energy (electronic thermal conductivity). As regards lattice conduction, we are concerned with the quantized lattice vibrations, the phonons, which carry only heat energy. In the case of thermoelectric effects, we consider both electrical and thermal currents.

In metallic systems, the electrons and holes are described by their Fermi surfaces. Thus we use the recent models developed for this Fermi surface in the discussion of the conduction phenomena. In acceptor graphite intercalation compounds, which are highly anisotropic systems, the Fermi surfaces are well approximated by circles for stage-1 compounds and by cylinders for higher-stage compounds [6.4]. For donor compounds, the situation is more complicated, since the anisotropy may vary considerably from one type of compound to another, and from one stage to another for a given compound. Also, despite the large amount of work devoted to band-structure studies of donor compounds, the situation is not yet clear.

The main interactions that limit the mean free path of the charge carriers are interactions with the phonons, the electrons or holes, the impurities, the static lattice defects, and the crystal boundaries. On the other hand, phonons are mainly scattered by other phonons (through normal or umklapp processes), impurities, lattice defects, crystal or grain boundaries, and possibly by the charge carriers.

Following the pioneering work of *Hennig* and *Ubbelohde* and their respective groups in the 1950s [6.3], the more recent work on the transport properties of GICs was stimulated by the promise of realizing electrical conductors with conductivities that could reach or even exceed that of copper. We thus speculate about the limits which could be attained on the basis of the available data. Also, HOPG for in-plane conduction is, with diamond, the best heat conductor at room temperature. Therefore, it is interesting to consider how intercalation would modify this property.

The low-temperature electrical resistivity, the so-called residual resistivity, reveals information about the lattice defects which are more effective in scattering the charge carriers, while the interpretation of the temperature variation of the ideal electrical resistivity yields information about electron–phonon and possibly electron–electron interactions. The thermal conductivity measurements also yield information about lattice defects and phonon scattering events.

Among the most remarkable features of acceptor GICs is the very high anisotropy of their electrical resistivity and, to a lesser extent, of their thermal conductivity. For some acceptor GICs, the ratio of the in-plane conductivity to that along the *c* axis may reach six orders of magnitude at room temperature [6.1]. This justifies our consideration of the in-plane hole system as a quasi-two-dimensional (2D) gas. On the other hand, the anisotropy of the thermoelectric power is very small compared with that of the other transport properties [6.5], and its temperature variation both in-plane and in the *c* axis direction is puzzling [6.6]. Recent findings on the peculiar behavior of stage-1 compounds have further complicated the situation [6.7]. We consider in some detail the thermoelectric power situation since it has not been previously reviewed extensively.

Today we are more and more frequently faced with new materials having properties that contradict the traditional classification of solids and that seem to violate what were once well-established rules concerning the nature of solids. Acceptor GICs belong to this category of materials, since they are "covalent metals" just as the high T_c ceramics may be labeled "iono-covalent superconductors", if we use the traditional terminology. In that context, we believe that transport measurements on GICs encompass more than a study of the particular case of a specific group of materials. *Dresselhaus* [6.8] recently pointed out that there were many similarities between GICs and other types of "deliberately structured materials". Acceptor GICs are indeed metallic superlattices with very small repeat distances and exceptionally sharp interfaces in the *c* axis direction due to the strong graphitic in-plane cohesive energy. On the other hand, all acceptor GICs and, to a lesser extent, high T_c superconductors have high electrical anisotropies. Very recently *Iye* presented a detailed comparison between high T_c superconductivity and superconductivity in GICs [6.9], which has only been observed in donor compounds. Because of the inherent 2D nature of their electronic structures, weak localization effects are also particularly interesting in these compounds [6.10].

In the following we are mainly concerned with low-stage acceptor compounds since we may then rely on a generally accepted 2D band structure [6.4]. We do not differentiate between the various host materials, provided they are well ordered, except when we consider lattice defects explicitly. As long as graphene planes host the carriers, the same 2D model can be applied. If charge transfer is effective in the conduction process, then we have a Fermi level whose magnitude is directly related to the amount of the charge transfer.

In Sect. 6.1 we analyze the recent in-plane electrical resistivity data obtained for low-stage acceptor GICs and discuss the effects of localization and of carrier–carrier interaction, while in Sect. 6.2 we analyze the thermal conductivity of intercalation compounds with various host materials. In Sect. 6.3 we review and discuss experimental observations on in-plane thermoelectric power, and briefly treat *c* axis transport in Sect. 6.4. We then discuss the particular situation of magnetic compounds in relation with transport properties in Sect. 6.5. We conclude with remarks emphasizing the contribution of GICs to the understanding of the transport properties of 2D systems.

6.1 The In-Plane Electrical Resistivity

6.1.1 General Considerations

In the 1950s, *Hennig* and his group and *Ubbelohde* and coworkers already realized that compounds of very high in-plane conductivities could be synthesized [6.3]. Thus, the first physical property to be investigated was naturally the in-plane electrical resistivity.

A simple picture considers the electrical resistivity ϱ as the sum of the contributions from a temperature-independent residual term ϱ_r and a temperature-sensitive ideal term ϱ_i:

$$\varrho = \varrho_r + \varrho_i \, . \quad (6.1)$$

This was considered as a general feature of 3D metals and in the beginning was also applied to GICs. However, it is now well established that, in the presence of weak disorder, we should consider in addition quantum effects due to localization [6.11–13] and/or interaction effects [6.14, 15]. It was recently shown that this is also necessary for acceptor GICs [6.10, 16–18]. These quantum effects, though they do not generally much affect – at least down to 1 K – the magnitude of the low-temperature resistivity, introduce new qualitative features in our understanding of low-temperature transport effects. Thus, after reviewing the classical ideal and residual resistivities in Sects. 6.1.2, 3, we discuss in Sect. 6.1.4 the contributions due to quantum localization and interaction effects. Note that the interpretation of these quantum effects confirms the 2D conduction characteristics in acceptor GICs [6.10, 18].

From an experimental viewpoint, the problems encountered in the measurement of the electrical resistivity are directly related to the special electronic properties of these compounds:

(1) The anisotropy of the electrical resistivity, which greatly complicates the measurement of the in-plane resistivity in HOPG-based acceptor GICs [6.19, 20]. This is due to the fact that samples with extremely high length to width ratios are needed to obtain homogeneous current densities in the sample cross section. HOPG-based samples with such high ratios are not practically available.

(2) The importance of the residual resistivity relative to the ideal one. In acceptor GICs, the residual resistivity ratio (RRR) – the ratio of the resistance at 300 K to that at 4.2 K – is rather small: roughly less than 10 for HOPG-based compounds [6.21–23], less than 5 for vapor-deposited carbon fiber (VDF) compounds [6.24], and less than 2 for mesophase fiber-based compounds (PDF) [6.17]. Because of the low RRR, it is difficult to extract the ideal term from the measured total resistivity using (6.1) – particularily in the low-temperature range, since ϱ_i results from subtracting two large numbers.

Unlike HOPG, four-probe dc electrical measurements may be performed on graphite fibers because of the favorable length to cross section ratios [6.25]. It is thus possible to perform high-resolution electrical resistivity measurements on graphite fibrous acceptor compounds and to separate the ideal resistivity from the residual resistivity despite their very low RRR [6.17, 24]. As a corollary, we can simultaneously measure the electrical and thermal conductivities on the same sample (Sect. 6.2.3). This allows a direct determination of the electronic component from the measured thermal conductivity in some temperature ranges using the Wiedemann-Franz relation (Sect. 6.2.3).

With regard to the interpretation of the electrical resistivity measurements, the particular situation of acceptor GICs when compared with 3D metals is interesting. In GICs, the in-plane lattice properties, and thus the phonon spectra, are governed by the strong in-plane covalent bonds of the carbon atoms. The electronic system instead is a direct consequence of the charge transfer from the intercalate. Thus we might expect substantial differences from the behavior of 3D metals, especially when treating the electron–phonon interactions, as is the case for the ideal electrical resistivity. This is particularly true when we compare the high in-plane Debye temperature of these compounds, which is higher than 2000 K, with those of the 3D metals, which lie around and below 300 K. This means that in acceptor GICs we might observe electron–phonon interactions, and thus temperature dependences of the resistivities, quite different from those observed in 3D metals. This point has recently been discussed in some detail [6.2, 20].

In light of the situation in acceptor GICs, a simple picture was proposed to interpret the in-plane resistivity data [6.2, 20]. It was thought of as a 2D hole gas, originating from the charge transfer from the intercalate, which is fully described once its Fermi energy has been experimentally determined. This hole gas interacts with scattering centers (mainly phonons and defects) in the host layers where the Fermi gas has been introduced through the charge transfer. As was previously done for AsF_5 compounds (Sect. 6.1.5), in a first approach we shall neglect the possible scattering from the intercalate phonons, for which there is not yet experimental evidence.

The expression for the electrical conductivity is given by

$$\sigma = \frac{q^2 N \tau}{m^*} , \tag{6.2}$$

where q is the electronic charge, N is the charge carrier density, τ is the relaxation time of the charge carriers, and m^* is their effective mass. Equation (6.2) holds independent of the model chosen for the charge carrier distribution, and it was recently demonstrated that it also applies for 2D conduction in acceptor GICs [6.2].

Since the carrier mean free path l is equal to $v_F \tau$, and $\hbar k_F$ corresponds to $m v_F$, we can write

$$\sigma = \frac{q^2 N l}{\hbar k_F} . \tag{6.3}$$

For stage-1 acceptor compounds, where the carrier dispersion is linear as described by the *Blinowski-Rigaux* model [6.4], the 2D Fermi surface is a circle and the conductivity can be expressed as [6.2]

$$\sigma_{2D} = q^2 v_F \tau (k_F^2/\pi)(\hbar k_F)^{-1} , \tag{6.4}$$

where v_F and k_F are the Fermi velocity and wave number, respectively.

For 2D systems, N is a carrier density per cm^2 (per cm^3 in 3D systems), and the conductivity is in units of $\Omega^{-1}(\Omega^{-1}\,cm^{-1}$ in 3D systems). The relation between the 2D resistivity and the measured electrical resistance R_m of a sample consisting of $1/I_c$ planes of 1 cm width and 1 cm length is

$$R_m = I_c \varrho_{2D} , \tag{6.5}$$

since what is measured is the electrical resistance of a stack of d/I_c parallel planes, where d is the sample thickness, I_c the c axis repeat distance, and $\varrho_{2D} = \sigma_{2D}^{-1}$ the 2D electrical resistivity. In that case, R_m is the equivalent of a 3D resistivity ϱ_{3D}.

There are two prerequisites for analyzing the electrical resistivity results for acceptor GICs consistently:

(1) A reliable model for the band structure and a knowledge of the Fermi energies.
(2) High-resolution resistivity measurements extending from the liquid-helium range and covering about two decades of temperature.

Expressions (6.2–4) are derived for a single-carrier valley. If the Fermi surface consists of more than one valley and/or of more than one band, as is the case for higher-stage GICs, the contribution of each valley and band should be taken into account. In that case, the total electrical conductivity is given by the sum of the conductivities σ_j for each group of carriers j:

$$\sigma = \sum \sigma_j . \tag{6.6}$$

Applying Matthiessen's rule (6.1) for each band and taking into account the various scattering events s, we write

$$\varrho_j = \sum \varrho_{js} . \tag{6.7}$$

In GICs, s may denote scattering by static defects ($s = r$), charge carriers, or phonons ($s = i$).

These formulas show that contributions to the conductivity from different carrier groups add up, while it is the resistivity due to various scattering mechanisms that add and Matthiessen's rule – (6.1) or (6.7) – is applied for each band.

6.1.2 Ideal Electrical Resistivity

For 3D metallic systems, when the electrical resistivity is due to electron–phonon scattering, the Bloch-Grüneisen relation [6.26, 27] predicts a variation of resistivity as T^n with $n = 5$ at very low temperature, with a gradual decrease in n with increasing temperature until it is equal to 1 around and above the Debye temperature. It was suggested instead from earlier experimental data that the temperature dependence of the ideal resistance in GICs could be expressed [6.1] as

$$R(T) = BT + CT^2 , \tag{6.8}$$

where B and C are constants.

In acceptor GICs the ideal resistivity is sensitive to the charge transfer from the intercalate and to the phonon spectrum of the host material [6.2, 20]. The charge transfer, in turn, depends on the nature of the intercalated species and the stage of the compound. For well-graphitized carbons, for which the defect structure is not very pronounced, the phonon dispersion relations do not vary significantly from one sample to another. Thus, when comparing various samples with different host materials, whether from the same precursor or not, we can reasonably assume that the differences in ideal resistivities should be ascribed only to the charge transfer, i.e., to the Fermi energy if the same 2D band model is assumed for all acceptor GICs. Since the Fermi energies for metal-chloride compounds are almost the same for a given stage and are not too different from one stage to the other for low stages, the ideal resistivities are not expected to vary much from one chloride compound to another. This is what was observed (Fig. 6.1).

The temperature dependence of the electrical resistivities observed in low-stage (stages $n \leqslant 4$) metal-chloride acceptor GICs, which are presented in Fig. 6.1, show that the ideal resistivities are almost independent of the intercalate, stage, or host material (HOPG; carbon fibers that are vapor deposited, pitch-derived, etc.), provided that the host material is of good crystalline perfection. Figure 6.1 also shows that to separate the two contributions of the resistivity, high-resolution measurements are needed. The electrical resistivity has to be measured to better than one part per 10^5 in the lowest temperature range. The residual resistivity ϱ_r is first determined by extrapolating to $T = 0\,\mathrm{K}$ the experimental values of the total classical resistivity ϱ, measured in the lowest temperature range. Then, using Matthiessen's rule (6.1), the temperature dependence of the ideal resistivity ϱ_i is determined.

Results obtained by *Meschi* and coworkers [6.28, 29] revealed the same behavior as that shown in Fig. 6.1. Although these authors did not separate the ideal and residual contributions, this can be roughly done from their data for some compounds reported. The same applies to other published data on the temperature dependence of the electrical resistivity [6.30, 31].

The high-resolution resistivity measurements [6.17, 20, 24] suggest that in the lowest temperature range, hole–hole interaction is important. Nevertheless, the main scattering mechanism for charge carriers that leads to a temperature-dependent relaxation time at high temperatures is scattering by acoustic phonons. We have already pointed out the complexity of a quantitative description of the electron–phonon interaction in GICs [6.2, 20]. Indeed, the effectiveness of a scattering event for a degenerate electron gas depends on the angle through which the charge carriers are scattered on the Fermi surface and, for a given scattering angle, on the number of phonons available. The main difference between a 3D free-electron system and the 2D hole system in acceptor GICs is that around room temperature the charge carriers do not interact with the same

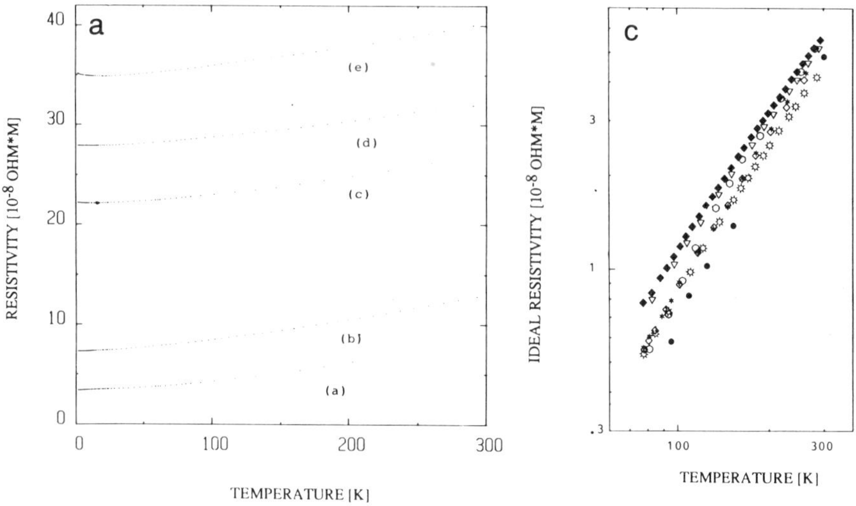

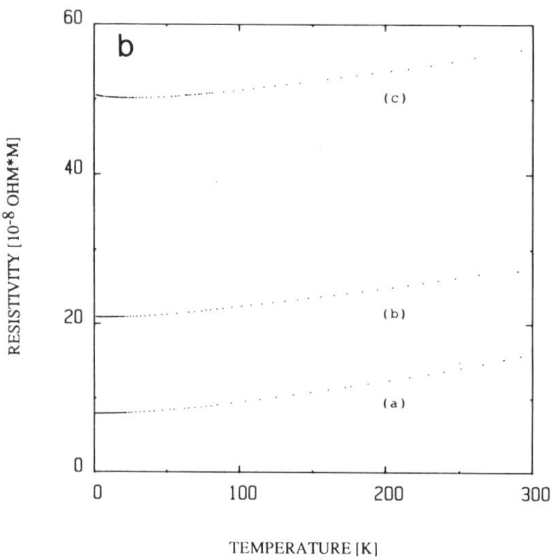

Fig. 6.1. Temperature dependence of the in-plane electrical resistivity of various low stage (for stages $n \leqslant 4$) GICs. Note that the resistivities are given as the equivalent of the 3D electrical resistivity, and not in 2D, to facilitate comparison with other 3D materials. In parts (**a, b**) the total measured resistivities are presented, while in (**c**) only the ideal contribution is shown. The curves may be fitted to relation (6.8) in the temperature range $1.5 < T < 300$ K; i.e., for the ideal contribution an almost linear variation of the ideal resistivity in the lowest temperature range and a T^n behavior with $n \approx 2$ around room temperature is observed.
(**a**) In-plane resistivity of $CuCl_2$ intercalates for different host materials and stage numbers, n. The traces are (a) BDF, $n = 2$; (b) PX5, $n = 4$; (c) VSC25, $n = 1$–2 (mainly 2); (d) P55 ($n = 3$); (e) P100-4, $n = 1$–2 (mainly 2).
(**b**) In-plane resistivity of stage-2 compounds with two other metal-chloride intercalates and three different host materials: (a) BDF-$SbCl_5$, (b) PX5-$SbCl_5$, (c) VSC25-$CoCl_2$.
(**c**) The ideal resistivities in the temperature range $77 < T < 300$ K for the $CuCl_2$ samples whose total resistivities are presented in (**a**). (★) BDF, $n = 2$; (◆) PX5, $n = 4$; (◇) VSC25, $n = 1$–2 (mainly 2); (✡) P55, $n = 3$; (▽) P100-4, $n = 1$–2 (mainly 2), and for two other samples reported in [6.36]: (○) HOPG, $n = 2$ and (●) PDF, $n = 2$–3–4. The ideal resistivities are almost the same in spite of the wide differences in total resistivities as shown in part (**a**), i.e., whatever the stage or host material. Note that the temperature-dependent part of the resistivity is a tiny fraction of the total resistivity in most of the compounds. The data for the intercalation compounds BDF and VSC25-$CuCl_2$ are from reference [6.20] and that for HOPG-$CuCl_2$ from reference [6.36]. All the other data are courtesy of L. Piraux (Ph.D. thesis)

class of phonons in the two systems. For a 3D free-electron system, the electrons are mainly scattered by the Debye phonons around and above the Debye temperature and by phonons of energy $\sim k_B T$ below the Debye temperature. This leads to small relaxation times around room temperature and to large ones at low temperatures. For stage-1 and stage-2 acceptor GICs, which can be considered as quasi-2D for electronic conduction, the holes are scattered through small angles by low-energy ($\sim k_B T$) thermal phonons well below room temperature. Above room temperature, more precisely, above a characteristic temperature θ^*, where [6.20]

$$\theta^* = \frac{2k_F v_s \hbar}{k_B}, \tag{6.9}$$

the holes are scattered by subthermal phonons – which are not the dominant ones at the temperature considered – through large angles. These *subthermal phonons* are the same that drag along the carriers in the phonon-drag situation and that we will call hereafter the *drag phonons* (Sect. 6.3.2b).

For graphite, the velocity of sound $v_S \sim 2.1 \times 10^6$ cm/s in-plane and for metal-chloride GICs, $k_F \sim 2 \times 10^7\ \mathrm{cm}^{-1}$. Thus we obtain for θ^* a value of ~ 540 K. The Fermi wave vector in acceptor compounds is an order of magnitude smaller than in ordinary metals. However, the velocity of sound is one order of magnitude higher in the graphene planes. The result is that θ^* is accidentally of the same order as in metals. However, in metals θ^* roughly corresponds to θ_D, the Debye temperature, whereas for the intercalated graphene layers $\theta^* \sim \theta_D/4$.

As a result, the electron-phonon interaction might be weaker in GICs than in 3D metals at a given temperature. Although the angle of scattering is not very different in both cases at a given temperature, the number of interacting phonons is smaller in acceptor GICs, since we are well below the in-plane Debye temperature. Thus, for acceptor GICs, at low and high temperatures, scattering should be less effective than in 3D metals.

Recent measurements performed on low-stage acceptor GICs confirmed the validity of (6.8) in the temperature range $1.5 < T < 300$ K [6.20, 24]. An almost linear variation of the ideal resistivity was observed in the low-temperature range and a T^n behavior with $n \approx 2$ around room temperature. Figure 6.1 shows these data for the various samples investigated where the temperature variation of the ideal resistivities of low-stage acceptor GICs are presented for various intercalates and host materials. It may be seen that, contrary to what is observed in a 3D metal, we find a coefficient n larger at higher temperatures than at lower temperatures. It is hard to explain such a behavior if we consider only electron–phonon (acoustic) interactions, which would then be described by a relaxation time varying as T^1 for small-angle scattering at low temperatures. The relaxation time for 2D hole–hole interactions instead is expected to vary as T^1 in the presence of weak disorder (Sect. 6.1.4). Indeed, the recent findings concerning the low-temperature behavior of these compounds suggest that strong hole–hole interactions might be effective at low temperatures. In this connection, we should consider what is the upper limit of temperature where weak disorder might have an effect on the temperature-dependent part of the resistivity.

The explanation of the temperature dependence of the resistivity remains puzzling, but the experimental facts are now well established. Indeed, Fig. 6.1 shows that the high-temperature ideal resistivity of a stage-2 $CuCl_2$ intercalation compound with vapor-deposited benzene-derived fibers (BDF) or PDF as host materials is exactly the same for both systems, while the residual resistivity is, as expected, quite different. This applies also to $FeCl_3$ with BDF and HOPG respectively and to other intercalates. It justifies our starting assumption: that the electronic system depends solely on the charge transfer from the intercalate (i.e., the Fermi energy), while the scattering mechanism in the ideal range at high temperature is mainly due to the phonon spectrum of the host graphite. Since, for a given stage, the Fermi energies are not very different for the various metal-chloride acceptor intercalates considered here, the electron–phonon interaction and thus the ideal resistivity should not be very different from one sample to another at high temperature. On the other hand, the residual resistivity depends on the defect structure of the pristine host material and on the additional defects introduced by intercalation. Thus it should vary from one host material to another and for the various intercalate species.

6.1.3 Residual Electrical Resistivity

Since electrical conduction is assumed to occur in the graphene layers, the residual resistivity of acceptor GICs will be mainly determined by the defect structure of these layers. This defect structure will depend on that of the pristine host material and on the defects possibly introduced during the intercalation process, if the scattering lengths in both cases are comparable. In pristine graphite fibers, for example, the defect structure may vary widely according to the type of fiber (vapor deposited, pitch-derived, pan-based, . . .), the heat treatment temperature, and the quality of the precursor for a given type of fiber. Thus carbon fibers provide a large variety of host structures of different crystalline perfection.

For acceptor GICs where the intercalation process occurs through the *Daumas-Hérold* [6.32] mechanism (i.e., for stages $n \geqslant 2$), some of the additional defects introduced during synthesis, one cannot avoid these large-scale 2D defects, the so-called Daumas-Hérold domain boundaries. We shall see that this is indeed the case in intercalated metal chlorides.

Thus, the residual resistivity of an acceptor GIC depends essentially on the following parameters:

(1) the quality of the pristine host material prior to intercalation
(2) the nature of the inercalate

(3) the intercalation process
(4) the stage of the intercalate.

Later we examine these factors in the light of recent experimental results.

Since the residual resistivity is due to scattering by static lattice defects, its value should provide, in principle, some information about these defects. The static defects usually considered are point defects and dislocations – and in 3D materials surface defects such as sample boundaries for single crystals and grain boundaries for a polycrystalline material. These are large-scale defects with respect to the carrier wavelengths. In a 2D material, a linear defect in the plane of the carrier propagation is equivalent to a surface defect in 3D solids. The Daumas-Hérold domain boundaries are therefore large-scale linear defects for holes. As we shall see later, this will also be the case for phonons (Sect. 6.2.2). One case of practical importance is when large-scale defect scattering dominates, which is believed to be the case for all acceptor GICs of stage $n \geqslant 2$ in the liquid-helium range [6.20]. In that case, the mean free path l_r is temperature and energy independent and we can use the residual resistivity to determine the size of these defects. We have recently shown [6.2, 20] that for a stage-1 compound, when $\varrho_r \gg \varrho_i$, the hole mean free path may be expressed as

$$l_r = 3\pi\hbar b\gamma_0 I_c (2q^2 R_r \varepsilon_F)^{-1} , \tag{6.10}$$

where b is the in-plane nearest-neighbor distance, γ_0 the overlap energy, ε_F the Fermi energy, and R_r the measured equivalent of a 3D residual resistivity.

Since in HOPG-acceptor GICs R_r is of the order of $10^{-6}\ \Omega$ cm, $I_c \sim 1$ nm, and $\varepsilon_F \sim 1$ eV, the mean free path for defect scattering is on the order of few 10^2 nm, a value much smaller than in the pristine material. This gives an upper limit for the size of large-scale defects in stage-1 HOPG intercalation compounds. For fibers, whose structure is generally more disordered than that of HOPG, the R_r should be higher and l_r correspondingly smaller. For intercalated BDFs heat treated at high temperatures, which have a structure close to that of HOPG, l_r should be larger than for intercalated PDFs and is indeed close to l_r for intercalated HOPG. In the pristine material, the mean-free path is of the order of the crystallite size, which is smaller in PDF than in BDF or HOPG. The relative decrease of l_r after intercalation is less pronounced for PDFs. Thus, in addition to the initial defect strucure of the host material, we should also take into account the effect of defects introduced by the intercalation process. However, the relative effect of these defects is more important when the host material is of higher crystalline perfection. This results from the combination of the mean free paths for various defects following Matthiessen's rule (6.7).

For a stage-2 compound in the residual range we have [6.2]

$$l_r = \frac{I_c}{R_r}\frac{\pi\hbar}{q^2}(k_{F1} + k_{F2})^{-1} , \tag{6.11}$$

where k_{F1} and k_{F2} are the Fermi wave vectors in the hole bands 1 and 2 respectively [6.2, 4].

Relations (6.10, 11) show that the measurement of the low-temperature resistance R_r and a knowledge of the Fermi wave vector or energy allow a direct estimate of the size of the large-scale defects. In Fig. 6.1 the values of R_r obtained for acceptor GICs with different host materials and intercalates may be compared.

6.1.4 Two-Dimensional Localization and Interaction Effects

A logarithmic increase of resistivity with decreasing temperature was observed in a variety of quasi-2D metallic systems. This was interpreted in terms of weak electron localization [6.11–13] and electron–electron Coulomb interactions [6.14, 15], which are both enhanced by defect scattering [6.33–35]. Logarithmic resistivities were also recently observed in acceptor GICs in the liquid-helium range [6.16–18], where changes in ϱ from 0.1% to 1% were detected over a decade of temperature using high-resolution resistivity measurements (Fig. 6.2). This resistivity behavior was accompanied by a negative magnetoresistance (Fig. 6.3).

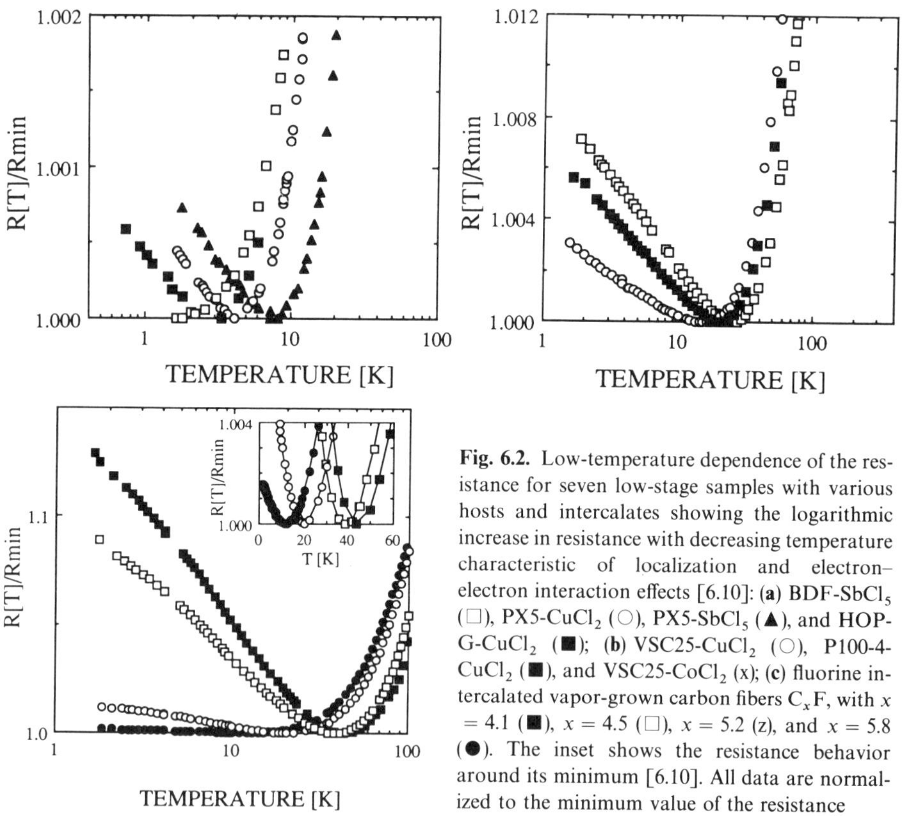

Fig. 6.2. Low-temperature dependence of the resistance for seven low-stage samples with various hosts and intercalates showing the logarithmic increase in resistance with decreasing temperature characteristic of localization and electron–electron interaction effects [6.10]: (**a**) BDF-$SbCl_5$ (□), PX5-$CuCl_2$ (○), PX5-$SbCl_5$ (▲), and HOPG-$CuCl_2$ (■); (**b**) VSC25-$CuCl_2$ (○), P100-4-$CuCl_2$ (■), and VSC25-$CoCl_2$ (x); (**c**) fluorine intercalated vapor-grown carbon fibers C_xF, with $x = 4.1$ (■), $x = 4.5$ (□), $x = 5.2$ (z), and $x = 5.8$ (●). The inset shows the resistance behavior around its minimum [6.10]. All data are normalized to the minimum value of the resistance

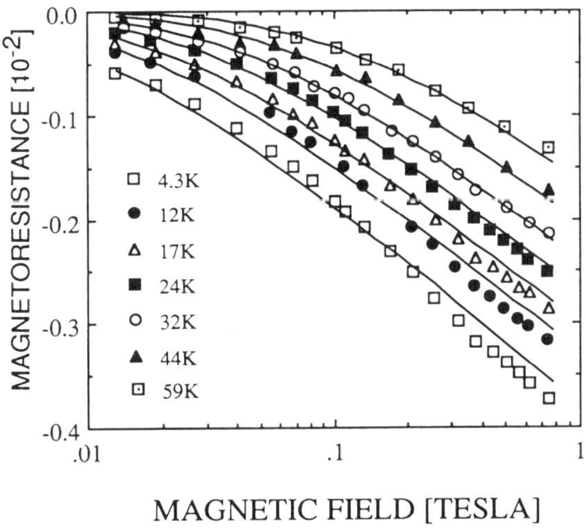

Fig. 6.3. Transverse negative magnetoresistance versus magnetic field for a C_xF fiber with $x = 4.1$ at various temperatures as indicated. The solid lines are calculated from a best fit using the weak localization theory [6.10]

Two mechanisms may contribute to the correction term that gives rise to the logarithmic increase in resistivity: a single-carrier weak localization effect produced by constructive quantum interference between elastically backscattered partial carrier waves [6.12], and charge carrier many-body Coulomb interactions [6.14, 15]. Both effects are enhanced by weak disorder, i.e., by defect scattering. A weak external magnetic field suppresses the phase coherence of the backscattered waves but does not influence the Coulomb interaction phenomenon. Thus magnetoresistance measurements allow separation of the two effects.

Acceptor GICs, and especially stage-1 compounds which have a circular Fermi surface, are natural 2D electronic systems since this character is inherent to their band structure. The 2D behavior results from the distribution of the charge carriers, which are strongly localized in the graphene planes and which may be considered as quasi-free carriers – although weakly localized – for motion parallel to the planes. This is in contrast to metallic films where the quasi-2D behavior results from strong scattering of the charge carriers at surfaces and interfaces. In metallic films, where localization effects were first observed, the 2D character is associated with the anisotropy of the mean free path, the latter becoming very small in the direction perpendicular to the film, thus confining the carrier motion to the plane of the film. The differences between the 2D electronic structure of acceptor GICs and that of other quasi-2D electronic systems originate from differences in the energy dependence of the density of states, which is linear in acceptor GICs [6.2].

Also, the possibility of varying the defect strucure of the host material over wide ranges in GICs allows extensive investigations of the phenomena of weak localization and electron–electron interactions. Indeed we can vary the nature of the precursor and the heat treatment temperature or cycle. In addition, to a lesser extent, the Fermi level can be modified by varying the nature of the intercalate and its concentration (stage). Acceptor GICs are thus among the best candidates to investigate 2D localization and interaction effects.

In the weak disorder limit, which is also the condition for transport in the Boltzmann approximation, i.e., when

$$k_F l \gg 1 . \tag{6.12}$$

Here k_F is the Fermi wave vector, and l is the mean free path of the carriers, a correction term $\delta\sigma_{2D}$ is added to the Boltzmann classical electrical conductivity. This then becomes

$$\sigma_{2D} = \sigma_{2D,\mathrm{Boltz}} + \delta\sigma_{2D} , \tag{6.13}$$

with $\sigma_{2D,\mathrm{Boltz}}$ as given in (6.2).

The term $\delta\sigma_{2D}$ accounts for weak localization and carrier-carrier interaction effects, which both predict a similar resistance-temperature anomaly with temperature.

Figure 6.2 shows that the logarithmic increase in resistivity is apparent in most acceptor GICs whatever the host material or intercalate. The temperature at which the minimum occurs in the $\varrho(T)$ curve was found to increase with the residual resistivity ϱ_r (Fig. 6.4), thus suggesting that for the more highly ordered

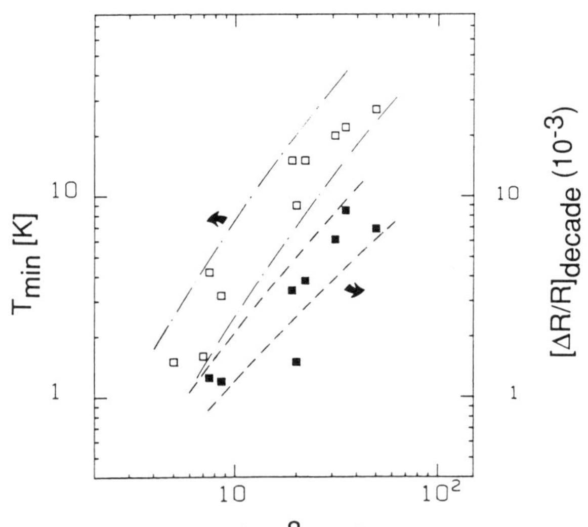

Fig. 6.4. Relation between the value of the residual resistivity $\varrho_{4.2\,\mathrm{K}}$ of carbon fibers intercalated with metal chlorides and the temperature at which the resistivity is minimum, $T_{\min}$ (open squares, left scale), and the logarithmic variation of resistance over a decade of temperature $[\Delta R/R]_{\mathrm{decade}}$ (dark squares, right scale). The dashed lines are calculated values [6.10]

carbons, decreasing the temperature might reveal the same behavior, but with a minimum of resistivity shifted below 1.5 K. Indeed, the fact that the most spectacular examples for localization and Coulomb interaction were observed on fiber-based compounds of low crystalline perfection does not mean at all that such effects do not exist in more highly graphitized fibers or in HOPG. More generally speaking, we think that localization effects are a feature of all metallic systems. However, these effects are more difficult to observe in the HOPG-based compounds for the following reason. The increase in resistivity is usually small above 1.5 K, the lowest temperature at which measurements were performed. The better the samples are, the smaller the increase in resistance due to localization (Fig. 6.4). Thus, for good samples, we need even higher resolution electrical resistivity measurements and lower temperatures to detect localization effects. For HOPG-based compounds, four-probe dc measurements are not reliable (Sect. 6.1.1), and the ac measurements are not accurate enough (1%) to detect the increase in resistance. Despite that, it seems that localization effects have been observed in compounds with HOPG of low quality as the host material [6.29, 36].

The results presented in Figs. 6.3, 4 were quantitatively interpreted using the theories of weak localization and electron–electron Coulomb interaction in 2D disordered systems [6.18]. It was shown that the amplitude of the effect and the temperature of the resistance minimum were indeed directly related to the residual resistivity of the compound as observed experimentally (Fig. 6.4). Also, the values of the temperature-dependent inelastic scattering time obtained from the temperature and field dependence of the magnetoresistance were found to agree well with those predicted by both theories of electron–electron Coulomb interaction and electron–phonon interaction in 2D disordered systems.

With $\delta R = -R^2\,\delta\sigma_{2D}$, the temperature dependence of the resistance of 2D disordered systems, due to the occurrence of both weak localization and Coulomb interaction effects, is given by [6.13, 14]

$$\frac{\delta R(T)}{R(T_0)} = -\frac{q^2}{2\pi^2\hbar}\,\varrho_{2D}(\alpha p + \gamma)\ln\left(\frac{T}{T_0}\right), \qquad (6.14)$$

where $\delta R(T) = R(T) - R(T_0)$ and ϱ_{2D} is the 2D resistivity at the temperaure T_0. The weak localization contribution is determined by the product αp, while the Coulomb interaction contribution is given by γ, where γ – a measure of the screening by other charge carriers – has a value close to 1 in the limit of weak screening [6.14, 37]. Here p is the exponent in the temperature dependence T^p of the inverse of the inelastic scattering time τ_i. The value of α depends on the effect of spin-dependent processes due to magnetic scattering and spin-orbit coupling. The characteristic scattering times for these last two processes are respectively τ_s and τ_{so}; both are assumed to be temperature independent.

The physical meaning of the different scattering times is that for times smaller than $\tau_k (k = \mathrm{i, s, so})$ the phase coherence of interfering complementary waves, which is the source of the weak localization effect, is maintained [6.12]. Therefore, weak localization occurs only when the scattering by static defects dominates, i.e., when $\tau_0 \ll \tau_k$, where τ_0 is the elastic scattering time due to defect scattering. In the limit where $\tau_i \ll \tau_{so}, \tau_s$, $\alpha = 1$. For $\tau_s \ll \tau_i, \tau_{so}$, $\alpha = 0$, and if $\tau_{so} \ll \tau_i, \tau_s$, $\alpha = -1/2$.

The weak localization theory for 2D systems [6.13] predicts that a uniform perpendicular magnetic field H introduces a phase shift between interfering partial carrier waves, thus weakening the localization effect when the characteristic magnetic time

$$\tau_H = \frac{\hbar}{4qDH} \tag{6.15}$$

becomes comparable to or smaller than the dominant scattering time τ_k. In (6.15) $D = \frac{1}{2}(v_F^2 \tau_0)$ is the 2D diffusion constant, and $v_F \sim 8 \times 10^5$ m/s in low-stage acceptor GICs is the Fermi velocity. Therefore, since the Coulomb interaction is not affected by the low magnetic field applied [6.35, 38], the measurement of the magnetoresistance at various temperatures does allow the determination of the relative contribution of both effects to the temperature dependence of the resistance. It was explicitely shown [6.18] that the dominant contribution to the low-temperature anomalous behavior is the hole–hole interaction (65%–85%).

In addition, note that the minimum of the electrical resistivity cannot be attributed to the Kondo effect. Indeed, near this minimum, there are no anomalies typical of the Kondo effect in the thermoelectric power, which is found to depend linearly on the temperature [6.39].

Altshuler et al. [6.40] calculated the scattering time τ_i due to electron–electron interaction in a strictly 2D electron gas and obtained

$$\left(\frac{1}{\tau_i}\right)_{ee} = \frac{k_B T}{2\varepsilon_F \tau_0} \ln\left(\frac{\varepsilon_F \tau_0}{\hbar}\right). \tag{6.16}$$

Using $\tau_0 = l_0/v_F \cong 2.0 \times 10^{-14}$ s and $\varepsilon_F \cong 0.8$ eV, we find $(1/\tau_i)_{ee} \cong 3.8 \times 10^9\, T$, where T is the absolute temperature. This was found to agree well with the values obtained in other experimental observations [6.18].

The expression for the electron–phonon lifetime in the dirty limit has been derived by *Takayama* [6.41], who proposed

$$\left(\frac{1}{\tau_i}\right)_{ep} = \frac{2\pi^2 a k_B}{k_F l_0 \hbar \theta_D} T^2 , \tag{6.17}$$

where a is a numerical factor of the order of unity, k_F is the Fermi wave vector, and θ_D is the Debye temperature. In the case of acceptor GICs, taking for θ_D the value of 540 K, the effective temperature for hole–phonon interaction θ^* (Sect. 6.1.2), it is found from (6.17) that $(1/\tau_i)_{ep} \sim 1.5\ 10^8\, T^2$. The relation between this value of τ_i and the one that may be obtained via the ideal resistivity measurements has been discussed [6.18].

It is worth adding that localization effects have been observed in all intercalated mesophase fibers investigated, as well as in vapor-deposited fibers and HOPGs of low structural perfection. All the compounds investigated were found to verify the weak disorder condition (6.12). The temperature dependence of the resistivity increase was found consistent with the 2D electronic structure, and the results obtained concerning the temperature and magnetic field dependences of the effect were correctly interpreted in terms of the existing theories derived for quadratic dispersions, which remain valid for linear dispersion relations [6.18].

6.1.5 Electrical Conductivity and Charge Transfer

From the analysis of the results on the temperature variation of the electrical resistivities of various metal-chloride GICs, we now speculate about the intrinsic (ideal) and extrinsic (residual) limits for their room-temperature resistivities. Also, we briefly discuss the effect of charge transfer on the resistivity.

Indeed, since the discovery of an acceptor GIC with electrical conductivities higher than that of copper, a question which has periodically been raised concerns the highest electrical conductivity limit that can be attained with GICs.

From the available experimental results and the phenomenological ingredients of the transport theory, we now can estimate this limit for the metal-chloride compounds. As described above, these compounds have been thoroughly investigated with HOPG and carbon fibers as host materials.

Let us consider a 3D free-electron system at a given temperature. If, by some artificial means, we were able to add more free electrons to the system keeping the other parameters unchanged, we would expect an increase in electrical conductivity. Indeed, for a 3D free-electron system (6.2), the carrier density N is proportional to k_F^3 [6.26]. The electron mass m^* is here by definition a constant, independent of k_F and equal to the free-electron mass m_0. Thus, the 3D electrical conductivity may be expressed as

$$\sigma_{3D} \propto k_F^3 \, \tau(k_F) \, . \tag{6.18}$$

Since the relaxation time $\tau(k_F)$ is expected to decrease with increasing wave number or carrier density, i.e.,

$$\tau(k_F) \propto k_F^{-a} \, , \tag{6.19}$$

from (6.18) the conductivity increases with increasing carrier density only if $a < 3$. This is the case for a 3D free-electron system where the conductivity is found to be proportional to $N^{2/3}$, and thus to k_F^2 [6.26].

However, it is not obvious at all that this conclusion should apply to any charge carrier distribution, and in particular to the special case of the 2D hole gas in acceptor GICs. The latter is far from being a free-electron system, and we might thus expect to see large deviations from the simple model.

In the 2D *Blinowski-Rigaux* model [6.4], N is proportional to k_F^2, and the effective mass m^* varies linearly with k_F [6.2]. Thus, from (6.2), the 2D electrical conductivity may be expressed as

$$\sigma_{2D} \sim k_F \, \tau(k_F) \, . \tag{6.20}$$

Then we see from relations (6.19, 20) that if $a > 1$ the electrical conductivity decreases when the Fermi wave vector, i.e., the carrier density, increases. The exact value of the exponent a is difficult to estimate since for such a calculation we should take into account the interaction of a 2D carrier system with a 3D phonon system, but around room temperature we expect a value of $a \geqslant 1$. Then the conductivity should be constant or decreasing with increasing charge transfer in the range wherever the hole dispersion relation is linear in k. If, as was previously assumed for AsF_5 compounds [6.42, 43], only scattering by in-plane graphitic phonons is considered, $a = 1$, then $\tau(k_F) \sim k_F^{-1}$. In this case we expect a conductivity independent of charge transfer. It would be interesting to check experimentally whether this is so. This could be done by measuring the room-temperature electrical resistivity of a stage-1 acceptor GIC as a function of charge transfer.

6.2 The In-Plane Thermal Conductivity

In contrast to the case of the electrical resistivity, little attention was paid to the thermal conductivity of GICs until the beginning of the last decade. The importance of this property as a method of sample characterization had not been given proper recognition. Indeed, not only is thermal conductivity the best way to investigate phonon–phonon interactions; phonons are also powerful means of probing the lattice defects in these compounds. On the other hand, thermal-conductivity measurements are time consuming and very delicate to perform [6.44]. This is particularly true when only samples of small dimensions are available, as with carbon fibers [6.45].

There are essentially two mechanisms for thermal conduction in solids around and below room temperature [6.46]. The first, the electronic thermal conductivity κ_E, involves the charge carriers, while the second, which involves the quantized lattice vibrations (the phonons), is called the lattice thermal conductivity κ_L. In electrical insulators, heat is exclusively carried by phonons, while in pure metals it is predominantly carried by the charge carriers. For some materials like highly doped semiconductors, metallic alloys, and group-V semimetals, the lattice thermal conductivity may be comparable to the electronic contribution in certain temperature ranges. This is also the case for GICs [6.47]. The total thermal conductivity is expressed as

$$\kappa = \kappa_E + \kappa_L \, . \tag{6.21}$$

In the case of magnetic compounds, magnons may also contribute to the transport of heat energy. We discuss this situation in Sect. 6.5.

To understand thermal transport in solids where many contributions to the thermal conductivity are comparable in magnitude, we must separate them (Sect. 6.2.3).

The temperature variation of the in-plane thermal conductivity in GICs based on HOPG [6.7, 47–53] and on highly graphitized fibers [6.39, 54] has been reported by several groups. *Issi* has also recently shown how intercalation modifies the thermal conductivity of graphites [6.2]. In the case of intercalated HOPG, both acceptor [6.7, 47–49, 51–53] and donor [6.47, 50] compounds have been investigated, while for intercalated fibers, only acceptor compounds have been studied [6.39, 54]. In contrast to the pristine material, the results of these studies have shown that in the intercalation compounds the lattice contribution is not the sole heat transport mechanism above liquid-helium temperature [6.2, 47, 55]. Since intercalation introduces lattice defects, it decreases the absolute magnitude of the lattice thermal conductivity around and above the dielectric maximum. On the other hand, the large increase in carrier concentration in GICs due to charge transfer adds a significant electronic contribution to the thermal conductivity, especially at low temperatures. The combination of these two effects leads to a decrease of the total thermal conductivity at high temperatures and to an increase at low temperatures with respect to that of the pristine material [6.2, 47, 55]. Figure 6.5 schematically represents the effect of intercalation on the thermal conductivity of pristine graphite [6.2]. Figure 6.6 presents some typical experimental results on the temperature and stage dependence of the thermal conductivity of HOPG-based GICs.

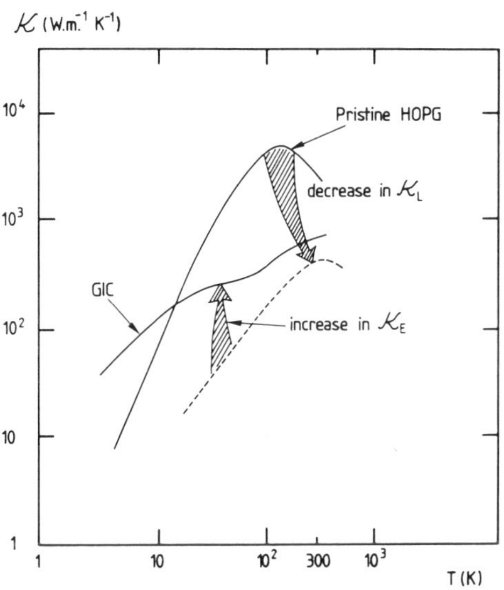

Fig. 6.5. Schematic representation of the effect of intercalation on the in-plane thermal conductivity of pristine graphites. The figure shows that intercalation has two main effects on the thermal conductivity of the pristine material. First, the defects introduced during intercalation decrease the in-plane lattice thermal conductivity κ_L through a reduction of the phonon mean path (lower dashed curve). Second, the main effect of intercalation is to increase the electrical conductivity as a result of the charge transfer: an increase in the carrier density accompanied by a decrease, but in a smaller proportion, of the electronic mobility. An increase in electrical conductivity will lead directly to an increase in the electronic thermal conductivity κ_E. So there are essentially two competing effects due to intercalation. The first tends to decrease the total conductivity, while the second has an opposite effect. The observed effect in intercalation compounds is that the thermal conductivity is decreased at high temperatures and increased at low temperatures with respect to the pristine material. This behavior was observed in all intercalation compounds investigated up to now, whatever the intercalate or the host material

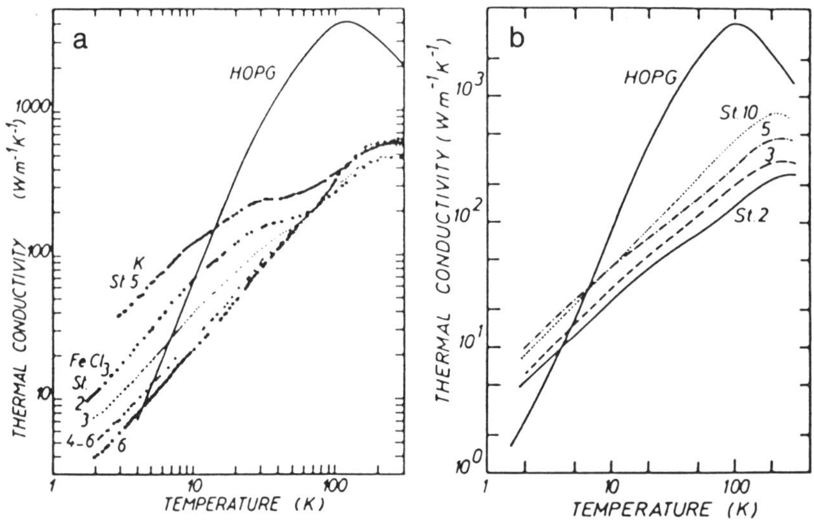

Fig. 6.6. The in-plane thermal conductivity of various HOPG GICs plotted as a function of temperature and compared with that of the pristine material: (**a**) pure stages 2, 3, and 6, and mixed-stage (4–6) HOPG-$FeCl_3$ compounds, and a stage-5 potassium (donor) intercalation compound [6.47]; (**b**) pure stage HOPG-$SbCl_5$ compounds [6.51]. The various compounds follow the general trend described in Fig. 6.5

6.2.1 Electronic Thermal Conductivity

Electronic thermal conductivity results from the transfer of energy due to the difference in the broadening of the Fermi-Dirac distribution around the Fermi level caused by a temperature gradient. The Wiedemann-Franz law directly relates the electronic thermal conductivity to the electrical conductivity σ:

$$\kappa_E = LT\sigma \; . \tag{6.22}$$

The Lorenz ratio L takes the value of the Lorenz number ($L_0 = 2.44 \times 10^{-8}\,V^2\,K^{-2}$) when both the electrical conductivity and electronic thermal conductivity are expressed in terms of the same relaxation time [6.26, 56]. This is true for a degenerate free-electron system that experiences elastic collisions. This relation holds in the low-temperature residual resistivity range when scattering is dominated by impurities and lattice defects. For 3D metals it is also applicable above the Debye temperature, i.e., when large-angle intra-valley electron–phonon (acoustic) scattering is the predominant scattering mechanism. Thus $L = L_0$ for metals over limited temperature ranges. For GICs, we should consider an effective temperature for electron–phonon interaction instead of the Debye temperature, as discussed in Sect. 6.1.2. Since this temperature was found to be around 540 K for low-stage acceptor GICs, we expect large-angle scattering to dominate the electron–phonon interaction in GICs above room temperature. Thus, only in the residual range and above room temperature can we consider that $L = L_0$ in low-stage acceptor GICs.

In pure metals, at low temperatures and within the residual electrical resistivity range, the electronic thermal conductivity varies linearly with temperature. At higher temperatures, κ_E reaches a maximum, which is sharper for purer samples than for samples with impurities and defects. At still higher temperatures, κ_E decreases and saturates to a constant value. The saturation takes place in a temperature range where the electrical resistivity should vary linearly with temperature, because of large-angle electron–phonon interactions. For samples with a high concentration of impurities or lattice defects, the linear increase may be directly followed by a temperature insensitive κ_E without any intermediate thermal-conductivity peak [6.46]. From the temperature variation of the electrical resistivity, we expect this last situation to prevail in GICs (Sect. 6.2.3). However, a linear variation of the ideal resistivity was never observed at high temperatures in acceptor GICs (Sect. 6.1.2).

The Wiedemann-Franz law (6.22) allows a determination of the electronic thermal conductivity from the knowledge of the electrical conductivity. However, κ_E can also be directly expressed in terms of electronic parameters [6.26]. Using elementary kinetic arguments, we can show that the electronic thermal conductivity is directly proportional to the electronic specific heat C_E, to the velocity of the electronic charges at the Fermi energy v_F and to the mean free path of the charge carriers l_E:

$$\kappa_E = \tfrac{1}{3} C_E v_F l_E . \tag{6.23}$$

Because of the large charge transfer upon intercalation, we expect a large increase in the Fermi energy. This is accompanied by a decrease in the mean free path of the charge carriers, because of increased defect scattering. This situation is similar to that of the electrical conductivity, where intercalation causes an increase in the carrier density along with a decrease in the electronic mobility. Since the carrier density increases faster than the mobility decreases, the net result of intercalation is an increase in electrical conductivity. In the same way, we should observe an increase in the electronic thermal conductivity. This follows directly from inspection of relation (6.22).

At low temperatures, the thermal conductivity of HOPG-derived GICs was first observed to vary almost linearly with temperature [6.48]. When applying the Wiedemann-Franz law, a reasonable value for the electrical resistivity was found [6.47]. This suggested that in the lowest temperature range the dominant mechanism for heat conduction is via the charge carriers. However, considering more recent results and a further analysis, this view must now be taken with some caution [6.57]. As we see in Sects. 6.2.3, 4, the charge carriers are the dominant mechanism only for very good samples. For samples of less structural perfection, the phonon conductivity may be comparable to the electronic conductivity in the lower temperature range. In fact, we have a contribution from three distinct types of carriers in the low-temperature range: the charge carriers, the graphite phonons, and the intercalate phonons.

Figure 6.7 shows the subtle balance between the effect of these three heat carriers that will determine the total low-temperature thermal conductivity [6.58].

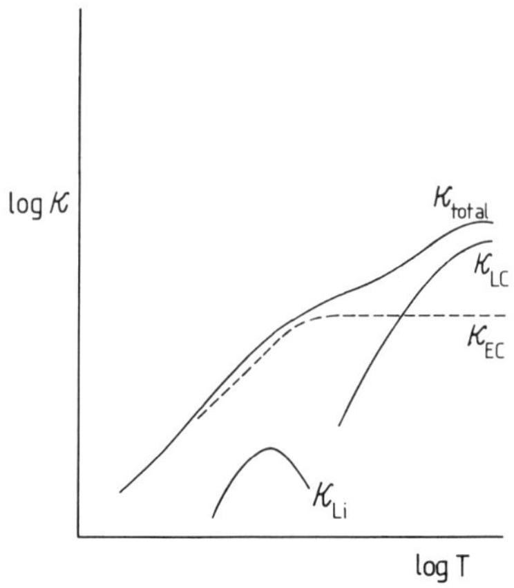

Fig. 6.7. Log-log schematic representation of the contribution of the three competitive heat carriers that will determine the total thermal conductivity in GICs below room temperature [6.58]. Two contributions are due to the graphene layers: electronic κ_{EC}, and lattice κ_{LC}. One is due to the phonons of the intercalate, κ_{Li}. As may be seen from Fig. 6.6, the maximum total thermal conductivity is on the order of a few hundred $Wm^{-1}K^{-1}$. In the lowest temperature range, the thermal conductivity increases almost linearily with temperature in some samples; then the additional contribution due to the intercalate, κ_{Li}, shows up. In other samples this contribution κ_{Li} can be very important, even in the lowest temperature range and hence no linear variation is observed (Sect. 6.2.4)

6.2.2 Lattice Thermal Conductivity

The lattice thermal conductivity is the transport of heat energy by phonons, which is induced by the difference in phonon concentration caused by a temperature gradient. The lattice or phonon thermal conductivity κ_L is often well approximated by the Debye formula [6.46]:

$$\kappa_L = \tfrac{1}{3} C_L v_s l_p , \tag{6.24}$$

where C_L is the lattice specific heat at constant volume, v_s is the phonon group velocity (the velocity of sound), and l_p is the phonon mean free path, which is directly related to the phonon relaxation time τ_p:

$$l_p = v_s \tau_p . \tag{6.25}$$

The Debye formula is based on an approximation, the dominant phonon mode approximation, which avoids the complexity of considering the entire phonon spectrum, with a velocity and mean free path associated with each phonon mode and polarization. Indeed, in a more rigorous approach we can consider the exact expression [6.46]:

$$\kappa_L = \sum \int_{\omega_0}^{\omega_D} C(\omega) v(\omega) l(\omega) d\omega , \tag{6.26}$$

where $C(\omega)$ is the heat capacity of a phonon of frequency ω, with velocity $v(\omega)$ and mean free path $l(\omega)$. The integration should be taken over all phonon modes and the summation for the three polarizations, and thus (6.26) takes into account the entire phonon spectrum.

Despite its simplicity, the Debye formula is very useful in the discussion of the essential features of the temperature variation of the lattice thermal-conductivity results. We know that the velocity of sound v_s is almost temperature independent, so it leaves us with an examination of the temperature variation of C_L and l_p to analyze κ_L. In the lowest temperature range, where the phonon mean free path is constant and limited by boundary scattering, the temperature variation of the lattice thermal conductivity directly reflects that of the lattice specific heat. For pristine graphite, phonons are scattered at the crystallite boundaries; thus knowing C_L, κ_L, and v_s, we can determine l_p, which is directly related to the crystallite size [6.59, 60].

At higher temperatures, the phonon mean free path decreases with increasing temperature due to resistive phonon–phonon umklapp scattering [6.46, 56]. Now the lattice specific heat is temperature independent and κ_L decreases when the temperature increases. Between these two extreme temperature ranges lies the dielectric maximum κ_{max}, whose magnitude and corresponding temperature depends on the lattice defects – mainly point defects – and sample size for single crystals, or the crystallite size for polycrystalline material. Therefore, from the analysis of the lowest temperature region and the region around the maximum κ_{max}, we can get a fair amount of information on lattice defects. Such an analysis was indeed carried out for GICs [6.47], and the results are discussed below.

To analyze quantitatively the low-temperature thermal-conductivity results of GICs and to determine the various phonon scattering mechanisms, we must take into account all the resistive mechanisms. These include scattering by lattice defects (mainly point defects and large-scale defects) and phonon–phonon scattering. Phonons may undergo umklapp processes and normal processes. Although the latter are not direct resistive processes, they lead to the creation of higher-frequency phonons, which are more apt to undergo umklapp and point defect scattering, both being resistive processes.

The large-scale lattice defects are those with much larger size than the phonon wavelength, such as crystal boundaries or possibly Daumas-Hérold domains (Sect. 6.1.3). These lead to relaxation times which are independent of phonon frequency:

$$\tau_B = L/v_s , \tag{6.27}$$

where the boundary scattering length L is, within a numerical factor close to unity, equal to the distance between such defects.

On the other hand, point defects are lattice defects with a size small compared with phonon wavelengths. The relaxation time for point defect scattering varies strongly with the phonon frequencies ω:

$$\tau_D = D\omega^r , \tag{6.28}$$

where D is a constant and $r = 4$ for 3D solids [6.46] and 3 for 2D solids [6.61].

For phonon normal and umklapp processes we shall denote the relaxation times by τ_N and τ_U respectively, which are both frequency and temperature dependent [6.46].

To take into account these various mechanisms in a quantitative way, *Callaway* [6.62] expressed the lattice thermal conductivity κ_L as a sum of two terms:

$$\kappa_L = \kappa_1 + \kappa_2 , \tag{6.29}$$

where κ_1 includes all relaxation mechanisms including normal processes as resistive mechanisms, and κ_2 is a correction term that allows for the fact that normal processes do not contribute directly to the thermal resistance.

The relaxation frequency τ_R^{-1}, which takes into account all resistive processes, is then

$$\tau_R^{-1} = \tau_U^{-1} + \tau_B^{-1} + \tau_D^{-1} , \tag{6.30}$$

and the relaxation frequency, which takes into account both resistive and normal processes, is

$$\tau_C^{-1} = \tau_N^{-1} + \tau_R^{-1} . \tag{6.31}$$

Using the *Callaway* formalism [6.62] and taking into account the particular anisotropy of phonon conduction in GICs, an analysis of the thermal conductivity of $FeCl_3$ GICs was carried out [6.47]. This allowed an estimate of the size of the point defects and the large-scale defects in these compounds [6.47].

6.2.3 Separation of the Electronic and Lattice Contributions

There are essentially two experimental methods for separating the electronic and lattice contributions to the thermal conductivity. The first and easiest way is to measure the electrical and thermal conductivities simultaneously on the same sample. Then, using the Wiedemann-Franz law (6.22), we may compute from the measured electrical resistivity the corresponding electronic thermal conductivity. By subtracting the latter from the total measured thermal conductivity (6.21), we obtain the lattice thermal conductivity for each temperature. However, this method is only applicable in the temperature range where the Wiedemann-Franz ratio is equal to L_0, the free-electron Lorenz number (Sect. 6.2.1).

Since most acceptor GICs have very low residual resistivity ratios (Sect. 6.1.3), the residual resistivity dominates up to rather high temperatures (Sect. 6.1). Thus the validity of the Wiedemann-Franz law in the residual range is expected to be maintained up to temperatures higher than those for pure 3D metals. This is particularly true in some fiber compounds, especially in pitch-derived fibers of poor structural perfection, where the electrical resistivity is very weakly temperature dependent. In that case the electronic thermal conductivity can be directly computed over a large temperature range from the measured electrical resistivity via the Wiedemann-Franz law using the free-electron Lorenz number [6.39, 54]. A recent report on the thermal conductivity of intercalated carbon fibers [6.57] has shown that for a P100-4-$CuCl_2$ compound, the electronic contribution may be computed in this way from liquid-helium temperature up to room temperature. This procedure can be applied up to only 130 K for PX5-$CuCl_2$ compound, where the host material is of higher structural perfection. At 300 K and 130 K respectively, the temperature-dependent part of the resistivity reaches almost 20% of the residual resistivity. Since this percentage decreases as the temperature is lowered, the separation procedure is justified. The separation of the total measured thermal conductivity of fiber-based compounds into the electronic and lattice contributions is shown in Sect. 8.5 of this volume. For earlier work please refer to [6.3].

If we compare these high-temperature results with those previously obtained for HOPG, BDF, and other pitch-derived carbon fibers, we see that the relative decrease in the lattice thermal conductivity due to intercalation is less pronounced as the lattice perfection of the pristine material decreases. This is akin to what is observed for the residual electrical resistivity discussed in Sect. 6.1.3 in terms of combination of the scattering lengths for large-scale defects of different sizes.

Another limitation of the separation method is specific to HOPG-based acceptor GICs. We have seen previously that in these intercalation compounds it is not possible to measure the dc electrical resistivity (Sect. 6.1.1). Thus, it is not possible to measure simultaneously on the same sample both the electrical resistivity and thermal conductivity.

So for the purpose of separating the two contributions to the thermal conductivity, the fibers have two major advantages as compared with HOPG. First, we can measure electrical and thermal conductivities on the same sample. Second, the Wiedemann-Franz law applies over larger temperature ranges.

The second way to separate κ into κ_E and κ_L is through the application of a large magnetic field. From the Wiedemann-Franz law (6.22), it may be seen that the application of a magnetic field to a sample with a large magnetoresistance will deplete the electronic thermal conductivity. At high magnetic fields $\kappa_E(H)$ may become negligible in comparison with κ_L. In the saturation region, $\kappa(H) \approx \kappa_L$, and κ_E and κ_L may be separated. In this analysis we must assume that κ_L is not affected by the magnetic field, which is generally the case. The magnetic field intensity required to reduce κ_E to insignificance as compared with κ_L depends on the carrier mobility: the higher the mobility the lower the field required. If the mobilities are too low, the separation of κ_L and κ_E might require higher magnetic fields than are presently available. For low-stage HOPG-based acceptor intercalated compounds the complete separation for low-stage compounds requires very high magnetic fields [6.51, 63].

6.2.4 The Extra Contribution due to the Intercalate

Recent work performed on intercalated graphitic fibers confirmed the existence of a low-temperature extra phonon contribution due to intercalation in addition to the electronic thermal conductivity [6.57]. This was previously discussed [6.64] and was already inferred from the first thermal-conductivity results [6.48]. Further studies of this extra contribution have shown a decrease of the effect with increasing stage number for low stages [6.57].

A small extra contribution is already apparent at low temperatures for the stage-2 HOPG-$FeCl_3$ intercalation compound as shown in Fig. 6.6a. This effect is, however, more pronounced in pitch-based carbon-fiber intercalation compounds [6.57, 64].

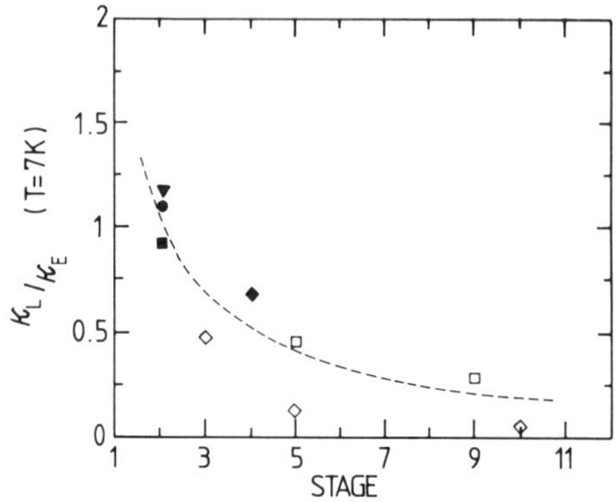

Fig. 6.8. Stage dependence of the low-temperature additional contribution to the electronic thermal conductivity at 7 K [6.57]. The results are presented in the form of the ratio of the lattice thermal conductivity (κ_L) to the electronic thermal conductivity (κ_E) as a function of stage for various intercalates and host structures: (□) HOPG-$FeCl_3$, (◇) HOPG-$SbCl_5$, (●) BDF-$CuCl_2$, (■) VSC25-$CuCl_2$, (◆) PX5-$CuCl_2$, and (▼) P100-4-$CuCl_2$. The dashed curve shows the variation expected from the sandwich model. The additional contribution due to intercalation is seen to decrease as the stage increases

In Fig. 6.8 we present the stage dependence of the low-temperature additional contribution to the electronic thermal conductivity for various acceptor GICs prepared with HOPG, benzene, and pitch-derived carbon fibers as host materials and metal chlorides as intercalates. The additional contribution due to intercalation decreases as the stage number increases. This is justified by the fact that as the compound becomes more dilute there are fewer of these extra phonons due to the intercalate per unit volume. The figure reveals that the ratio κ_L/κ_E decreases with increasing stage index, as expected from the sandwich model applied to thermal conductivity [6.47]. For very dilute compounds, the situation is qualitatively different. The relative contribution of the electronic term as well as the extra lattice term should become negligible with respect to that of the lattice thermal conductivity of graphite. The ratio κ_L/κ_E should increase with increasing stage and should tend to a very large value for an infinite stage, i.e., for the case of pristine graphite. In that case, κ_L is due only to the graphite phonons. Thus the above considerations should apply only for low stages.

6.3 The In-Plane Thermoelectric Power

Before discussing the thermoelectric power data obtained on GICs, we should review the elementary physics underlying this transport phenomenon and the status of our understanding of this property for other types of materials. Although the theory of the thermoelectric power is well established and this property has been widely measured on a large variety of solids, theory and experiments are almost always at variance, even in materials that are described by the simplest solid-state models corresponding to nearly free-electron systems. So the thermoelectric power is one of the easiest thermal parameters to measure but the most delicate to interpret correctly. However, the results obtained so far on GICs are fascinating, and some challenge our understanding.

In the following we first review the experimental data available on GICs, then describe the mechanisms responsible for the thermal emf generation in solids in general. Finally, we show how these models apply to GICs, pointing out some particular features of GICs in this respect.

6.3.1 Experimental Results

Since the pioneering work of *Blackman* et al. [6.65] followed by that of *Ubbelohde* [6.66], many papers have been published on the temperature dependence of the in-plane [6.6, 7, 39, 49, 51–54, 67–70] and the c-axis [6.5, 71, 72] thermoelectric power of GICs. These results show a major departure of the temperature dependence of the thermoelectric power for GICs as compared with that of the pristine host material. All the results for GICs were obtained with HOPG as host material, except for a few that involved the use of graphitic fibers [6.39, 54, 70]. Most of the data pertain to acceptor compounds, though two donor compounds have also been investigated [6.50, 69]. If we ignore the expected difference in sign – i.e., that donors have negative thermoelectric powers and acceptors positive ones – at *high temperatures* the temperature dependence was found to be almost the same for all samples investigated. In addition, the magnitudes of the thermoelectric power appear quite insensitive to the stage of the compounds [6.51, 49, 68], except for very dilute compounds [6.51], and to the type of intercalate [6.6]. For stages higher than 1, the in-plane thermoelectric power of GICs is a linear function of the temperature in the lowest temperature range, then increases monotonically with increasing temperature, reaching a broad maximum at about 200 K. For stage-1 compounds, it was recently shown that the high-temperature behavior is qualitatively the same as that of higher-stage compounds, while the low-temperature behavior is totally different [6.6, 7, 53].

Figures 6.9, 10 present the temperature dependence of the in-plane thermoelectric power of various samples of low-stage acceptor GICs with different intercalates, stages, and host materials. Figure 6.9b shows the results obtained for stage-2 P100-4-$CuCl_2$ and stage-4 PX5-$CuCl_2$ intercalation compounds. Figure 6.10 presents the thermoelectric power of stage-1 HOPG-$CuCl_2$ and stage-1 HOPG-$CdCl_2$ intercalation compounds. These figures show that in the higher temperature range the in-plane thermoelectric power follows the general trend described above. However, if we consider all the published data, it appears that the magnitude of the maxima may be slightly different from one compound to another. Indeed, for stage numbers $n > 1$ the room-temperature value is 25 μV/K for HOPG-$SbCl_5$ [6.51] intercalation compounds and intercalated graphite fibers [6.39, 54] and 30–40 μV/K for HOPG-$FeCl_3$ intercalation compounds [6.49, 68], while it is in the range 15–20 μV/K for all stage-1 HOPG-based intercalation compounds measured.

In the *lowest temperature range*, we observe for the in-plane thermoelectric power a totally different behavior according to whether the stage is higher than or equal to 1. Indeed, for stages higher than stage-1, a stage-dependent linear

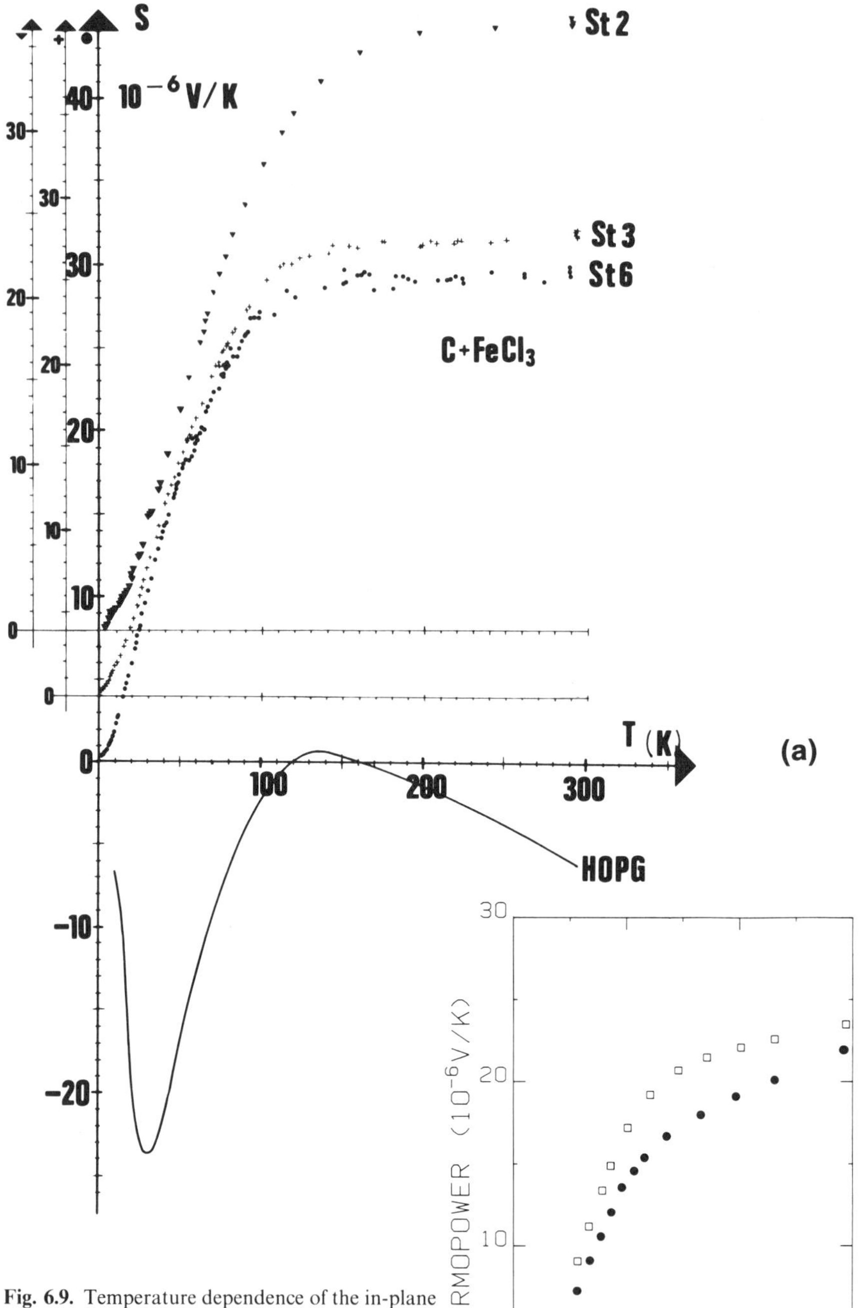

Fig. 6.9. Temperature dependence of the in-plane thermoelectric power below room temperature of HOPG-based GICs: (**a**) $FeCl_3$ compounds as a function of stage for $n \geqslant 2$, compared with pristine HOPG [6.49] and (**b**) stage-2 P100-4-$CuCl_2$ (●) and stage-4 PX5-$CuCl_2$ (□) intercalation compounds [6.6]

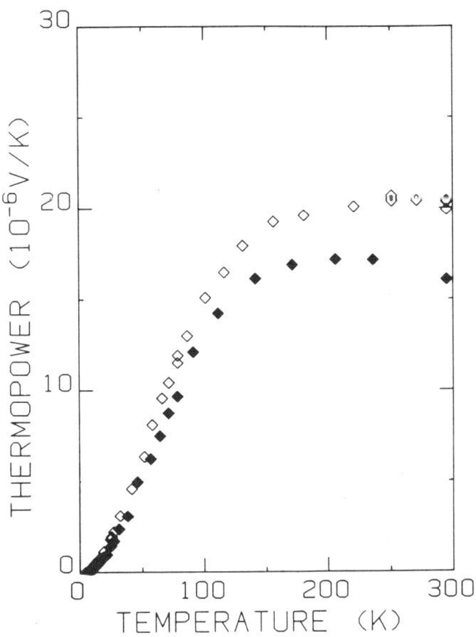

Fig. 6.10. Temperature dependence of the in-plane thermoelectric power of stage-1 HOPG-$CuCl_2$ (◇) and stage-1 HOPG-$CdCl_2$ (◆) intercalation compounds [6.6, 7]. In the higher temperature range the in-plane thermoelectric power follows the same general trend as for higher stages

function of the temperature is observed. As Fig. 6.11b shows, the linear coefficient for the stage-4 compound ($S/T = 10^{-7}$ V/K^2) is higher than for the stage-2 compounds ($S/T = 6$ to $7\ 10^{-8}$ V/K^2). This is in qualitative agreement with the results previously obtained on a few samples of HOPG-$SbCl_5$, whose stage ranged from 2 to 10 [6.51] (Fig. 6.11a). However, for the stage-2 HOPG-$SbCl_5$ sample, the coefficient of the linear variation is lower than for the corresponding intercalated graphite fibers. On the other hand, for stage-1 compounds, the thermoelectric power drops rapidly to a negligibly small value below a characteristic temperature that lies around 7 K (Fig. 6.11b) [6.6, 7, 53].

It was also recently found that the low-temperature thermoelectric power of magnetic acceptor GICs displays an anomalous behavior [6.53]. Indeed, for first-stage compounds, the thermoelectric power is no longer zero and for higher stages, it has a higher value than in nonmagnetic compounds (Sect. 6.5.3).

We can summarize the behavior observed for the in-plane thermoelectric power in the *liquid-helium temperature range* as follows:

(1) For stages higher than stage 1, it is a stage-dependent linear function of the temperature.

(2) For stage-1 compounds, two types of behavior are observed, according to whether the compound is magnetic or not. For *nonmagnetic compounds*, the thermoelectric power drops rapidly to a negligibly small value below a characteristic temperature around 7 K. For *magnetic compounds* the thermoelectric power does not vanish and may be even higher than for stage $\geqslant 2$ compounds (Sect. 6.5.3).

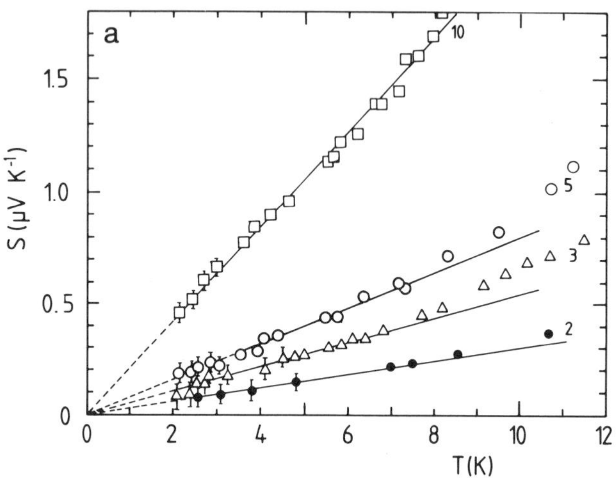

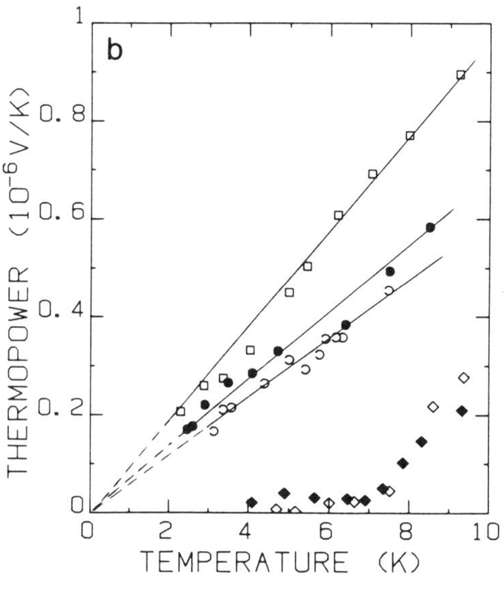

Fig. 6.11. Low-temperature thermoelectric power of various HOPG-based acceptor nonmagnetic GICs showing the linear variation (**a**) for $SbCl_5$ compounds as a function of stage [6.51], (**b**) for PX5-$CuCl_2$, $n = 4$ (□); P100-4-$CuCl_2$, $n = 2$ (●); BDF-$CuCl_2$, $n = 2$ (○); HOPG-$CuCl_2$, $n = 1$ (◇); and HOPG-$CdCl_2$, $n = 1$ (◆). The solid lines are guides to the eye. For stage-1 compounds, the thermoelectric power drops rapidly to a negligibly small value below about 7 K [6.6, 7, 53]

6.3.2 Mechanisms for Thermoelectric Power Generation in Solids

There are essentially three mechanisms for thermoelectric power generation. The *diffusion thermoelectric power* S_d results from the diffusion of the charge carriers from hot to cold caused by the redistribution of the carrier energies arising from the temperature gradient. Charge carriers tend to pile up at the low-temperature end of the sample, giving rise to an electric field that tends to counterbalance the stream of diffusing carriers until a steady state is reached. The exact physics of the process is not obvious at all, since it requires an understanding of all the subtleties of the scattering processes, but the origin of the diffusion is straightforward.

The second mechanism is the so-called *phonon-drag thermoelectric power* S_g which involves a transfer of momentum from the phonon system. Transport effects are usually analyzed under Bloch's conditions, i.e., assuming that the electron and phonon systems can be treated independently. This is certainly not the case when thermoelectric effects are considered. If the coupling between electron and phonon systems is strong, there might be an anisotropic transfer of momentum from one system to the other resulting in a drag on the charge carriers. The resulting extra electronic motion requires an additional electric field to counterbalance it, giving rise to the phonon-drag or lattice thermoelectric power. Conversely, when an electric current flows in a solid, the electrons may drag along the phonons, leading to a possible enhancement of the electrical conductivity. To determine the magnitude of the phonon-drag effects, we need to know about the strength of the electron–phonon coupling in the temperature range considered. Thermal-conductivity measurements might help in this context.

In magnetic conductors, or in nonmagnetic conductors containing magnetic impurities, anomalous low-temperature thermopowers have often been observed [6.73]. This was attributed to the presence of an additional thermoelectric power mechanism in which electrons are driven along the temperature gradient by spin-wave excitations. This *magnon-drag thermoelectric power* S_m bears some resemblance to the phonon-drag effect. Since in ferromagnetic materials the temperature dependences of S_g and S_m are different [6.73], the two effects can in principle be separated. However, such a separation is generally very difficult to achieve, because S_m is expected to be extremely small as compared with S_g at temperatures where magnetic effects are most prominent, i.e., generally above 50 K. In antiferromagnetic materials, the situation is still more complicated, since the temperature dependences of S_g and S_m are expected to be the same [6.73]. As we shall see in Sect. 6.5, in GICs we are again in a special situation.

The mechanisms invoked for thermoelectric power generation concern a given group of charge carriers, and we shall call the thermoelectric power of this group the partial thermoelectric power S_i, where i denotes the group of carriers considered. In a many-valley system a partial thermoelectric power should be ascribed to each of the valleys of a given band.

For a stage higher than stage 1, the various bands should be considered. The total thermoelectric power is obtained by considering the different groups of carriers with partial thermoelectric powers that contribute to the total thermoelectric power, S, as emfs in parallel. The total thermoelectric power is thus expressed as

$$S = \frac{\sum \sigma_i S_i}{\sum \sigma_i}, \qquad (6.32)$$

where σ_i is the partial electrical conductivity of carriers i.

In addition, each group of charge carriers contributing to the phonon-drag thermoelectric power may have two contributions opposite in sign, according to whether they experience normal or umklapp collisions with phonons [6.26, 27]. Indeed, contrary to the case of the electrical resistivity, where normal and umklapp processes add their contribution by increasing the scattering of the charge carriers, in the phonon-drag thermoelectric power, their contributions have opposite effects. In the case of the umklapp phonon–electron interaction, the electron is driven after scattering to the side of the Fermi surface opposite to that on which it is driven in a normal interaction. Thus, whereas with a normal process the charge carrier is dragged along the temperature gradient, with an umklapp process, it is dragged in the opposite direction, resulting in a thermoelectric power of opposite sign. Since in GICs the sign of the thermoelectric power corresponds to that of the majority carriers, we may reasonably assume that normal processes should dominate the phonon-drag thermoelectric power, where this mechanism is operative.

This shows that considering more than one group of carriers in calculating the thermoelectric power is a formidable task. To explain even qualitatively the physics of thermal emf generation in GICs, we must work first on a single group of carriers, i.e., on first-stage acceptor compounds. This has been done recently [6.6, 7, 53].

(a) Diffusion Thermoelectric Power

The general expression for the diffusion thermoelectric power is given by [6.26, 27]

$$S_{\mathrm{d}} = \frac{\pi^2 k_{\mathrm{B}}^2}{3q} T \left[\frac{\partial \ln \sigma}{\partial \varepsilon} \right]_{\varepsilon_{\mathrm{F}}} , \tag{6.33}$$

where ε_{F} is the Fermi energy, T is the absolute temperature, σ is the electrical conductivity, and the derivative is evaluated at the Fermi level.

Equation (6.33) implies that for a given scattering mechanism, the diffusion thermoelectric power depends solely on the Fermi energy, i.e., the energy from the band extremum to the Fermi level. This may be understood from simple arguments. The diffusion thermoelectric power results from the difference in the broadening of the Fermi distribution around the Fermi level due to the temperature difference between the hot and cold ends of a sample in a temperature gradient. The smaller the Fermi energy, the greater the relevant fraction of the charge carriers of the total Fermi gas. This explains why semimetals have higher thermoelectric powers than metals [6.74]. In GICs the diffusion thermoelectric power will thus depend on the charge transfer, since it determines the magnitude of the Fermi energy.

Note also that the partial diffusion thermoelectric power S_{di} is essentially isotropic since it depends on the Fermi energy, provided the relaxation time dependence on energy is isotropic too. This is not the case for the partial phonon-drag thermoelectric power S_{gi}, since it depends on the electron-phonon interaction, which might be highly anisotropic in such materials as GICs.

(b) Phonon-Drag Thermoelectric Power

In a metal, the phonon-drag thermoelectric power for a given valley of charge carrier density N is written as

$$S_g = \frac{C_L}{3Nq}\left(\frac{\tau_{p-x}}{\tau_{p-x} + \tau_{p-e}}\right), \tag{6.34}$$

where C_L is the lattice specific heat at constant volume and q is the electronic charge. Here τ_{p-e} is the phonon–electron relaxation time, while τ_{p-x} is the phonon relaxation time for all other scattering events.

In the simplest case, if we assume that we are at low temperatures and that only phonon–electron scattering is important (i.e., that $\tau_{p-x} \gg \tau_{p-e}$), relation (6.34) becomes

$$S_g = C_L/Nq \,. \tag{6.35}$$

Thus, we expect the temperature variation of the phonon-drag thermoelectric power to follow that of the specific heat. This would mean that, if it is the thermal phonons that drag along the charge carriers, the mean free path of these phonons should be limited by electron scattering. And since the lattice thermal conductivity is given by (6.24), we expect a behavior similar to that of the lattice thermal conductivity in the same temperature range. As we shall see in the next section (Sect. 6.3.3), this is exactly what was recently observed (Fig. 6.12) [6.75]. Indeed, contrary to the relaxation time for electron-phonon scattering, which is operative in the electrical conductivity and is essentially temperature dependent,

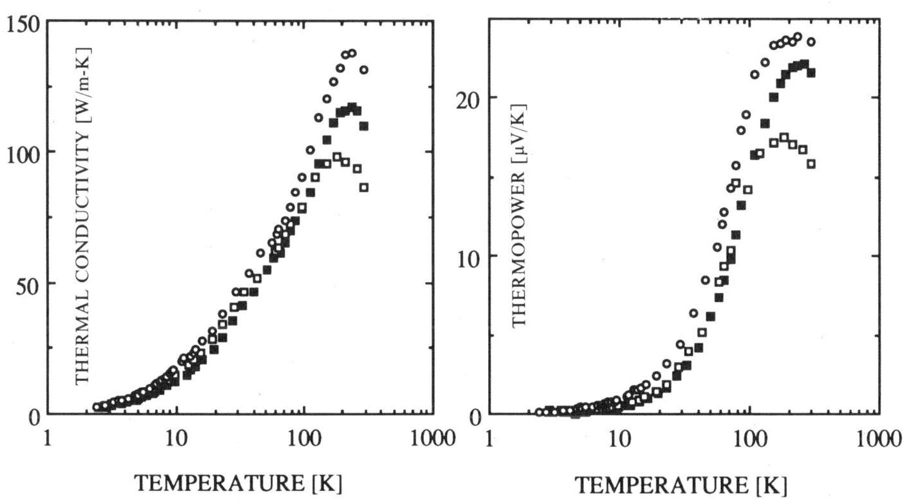

Fig. 6.12. Comparison of the temperature dependence of the thermal conductivity and the thermoelectric power for three different samples of first-stage HOPG-$CoCl_2$ compounds showing clearly a parallel behavior in the higher temperature range [6.76]

the relaxation time for phonon–electron scattering is temperature insensitive at high temperatures, since in a degenerate system the carrier density does not depend on temperature.

In pure ordinary metals, this parallel behavior is never observed, since at all temperatures the electronic thermal conductivity dominates. In metallic alloys, where the lattice thermal conductivity may become important, the phonon-drag thermoelectric power is greatly reduced because of alloy scattering of the dragging phonons.

In semiconductors or semimetals with small Fermi energies, where the lattice thermal conductivity usually dominates, the phonons that mainly contribute to the conductivity are those that dominate the spectrum, i.e., the more energetic ones. We shall call these phonons the *thermal phonons*. The thermal phonons are not necessarily those that exert drag on the charge carriers. Indeed, because of energy and momentum conservation requirements, the phonons that are apt to drag the charge carriers, and that we shall call hereafter the *drag phonons*, are generally of much lower energy than the thermal phonons in semimetals and semiconductors [6.75]. Although the Fermi surface in GICs might be smaller than in an ordinary metal, it is still much larger than in semimetals in stage-1 acceptor compounds. This leads to the situation where it is the thermal phonons that drag the charge carriers in acceptor GICs up to about 540 K (Sect. 6.1.2).

So here again with GICs we are in a unique situation where there is a significant contribution from the lattice thermal conductivity and where, if operative, the phonon-drag thermoelectric power could show a parallel behavior. Indeed, then it is the thermal phonons that would exert drag on the charge carriers up to at least room temperature. The only problem is that the lattice thermal conductivity, though dominant in some temperature ranges, is not the sole mechanism, since there remains a significant electronic contribution (Sect. 6.2).

6.3.3 Discussion of the GIC Results

For a stage-1 compound, applying (6.33) and using the Blinowski-Rigaux model [6.4], we find [6.7]

$$S_d = 2.45 \times 10^{-8}\, T(1+p)\varepsilon_F^{-1}\,, \tag{6.36}$$

where S_d is expressed in V/K and p is the scattering parameter, which expresses the energy dependence of the relaxation time:

$$\tau = \tau_0 \varepsilon^p\,. \tag{6.37}$$

If we assume defect scattering in the low-temperature regime, with an energy-independent relaxation time, as is the case for boundary scattering ($p = 0$ in the Blinowski-Rigaux model), and a Fermi energy of 1 eV, we obtain for a stage-1 compound

$$S_d = 2.45 \times 10^{-8}\, T\,. \tag{6.38}$$

For a stage-2 compound, taking a Fermi energy $\varepsilon_F = 0.8$ eV and $\gamma_1 = 0.38$ eV, we obtain for the hole bands 1 and 2 [6.7] a diffusion thermoelectric power

$$S_{d1} = 2.45 \times 10^{-8} T(1.1 + 1.25 p_1) \tag{6.39}$$

and

$$S_{d2} = 2.45 \times 10^{-8} T(2.0 + 2.38 p_2) . \tag{6.40}$$

Since scattering in these compounds at low temperatures is mainly on large-scale defects [6.2, 47], we may assume that $p_1 = p_2 = 0$ in the residual resistivity range and thus that $S_{d1} = 2.7 \times 10^{-8}\ T$ and $S_{d2} = 4.9 \times 10^{-8}\ T$. Since it was shown that $\sigma_1 \sim 2\sigma_2$ [6.7], using (6.32) we find

$$S_d \sim 3.5 \times 10^{-8} T . \tag{6.41}$$

From the linear temperature dependence of the thermoelectric power of stage-2 compounds at low temperatures, *Piraux* and coworkers have estimated the Fermi energies of their compounds [6.7]. Assuming that carriers from the two hole bands were scattered at the boundaries of the Daumas-Hérold domains – i.e., that $p = 0$ – they obtained $\varepsilon_F = 0.9$ eV, 0.55 eV, and 0.40 eV for the stage-2 HOPG-$SbCl_5$, BDF-$CuCl_2$, and P100-4-$CuCl_2$ intercalation compounds, respectively. They tentatively ascribed the lower values obtained for intercalated fibers with respect to HOPG-based compounds to a nonuniform intercalation in the fibers and/or to a lower charge transfer from the $CuCl_2$ intercalate.

For the stage-1 compounds, (6.38) shows that at 5 K the diffusion thermoelectric power should take a value around 10^{-7} V/K, while the experimental results show that it does not exceed 10^{-8} V/K (Fig. 6.11b), showing clearly that the diffusion thermoelectric power is negligibly small. This is contrary to what is observed for higher stages. To have a zero thermoelectric power in stage-1 compounds, the scattering parameter p in (6.36) should be equal to -1. Since the electrical resistivity data clearly show that we are in the residual range at this temperature, this means that we should invoke a low-temperature defect scattering mechanism, which is strengthened as the energy increases. This excludes large-scale defect scattering like boundary scattering, which is energy independent. This also contrasts with what is observed for higher stages where the mean free path at low temperatures is mainly limited by boundary scattering at the limits of the Daumas-Hérold domains.

In the light of the existing theories for the diffusion thermoelectric power, we can conclude that the unusual negligibly small effect observed at low temperatures in first-stage compounds indicates a scattering mechanism qualitatively different from that of higher-stage compounds.

If we now compare the predicted diffusion thermoelectric power at higher temperatures with that observed experimentally (Figs. 6.9, 10), we see that the situation is reversed and that the observed values are higher than the predicted ones. At 100 K, for example, the calculated thermoelectric power is 2.45×10^{-6} V/K, while the experimental value is more than six times higher.

It was early realized that it would be hard to explain the magnitude of the high-temperature thermoelectric power and the presence of a maximum in terms of a diffusion mechanism [6.55, 76]. This suggested that there is an additional mechanism for thermal emf generation.

The only attempt to interpret the thermoelectric power data along these lines was made by *Sugihara* [6.77]. He suggested an important phonon-drag contribution above 10 K, associated with Rayleigh scattering, to explain the peculiar temperature dependence of the thermoelectric power, and computed the thermoelectric power for stage-2 acceptor compounds. However, the occurrence of Rayleigh scattering that he invoked has not been demonstrated. With the new results available now on stage-1 compounds [6.6, 7, 53], a clearer picture emerges, leading to a straightforward interpretation of the phonon-drag contribution [6.75].

At high temperatures, the charge carrier mobility is mainly limited by scattering by the phonons of the graphene layers. Previous calculations [6.42, 43] of the electron-phonon contribution to the resistivity of low-stage GICs have shown that the energy dependence of the relaxation time follows an ε^{-1} behavior, i.e., that $p = -1$. Therefore, for stage-2 compounds, the diffusion thermopower is expected to be small around room temperature, while for stage-1 nonmagnetic compounds, we are in a unique situation where over the entire temperature range below room temperature the diffusion thermopower is negligibly small and the phonon-drag component dominates the scene [6.75]. For both stage-1 and stage-2 magnetic compounds, there is an additional contribution from magnon drag at low temperatures.

The effectiveness of the phonon-drag mechanism is accentuated as the charge carriers with wave vectors initially in the direction opposite to that of the dragging phonons are driven in the dragging direction through their interaction, i.e., from hot to cold. This would require phonons with wave vectors on the order of k_F, i.e., temperatures on the order of $\frac{1}{2}\theta^*$, where θ^* is given by (6.9). This yields a temperature of 270 K, which is roughly the temperature at which the maximum thermopower is observed experimentally.

Moreover, as emphasized in the previous section, the phonon-drag thermopower and the lattice thermal conductivity should show a parallel behavior in some restricted temperature range. This is what we observe experimentally, as shown in Fig. 6.12. The temperature dependence of the thermal conductivity of three stage-1 HOPG-$CoCl_2$ compounds are compared with their thermopower over the same temperature range. In the high-temperature range, it is reasonable to assume that the lattice contribution to the thermal conductivity dominates the electronic contribution. The close relationship shown is expressed not only in the temperature dependence of both properties but also in their relative magnitudes. These correlations strongly suggest a dominant phonon-drag contribution to the thermopower of low-stage acceptor GICs in the high-temperature range.

In agreement with the present observations, it was predicted that for a 2D electron gas coupled with 3D phonons, the phonon-drag thermopower should be large as compared with the diffusion thermopower [6.78]. This theory was applied with success to heterostructures [6.79].

6.4 *c* Axis Transport and Anisotropy

Among the most intriguing properties of GICs are the transport phenomena in the direction perpendicular to the graphitic planes. This applies to the electrical conductivity as well as to the thermal conductivity and thermoelectric power.

6.4.1 *c* Axis Electrical Resistivity

A comprehensive review has recently been provided by *McRae* and *Marêché* [6.80], who examine the stage and temperature dependence of the *c*-axis electrical resistivity in light of all published experimental data and proposed theoretical models. Therefore, here we only briefly describe the situation, referring for more detailed discussions to [6.80] and its wide bibliographical survey.

From exploratory measurements performed on two compounds and from inspection of a few scattered data in the literature, it was suggested that the temperature dependence of the *c* axis resistivity ϱ_c of both donor and acceptor compounds follows a general trend [6.5]. For low stages, the resistivity was found to increase with increasing temperature, akin to the "usual" metallic behavior, whereas for higher stages a decrease in resistivity with increasing temperature was observed. This was independently verified first on a donor [6.81] and later on an acceptor compound [6.82], both measured as a function of stage. Further investigations have confirmed this general trend [6.80]. So, at least with regards to the experimental data, the situation is clear. The general trend for donor as well as acceptor compounds is qualitatively the same and is well illustrated by the behavior of the trichloride compounds presented in Fig. 6.13. This holds whether the compound contains one or more intercalated species, or whether the intercalate has a mono or multilayered arrangement [6.80].

Essentially two models have been proposed to interpret *c*-axis dc electrical conductivities observed. They both imply two additive mechanisms. The first model, proposed by *Sugihara* [6.83], involves a phonon-assisted hopping and an impurity-assisted mechanism. These impurities are assumed to be localized charged centers such as charged defects in the intercalate planes. The second model, proposed by *Shimamura* [6.84], also invokes a phonon-assisted hopping mechanism. However, as a second mechanism he proposed a conduction mechanism via conducting paths allowing the charge carriers to be transferred from one graphene plane to another. These conducting paths were attributed to structural defects. For stages $n < 3$, the impurity-assisted or the conducting path mechanism should be favored at all temperatures according to Sugihara and

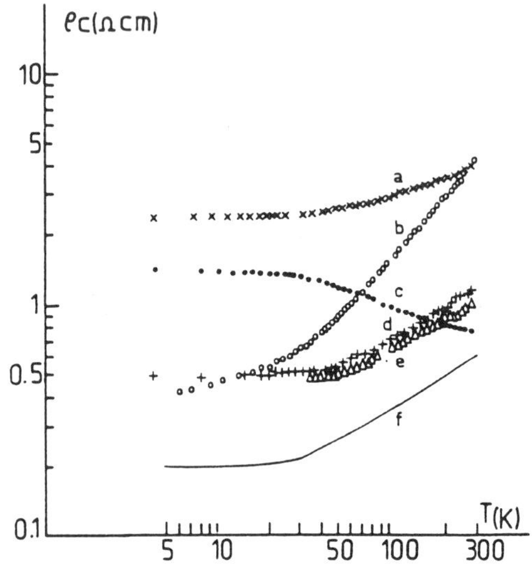

Fig. 6.13. Temperature dependence of the c axis electrical resistivity of two trichloride acceptor GICs showing the general trend for this property, which is the same in both donor and acceptor compounds [6.82]: (**a**) $GaCl_3$, $n = 4$; (**b**) $GaCl_3$, $n = 1$; (**c**) $FeCl_3$, $n = 4$; (**d–f**) three samples of $FeCl_3$, $n = 2$, of various origins

Shimamura, respectively. For stages $n > 3$ both authors agree that a phonon assisted hopping mechanism is dominant.

In spite of the two models proposed to explain the temperature variation of ϱ_c invoking different physical mechanisms, we believe that an interpretation that coherently accounts for the experimental findings is still lacking. The problem with the models proposed is that they invoke additive mechanisms and many adjustable parameters to account for the observed facts, none of which are supported by other experiments. Also, any model describing the electrical conduction in the c-axis direction should take into account the observation that the anisotropy of the thermoelectric power is less than 10, thus many orders of magnitude smaller than that of the resistivity of acceptor compounds.

Whether or not we retain the interpretations proposed, it is interesting to note that for GICs the conduction mechanisms in the two main directions are qualitatively different. While the in-plane conductivity definitely has to be interpreted in terms of ordinary band conduction models, provided the appropriate dispersion relations are applied, in the c axis direction other types of conduction mechanisms need to be considered.

6.4.2 *c* Axis Thermal Conductivity and Thermoelectric Power

Using the Wiedemann-Franz law (Sect. 6.2), the c-axis electronic thermal conductivity was found to be negligibly small in all the compounds studied [6.47, 55]. Therefore, the observed thermal conductivity was entirely ascribed to phonons. As Fig. 6.14 shows, the thermal conductivity in the c-axis direction was found to be much smaller than in-plane, and the temperature dependences were found to be different in the two directions [6.47]. No detailed analysis of the data in the c axis direction was performed.

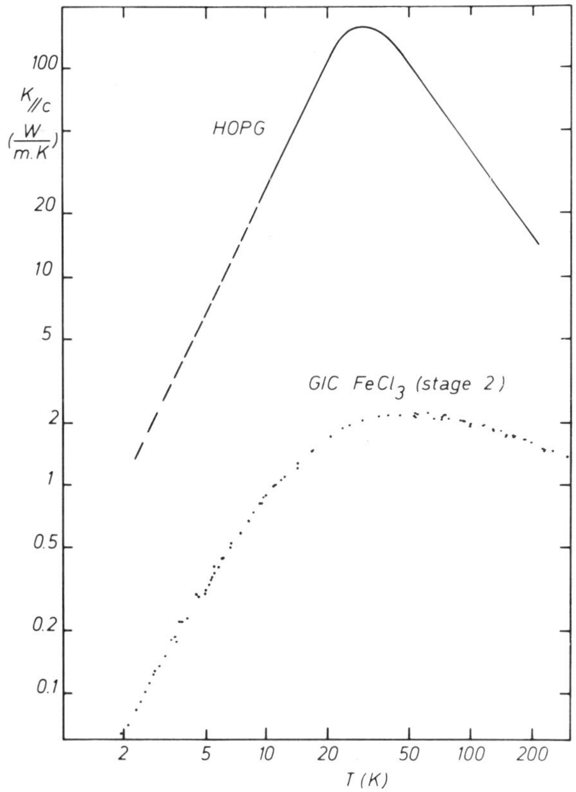

Fig. 6.14. The temperature dependence of the thermal conductivity in the c axis direction of a stage-2 HOPG-$FeCl_3$ GIC compared with that of pristine HOPG in the same direction [6.47]

The anisotropy of the thermoelectric power shown in Fig. 6.15 has been totally ignored in the tentative interpretations of the c-axis conductivity. We think that this anisotropy may be important in our quest for understanding the puzzling conduction mechanisms in the c axis direction.

In Sect. 6.3 we analyzed the in-plane thermoelectric power. Now, if we look at the temperature variation of the thermoelectric power in the c-axis direction (Fig. 6.15), we find that it is not very different from the in-plane thermoelectric power for a given compound with stage $\geqslant 2$ in the lowest temperature range [6.5]. This is what we expect for a single band in a 3D anisotropic solid, since the thermoelectric power mainly depends on a scalar quantity, the Fermi energy. Thus, the same charge carriers, the holes, exhibit very different directional behavior in relation to the electrical conductivity and the thermoelectric power.

For the higher temperature range, the situation is quite different and more delicate to analyze. Although the magnitude is several orders less than for the electrical resistivity, the thermoelectric power is anisotropic and the anisotropy

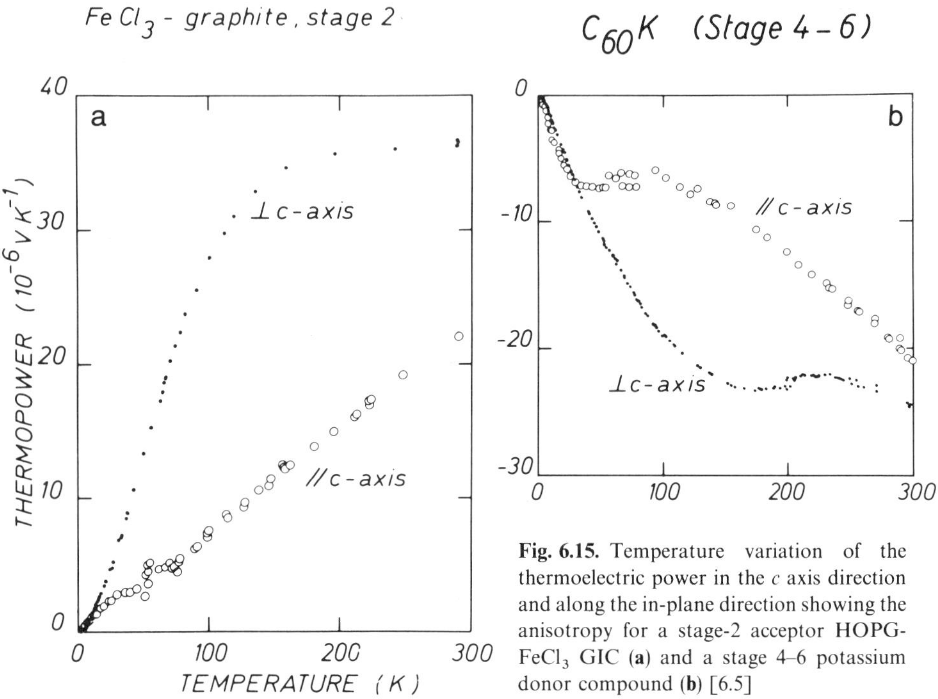

Fig. 6.15. Temperature variation of the thermoelectric power in the c axis direction and along the in-plane direction showing the anisotropy for a stage-2 acceptor HOPG-$FeCl_3$ GIC (**a**) and a stage 4–6 potassium donor compound (**b**) [6.5]

varies from one compound to another. However, if we assume a phonon drag mechanism to be dominant, the partial phonon-drag thermoelectric power is not expected to be isotropic as for the diffusion case. Indeed, the phonon-drag mechanism depends strongly on the electron–phonon interaction, which is strongly anisotropic in these compounds.

6.4.3 Anisotropy and Dimensionality

Because of the exceptionally high anisotropy of the electrical conductivity in acceptor GICs ($\approx 10^6$ at room temperature), for most practical situations a 2D model is justified for the analysis of the results. The 2D model is also applicable to the electronic thermal conductivity. However, the situation is quite different as regards the lattice thermal conductivity. For lattice waves that belong to the crystal as a whole, the graphite layers cannot be regarded as individual systems. The graphene layers must be coupled in some way to secure the mechanical cohesion in the c axis direction. In addition, the anisotropy of the lattice thermal conductivity of pristine HOPG reaches $\approx 10^2$ around room temperature and decreases down to a factor of ≈ 3 in the liquid-helium range. In GICs the situation is almost the same, but the anisotropy of the total thermal conductivity remains around 10^2 in the liquid-helium range. However, this high anisotropy at low temperature is due to the electronic contribution κ_E and not to κ_L [6.2, 55].

For the thermoelectric power, the anisotropy is even much less pronounced than that of the thermal conductivity. So for a given sample we may see that the anisotropy varies by orders of magnitude according to the transport coefficient considered. Also, as regards the electrical resistivity, the anisotropy varies widely with stage and temperature. So the generally accepted picture that acceptor compounds exhibit higher anisotropy than donor compounds is not always true.

6.5 Transport in Magnetic GICs

Anomalous behavior in the transport properties of magnetic acceptor compounds was first observed in their low-temperature electrical resistivities [6.85, 86]. Later on, anomalies in their thermal conductivity and thermoelectric power were also reported [6.53, 87].

More generally, it is known that the presence of magnetic species might affect the electrical resistivity and the thermal conductivity [6.88, 89], as well as the thermoelectric power of solids [6.73]. On one hand, the spins may contribute to the scattering of electrons and phonons. Thus they are liable to generate an increase of electric, electronic thermal, and lattice thermal resistivities and to modify the thermoelectric power. On the other hand, since magnons may also carry energy, they may enhance the thermal conductivity and also contribute to the thermoelectric power (Sect. 6.3.2).

In particular, for the thermoelectric power, both effects due to the spins have been proposed from time to time to explain the low-temperature anomalies observed in magnetic materials. However, the observed behavior was usually complex and could hardly be interpreted in terms of the existing theoretical models. As a result, the presence of magnon drag was invoked only to justify the anomalous behavior of the measured thermoelectric power around the magnetic phase transition temperatures, but the effect has not been clearly demonstrated. We show below that from the results obtained on GICs a comprehensive study of the contribution of magnons to the thermal conductivity and thermoelectric power of solids may be foreseen.

Contrary to the electronic properties, the 2D magnetic behavior is expected to be more pronounced for higher-stage acceptor GICs, since it is the properties of the intercalate layers that are considered. The transition from 3D to 2D magnetic coupling occurs with increasing stage from $n = 1$ to $n \geqslant 2$, according to the intercalate. This is obvious since the increase in stage leads to a larger distance between the intercalate layers and thus to a reduction of interlayer coupling.

The $FeCl_3$-GIC and $CoCl_2$-GIC systems are usually believed to be typical Heisenberg spin systems and 2D-XY systems, respectively [6.8]. In the low-temperature phase, the spins in the $CoCl_2$-GIC system exhibit in-plane ferromagnetic ordering with antiferromagnetic coupling along the c axis. For the

$FeCl_3$-GIC system, the in-plane and out-of-plane correlation lengths as determined by neutron-diffraction experiments indicate that the magnetic transition occurs with a simultaneous 2D-to-3D crossover.

6.5.1 Electrical Resistivity

Electrical resistivity measurements performed on first- and second-stage HOPG-$CoCl_2$ compounds revealed anomalies in the temperature dependences at low temperatures [6.85, 86]. The stage-1 behavior was found to be qualitatively different from the stage-2 behavior. For the stage-1 compound presented in Fig. 6.16, a sharp resistivity increase of nearly 10% was observed with decreasing temperature below the magnetic transition temperature T_c (9.7 K). The stage-2 compound instead showed a very small decrease ($\sim$ 1%) below this temperature. It was also found that a magnetic field of 0.1 T applied in-plane destroyed the effect [6.85, 86]. On the other hand, the same qualitative behavior was observed by other workers [6.87] for a stage-1 $CoCl_2$ compound, but the effect was of smaller magnitude (1–3%) and occurred at 7.5 K. However, these results obtained on a stage-2 HOPG-$CoCl_2$ [6.87] were at variance with those previously obtained [6.85, 86], since an increase in resistivity was found below the transition temperature (0.3–0.6%). The origin of the discrepancy is not yet understood.

The jump in electrical resistivity was explained by invoking two competing mechanisms [6.85, 86]. The first one consisted of a zone-folding effect in the c direction due to a doubling of the unit cell in this direction, resulting from the

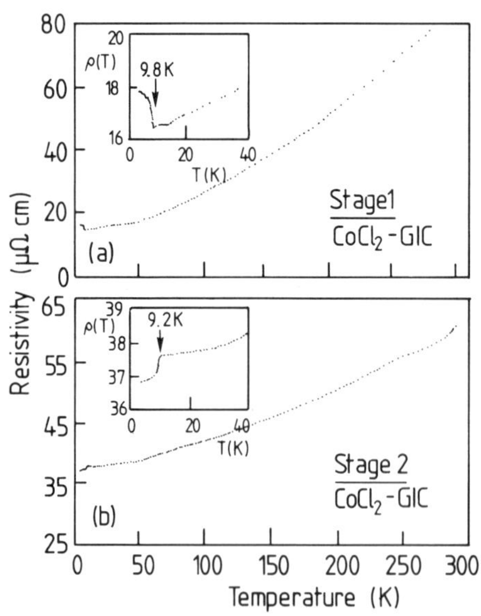

Fig. 6.16. Temperature dependence of the zero-field electrical resistivity of stage-1 and stage-2 HOPG-$CoCl_2$ compounds showing anomalies at 9.8 K and 9.2 K respectively (insets) [6.86]. The behaviors of stages 1 and 2 are qualitatively different. For the stage-1 compound, a sharp resistivity increase of nearly 10% is observed, with decreasing temperature below the magnetic transition temperature. For the stage-2 compound, however, a very small decrease ($\sim$ 1%) is observed below this temperature

presence of the magnetic phase. This resulted because in HOPG-$CoCl_2$ compounds the magnetic intercalate layers are stacked antiferromagnetically, while within the layers a ferromagnetic order prevails. The modification of the Fermi surface resulting from the zone folding should lead to an increase in resistivity [6.90]. The second effect, which causes a decrease in resistivity as T is lowered, was ascribed to the order-disorder transition of the spins in the low-temperature magnetic phase.

The different behavior in the resistivity anomaly between stage-1 and stage-2 HOPG-$CoCl_2$ compounds observed by the MIT group was attributed to a larger correlation length for the antiferromagnetic stacking in stage-1 compounds as compared with stage-2 compounds [6.85, 86].

Figure 6.17 shows the low-temperature dependence of the resistivity of a stage-1 HOPG-$FeCl_3$ sample. Around 5–10 K, a resistivity anomaly is observed. Such an anomaly may be correlated with a magnetic phase transition revealed by magnetic-susceptibility measurements [6.91], for which a peak was observed at around 5.5 K on the first-stage $FeCl_3$-GIC system. On the basis of a neutron-diffraction experiment, *Simon* et al. [6.91] attributed the magnetic phase transition around 5.5 K to a 2D to 3D crossover. Moreover, since no antiferromagnetic stacking exists in these compounds, no modification of the Fermi surface is expected. The anomaly observed in the resistivity curve is thus consistent with a spin-disorder effect, leading to a decrease of the resistivity in the low-temperature regime.

So, as regards the electrical resistivity data, the $CoCl_2$-GIC and the $FeCl_3$-GIC systems show contrasting behaviors.

6.5.2 Thermal Conductivity

Several authors obtained specific-heat data on magnetic GICs. A recent overview of these data obtained on low-stage $CoCl_2$ and $FeCl_3$ GICs [6.91] suggested that the magnetic contribution to the specific heat dominates in the low-temperature range. *Simon* et al. [6.91] found a peak in the specific-heat curve in stage-1 $CoCl_2$ and $FeCl_3$ GICs at the magnetic phase transition temperature.

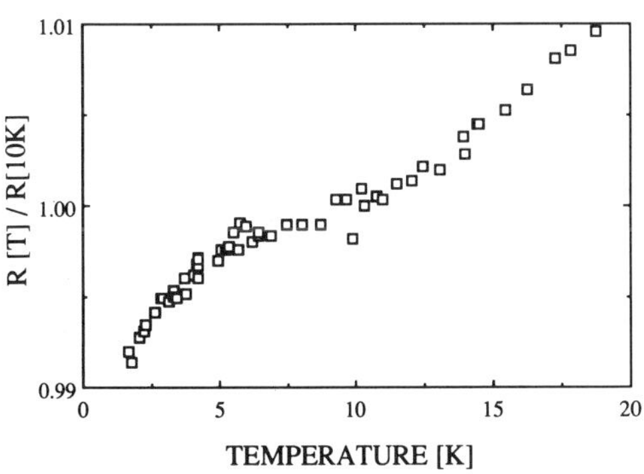

Fig. 6.17. In-plane electrical resistivity of a stage-1 HOPG-$FeCl_3$ sample at low temperatures. A resistivity anomaly is observed below 10 K

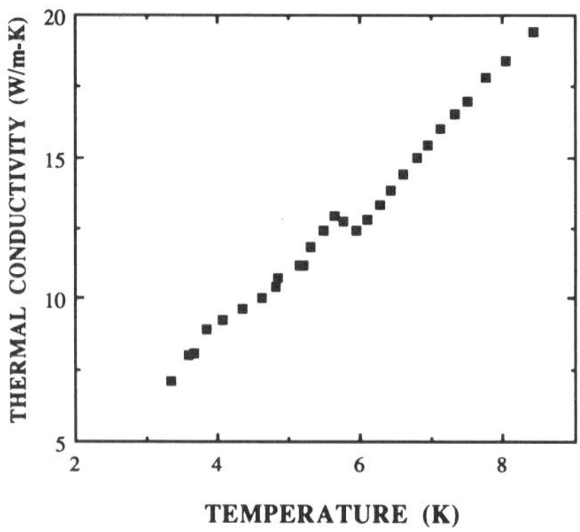

Fig. 6.18. In-plane thermal conductivity at low temperatures of a stage-1 HOPG-$FeCl_3$ compound around the magnetic transition temperature. A small hump revealing an enhancement of the thermal conductivity is observed around 5.5 K

On the other hand, a recent neutron scattering study on $CoCl_2$ GICs [6.92] provided their magnon dispersion, $\omega = q^2$. From this relation a $T^{3/2}$ variation for the specific heat of 3D magnons could be derived. It contradicts the T^3 variation claimed by *Simon* et al. [6.91] from their specific heat data. However, this T^3 variation was observed over too narrow a temperature range (3 K) to be convincing.

The temperature dependences of the in-plane thermal conductivity of stage-1 HOPG-$FeCl_3$ and HOPG-$CoCl_2$ intercalation compounds and of an HOPG-$CoCl_2$-$AlCl_3$ bi-intercalation compound have recently been reported [6.53, 87]. As is the case for 3D ferromagnetic and antiferromagnetic materials [6.88, 89], spin-wave excitations were expected to carry heat currents in magnetic GICs. However, since in GICs other quasiparticles of nonmagnetic origin already contribute to the thermal conductivity (Sect. 6.2), it was difficult to isolate a magnetic contribution in these compounds. Nevertheless, for the $FeCl_3$ compound, a small hump revealing an enhancement of the thermal conductivity was observed around 5.5 K (Fig. 6.18). Oddly enough, this temperature coincides with that of the magnetic phase transition temperature determined by magnetic-susceptibility measurements, but not with that reported for the specific heat that shows up at a lower temperature, i.e., around 3.7 K. Anyway, since the hump was observed in a temperature range where there is an excess specific heat of magnetic origin, it was reasonable to assume that the existence of an additional contribution to the thermal conductivity is due to magnons. Indeed, a magnetic scattering of phonons or charge carriers should lead to a decrease in thermal conductivity, which is contrary to what was observed. The hump in the thermal conductivity resulting from magnetic phase transitions shown in Fig. 6.18 was observed for the first time in a solid.

6.5.3 Thermoelectric Power

Figure 6.19 presents results on the temperature dependence of the in-plane thermoelectric power of two stage-1 compounds and of a bi-intercalation compound at low temperatures. The measured values of the in-plane thermoelectric power at higher temperatures were close to those observed for nonmagnetic first-stage acceptor GICs (15–20 μV/K at room temperature), and followed the same general trend observed in all GICs (Figs. 6.9, 10). In the lowest temperature range, the situation is quite different. While for stage-1 nonmagnetic acceptor compounds, the in-plane thermoelectric power was found to drop rapidly to negligibly small values below ~ 7 K (Sect. 6.3), the thermoelectric powers of the HOPG-$FeCl_3$ and HOPG-$CoCl_2$ intercalation compounds exhibit large values, as Fig. 6.19 shows. Thus, there is a thermoelectric power enhancement in magnetic compounds that should be associated with the presence of the magnetic species in the intercalate layers. This is supported by the fact that the thermoelectric power of the bi-intercalation compounds, also shown in Fig. 6.19, is significantly lower than that of the corresponding stage-1 magnetic compounds and is instead close to that of stage-1 nonmagnetic acceptor GICs. Indeed, in the bi-intercalation compound the $CoCl_2$ magnetic layer is sandwiched between two $AlCl_3$ nonmagnetic layers, thus reducing the influence of the $CoCl_2$ layer on the graphitic planes where conduction takes place.

As we saw above, a magnetic effect on the thermoelectric power may originate either from the contribution of the magnon-drag effect or from a modification of the scattering parameter due to magnetic scattering of the charge carriers. In light of the results available now, it is premature to decide definitely which of the two effects dominate, or whether both effects act simultaneously.

However, since the magnitude of the residual resistivity of a magnetic compound, though usually somewhat higher, is not much different from that of a nonmagnetic compound with the same host material, we might reasonably assume that we have a magnon-drag mechanism. Otherwise, if magnetic scattering were determining the thermoelectric power, it should also dominate

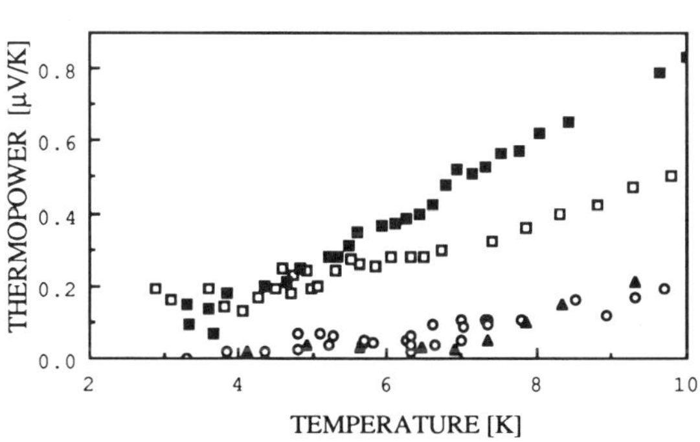

Fig. 6.19. Low-temperature behavior of the in-plane thermopower of stage-1 HOPG-$FeCl_3$ (■) and HOPG-$CoCl_2$ (□) intercalation compounds compared with that of HOPG-$CoCl_2$-$AlCl_3$ (○) bi-intercalation compounds and of a nonmagnetic GIC (▲) [6.6]

the electrical resistivity in the residual range. To explain the results in that case, we should assume that the relaxation time has almost the same magnitude for both magnetic scattering and scattering by static defects. Also, we would expect a large variation of electrical resistivity at the transition temperature because of spin alignment. But this is not what is observed.

An interesting feature of 2D magnetic GICs is that magnetic phase transitions are observed at lower temperatures than in the pristine 3D magnetic material, since the interplanar magnetic interactions are reduced by the stacking structure. In the $FeCl_3$-GIC and $CoCl_2$-GIC systems, the magnetic transition temperatures are below 10 K. Also, because of the higher Debye temperature of the graphitic planes, phonon drag is expected to occur at higher temperatures than in 3D metals. These two effects should thus favor a separation of S_m and S_g (Sect. 6.3). Thus, magnetic GICs may turn out to be choice materials for the study of magnon drag and of the magnon contribution to thermal conductivity in general.

6.6 Concluding Remarks

In this chapter we have mainly reviewed the experimental work performed during the last decade on the zero-field transport properties of acceptor metal chloride GICs. This work concerns the temperature variation of the electrical resistivity, the thermal conductivity, and the in-plane thermoelectric power, with some reference to the remarkably high anisotropy of the compounds. Emphasis has been placed on the situation of acceptor GICs among other solids and its implications for our general understanding of the transport properties of systems with reduced dimensionality.

We first analyzed the ideal and residual in-plane electrical resistivity data, relative to low-stage acceptor GICs, and the effects of 2D weak localization and of carrier–carrier interaction. We saw how the residual resistivity allows an estimate of the size of the large-scale lattice defects, which were found to be on the order of a few 10^2 nm in HOPG acceptor GICs, a value much smaller than in the pristine material. The temperature variation of the ideal electrical resistivity is not yet fully understood, but the low-temperature behavior is probably due to hole–hole interaction in the presence of weak disorder, while electron–phonon scattering dominates at higher temperatures. The model of a 2D hole gas introduced by the intercalate and interacting with the phonons and defects of the host layers gives a good picture of electronic transport in the Boltzmann conductivity range.

Because of the inherent 2D nature of the acceptor GIC electronic structure, weak localization effects and electron–electron interactions are particularily interesting. The possibility of varying the defect structure of the host material over wide ranges allows large experimental possibilities for the investigation of these phenomena.

We also saw that in the case of the particular 2D electronic structure of acceptor GICs, we should expect a conductivity independent of charge transfer for a given stage if we consider only scattering by in-plane graphitic phonons. This, together with the experimental data available on metal-chloride GICs, indicates the limits of conductivity that could be attained in these compounds.

The thermal conductivity of intercalation compounds with various host materials, intercalates, and stages was also analyzed. The general trend is a decrease of the total thermal conductivity at high temperatures and an increase at low temperatures with respect to that of the pristine material. We saw that phonons are powerful and direct tools for probing lattice defects. The interpretation of the low-temperature lattice thermal conductivity allows estimates of the size of the large-scale defects and the concentration of point defects. Also, the extra contribution from the intercalate phonons is clearly shown.

The experimental observations concerning the in-plane thermoelectric power were also particularly interesting. While at high temperatures all compounds show the same temperature variation, in the low-temperature region the behavior varies widely, according to the stage of the compound. At low temperatures, the diffusion thermopower dominates in nonmagnetic compounds. For stages higher than stage 1, a stage-dependent linear function of the temperature is observed for the diffusion thermoelectric power. For stage 1 itself, two types of behaviors are observed, according to whether the compound is magnetic or not. For nonmagnetic compounds the thermoelectric power drops rapidly to a negligibly small value below a characteristic temperature around 7 K, while for magnetic compounds the thermoelectric power does not vanish and may be even higher than for stage $\geqslant 2$ compounds. The unusual negligibly small diffusion thermoelectric power in nonmagnetic stage-1 GICs indicates that for the first-stage compounds the scattering mechanism in the lowest temperature range is qualitatively different from that for higher stages. At higher temperatures, we have a unique situation where the phonon-drag thermoelectric power dominates the scene, while the diffusion contribution is negligibly small.

As regards c axis transport, the behaviors of the electrical conductivity, thermal conductivity, and thermoelectric power are puzzling. For low stages, the electrical resistivity, which may be six orders of magnitude larger than in-plane, increases with increasing temperature, while for higher stages the reverse trend is observed. The thermal conductivity, which is entirely ascribed to phonons in this direction, is much smaller than in-plane, and the temperature dependence is different in both directions. The anisotropy of the total thermal conductivity is around 10^2 from the liquid-helium range up to room temperature. For the thermopower, the anisotropy is even much less pronounced than that of the thermal conductivity. So for a given sample the anisotropy varies by orders of magnitude according to the transport coefficient considered.

The magnetic compounds showed anomalies in electrical resistivity and thermal conductivity, and a magnon-drag mechanism is very likely responsible for the observed enhancement of the thermoelectric power with respect to the nonmagnetic compound.

Table 6.1. Summary of the features and advantages of GICs with regard to the solid-state properties discussed in this chapter

Property	advantage of GICs or specific situation	layers concerned
2D weak localization	– 2D inherent in the band structure – control of disorder – control of de Broglie wavelength	graphite
magnons	– lower transition temperatures	intercalate
thermal conductivity	– control of relative lattice and electronic contributions (charge transfer) – relative contribution of intercalate phonons	graphite intercalate
thermopower	– phonon drag dominant – possible study of exclusive phonon drag contribution – possible study of magnon contribution	graphite

Table 6.1 summarizes the situations of acceptor GICs with regard to the various solid-state phenomena discussed in this chapter.

Acknowledgements. I am much indebted to Professors M.S. Dresselhaus and J.-P. Michenaud and to Dr. L. Piraux for very interesting discussions and for pertinent comments and suggestions. I also thank Dr. E. McRae for enlightening discussions concerning the *c*-axis electrical resistivity.

References

6.1 M.S. Dresselhaus, G. Dresselhaus: Adv. Phys. **30**, 139 (1981)
6.2 J.-P. Issi: Proc. 10th Erice Summer School, International School of Materials Science and Technology, Erice Sicily July 5–15, Nato ASI Series B (Plenum, New York 1986) p. 347
6.3 For reference to early work on GICs, see for example, G.R. Hennig: J. Chem. Phys. **43**, 1202 (1965); A.R. Ubbelohde: Proc. Roy. Soc. **A327**, 289 (1972) and references therein
6.4 J. Blinowski, Nguyen Hy Hau, C. Rigaux, J.-P. Vieren, R. Le Toullec, G. F urdin, A. Hérold, J. Mélin: J. Phys. **41**, 47 (1980)
6.5 J.-P. Issi, B. Poulaert, J. Heremans, M.S. Dresselhaus: Solid State Commun. **44**, 449 (1982)
6.6 J.-P. Issi, M. Kinany-Alaoui, J.-P. Michenaud, L. Piraux: Extended Abstracts of the Symposium on Graphite Intercalation Compounds: Science and Applications, ed. by M. Endo, M.S. Dresselhaus, G. Dresselhaus (Mater. Res. Soc. 1988) p. 117
6.7 L. Piraux, M. Kinany-Alaoui, J.-P. Issi, A. Pérignon, P. Pernot, R. Vangelisti: Phys. Rev. **B38**, 4329 (1988)
6.8 M.S. Dresselhaus: Proc. 10th Erice Summer School, International School of Materials Science and Technology, Erice Sicily July 5–15, Nato ASI Series B (Plenum, New York 1986) p. 1
6.9 Y. Iye: Extended Abstracts of the Symposium on Graphite Intercalation Compounds: Science and Applications, ed. by M. Endo, M.S. Dresselhaus, G. Dresselhaus (Mater. Res. Soc. 1988) p. 49
6.10 L. Piraux: J. Mater. Res. **5**, 1285 (1990)

6.11 E. Abrahams, P.W. Anderson, D.C. Licciardello, T.V. Ramakrishnan: Phys. Rev. Lett. **42**, 613 (1979)
6.12 G. Bergmann: Phys. Rev. B**28**, 2914 (1983)
6.13 S. Hikami, A.I. Larkin, Nagaoka: Prog. Theor. Phys. **63**, 707 (1980)
6.14 B.L. Altshuler, A.G. Aronov, P.A. Lee: Phys. Rev. Lett. **44**, 1288 (1980)
6.15 H. Fukuyama: J. Phys. Soc. Jpn **48**, 2169 (1980)
6.16 L. Piraux, J.-P. Issi, J.-P. Michenaud, E. McRae, J.F. Marêché: Solid State Commun. **56**, 567 (1985)
6.17 L. Piraux, V. Bayot, J.-P. Michenaud, J.-P. Issi, J.F. Marêché, E. McRae: Solid State Commun. **59**, 711 (1986)
6.18 L. Piraux, V. Bayot, X. Gonze, J.-P. Michenaud, J.-P. Issi: Phys. Rev. B**36**, 9045 (1987)
6.19 G.M.T. Foley, C. Zeller, E.R. Falardeau, F.L. Vogel: Solid State Commun. **24**, 371, (1977)
6.20 J.-P. Issi, L. Piraux: Annales de Phys. **11**(2), 165 (1986)
6.21 E. McRae: Thèse d'Etat, Nancy (1982) [in French]
6.22 J.-F. Marêché, E. McRae, N. Nadi, R. Vangelisti: Synth. Met. **8**, 163 (1983)
6.23 L.A. Pendrys, T.C. Wu, C. Zeller, H. Fuzellier, F.L. Vogel: Extended Abstracts of the 14th Biennal Conf. Carbon (1979) ed. by P.A. Thrower, p. 306
6.24 L. Piraux, J.-P. Issi, L. Salamanca-Riba, M.S. Dresselhaus: Synth. Met. **16**, 93 (1986)
6.25 T.C. Chieu, M.S. Dresselhaus, M. Endo: Phys. Rev. B**26**, 5869 (1982)
6.26 J.M. Ziman: *Electrons and Phonons* (Clarendon, Oxford 1960)
6.27 F.J. Blatt: *Physics of Electronic Conduction in Solids* (McGraw-Hill, New York 1968)
6.28 A. Ansart, C. Meschi, S. Flandrois: Synth. Met. **23**, 455 (1988) Proc. 4th Int. Symp. on GICs, Jerusalem
6.29 C. Meschi: Ph.D thesis, Université de Bordeaux I (1988) [in French]
6.30 C. Manini, J.-F. Marêché, E. McRae: Synth. Met. **8**, 261 (1983)
6.31 P.D. Hambourger, D.A. Jaworske, J.R. Gaier: Extended Abstracts of the Symposium on Graphite Intercalation Compounds, ed. by P.C. Eklund, M.S. Dresselhaus, G. Dresselhaus (Mater. Res. Soc. 1984) p. 208
6.32 N. Daumas and A. Hérold, C.R. Hebd: Séance Acad. Sci. Paris C **268**, 373 (1969)
6.33 B. Kramer, G. Bergmann, Y. Bruynseraede (eds.): *Localization, Interaction and Transport Phenomena* Springer Ser. Solid-State Sci. Vol. 61 (Springer, Berlin, Heidelberg 1985)
6.34 G. Bergmann: Phys. Rep. **107**, 1 (1984)
6.35 P.A. Lee, T.V. Ramakrishnan: Rev. Mod. Phys. **57**, 287 (1985)
6.36 C. Meschi, J.P. Manceau, S. Flandrois, P. Delhaes, A. Ansart, L. Deschamps: Ann. de Phys. **11**, colloq. 2, suppl. 2, 199 (1986)
6.37 H. Fukuyama: In *Percolation, Localization and Superconductivity*, ed. by A.M. Goldman, S.A. Wolf (Plenum, New York 1984) p. 161
6.38 P.A. Lee, T.V. Ramakrishnan: Phys. Rev. B**26**, 4009 (1982)
6.39 L. Piraux, B. Nysten, J.-P. Issi, J.F. Marêché, E. McRae: Solid State Commun. **55**, 517 (1985)
6.40 B.L. Altshuler, A.G. Aronov, D.E. Khmelnitzkii: J. Phys. C **15**, 7367 (1982)
6.41 H. Takayama: Z. Phys. **263**, 329 (1973)
6.42 L. Pietronero, S. Strässler: Synth. Met. **3**, 213 (1981)
6.43 T. Inoshita, H. Kamimura: Synth. Met. **3**, 223 (1981)
6.44 J.-P. Issi, J. Boxus, B. Poulaert, J. Heremans: Thermal Conductivity Vol. 17, ed. by J.G. Hust (Plenum, New York 1982) p. 537
6.45 L. Piraux, J.-P. Issi, P. Coopmans: Measurement **5**, 2 (1987)
6.46 R. Berman: *Thermal Conduction in Solids* (Clarendon, Oxford 1976)
6.47 J.-P. Issi, J. Heremans, M.S. Dresselhaus: Phys. Rev. B**27**, 1333 (1983)
6.48 J. Boxus, B. Poulaert, J.-P. Issi, H. Mazurek, M.S. Dresselhaus: Solid State Commun. **38**, 1117 (1981)
6.49 B. Ploulaert, J. Heremans, J.-P. Issi, I. Zabala Martinez, H. Mazurek, M.S. Dresselhaus: Extended Abstracts 15th Biennial Carbon Conf. (1981) p. 92

6.50 J. Heremans, J.-P. Issi, I. Zabala Martinez, M. Shayegan, M.S. Dresselhaus: Phys. Lett. **84A**, 387 (1981)
6.51 M. Elzinga, D.T. Morelli, C. Uher: Phys Rev B**26**, 3312 (1982)
6.52 F.J. Blatt, I. Zabala Martinez, J.-P. Issi, M. Shayegan, M.S. Dresselhaus: Phys. Rev. B**27**, 2558 (1983)
6.53 M. Kinany-Alaoui, L. Piraux, J.-P. Issi, P. Pernot, R. Vangelisti: Solid State Commun. **68**, 1065 (1988)
6.54 L. Piraux, B. Nysten, J.-P. Issi, L. Salamanca-Riba, M.S. Dresselhaus: Solid State Commun. **58**, 265 (1986)
6.55 J.-P. Issi: Mater. Res. Soc. Symp. Proc. **20** (Elsevier, Amsterdam 1983) p. 147
6.56 P.G. Klemens: in *Encyclopedia of Physics*, ed. by S. Flügge (Springer, Berlin, Heidelberg 1956) Vol. XIV, p. 198
6.57 L. Piraux, J.-P. Issi, J.-F. Marêché, E. McRae: Synth. Met. **30**, 245 (1989)
6.58 J.-P. Issi, J. Heremans, M.S. Dresselhaus: In *Physics of Intercalation Compounds*, ed. by L. Pietronero, E. Tosatti (Springer, Berlin, Heidelberg 1981) p. 310
6.59 B.T. Kelly: *Physics of Graphite* (Applied Science Publishers, London 1981)
6.60 B. Nysten, L. Piraux, J.-P. Issi: Thermal Conductivity, Vol. 19, ed. by D.W. Yarbrough (Plenum, New York 1985) p. 341
6.61 B. Dreyfus, R. Maynard: J. de Phys. **28**, 955 (1967)
6.62 J. Callaway: Phys. Rev. **113**, 1046 (1959)
6.63 J. Heremans, M. Shayegan, M.S. Dresselhaus, J.-P. Issi: Phys. Rev. B**26**, 3338 (1982)
6.64 B. Nysten, L. Piraux, J.-P. Issi: Synth. Met. **12**, 505 (1985)
6.65 L.C.F. Blackman, J.F. Mathews, A.R. Ubbelhode: Proc. Roy. Soc. **258**A, 339 (1960)
6.66 A.R. Ubbelohde: Carbon **6**, 177 (1968)
6.67 F. Maeda, H. Oshima, K. Kawamura, T. Tsuzuku: In Extended Abstracts of the Fourteenth Biennial Carbon Conference, State College, PA, 1979, ed. by P.A. Thrower, p. 197
6.68 J.-P. Issi, J. Boxus, B. Poulaert, H. Mazurek, M.S. Dresselhaus: J. Phys. **14**, L307 (1981)
6.69 T. Enoki, K. Imaeda, H. Inokuchi, S. Miyajima, T. Chiba, M. Sano: Extended Abstracts Mater. Res. Soc. Symp. (1986) p. 63
6.70 M. Endo, T. Koyama, M. Inagaki: Synth. Met. **3**, 177 (1981)
6.71 L. Piraux, J.-P. Issi, P. Eklund: Solid State Commun. **56**, 413 (1985)
6.72 C. Uher, D.T. Morelli: Synth. Met. **12**, 91 (1985)
6.73 F.J. Blatt, P.A. Schroeder, C.L. Foiles, D. Greig: *Thermoelectric Power of Metals* (Plenum, New York 1976)
6.74 J.-P. Issi: Aust. J. Phys. **32**, 585 (1979)
6.75 L. Piraux, M. Kinany-Alaoui, J.-P. Issi: To be published
6.76 J.-P. Issi: Revue de Chimie Minerale **19**, 394 (1982)
6.77 K. Sugihara: Phys. Rev. B**28**, 2157 (1983)
6.78 D.G. Cantrell, P.N. Butcher: J. Phys. C **19**, L429 (1986)
6.79 D.G. Cantrell, P.N. Butcher: J. Phys. C **20**, L1993 (1987)
6.80 E. McRae, J.-F. Marêché: J. Mater. Res. **3**, 75 (1988)
6.81 K. Phan, C.D. Fuerst, J.E. Fischer: Solid State Commun. **44**, 1351 (1982)
6.82 D.T. Morelli, C. Uher: Phys. Rev. B**27**, 2477 (1983)
6.83 K. Sugihara: Phys. Rev. B**29**, 5872 (1984); Phys. Rev. B**37**, 4752 (1988)
6.84 S. Shimamura: Synth. Met. **12**, 365 (1985)
6.85 N.C. Yeh, K. Sugihara, J.T. Nicholls, G. Dresselhaus: In Proc. 10th Erice Summer School, International School of Materials Science and Technology ed. by M.S. Dresselh aus, G. Dresselhaus, Erice Sicily July 5–15, Nato ASI Series B (Plenum, New York 1986) p. 429
6.86 M.S. Dresselhaus, N.C. Yeh, K. Sugihara, J.T. Nicholls, G. Dresselhaus: Int. Coll. Layered Compounds, Pont-à-Mousson (France) (1988) p. 199; K. Miura, Y. Iye, J .T. Nicholls, G. Dresselhaus: In Extended Abstracts of the Symposium on Graphite Intercalation Compounds, ed. by M. Endo, M.S. Dresselhaus, G. Dresselhaus (Mater. Res. Soc., Pittsburgh 1988) p. 69

6.87 M. Kinany-Alaoui, L. Piraux, V. Bayot, J.-P. Issi, P. Pernot, R. Vangelisti: Proc. 5th International Conference on Graphite Intercalation Compounds (Berlin, 1989)
6.88 S.A. Friedberg, E.D. Harris: In Proc. 8th International Conference on Low Temperature Physics (1962) ed. by R.O. Davies (Butterworths, London 1963) p. 302
6.89 H.W. Gronert, D.M. Herlach, E.F. Wassermann: Europhys. Lett. **6**(7), 641 (1988)
6.90 R.J. Elliott, F.A. Wedgwood: Proc. Phys. Soc. **81**, 846 (1963)
6.91 Ch. Simon, F. Battalan, I. Rosenman, C. Ayache, E. Bonjour: Phys. Rev. B**35**, 5816 (1987)
6.92 H. Zabel, S.M. Shapiro: Phys. Rev. B**36**, 7292 (1987)

7. Magnetic Intercalation Compounds of Graphite

By Gene Dresselhaus, James T. Nicholls and Mildred S. Dresselhaus

With 32 Figures

Magnetic graphite intercalation compounds are layered magnetic materials in which the ratio of the intraplanar to interplanar exchange coupling can be varied by several orders of magnitude through intercalation and staging. In this chapter, the structure and magnetic properties of magnetic acceptor and donor graphite intercalation compounds (GICs) are reviewed. The relation between the GIC and the pristine intercalate makes it possible to measure the effect of staging in the GIC, including the crossover to lower dimensionality. The theory of two-dimensional (2D) magnetism is reviewed and compared with recent evidence for 2D magnetism in acceptor graphite intercalation compounds. The transition-metal dichloride acceptor compounds show larger magnetic anisotropies than the donor GICs and provide instructive examples of quasi-2D magnetic systems. The prototype 2D-*XY* system is the $CoCl_2$ GIC, for which a large body of experimental data is presented to infer the 2D behavior. Experimental results for the commensurate stage-1 donor compound C_6Eu are presented as an example of an anisotropic RKKY magnetic interaction. In addition, progress in the study of many magnetic acceptor GICs is reviewed.

7.1 Background

The study of magnetism in layered materials has long been considered an attractive area for investigating the effect of lower dimensionality in magnetic systems. Because of this interest in low-dimensional magnetism, a number of layered transition-metal chlorides (e.g., $CoCl_2$ [7.1], $NiCl_2$ [7.2], $FeCl_2$ [7.3]) were studied as potential experimental candidates for studying such phenomena because the *interplanar* exchange interactions in these materials are approximately an order of magnitude smaller than the *intraplanar* exchange interactions. However, subsequent experiments showed that these materials were all 3D magnets. With the development of experimental techniques for intercalating these layered magnetic transition-metal chlorides into graphite, interest in low-dimensional magnetic systems was recently renewed, because it is now possible through intercalation to decrease the interplanar interactions by several more orders of magnitude. In addition, the special nature of the intercalation process for these acceptor compounds retains most of the structural integrity of the intercalate layer, and thus the results of earlier studies on the pristine layered

Springer Series in Materials Science, vol. 18
H. Zabel · S.A. Solin (eds.): Graphite Intercalation Compounds II

materials can now be applied to the graphite intercalation compounds (GICs) with only minor changes to the in-plane parameters in the magnetic Hamiltonian.

Theoretical interest in 2D magnetism was stimulated when *Onsager* [7.4] solved the statistical mechanics problem for the 2D Ising model (spin dimension 1) exactly, and showed theoretically that such a system undergoes a continuous phase transition at a finite temperature. Much later, a new kind of phase transition, unlike the familiar order/disorder type shown by the 2D Ising model, was suggested in the context of spin systems with spin dimension 2 or 3 (as in the *XY* and Heisenberg models) [7.5–7]. The new type of phase transition was later established for the 2D *XY* model (Sect. 7.2.3) [7.8–13]; in particular, *Kosterlitz* and *Thouless* [7.11–13] gave new insights into the nature of the transition, which is customarily referred to as the Kosterlitz-Thouless transition, enabling quantitative calculations of critical properties.

The Kosterlitz-Thouless (KT) theory showed that there was another degree of freedom associated with the spins, i.e., their ability to form positive and negative spin vortices that would show a phase transition when vortex pairs $(+/-)$ undergo an unbinding transition at a finite temperature. Though most experimental magnetic systems are 3D, the 2D-*XY* model represents a system with a critical dimension [7.14–16], where fluctuations in the order parameter in the critical region near the phase transition temperature T_c determine the functional form of all observables near the transition [7.14]. The application of renormalization group techniques [7.16, 17] has enabled one to deduce the functional forms for the temperature and magnetic field dependence of a number of thermodynamic variables [7.13]. In addition, the advent of large computers and refined sampling algorithms has made possible Monte Carlo calculations on low-dimensional systems with reasonably complex magnetic interactions and anisotropies [7.18]. Theoretical developments connected with the KT transition stimulated a great deal of interest in identifying experimental systems with sufficiently anisotropic magnetic properties so that they could be approximated as quasi-2D systems.

Because of their very high structural anisotropy, magnetic graphite intercalation compounds [7.19] appeared to provide a framework to study the fundamental physics of magnetism of quasi-2D systems. This follows from the unique properties of GICs, namely that the host graphite material is nonmagnetic, but allows (through the intercalation mechanism shown schematically for a stage-1 $CoCl_2$ GIC in Fig. 7.1) the introduction of single layers of magnetic species separated by controlled distances, usually the c axis repeat distance I_c of the GIC. As the stage index n for the magnetic acceptor GICs increases (and, consequently, I_c also increases), each magnetic layer remains essentially like a layer of the parent bulk magnetic solid, but is now separated in the GIC by larger and larger distances, thereby approaching 2D magnetic behavior at some high stage (the limit $n \to \infty$ is clearly 2D behavior). As discussed below, other magnetic layered compounds also offer the possibility of studying anisotropic magnetic phenomena in systems of reduced dimensionality.

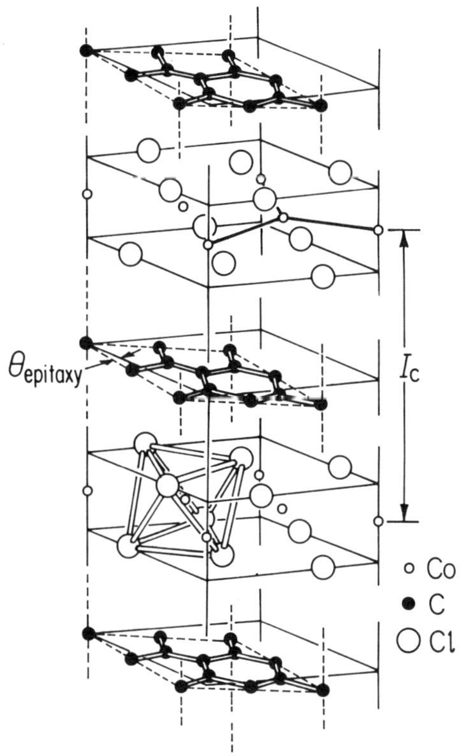

Fig. 7.1. Schematic crystal structure of stage-1 $CoCl_2$ GIC. The $CoCl_2$ intercalate layer is incommensurate with respect to the adjacent graphene layers but exhibits rotational locking of $\sim 0°$. The rotational epitaxy angle is indicated by $\theta_{epitaxy}$ (Sect. 7.1.3a). The intercalate layers in the actual stage-1 compounds are usually stacked in an . . . $\alpha\beta\gamma\alpha\beta\gamma$ . . . arrangement, in contrast to the schematic diagram showing . . . $\alpha\alpha$ . . . stacking

The possibility of preparing single isolated magnetic layers either by intercalation or by atomic layer-by-layer deposition processes has stimulated a great deal of experimental activity in the observation of highly anisotropic and quasi-2D magnetic phenomena. Likewise, the intercalation process for different magnetic intercalate species and for different host materials also provides complementary information with regard to quasi-2D magnetism (Sect. 7.2.3). At present, progress in experimental 2D magnetic studies is largely limited by the types of materials that can be synthesized, and by their structural perfection.

There have been several recent reviews of magnetic GICs [7.20–23] and one review of the stage-2 $CoCl_2$ GIC [7.24]. Most of the interest from the physics community has been related to the issue of low dimensionality and this view point is reflected in these reviews. The recent studies of the magnetic properties of the high-T_c metal-oxide superconductors [7.25] (which have also been described as intercalation compounds [7.26], because of the building block structure of the high-T_c oxides) have further enhanced the interest in 2D magnetism.

This review focuses on the structure and properties of the magnetic graphite intercalation compounds. Some of the most promising work is either quite recent or is still in progress. Thus the present review should be considered as a progress report on the potential of GICs for the study of highly anisotropic and quasi-2D magnetic phenomena.

7.1.1 Theoretical Considerations

The Heisenberg exchange Hamiltonian for a magnetic system is generally written as

$$\mathscr{H}_{\mathrm{ex}} = -2 \sum_{i<j} J_{ij} \boldsymbol{S}_i \cdot \boldsymbol{S}_j , \tag{7.1}$$

where the sum is over all the spins in the system and only nearest-neighbor exchange interactions are considered. The spatial dimensionality of the system is defined by geometrical considerations of the various neighbor distances between the magnetic species. If the magnetic species are arranged on layers separated by large distances so that the first few near-neighbor distances are within the basal plane, the magnetic system can be approximately described by a 2D spatial system whereby the sum in (7.1) is dominated by spins on a single layer. For actual GICs, the weak interplanar coupling J'_{ij} (where $J'_{ij} \ll J_{ij}$) leads to quasi-2D behavior, but however small the magnitude of J'_{ij}, 3D ordering is eventually established as $T \to 0$.

With regard to the spin dimensionality, the exchange interaction can be isotropic (three dimensional), and given by the Heisenberg Hamiltonian of (7.1). If there is an anisotropy that causes the spins to lie in the basal plane, the XY model is valid:

$$\mathscr{H}_{XY} = -2 \sum_{i<j} J_{ij} (S_i^x S_j^x + S_i^y S_j^y) . \tag{7.2}$$

Finally, when the spins are confined along a unique direction (e.g., normal to the basal plane), the Ising model, written as

$$\mathscr{H}_{\mathrm{Ising}} = -2 \sum_{i<j} J_{ij} S_i^z S_j^z , \tag{7.3}$$

is appropriate. All three of these cases (7.1–3) can be represented by the general Hamiltonian

$$\mathscr{H} = -2 \sum_{i<j} J_{ij} \boldsymbol{S}_i \cdot \boldsymbol{S}_j + 2 \sum_{i<j} (J_{\mathrm{A}})_{ij} S_i^z S_j^z , \tag{7.4}$$

where the anisotropy term $(J_{\mathrm{A}})_{ij} = 0$ for the Heisenberg Hamiltonian, $(J_{\mathrm{A}})_{ij} = -J_{ij}$ for the XY Hamiltonian, and $J_{ij} = 0$ for the Ising Hamiltonian. In practice, actual experimental systems do not correspond to these idealized Hamiltonians, but rather only represent approximations to these idealized limits. Furthermore, because of the different conventions used in writing the exchange terms in the Hamiltonian, and often the absence of clear definitions in the experimental literature, there are often ambiguities in factors of 2 in comparing the results from different groups.

For the GICs, the spatial dimensionality is either $d = 2$ if $J'_{ij} \ll J_{ij}$ or $d = 3$ if the interplanar coupling is important. From a theoretical standpoint, the most interesting GIC systems are those that can be approximately described by the 2D-XY model [7.21, 22], because 2D-XY systems have a critical dimensionality where fluctuations play an important role very close to the thermodynamic transition temperature T_c. Though developed for magnetic systems, the 2D-XY model has also been widely used to model a variety of structural phase transitions [7.27], Josephson junction arrays [7.28], electrons on the surface of a liquid-helium film [7.29], and superfluid helium ^{4}He films [7.30]. In Sects. 7.2.1, 2 we discuss in more detail the magnetic Hamiltonians for magnetic acceptor and donor GICs, while Sect. 7.2.3 is devoted to further discussion of the 2D-XY model.

7.1.2 Magnetism in Layered Compounds

From an experimental standpoint, the materials that show the most promise as model 2D magnetic systems are layered compounds with a high degree of spatial anisotropy. Beyond that, various investigators have sought specific physical systems in layered compounds that would emulate the 2D-Heisenberg, 2D-XY, and 2D-Ising interactions described in Sect. 7.1.1. In these layered materials the magnetic ions are arranged in 2D layers, where there is a strong magnetic interaction within the layers and there is a much weaker magnetic interaction between the layers due to the larger physical separation between them [7.31].

Naturally occurring layered magnetic compounds (such as layered transition-metal chloride compounds) are of interest as anisotropic magnetic materials in their own right. An attractive experimental approach to achieving quasi-2D magnetic systems is to have a mechanism by which the physical separation between the magnetic layers can be increased. The intercalation of a magnetic intercalate into a nonmagnetic host is one such mechanism. When applied to graphite or transition-metal dichalcogenide host materials, we refer to the pristine magnetic intercalates as the zero-stage limit of magnetic intercalation compounds. Other experimental approaches to the synthesis of quasi-2D systems include layer-by-layer synthesis of magnetic superlattices [7.32, 33], preparation of highly anisotropic layered compounds such as $(C_nH_{2n+1}NH_3)_2CuCl_4$ [7.31, 34], and preparation of magnetic Langmuir–Blodgett films where the chain length is variable [7.35] so that the separation between the layers of magnetic ions can be controlled.

For all magnetic intercalation compounds, the magnetic layers consist of ions with unfilled d or f shells. In this context, we summarize in Table 7.1 the magnetic crystal field properties associated with various M^{2+} and M^{3+} ions with partly filled $3d$ shells, including the ground-state configuration of the free ion, the spin–orbit splitting, and the ground-state configuration of the ion in an octahedrally coordinated cubic field. These crystal-field splittings are important in determining the electronic states within a few electron volts of the Fermi level, and therefore the type of magnetic interactions governing the electronic ground

Table 7.1. Crystal-field parameters for transition-metal ions[a]

Ion	Free ion[b] Ground state $^{2S+1}\mathscr{L}_J$	Spin-orbit (cm^{-1})	Octahedral coordination[c] Cubic field, O_h Ground state	Reference
Sc^{2+}	$^2D_{3/2}$	197.5	$^2T_{2g}$	[7.36]
Ti^{2+}	$^3F_2$	184.7	$^3T_{1u}$	[7.37]
$V^{2+}(Cr^{3+})$	$^4F_{3/2}$	145.0	$^4A_{2u}$	[7.37]
Cr^{2+}	$^5D_0$	59.9	^5E_g	[7.37]
Mn^{2+} (Fe^{3+})	$^6S_{5/2}$	—	$^6A_{1g}$	[7.38]
Fe^{2+}	$^5D_4$	436.2	$^5T_{2g}$	[7.39]
Co^{2+}	$^4F_{9/2}$	841.2	$^4T_{1u}$	[7.40]
Ni^{2+}	$^3F_4$	1360.7	$^3A_{2u}$	[7.40]
Cu^{2+}	$^2D_{5/2}$	2071.8	^2E_g	[7.37]

[a] Not all 3*d* transition-metal ions have yet been synthesized into GICs.
[b] Free-ion ground state from *Moore* [7.41].
[c] Crystal-field splittings of the orbital degeneracy in a cubic field are as follows: $S \to A_{1g}$; $P \to T_{1u}$; $D \to E_g + T_{2g}$; $F \to A_{2u} + T_{1u} + T_{2u}$.

state. In general, if the lowest crystal-field-split level is nondegenerate, then the anisotropy terms might be expected to be small and the spin Hamiltonian is considered to be of the Heisenberg type, as for example in the case of $FeCl_3$. If on the other hand, the lowest crystal field state is degenerate and the spin–orbit coupling is large, then the spin Hamiltonian will have large anisotropy terms and the Hamiltonian could be of the Ising or *XY* type, depending on the sign and magnitude of the anisotropy. We give several examples of such materials below. We note that no GICs have yet been synthesized for some of the entries in Table 7.1. The 4*d* (Y, Zr, Nb, Mo, Tc, Ru, Rh, Pd, Ag) and 5*d* (La, Hf, Ta, W, Re, Os, Ir, Pt, Au) transition-metal ions are expected to show behaviors analogous to the 3*d* compounds as the outer *d* shell is filled, except for the greater role played by the spin–orbit interaction relative to the crystal-field interaction as the atomic number of the ion increases.

(a) Comparison of Magnetic Superlattices

Two approaches to achieve quasi-2D behavior are the intercalation of magnetic species into nonmagnetic layered host materials and the layer-by-layer synthesis of magnetic superlattices (deliberately structured materials). In this section we briefly review these two approaches and compare their relative advantages. We then contrast (Sect. 7.1.3) the variety of magnetic intercalation compounds that can be prepared, depending on the choice of host material.

The greatest advantage of the layer-by-layer atomic deposition process relates to the flexibility in the choice of materials and periodicities. The greatest control of the layer thickness and in-plane homogeneity is provided by molecular-beam epitaxy (MBE), though metal-organo chemical vapor deposition (MOCVD) offers almost as much control and greater sample throughput. With

MBE, it is possible to prepare a superlattice with layers of magnetic species A and nonmagnetic species B, or even with layers of two different magnetic species. Layers of almost any magnetic element or compound can be deposited, thereby allowing in principle the study of 2D ferromagnetic, antiferromagnetic, ferrimagnetic, and spin-glass behaviors. Likewise, the nonmagnetic species can be chosen to be insulating, semiconducting, or metallic. With MBE, the layer thicknesses d_A and d_B can be controlled over wide ranges ($3\,\text{Å} \leqslant d \leqslant 200\,\text{Å}$), and samples of several hundred superlattice unit cells with repeat distances $d = d_A + d_B$ can be fabricated.

One fascinating application of the MBE technique to the study of size effects in magnetic materials has been in a superlattice of magnetic Dy and nonmagnetic Y, where the layer thickness d_A of the magnetic Dy layers was chosen to be less than the period of the spin spiral along the c axis of bulk metallic Dy [7.42]. Also of interest has been the magnetic properties of the Gd–Y superlattice grown by MBE techniques [7.43, 44]. Although sputter deposition techniques have been widely used to prepare magnetic multilayers [7.45], the constituent layers d_A and d_B tend to be thick compared with the range of the magnetic exchange interactions, and therefore are not suitable for the quantitative study of 2D properties.

The unique advantage of intercalation compounds with regard to studies of 2D magnetism is the possibility of preparing isolated magnetic layers and atomically sharp interfaces between materials and the macroscopic ($\sim 1000\,\text{Å}$) lateral distances over which these sharp interfaces are maintained [7.46]. Thus it is possible to accurately control the interlayer exchange interaction J' by intercalation; this level of control of J' is not possible by any of the atomic deposition processes because J' falls off so rapidly with distance. The magnetic behavior of atomically sharp isolated monolayers is qualitatively different from that of multiple magnetic layers with regard to the observation of 2D magnetic phenomena. Thus, intercalation compounds provide an excellent opportunity for experimental study of the magnetic properties of isolated magnetic layers (in high-stage compounds) and of the effect of weak interplanar exchange coupling on these properties (as the stage index is decreased).

For GICs the c-axis repeat distance I_c between sequential intercalate layers follows the simple relation $I_c = d_s + (n - 1)c_0$, where $c_0 = 3.35\,\text{Å}$ is the interlayer separation in the pristine graphite host material, n is the stage number, and d_s is the separation between two graphene layers between which the intercalate is sandwiched (as, for example, the thickness of the layers C–Cl–Co–Cl–C in a $CoCl_2$ GIC, where C denotes a graphene layer, defined as an isolated single layer of the graphite structure). Typical values of d_s for magnetic intercalates are 4.87 Å and 9.38 Å for the Eu donor and $CoCl_2$ acceptor intercalates, respectively. (See Table 7.2 for other representative d_s values.) Magnetic intercalation compounds with stages as high as 8 [7.56] have been synthesized, corresponding to $I_c \cong 33\,\text{Å}$. From entropy considerations, high stage compounds ($n \geqslant 4$), however, are more likely to be inhomogeneous and to have admixtures of adjacent stages. For the layered acceptor magnetic materials normally investi-

Table 7.2. Parameters for several magnetic materials and their magnetic GICs

			Structure[a]		T_c(K)	
Intercalate	d_s^b (Å)	Stages[c]	Lattice[a]	a_0[a] (Å)	Pristine	GIC
Eu	4.87[d]	1	bcc[e]	4.58	90[e]	40
Sm	4.68[d]	1	hexagonal[e]	3.62	14.8[e]	—
Tm	4.62[d]	1	hcp[e]	3.54	58[e]	—
$CoCl_2$	9.38	1–3	$CdCl_2$	3.54	24.9[f]	8–10
$CrCl_3$	9.45	2, 3	$FeCl_3$	6.00	14.5[g]	~10[h]
$CuCl_2$	9.40	1, 2	monoclinic	3.30	23.9[i]	—
$FeCl_2$	9.51	1, 2	$CdCl_2$	3.58	23.6[j]	15.5[k]
$FeCl_3$	9.37	1–11	$FeCl_3$	6.06	8.83[l]	1.7[m], 3.6[n]
$MnCl_2$	9.47	1, 2	$CdCl_2$	3.69	1.96[o]	1.2[p]
$MoCl_5$	9.32	1–5	monoclinic	17.31	22	1.6[q]
$NiCl_2$	9.42[r]	1, 2	$CdCl_2$	3.54	52.3[r]	18–23[s]

[a] Lattice structure and lattice constant a_0 refer to that of the pristine intercalate [7.47].
[b] The separation of magnetic layers in Å is $d_s + 3.35(n-1)$ where n denotes the stage index. Error bars are typically ±0.03 Å.
[c] Results are given only for well staged GICs.
[d] Ref. [7.48].
[e] Ref. [7.49] and references cited therein.
[f] Ref. [7.1].
[g] Ref. [7.50].
[h] Ref. [7.51].
[i] Ref. [7.52, 53].
[j] Ref. [7.3].
[k] Ref. [7.54].
[l] Ref. [7.55].
[m] Ref. [7.56].
[n] Ref. [7.57].
[o] Ref. [7.58].
[p] Ref. [7.59].
[q] Ref. [7.60].
[r] Ref. [7.2].
[s] Ref. [7.61, 62].

gated, the interplanar exchange interaction J' is small compared with that for the intraplanar interaction J, and the J'/J ratio decreases rapidly with increasing stage index (Sect. 7.2.1).

A second major advantage of GICs for the study of 2D magnetism is the high degree of in-plane order that can be achieved. GICs have been prepared with in-plane crystalline coherence distances in excess of 1 μm [7.27, 63], though typical magnetic in-plane coherence distances in magnetic GICs may be an order of magnitude smaller [7.64]. For typical magnetic acceptor GICs, the magnetic intercalate layer is incommensurate with respect to the two adjacent graphite layers on either side. Furthermore, many acceptor magnetic intercalate layers exhibit lattice constants that are equal to their values in the parent bulk compounds to within ±1.5% [7.65–67]. Since the in-plane intercalate lattice constants are almost identical to their bulk values, so are the in-plane magnetic exchange interactions J. A priori knowledge of the in-plane exchange interactions and anisotropies in the magnetic superlattices leads to a major simplification in the interpretation of quasi-2D magnetic phenomena. In contrast, the intercalate layers in the rare-earth magnetic donor compounds typically are commensurate with the graphite layers [7.48], leading to other simplifications in the analysis of the magnetic properties, though in the case of donors, all exchange interactions must be determined from the observed magnetic properties in the GIC. Higher-stage GIC samples do, however, present experimental problems due to staging inhomogeneities, as mentioned above.

To realize 2D magnetic behavior, it is necessary to maximize the separation between sequential magnetic layers. This is most readily accomplished through the preparation of high-stage intercalation compounds. A complementary approach to this problem has been the preparation of heterostructure or bi-intercalation compounds, whereby the magnetic intercalate is alternated with a nonmagnetic intercalate, thereby further increasing the separation between the magnetic layers. For example, the stage-1 $CoCl_2$ GIC has a c-axis repeat distance of 9.38 Å, whereas the acceptor/acceptor bi-intercalation compound C–$CoCl_2$–C–$GaCl_3$–C has a repeat distance and separation between magnetic layers of 19.12 Å [7.68]. Other examples of magnetic heterostructures or bi-intercalation compounds are the magnetic acceptor/nonmagnetic donor compound C–$CoCl_2$–C–K–C with a repeat distance of $I_c = 14.65$ Å [7.69], and the double magnetic acceptor bi-intercalation compound C–$CoCl_2$–C–$FeCl_3$–C with a repeat distance $I_c = 18.60$ Å [7.70]. The synthesis of such bi-intercalation compounds is carried out by first preparing a stage-2 $CoCl_2$ GIC and then intercalating the second intercalate species between the remaining adjacent graphite layers at a lower reaction temperature [7.71]. Since the actual bi-intercalation compounds often contain an admixture of the precursor stage-2 $CoCl_2$ GIC and the desired stage-1 heterostructure, it has been difficult to carry out quantitative magnetic studies on magnetic bi-intercalation compounds, though some qualitative studies have been carried out successfully [7.72, 73]. Bi-intercalation compounds (Sect. 7.4.1) are potentially of great interest for the preparation of 2D magnetic materials.

Another very useful method for varying the interlayer exchange interaction in GICs is through the application of pressure. Magnetic studies under pressure have been carried out for pristine $CoCl_2$ [7.74] and for stage-1 and stage-2 $CoCl_2$ GICs [7.75]. This work shows that the interlayer exchange interaction J' of the stage-1 $CoCl_2$ GIC can be increased by a factor of ~ 2 through the application of 15 kbar pressure [7.75].

In contrast, synthesis techniques for the atomic deposition process for metallic superlattices are not sufficiently developed to form crystalline in-plane structures with long-range order, except for the case of the epitaxial growth of materials with similar crystal structures. The flexibility of the materials selection in the superlattice growth of the deliberately structured materials in the z direction generally leads to problems with the in-plane crystal quality, arising from mismatches in crystal structures between constituents A and B of the superlattice. Even when the crystal structures are identical, large mismatches in the lattice constants of constituents A and B often occur, and the resulting strained layer superlattice forms misfit dislocations to relieve the large stresses that are thus introduced.

One type of MBE magnetic superlattice with promise for long range in-plane coherence is that based on magnetic semiconductor superlattices [7.76], as, for example, those prepared from constituents A and B in the diluted magnetic semiconductor (DMS) family $Cd_{1-x}Mn_xTe/CdTe$. In this case, strained layer superlattices [7.77, 78] are formed, so that the lattice constants are weakly dependent on the ratio of the layer thicknesses d_A/d_B. This dependence on d_A/d_B implies that the lattice constants must be measured for each sample, but this can readily be done using X-ray diffraction techniques. On the other hand, this dependence offers the flexibility of varying the lattice constants over small ranges. Because of the substitutional nature of the dilute magnetic cations, the magnetic interactions in DMS superlattices are mainly random, and spin-glass ground states are observed. To achieve a magnetic state with long-range in-plane order, it is necessary to tailor the lattice constant of the nonmagnetic constituent to match that of the pure magnetic species, or to prepare sufficiently thin layers d_A and d_B to minimize misfit dislocations at the interfaces.

(b) Magnetic Intercalation into Various Hosts

A variety of host materials are available for the synthesis of magnetic intercalation compounds, each exhibiting complementary properties. As discussed above, graphite intercalation compounds offer the possibility of intercalating single magnetic layers with significant control of the distance between the magnetic layers through the staging phenomenon. A good deal of detailed information is often available about the structure and magnetic properties of the parent magnetic materials (e.g., $CoCl_2$ or Eu), and this information can readily be utilized in the study of magnetic GICs. The graphite host material has the advantage of permitting the preparation of both donor and acceptor magnetic intercalation compounds. Unfortunately, only a few magnetic species have been successfully intercalated into graphite host materials (Table 7.2). As mentioned above, the intercalation of magnetic intercalates results in isolated magnetic layers, since it is not possible to intercalate more than a single consecutive magnetic layer between two graphene layers.

The clay intercalation compounds (CICs), in contrast to GICs, permit the formation of magnetic multilayers (e.g., aluminum silicates), which allow the intercalation of a larger number of guest species [7.79, 80]. To date there have been very few magnetic studies of intercalated clays, though in principle it should be possible to prepare a variety of anisotropic magnetic materials in clay hosts. For this type of host material, either the intercalate or the host material could have magnetic layers. Although there is no reported charge transfer in the intercalation of clays, the intercalate in this case consists of either cations or neutral species [7.81]. As noted above, clays are of particular interest because they permit the intercalation of multiple consecutive intercalate layers, and thus multiple consecutive *magnetic* intercalate layers may be possible. For the clay intercalation compounds (CICs), it is not generally possible to achieve good *c*-axis periodicity, with respect to either the number of sequential magnetic layers in each gallery or the separation between sequential groups of magnetic layers, since CICs do not generally exhibit good staging behavior. However, within the stage-1 compound it has been reported that in intercalated vermiculites the number of layers of water of hydration can be varied between 0 and 2 [7.82]. Because of the interesting structures that can be prepared with CICs, detailed studies of the magnetic properties of these materials should be rewarding.

In the case of the transition-metal dichalcogenides, magnetic species can also be intercalated, particularly transition-metal species, and a variety of magnetic phases have been reported [7.83]. For this class of intercalation compounds as well as for clays, both the transition-metal host material and the intercalate can exhibit magnetic behavior. The transition-metal dichalcogenides also permit intercalation of single magnetic layers, though there is normally no well-developed staging periodicity in these host materials. Some magnetic systems that have been studied include the $LiCrS_2$ and $AgCrS_2$ where the Cr^{3+} ion forms the magnetic species [7.84–86]. Particularly interesting magnetic systems are synthesized by intercalating magnetic transition metals such as Fe into TiS_2, TaS_2, and $NbSe_2$. For these systems, commensurate in-plane structures can be achieved for specific filling factors of the intercalate, such as $Fe_{1/4}TaSa_2$ or $Fe_{1/3}TaS_2$ [7.83, 87]. No corresponding effect has been reported for GICs. Large separations between layers of cations can be achieved by the introduction of large organic molecules, and in this way study of quasi-2D magnetism could be carried out in principle. For example, TaS_2 intercalated with *n*-octadecylamine yields a separation of ~ 60 Å between the sequential Ta layers, and this concept has been applied to study quasi-2D superconductivity [7.88]. A similar approach should be applicable to the study of quasi-2D magnetism. One method of using organic groups to separate the magnetic ions has been used to prepare cobaltocene compounds [7.89], but in this case all the magnetic species are separated by distances too large to achieve collective magnetic behavior.

Thus the various host materials (graphite, clays, dichalcogenides, etc.) permit synthesis of a variety of magnetic intercalation compounds with complementary but different properties. Within the class of intercalation compounds, the magnetic GICs provide a unique source of magnetic materials with single isolated magnetic monolayers embedded in a matrix with a well-established *c*-axis periodicity, thus providing a natural connection to the well-studied magnetic parent materials.

7.1.3 Structure of Magnetic Graphite Intercalation Compounds

(a) Structure of Acceptor Compounds

In contrast to the small number of magnetic donor compounds, a large number of magnetic acceptor compounds have been prepared by the intercalation of magnetic salts (mostly transition-metal halides [7.90]). Of these, the transition-metal chlorides compounds have been studied most extensively.

From the structural standpoint, the transition-metal (M) chloride in magnetic acceptor GICs form trilayer Cl–M–Cl intercalate structures with in-plane lattice constants very close (within 1.5%) to the corresponding values in the parent metal-chloride compounds (Fig. 7.1 and Table 7.3). Thus the intercalates are incommensurate with respect to the adjacent graphite layers (only the nonmagnetic $InCl_3$ exhibits a commensurate phase [7.99]). On the other hand,

Table 7.3. Lattice constants for selected transition-metal chlorides and their intercalation compounds at room temperature

GIC	Stage	Site symmetry	Intercalate stacking	a (Å)	I_c (Å)	c(Å)	$\theta_{epitaxy}$ [a]
$CoCl_2$	0	$\bar{3}2/m$	—	3.544	—	34.86[b]	—
$CoCl_2$[c]	1	$\bar{3}2/m$	$\alpha\beta\gamma$	3.572	9.38	28.15	$0° \pm 2.5°$
$CoCl_2$[c]	2	$\bar{3}2/m$	random	3.587	12.73	12.73	$28.5° \pm 3.0°$
$CrCl_3$[d]	3	$\bar{3}2/m$	random	5.95	—	16.15	30°
$CuCl_2$[e]	0	$2/m$	—	3.30[e]	6.70[e]	—	—
$CuCl_2$[c]	1	$2/m$	$\alpha\beta$	3.317[f]	9.40	18.80	$0° \pm 2.5°$
$CuCl_2$[c]	2	$2/m$	random	3.326[f]	12.71	12.71	$0° \pm 2.0°$
$FeCl_3$[g]	0	$\bar{3}2/m$	—	6.06	17.41	17.41	—
$FeCl_3$[h]	1	$\bar{3}2/m$	—	6.06	9.41	9.41	30°
$MnCl_2$	0	$\bar{3}2/m$	—	3.686	—	34.92[b]	—
$MnCl_2$[c]	1	$\bar{3}2/m$	$\alpha\beta\gamma$	3.702	9.47	28.40	$0° \pm 3.5°$
$MnCl_2$[c]	2	$\bar{3}2/m$	random	3.705	12.69	12.69	$30° \pm 4.0°$
$MoCl_5$[i]	4	hexagonal	random	6.18[i]	19.37	19.37	30°
$NiCl_2$	0	$\bar{3}2/m$	—	3.468[j]	—	34.40[b,j]	—
$NiCl_2$[c,k]	1	$\bar{3}2/m$	$\alpha\beta\gamma$	3.496	9.42	28.26	$0° \pm 5.5°$
$NiCl_2$[c]	2	$\bar{3}2/m$	random	3.498	12.75	12.75	$0° \pm 6.0°$

[a] The $\pm$ do not indicate errors of the angle, but rather that the lock-in angle is distributed from $(\theta+\delta\theta)$ to $(\theta-\delta\theta)$ (Fig. 7.1).

[b] Nonprimitive hexagonal cell (rather than the primitive rhombohedral) is used to emphasize the layered nature of these compounds [7.1].

[c] Ref. [7.91], see also [7.92, 93].

[d] Ref. [7.94].

[e] The pristine structure is monoclinic and has parameters [7.95] $a=6.7$ Å, $b=3.30$ Å, $c=6.85$ Å, $\beta=121°$. The pseudohexagonal parameters are given in the table.

[f] $CuCl_2$ forms a centered rectangular (001) net with $a=3.34$ Å and $b=6.85$ Å. We choose to use the primitive equiaxed oblique monoclinic unit cell, in which case $\gamma=128°$. The 110 reciprocal-lattice vectors of the intercalate and carbon are aligned [7.91].

[g] Ref. [7.96].

[h] Ref. [7.97].

[i] Hexagonal in-plane structure corresponds to a dimer of the intercalate [7.98].

[j] Ref. [7.2].

[k] Ref. [7.62].

the intercalate layers are orientationally locked to the adjacent graphene layers with the intercalate and graphene basis vectors oriented at either 0° or 30° (within $\pm 2°$) with respect to one another. This range of misorientation ($\pm 2°$) accounts for the circumferential arcing of the intercalate (*h*, *k*, 0) reflections shown in Fig. 7.2a [7.91]. The lock-in angle θ_{epitaxy} is found to depend on the intercalate species and the stage [7.91, 100–102] as shown in Table 7.3. Thus the in-plane diffraction patterns obtained from single-crystal regions of magnetic GIC acceptor compounds are characterized by sharp diffraction spots, with the

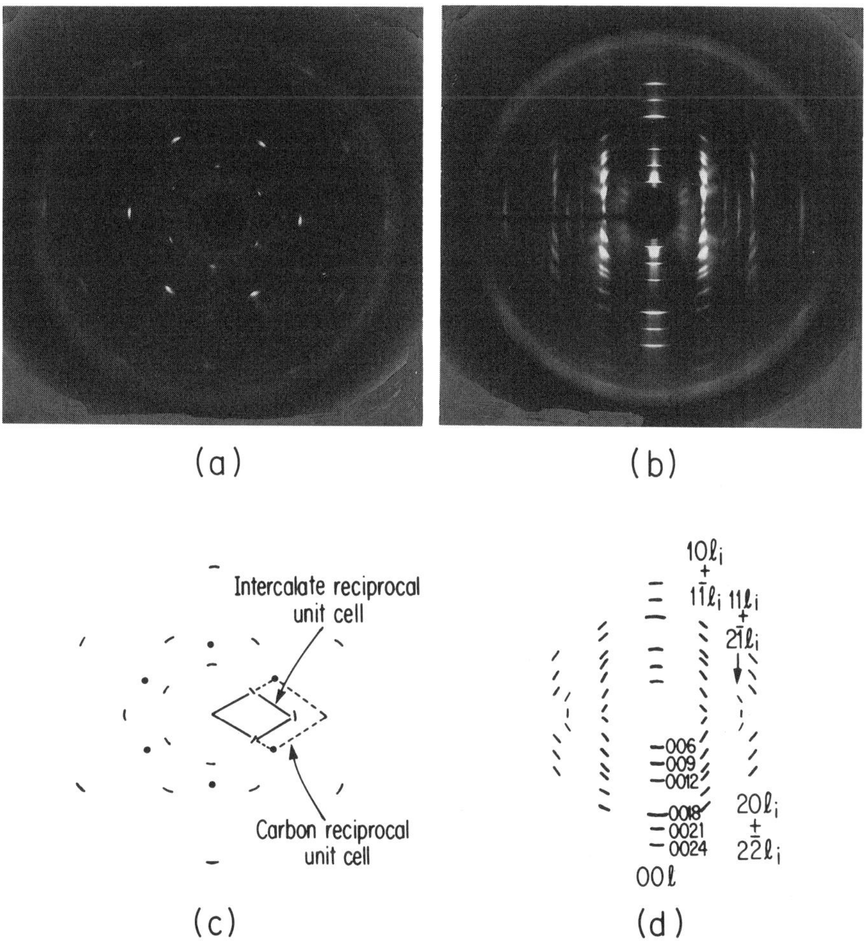

Fig. 7.2. Zero-level X-ray precession photograph from a stage-1 $CoCl_2$ GIC showing the (**a**) zone axis along *c* and the (**b**) zone axis along *a*. (**c**) A schematic interpretation of the photograph (**a**) showing both the graphitic and intercalate reciprocal unit cells. (**d**) Schematic interpretation of (**b**) showing *hkl* reciprocal-lattice point identifications. The subscript *i* refers to the intercalate sublattice. The indexing scheme used is that of the full lattice; hence the *l* values are three times larger than expected [7.91]

in-plane reciprocal-lattice vector of the intercalate being incommensurate but orientationally locked with that of the graphene layers (as shown in Fig. 7.2a for a stage-1 $CoCl_2$ GIC flake) [7.91]. With regard to the *c*-axis spacing, the interplanar distance between the metal and Cl layers in the GIC is essentially the same as in the pristine compound.

Zero-level X-ray precession photographs such as Fig. 7.2a recorded along ***c*** represent 2D "slices" of the reciprocal lattice (normal to the real-space zone axis), which includes the origin of reciprocal space. The stacking sequence is determined by indexing precession photographs along the in-plane axes (Fig. 7.2b). To successfully index the *a*-axis precession photographs, we note that the 3D intercalate structure possesses lower symmetry ($R\bar{3}m$) than an individual graphene layer (6*mm*). Assuming that the $CoCl_2$ layers have $R\bar{3}m$ symmetry, all of the intercalate reflections satisfy the condition $-h + k + l = 3m$, where m is an integer. From this result we conclude that the stacking for the intercalate layers (neglecting spin) is rhombohedral ($\ldots\alpha\beta\gamma\alpha\beta\gamma\ldots$), similar to the stacking in pristine $CoCl_2$ shown in Fig. 7.3. The diffuse nature of the intercalate (h, 0, l) or (0, k, l) reflections is attributed to both the large mosaic spread in the GIC crystal and the small 3D crystallographic coherence length along the *c* axis. For the dichloride compounds (e.g., the intercalates $CoCl_2$, $NiCl_2$, $MnCl_2$), the magnetic cation layers form a triangular lattice, while the trichloride compounds (e.g., the intercalates $FeCl_3$, $CrCl_3$) exhibit a honeycomb lattice, with a more open structure and larger in-plane lattice constants. $MoCl_5$ GICs are of special interest magnetically ($T_c \sim 1.6$ K) [7.60] because they can be prepared for stages up to 5. However, their in-plane structure has monoclinic symmetry

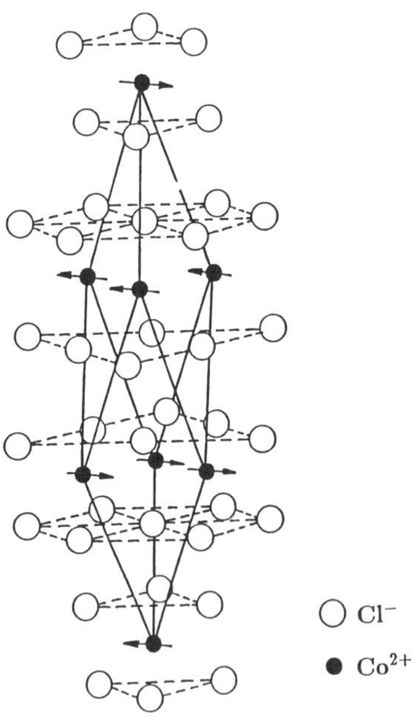

Fig. 7.3. Antiferromagnetic layer stacking of ferromagnetically aligned spin sheets in $CoCl_2$ at low temperatures ($T < T_N$). The low-temperature magnetic unit cell is double the rhombohedral structural unit cell shown in the figure. The $\alpha\beta\gamma$ stacking of the $CoCl_2$ layers is maintained in the stage-1 compounds, but this correlation is lost in the higher-stage compounds. The same low-temperature spin arrangement is found for stage-1 and stage-2 $CoCl_2$ GICs, as confirmed by elastic magnetic neutron scattering measurements (Sect. 7.3.4) [7.103].

[7.98]. Likewise $CuCl_2$ GICs, which can also be readily prepared for several stages [7.104], exhibit a monoclinic in-plane structure, where the triangular lattice that occurs in the other transition-metal dichlorides is distorted into an isosceles triangle [7.104]. The $OsCl_6$ GICs have also been studied magnetically [7.105, 106], though no ordered low-temperature magnetic phase has been reported.

Other GICs of potential interest for their magnetic properties are $EuCl_3$ [7.107], $ReCl_4$ [7.108], $RuCl_3$ [7.108], $OsCl_3$ [7.90], OsF_6 [7.106], $FeBr_2$ [7.109], $FeBr_3$ [7.109], and the rare-earth chlorides $ReCl_3$ [7.90]. Some of the structural information on the pristine and magnetic GICs (including compounds whose magnetic properties have not yet been studied) are summarized in Table 7.3. We further note that for a triangular lattice of magnetic ions (such as the Co^{2+} ions in the $CoCl_2$ GICs), the size of the reciprocal-lattice vector is $K = 4\pi/(\sqrt{3}a)$ which for the case of $CoCl_2$ GICs is 2.04 Å^{-1}.

Because of the possibility of preparing a variety of stages for magnetic acceptor compounds, low-dimensional studies of magnetic GICs have focused primarily on the acceptor compounds, particularly those with desirable magnetic properties. Unfortunately, the magnetic intercalate planes for the acceptor compounds do not extend throughout the sample, but rather form structural domains which influence the formation of magnetic domains. A comparison between the ideal stoichiometry (100% filling of the intercalate layer) and the actual stoichiometry (as, for example, measured chemically or by weight-uptake methods) yields filling factors often no larger than $\sim$ 70% [7.110]. When the stage-1 samples were prepared under high chlorine gas pressure (e.g., 10 atmospheres at 660 °C) filling factors as large as 90% have been claimed [7.111].

The in-plane microstructure of the intercalate layer in the acceptor compounds has been studied by a number of authors [7.112–114], and there has been much confusion in the literature on this topic. Historically, the measured in-plane structural coherence length has tended to increase with time as subsequent authors realized the deficiencies of earlier experiments and as the sample quality improved. Early theoretical arguments proposed that the charge transfer in transition-metal chloride acceptor compounds resulted in intercalate islands of $\sim$ 100 Å diameter with excess chlorine (negative charge) distributed on the island perimeter [7.92, 113, 115]. Experimental evidence for structural domains of this magnitude was then obtained from small-angle neutron scattering by *Flandrois* et al. [7.114], lending support to the island model and the chemical measurements of stoichiometric imbalance. Subsequently, TEM dark field images [7.116] were interpreted in terms of ellipsoidal islands of average size 100 Å × 500 Å, and approximately similar results were obtained by other groups [7.117]. However, later detailed work showed that these TEM images resulted from complicated moiré fringe patterns [7.100].

A much larger estimate of the in-plane intercalate ordering length in stage-1 $CoCl_2$ GICs later came from small-angle neutron scattering measurements (540–560 Å) [7.118], consistent with an earlier experiment showing in-plane coherence lengths (of ~ 1000 Å), an order of magnitude larger than the islands discussed above [7.114]. For example, quasielastic neutron scattering at (q_a, 0, 0.075) indicated [7.64] an in-plane spin correlation length of 900 ± 150 Å. Most recently, *Speck* [7.93, 100] measured an in-plane structural coherence length of ~ 1 µm using high-resolution transmission electron microscopy in small powder stage-1 $CoCl_2$ GIC flakes, thin enough so that only a single intercalate domain is observed and thus the moiré fringe interference patterns are not observed. These large structural coherence distances are comparable to those found in well-staged nonmagnetic commensurate acceptor and donor GICs [7.46].

From the various reports cited we conclude that there are several different length scales that are measured by the various experiments when probing the in-plane magnetic coherence length. Some measurements suggest that the magnetic coherence length may be as large as the structural coherence length, arguing that the magnetic spin boundaries are correlated with the microstructural features. With this interpretation, the structural coherence lengths in the intercalate (e.g., $CoCl_2$) domains are large enough (> 1000 Å in size) to contain a statistically meaningful number of spins (~ 10^4 spins/domain). To account for the low filling factors (70%–80%), pores or voids (of coalesced vacancies) must occur in the intercalate layer. An average size of these coalesced vacancies (or lakes) may be on the order of ~ 100–330 Å [7.91], thereby accounting for observation of this length scale by various techniques. A detailed study, attempting to clarify and interrelate the various studies of the magnetic and structural domain sizes on a specific (and representative) sample, and perhaps relating these results to the stoichiometric imbalance, would contribute much to further progress in this area.

(b) Structure of Donor Compounds

The most widely studied GIC donor compound is the stage-1 C_6Eu rare-earth compound, which on the basis of X-ray measurements [7.48] exhibits a $(\sqrt{3}\times\sqrt{3})R30°$ commensurate in-plane Eu superlattice (Fig. 7.4), and $A\alpha A\beta A\alpha\ldots$ layer stacking, where α, β refer to the intercalate layers [7.48] and A refers to the graphene layers. (Raman experiments have been proposed to look for the K-point folded phonon modes [7.120, 121] associated with the commensurate $(\sqrt{3}\times\sqrt{3})R30°$ structure of C_6Eu [7.122], but until now, no such zone-folded phonon modes have been reported.) Each Eu atom has six nearest-neighbor Eu atoms at a distance of 4.31 Å within the layer plane and six next-nearest neighbor Eu atoms lying at a distance of 5.47 Å in adjacent planes. The next-nearest neighbor Eu–Eu in-plane distance is 7.46 Å, indicating that higher-stage compounds are needed to demonstrate quasi-2D magnetic behavior for this family of donor GICs.

In the case of the lanthanides, only stage-1 C_6Eu binary intercalation compounds have been prepared of adequate size and quality for quantitative

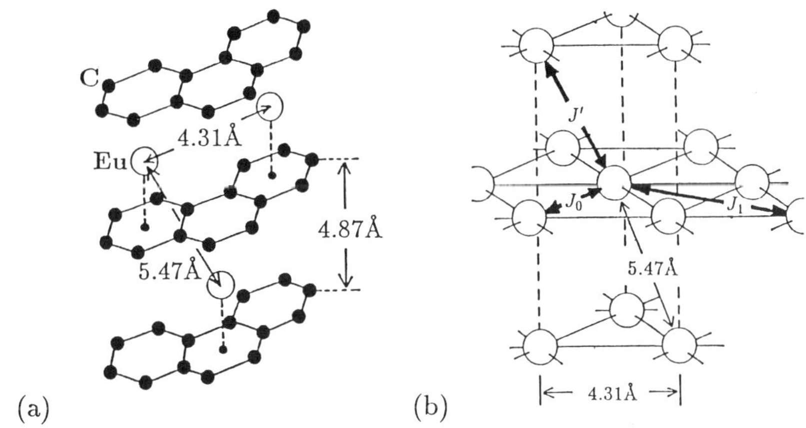

Fig. 7.4. Crystal structure of stage-1 C_6Eu (**a**) In-plane structure of Eu layers show the $(\sqrt{3} \times \sqrt{3})R30°$ in-plane superlattice commensurate with the adjacent graphene layers. (**b**) Structure of Eu ions in C_6Eu without explicitly showing the graphene layers, but showing instead the $\alpha\beta$ stacking of the Eu layers, and the various pertinent exchange interactions [7.119]

magnetic measurements. Very recently, higher-stage (i.e., stages 2–4) ternary donor magnetic compounds have been prepared by substitution of Eu for K to obtained $Eu_{1-x}K_x$ GICs, and these compounds have been studied magnetically for $x \sim 0.25$ [7.123]. The same approach has been applied to the synthesis of stage-2 through stage-4 $Sm_{1-x}K_x$ GICs [7.123]. This work shows promise for achieving a high-stage quasi-2D magnetic donor compound. Because of the smaller c axis repeat distance (I_c) in the donor GICs and because of their stronger interplanar coupling, the crossover to quasi-2D behavior in the donor compounds is expected to occur at a higher stage index, in comparison to acceptor GICs.

Though not of sufficient quality for magnetic measurements, small single crystals of C_6Yb have also been prepared, with X-ray diffraction patterns showing stage-1 ordering along the c axis and $(\sqrt{3} \times \sqrt{3})R30°$ ordering in the basal plane [7.48]. With regard to other lanthanides, some success has been achieved in the intercalation of graphite powders with the rare earths Ce, Pr, Nd, Sm, Eu, and Yb, as confirmed by their $(0, 0, l)$ X-ray reflections showing stage-1 ordering (Table 7.2) [7.48]. From these efforts, we conclude that studies of magnetic donor GICs are hindered primarily by the lack of suitable samples.

Thus far it has not been possible to prepare magnetic donor GICs by the direct intercalation of magnetic transition metals (e.g., Fe or Ni). However, it is claimed that transition metal donor compounds can be prepared by chemical reduction of transition metal chloride GICs [7.124]. However, no information on the in-plane structure of such compounds is presently available, nor are electronic structure or transport data available to verify that these compounds are donor compounds.

7.2 Origin of Magnetic Interactions

Magnetic interactions in solids are either of the quantum-mechanical exchange type or of a classical dipole–dipole nature. The layered systems that are the subject of this chapter involve both of these types of interactions to some degree. The strong in-plane interactions are always of the exchange type, whereas the interplanar interactions are of the exchange type for low-stage compounds and become dipolar for the higher-stage compounds. Pressure-dependent studies of the magnetic susceptibility [7.75] show that there is a change in the interplanar interaction mechanism in going from stage-1 to stage-2 $CoCl_2$ GICs.

Within the hierarchy of quantum-mechanical exchange interactions, the exact calculations depend on the wave functions of the electrons in the interacting system. For example, direct exchange interactions involving the d or f wave functions on neighboring ions do not occur normal to the layer planes in any of the GICs studied thus far. Instead, the interaction may take place through another electronic state, which can be an itinerant electron and goes by the name of the RKKY (Ruderman-Kittel-Kasuya-Yosida) interaction, or it can go through an intervening closed-shell ion (the superexchange interaction) as modified by the presence of a sheet of itinerant π electrons on the graphene planes (indirect exchange interaction).

In this section we consider specific Hamiltonians for the most common acceptor and donor GICs that have been studied for their magnetic properties. This is then followed by a discussion of the special magnetic properties of 2D-XY systems.

7.2.1 Magnetic Hamiltonians for Acceptor Compounds

The *Goodenough-Kanamori* rules [7.36, 125] provide a valuable guide for predicting the sign and order of magnitude of magnetic in-plane and c-axis exchange interactions for the pristine transition metal chloride compounds such as the MCl_2 and MCl_3 family of compounds. Studies of the $CoCl_2$ GIC system are consistent with extending these rules to acceptor GICs [7.126]. However, further work is needed to test the applicability of these rules to magnetic acceptor GICs in general.

The magnetic Hamiltonian for the acceptor GICs is deduced primarily from that for the pristine intercalate materials. The sign of the in-plane and interplanar exchange interactions in general remains unchanged from that of the pristine magnetic intercalate and seems to follow from the *Goodenough-Kanamori* rules [7.36, 125]. Since for many magnetic acceptor intercalates the in-plane distances and angles between magnetic ions and from the cations to the anions remain essentially unchanged by intercalation into the graphite host, the magnitudes of the in-plane exchange (J) and anisotropy constants (J_A) are also essentially unaffected by intercalation. In contrast, the interplanar exchange energy J' greatly decreases in magnitude, because of the increased interplanar separation between the magnetic layers after intercalation.

All the magnetic acceptor intercalation compounds that have been extensively studied thus far are related to insulating transition metal chloride intercalates which have magnetic ions subjected to a crystal field. The crystal-field interactions that determine the form of the magnetic Hamiltonian, and give rise to crystal-field splittings of the energy levels of the free ion, are large compared with the exchange interactions. For this reason, the effective spin states in acceptor magnetic GICs may differ from those of an isolated ion in vacuum and vary from one compound to another containing the same magnetic ion (Table 7.1). But they do not depend on the stage of the GIC or the magnitude of the intercalate filling factor.

In many cases, the crystal-field problem for the magnetic ions in magnetic acceptor GICs has already been solved, since magnetic acceptor intercalates generally retain their in-plane crystal structure upon intercalation, and the magnetic properties of the anhydrous pristine magnetic compounds have been previously studied. Such is the case for $NiCl_2$ [7.62], $CoCl_2$ [7.91], and $FeCl_2$ [7.39], where the local environment about each metal ion remains essentially unchanged upon intercalation. For other magnetic acceptor GICs such as $CuCl_2$ GICs, the crystal structure in the immediate neighborhood of the magnetic ion is only slightly distorted, in which case perturbation theory can be used to obtain the magnetic interactions in the GIC from what is known about pristine $CuCl_2$. Finally, we can, in principle, have magnetic acceptor GICs with crystal structures that are very different from their parent magnetic compounds, in which case it is necessary to consider the crystal-field problem explicitly for the magnetic acceptor GICs, as must also be done in cases where the magnetic properties of the pristine magnetic materials have not been previously investigated.

More generally, for magnetic intercalates where the trigonal interactions might be strong, the effect of a trigonal crystal field should be considered before the effect of the spin-orbit interaction. For $4d$ and $4f$ magnetic acceptor compounds, however, the spin-orbit interaction becomes greater than the crystal-field splittings. Thus the crystal-field problem has to be solved under conditions appropriate to the relative magnitudes of the physical interactions.

Most of the magnetic acceptor GICs show a threefold site symmetry at the position of the magnetic ion (Table 7.3). The Zeeman term in the magnetic Hamiltonian

$$\mathscr{H}_{\text{Zeeman}} = -\mu_B \sum_i \boldsymbol{S}_i \cdot \boldsymbol{g} \cdot \boldsymbol{H} \tag{7.5}$$

should thus reflect this crystal-field symmetry through the anisotropy of the g factor. The g factor measured for an arbitrary orientation of the magnetic field is related to the principal axis components by

$$g = g_{\perp} \sin^2\theta + g_{\|} \cos^2\theta \, , \tag{7.6}$$

where $g_{\|}$ and $g_{\perp}$ are respectively the g factor for $\boldsymbol{H} \parallel \hat{c}$ and $\boldsymbol{H} \perp \hat{c}$, and where θ is the angle between the magnetic field direction and $\hat{c}$. For many acceptor compounds, the g factors in the high-temperature region are close to the free-electron value (i.e., 2.0023), and the differences between $g_{\parallel}$ and $g_{\perp}$ are not significant. However, at low temperatures where the magnetic ordering establishes a preferred spin direction, the effective g factor will in general be very anisotropic (Sect. 7.3.7).

We illustrate the crystal field effects in this section for three representative cases to demonstrate the procedures used in constructing magnetic Hamiltonians:

(1) Ferromagnetic nearest-neighbor exchange for a magnetic ion with an effective spin of 1, which is unchanged from that of the free ion, illustrated by $NiCl_2$ GICs (Sect. 7.2.1a).

(2) Ferromagnetic nearest-neighbor exchange for a magnetic ion with an effective spin 1/2, illustrated by $CoCl_2$ GICs (Sect. 7.2.1b).

(3) Antiferromagnetic nearest-neighbor exchange with an effective spin of 5/2, illustrated by $MnCl_2$ GICs where the $3d$ shell is half filled (Sect. 7.2.1c).

In the pristine magnetic transition-metal salts, the in-plane and interplanar exchange interactions are dominated by the superexchange mechanism. The superexchange interaction between two magnetic ions occurs through the diamagnetic or closed-shell ions in magnetic insulators and involves the interplay between the potential exchange interaction and the kinetic exchange interaction [7.36, 127]. The potential exchange interaction gives rise to an interaction energy between the orthogonal orbitals of the magnetic ions, which involves the Coulomb exchange interaction under the Hartree-Fock approximation, and is in general ferromagnetic. The kinetic exchange interaction is determined by a hopping effect between the nonorthogonal orbitals of the magnetic ions through the intervening nonmagnetic anions. The hopping effect results in two important consequences: one is the pairing of electrons between one set of magnetic d electron orbitals and the s, p orbitals of the anions, thereby forming a partially covalent bond and lowering the total energy of the complex; the other is the formation of magnetic coupling between the other set of unpaired anion p orbitals and the magnetic d electrons. (This concept is schematically illustrated in Fig. 7.5a for pristine $CoCl_2$, where J_{Cl-Cl} denotes the exchange energy between Cl^- ions on adjacent layers, while $b_{mm'}$ is the transfer energy associated with an electron transfer from a Cl^- ion to a neighboring Co^{2+} cation, and U is the Coulomb repulsion energy of two d electrons in the same orbital.) In this context, the sign of the kinetic exchange energy is determined by both types of interacting orbitals as well as the bonding angles between the cations and anions. The resulting sign of the superexchange interaction in the magnetic insulators is determined by the competition between the potential exchange and kinetic exchange contributions. A set of empirical rules (the Goodenough-Kanamori rules) for determining the sign of the superexchange interactions by considering the symmetry of the anion and cation orbitals (in the crystal field) is well established [7.36], and has been explicitly applied to transition-metal chlorides.

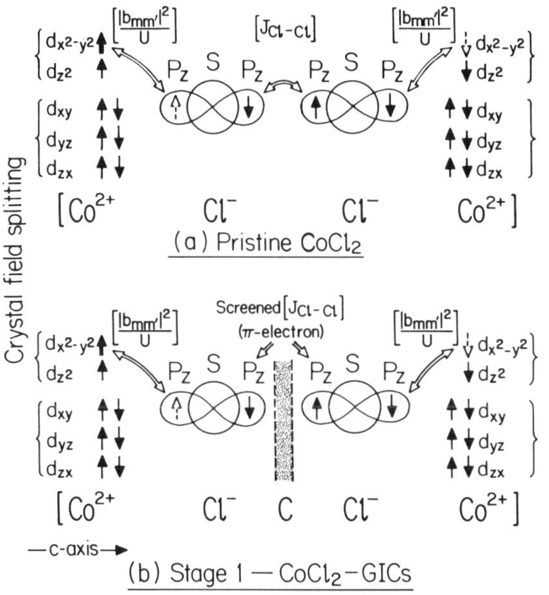

Fig. 7.5. Schematic diagram of the superexchange interaction in (**a**) pristine $CoCl_2$ and (**b**) as modified in the stage-1 $CoCl_2$ GICs by the presence of a conducting graphene layer. The magnitude of the Co^{2+}–Cl^- interaction is written as $|b_{mm'}|^2/U$ [7.128]

As shown in Fig. 7.5a, the pristine transition-metal chlorides have two layers of Cl^- ions between consecutive cation layers. Therefore, the superexchange interactions have to go through modified electron wave functions between Co^{2+} and Cl^- as well as through Cl^-–Cl^- exchange interactions. Since Cl^-–Cl^- can form a covalent p_σ bond through the Coulomb exchange interaction, an antiferromagnetic coupling for J_{Cl-Cl} is expected to lower the energy of the system.

Since the kinetic exchange energy in pristine $CoCl_2$ is not significantly changed after intercalation, while the Coulomb-type exchange energy J_{Cl-Cl} between Cl^-–Cl^- is greatly weakened by the insertion of sheets of π conduction electrons after intercalation (Fig 7.5b), *Yeh* et al. have argued [7.128] that the interplanar antiferromagnetic coupling in $CoCl_2$ GICs is reduced in a way resembling the screened Coulomb potential for charged impurities in a free-electron gas (see Fig. 7.5b). A detailed theory for this phenomenon remains to be carried out. Using this model, the stage dependence of the interplanar exchange coupling has been estimated for the $CoCl_2$ GICs [7.128]. The results indicate a rapid decrease in the interplanar exchange coupling J' with increasing stage. An experimental determination of J' as a function of interlayer separation I_c has been made for stage-1 $CoCl_2$ GICs by application of pressures up to 15 kbar (Fig. 7.6), yielding a value of $\partial J'/\partial I_c = -0.12 \pm 0.02$ K/Å [7.75].

The dominant c-axis magnetic exchange mechanism for pristine $CoCl_2$ is the superexchange interaction. In the case of stage-1 $CoCl_2$ GICs this superexchange interaction through the Cl^- ion layers is modified by an increased Cl^-–Cl^- separation, and by the presence of a sheet of itinerant π electrons on the graphene layers (Fig. 7.5), thereby significantly reducing J'. The difference in the functional dependence between the superexchange mechanism in the pristine

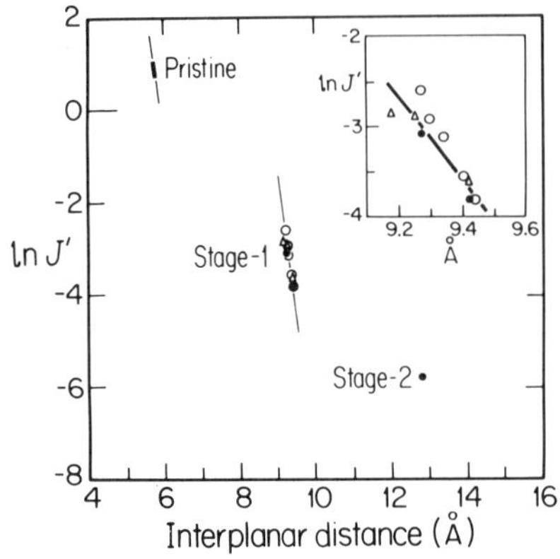

Fig. 7.6. The interlayer antiferromagnetic exchange $\ln J'$ as a function of the c axis interplanar distance I_c for pristine $CoCl_2$ [7.74], and stage-1 and stage-2 $CoCl_2$ GICs. The fits are extrapolated beyond the experimental points with thin solid lines. The inset shows the $\ln J'$ data points for the stage-1 compound in detail, as derived from resistivity data [7.129] (△), the magnitude of $1/\chi'$ at 4.2 K (○), and $\chi'(H)$ measurements (●) [7.75]

compound and the modified interaction in the stage-1 compound is demonstrated by the different $\partial J'/\partial I_c$ slopes observed for the two kinds of compounds when subjected to external pressure (Fig. 7.6) [7.75].

In the case of stage-2 $CoCl_2$ GICs, *Yeh* et al. argued that the dipole–dipole and the π-electron modified superexchange interactions become equally important [7.128]. In the case of higher-stage $CoCl_2$ GICs ($n \geqslant 3$), the dipole–dipole interaction dominates because of the rapid decrease of the superexchange and indirect exchange interactions with increasing stage index. Although the estimates have been made explicitly for the $CoCl_2$ GICs, it is believed that this discussion of the exchange mechanisms for interplanar coupling is generally applicable to all the magnetic transition-metal halide acceptor GICs [7.130].

Although dipole–dipole interactions are generally not as important in the low-stage magnetic GICs as the π-electron modified superexchange mechanism in determining the interplanar exchange coupling, the relatively long range of the dipole–dipole interaction (power-law dependence so that $J_{\mathrm{d-d}} \propto 1/I_c^3$) causes this mechanism to become important at large distances (high-stage GICs), above which the much shorter range (exponential dependence $\propto e^{-r/a}$) π-electron modified superexchange mechanism has fallen off substantially.

Because of the relatively long distance separating the graphite π-band carriers from the magnetic species in the magnetic acceptor GICs (Fig. 7.5), the RKKY [7.131] interaction is small for all the magnetic acceptor GICs, in contrast with the magnetic donor compound C_6Eu [7.132], where the π electrons associated with the graphene layers are strongly hybridized with the orbitals on the Eu^{2+} ions, giving rise to a strong RKKY coupling (Sect. 7.2.2). In the following subsections we consider explicit illustrative Hamiltonians for three transition-metal acceptor GICs.

(a) Magnetic Hamiltonian for $NiCl_2$ GICs

For the case of the N_1^{2+} ion, the ground state of the free ion is $^3F_4$ and has the same orbital state as $CoCl_2$ (Table 7.1). The $L = 3$ states common to both $NiCl_2$ and $CoCl_2$ are split by a cubic crystal field into three sets of levels (T_1, T_2, A_2). In a cubic field, the crystal-field parameter in $NiCl_2$ is of the opposite sign [7.40]

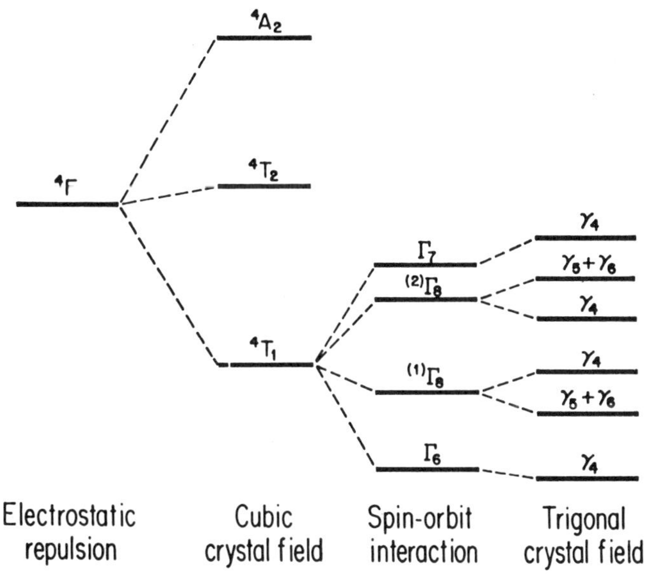

Fig. 7.7. Level order in crystal field for Co^{2+} [7.133]. The cubic crystal-field interaction causes the largest splitting and therefore is the first interaction considered. Further splittings due to the spin-orbit interaction and the trigonal crystal field are also indicated.

to that in $CoCl_2$ (Fig. 7.7), so that the lowest level for $NiCl_2$ is the $^3A_2$ state, while the threefold degenerate $^3T_1$ and $^3T_2$ states are much higher in energy. Since the orbital angular momentum for the $^3A_2$ state is fully quenched, there is no additional splitting introduced by the spin-orbit interaction. The ground state $^3A_2$ for $NiCl_2$ is threefold spin degnerate, and this suggests that $NiCl_2$ should be described by a $S = 1$ Heisenberg model. Thus, in the case of $NiCl_2$, the effective spin S for the ground state remains at $S = 1$, as in the free ion.

The magnetic Hamiltonian for pristine $NiCl_2$ was written down and experimentally determined by *Lindgard* et al. as [7.2]

$$\mathcal{H}_{\mathrm{NiCl_2}} = -2J\sum_{i<j} \boldsymbol{S}_i\cdot\boldsymbol{S}_j + 2J'\sum_{i<k} \boldsymbol{S}_i\cdot\boldsymbol{S}_k + D\sum_i (S_{iz}^2 - S_{ix}^2 - S_{iy}^2) - \mu_{\mathrm{B}}\sum_i \boldsymbol{S}_i\cdot\boldsymbol{g}\cdot\boldsymbol{H}\,, \tag{7.7}$$

where J (21.7 K) is the in-plane ferromagnetic coupling, and J' (0.77 K) is a much weaker interplanar antiferromagnetic coupling. The third term describes the single-ion anisotropy, and the positive sign of D determines that the spins will lie in the hexagonal plane. The magnitude of $D = 0.4$ K can be explained by a dipole–dipole interaction [7.2]. A sixfold anisotropy field H_6 is probably also present in pristine $NiCl_2$; however, neutron scattering experiments could not determine the orientation of the spins within the xy plane [7.2], and no experimental information about H_6 is available. The last term in (7.7) describes the effect of an external magnetic field where the anisotropy of the g factor has been included. For the case of $NiCl_2$ and its intercalation compounds, the

Table 7.4. Comparison between the magnetic properties of stage-1 and stage-2 $NiCl_2$ GICs and pristine $NiCl_2$

Property	Pristine $NiCl_2$	Stage-1 $NiCl_2$ GIC	Stage-2 $NiCl_2$ GIC
T_N (K)	52.3[a]	22.0 ± 0.5[b]	$T_{cl} = 17.3$, $T_{cu} = 19.4$[c]
$H_t(0)$ (Oe)	—	1200[b]	35[c]
$g_{\parallel}$	2.23[d], 2.25[e]	2.25 at 300 K[f]	2.096 at 300 K[g]
$g_{\perp}$	2.23[d], 2.25[e]	2.25 at 300 K[f]	2.156 at 300 K[g]
$C_{\parallel}$ (emu K/mole)	1.33[e]	—	—
$C_{\perp}$ (emu K/mole)	1.33[e]	1.36[f]	1.354[h]
$\Theta_{\parallel}$ (K)	67[e]	—	—
$\Theta_{\perp}$ (K)	67[e]	60[f]	70 ± 1[h]
J (K)	21.7[a]	—	17.5[h,i]
J' (K)	0.77[a], 0.79[d]	0.012[b,j]	0.007[c], 0.001[j]
D (K)	0.4[a], 0.40[d]	—	—
H_{sat} (kOe)	129[k]	—	—
H_6 (Oe)	~0[l]	—	—

[a] Ref. [7.2].
[b] Ref. [7.136].
[c] Ref. [7.61].
[d] Ref. [7.137].
[e] Ref. [7.138].
[f] Ref. [7.135].
[g] Ref. [7.134].
[h] Ref. [7.139].
[i] From high-temperature susceptibility data (Sect. 7.4.1a).
[j] Calculated from $H_t(0) = 25$ Oe using (7.22) (Sect. 7.3.1).
[k] Ref. [7.140].
[l] Believed to be small [7.2].

g factor is found from ESR measurements to be nearly isotropic [7.134, 135] (Table 7.4).

In discussing the magnetic properties of $NiCl_2$ GICs, the same form of the Hamiltonian (7.7) is used; J' decreases rapidly with increasing stage, and the g factor (e.g., for stage-2 compound) exhibits a very small amount of anisotropy as seen in Table 7.4 where $g_{\parallel}$ and $g_{\perp}$ respectively denote the g factors for $\boldsymbol{H} \parallel \hat{c}$ and $\boldsymbol{H} \perp \hat{c}$ (Sect. 7.2.1).

(b) Magnetic Hamiltonian for $CoCl_2$ GICs

In the case of $CoCl_2$ GICs, the in-plane crystal structure of the $CoCl_2$ (including the trilayer intercalate sandwich) is also preserved upon intercalation, so that we can make use of our knowledge of the magnetic properties of pristine $CoCl_2$ to describe those of the corresponding intercalation compounds. In vacuum, the free Co^{2+} ion has a $3d^7$ configuration leading to a $^4F_{9/2}$ ground state [7.41] with spin 3/2 (Fig. 7.7). We show below that in the crystal-field environment appropriate to $CoCl_2$ GICs, the effective spin of the Co^{2+} ion in the lowest crystal-field level is spin 1/2. When the crystal-field splittings give rise to an effective spin

Table 7.5. Comparison between the magnetic properties of stage-1 and stage-2 $CoCl_2$ GICs and pristine $CoCl_2$

Property	Pristine $CoCl_2$	Stage-1 $CoCl_2$ GIC	Stage-2 $CoCl_2$ GIC
T_N (K)	24.9[a]	9.9 ± 0.1[b]	$T_{cl} = 8.0$[c]
$H_t(0)$ (Oe)	—	380	10–15
$g_{\|\|}$	2.64[d]	2.80[e], 4.77[f,g]	2.86[e], 4.75g
$g_\perp$	4.98[d]	4.60[e], 5.94[g]	4.02[e], 6.40[g]
$C_{\|\|}$ (emu K/mole)	3.25[h]	3.78[i]	—
$C_\perp$ (emu K/mole)	3.65[h]	3.30[g], 3.36[i]	3.84[g], 3.41[i]
$\Theta_{\|\|}$ (K)	−35.2[h]	−68.3[i]	—
$\Theta_\perp$ (K)	27.2[h]	31.1[g]	23.2[g]
J (K)	28.5[a]	41.5[g]	31.0[g], 29.2 ± 4[j]
J_A (K)	16.0[a]	15.8[g], 20.8[e]	14.6[j]
J_A/J	0.56[a]	0.38[g]. 0.50[e]	0.35[e], 0.50[j], 0.48[g]
J' (K)	2.16[a]	0.25[k]	0.007[g]
J'_A (K)	3.3[a]	—	—
ξ_c (0) (Å)	LRO[l]	200[b]	22[m], 70[b]
H_{sat} (kOe)	32[n]	—	—

[a] Ref. [7.1].
[b] Ref. [7.103].
[c] $T_{cu} = 9.1$ K Ref. [7.141]. A table of values for T_{cl} and T_{cu} by various authors is given in Ref. [7.24].
[d] Ref. [7.142].
[e] Ref. [7.143].
[f] Ref. [7.24].
[g] Ref. [7.144].
[h] Ref. [7.138] (see also Ref. [7.1]).
[i] Ref. [7.145].
[j] Ref. [7.146]. The values quoted here are in terms of an unrestricted sum in the Hamiltonian of [7.146]. Because of the small dispersion along k_z for the magnon dispersion relation, the value for J' [7.146] is not accurate and has therefore been omitted from this table.
[k] Ref. [7.75].
[l] Long-range order, $\xi_c(0) \to \infty$.
[m] Ref. [7.147].
[n] Ref. [7.148].

1/2 ground state, the effective Hamiltonian looks similar to that of an isolated free ion, except that the effective g factor is strongly modified by the crystal field. Thus the splitting of the spin-up and spin-down states in a magnetic field governed by the effective g factor can be quite different from that for the isolated free ion, and this effect is observed for $CoCl_2$ GICs (Table 7.5).

The Co^{2+} ion in $CoCl_2$ and in the related GICs is surrounded by six nearest-neighbor Cl^- ions in an approximately octahedral cubic crystal environment with a small trigonal distortion (Fig. 7.3). The distortion can be handled in perturbation theory, since the interaction energies for this distortion are smaller in magnitude than the spin-orbit interaction (Table 7.5 and Fig. 7.7). In a cubic crystal field, the $^4F_{9/2}$ ground state of the Co^{2+} ion splits into three sets of levels

($^4T_1$, $^4T_2$, $^4A_2$) with the threefold orbitally degenerate $^4T_1$ state having the lowest energy (see Fig. 7.7). The spin-orbit coupling splits the twelve-fold degeneracy of the $^4T_1$ level (fourfold spin degeneracy and threefold orbital degeneracy) into a lowest doublet Γ_6, a fourfold level $^{(1)}\Gamma_8$, and a sixfold level ($^{(2)}\Gamma_8$ and Γ_7). Finally a trigonal field splits the fourfold and sixfold levels so that all the remaining levels are twofold degenerate and are well separated with the approximate splittings shown in Fig. 7.7.

Therefore, the ground state is twofold degenerate, and from Kramers' theorem the system has an effective spin of 1/2. Since the spin 1/2 state results from crystal-field interactions, the effective g factor need not be close to 2 (as for a free ion with spin 1/2) but can be highly modified by the crystal field. In the case of the Co^{2+} ions in $CoCl_2$, the calculated g values are large and are anisotropic (Table 7.5) [7.40]. This anisotropy is confirmed experimentally in pristine $CoCl_2$ [7.142] and in the related intercalation compounds [7.111, 143], either by direct measurements of the g factor by the electron-spin resonance (ESR) technique or by high-field magnetization studies (Sect. 7.3.2).

The Hamiltonian for pristine $CoCl_2$ has been written by *Lines* as [7.40]

$$\mathscr{H}_{\mathrm{CoCl_2}} = -J\sum_{i<j} \boldsymbol{S}_i\cdot\boldsymbol{S}_j + J_{\mathrm{A}}\sum_{i<j} S_{iz}S_{jz} + J'\sum_{i<k} \boldsymbol{S}_i\cdot\boldsymbol{S}_k - J'_{\mathrm{A}}\sum_{i<k} S_{iz}S_{kz} - \mu_{\mathrm{B}}\sum_i \boldsymbol{S}_i\cdot\boldsymbol{g}\cdot\boldsymbol{H}\,, \tag{7.8}$$

where the values for the various intraplanar (J, J_A) and interplanar (J', J'_A) exchange interactions [7.1] are given in Table 7.5. Note that others in the literature have modified Lines' Hamiltonian (7.8) by inserting factors of 2 into the first four terms of (7.8) or by using unrestricted sums in this equation. When so doing, attention must be given to clearly defining the convention used, so that values for the exchange couplings J, J', J_A, and J'_A can be uniquely defined. Such a protocol would allow work in different laboratories to be intercompared. To facilitate such comparisons it is suggested that workers in the field use (7.8) consistently. With regard to the intraplanar exchange coupling, J in (7.8) is ferromagnetic and the intraplanar exchange J' is antiferromagnetic, in agreement with predictions by the Goodenough-Kanamori rules. The sixfold intraplanar anisotropy field H_6 is represented by a term which is higher order in the spin (i.e., $(S_x + iS_y)^6$) and is much smaller than the terms given in (7.8). The magnitude of H_6 has been estimated to be small [7.24, 57].

The same basic magnetic Hamiltonian (7.8) has been applied to the $CoCl_2$ GICs, with the same J and J_A values for the in-plane magnetic interactions, but much smaller values for J' and J'_A arising from the large interplanar distances, that are also stage-dependent (Table 7.5) [7.130]. Although J'_A has never been determined experimentally for the $CoCl_2$ GICs, *Lines* [7.40] assumes that the relation $J_A/J = J'_A/J'$ holds for $CoCl_2$, which implies that J'_A will also decrease with increasing stage index.

Some comments are in order about the general methods that have been used for the determination of the exchange interaction energies appearing in (7.8).

The dominant interaction J is found from either the high-temperature susceptibility measurements by fitting to a Curie-Weiss law (Sect. 7.3.1) or from a fit to the magnon dispersion curves measured by inelastic neutron scattering (Sect. 7.3.4). At present there is considerable disagreement about the magnitude of J as measured by different techniques. The in-plane anisotropy J_A is determined in terms of the ratio J_A/J through measurement of the anisotropy of the g factor and use of *Lines*' theory [7.40]. The anisotropy in the g factor is found either directly by ESR measurements (Sect. 7.3.7), or from fitting the high-temperature dc susceptibility to the Curie-Weiss law (Sect. 7.3.1), or from the high-field magnetization for $\boldsymbol{H} \parallel \hat{c}$ and $\boldsymbol{H} \perp \hat{c}$ (Sect. 7.3 .2). Finally, the interplanar exchange J' can be found either from the c axis magnon dispersion relations determined from the inelastic neutron scattering measurements (which for many GICs are too flat to be measured by this method) or from measurements of saturation effects in the low-temperature magnetic field dependence of the in-plane susceptibility $\chi'(H)$, the in-plane and c axis resistivities $\varrho_a(H)$ and $\varrho_c(H)$, and the magnitude of the $(00\frac{1}{2})$ magnetic Bragg peak.

(c) Magnetic Hamiltonian for $MnCl_2$ GICs.

Like $NiCl_2$ and $CoCl_2$, pristine $MnCl_2$ has the rhombohedral $CdCl_2$ structure (space group D_{3d}^5). The ground state of the free Mn^{2+} ion is $^6S_{5/2}$ where the outer d shell is half filled (d^5). Since the orbital angular momentum is zero, there is no spin-orbit coupling and in a cubic crystal field, the symmetry assignment of the ground state remains $^6A_{1g}$, as shown in Table 7.1. The effect of the trigonal field on this level is expected to be small since the $^6A_{1g}$ level is orbitally nondegenerate, and therefore the Mn^{2+} ion in $MnCl_2$ has a spin $S = 5/2$, just as in the free ion. The Goodenough-Kanamori rules (Sect. 7.2.1) are useful in determining the sign of the in-plane exchange interaction for the half-filled $3d$ orbitals of $MnCl_2$. The rules predict that the interplanar exchange coupling between two in-plane Mn^{2+} ions via a Cl^- ion (making a 90° angle for the Mn^{2+}–Cl–Mn^{2+} bonds) is antiferromagnetic, in contrast to $NiCl_2$ ($3d^8$) and $CoCl_2$ ($3d^7$), where the d levels are more than half filled and the intraplanar interactions are predicted to be ferromagnetic [7.36].

The Hamiltonian for pristine $MnCl_2$ is therefore of the Heisenberg type and is given by [7.149]

$$\mathscr{H}_{\mathrm{MnCl_2}} = 2J \sum_{i<j} \boldsymbol{S}_i \cdot \boldsymbol{S}_j + D \sum_i (S_i^z)^2 - 2J' \sum_{i<k} \boldsymbol{S}_i \cdot \boldsymbol{S}_k - \mu_B \sum_i \boldsymbol{S}_i \cdot \boldsymbol{g} \cdot \boldsymbol{H}\,, \quad (7.9)$$

where the single-ion anisotropy term D was introduced by *Pryce* [7.150] and the values of J and D (for the single-ion anisotropy) have been estimated from the high-temperature dc susceptibility (Table 7.6) [7.144, 151]. Pristine $MnCl_2$ has a planar spin arrangement with in-plane antiferromagnetic interactions that show complicated antiferromagnetic phases below the Néel temperature ($T_N = 1.96$ K) [7.58]. As is usually the case, intercalation reduces the transition temperature so that the $MnCl_2$ GICs exhibit magnetic transitions at very low temperatures (~ 1 K) (Table 7.6).

Table 7.6. Comparison between the magnetic properties of stage-1 and stage-2 $MnCl_2$ GICs and pristine $MnCl_2$

Property	Pristine $MnCl_2$	Stage-1 $MnCl_2$ GIC	Stage-2 $MnCl_2$ GIC
T_N (K)	1.96[a]	1.2[b]	1.1[c] 1.2[b]
$g_{\parallel}$	2.04 ($T>50$ K)	1.91[d]	1.97[d], 1.912 ± 0.005[c] ($T=300$ K)
$g_{\perp}$	2.04 ($T>50$ K)	1.94[d]	1.97[d], 1.977 ± 0.005[c] ($T=300$ K)
$C_{\parallel}$ (emu K/mole)	—	4.01[d]	4.25[d]
$C_{\perp}$ (emu K/mole)	—	4.40[e] 4.12[d]	4.30[e] 4.26[d]
$\Theta_{\parallel}$ (K)[f]	−3.3	−9[e] −7.2[d]	−9[e] −9.1[d]
$\Theta_{\perp}$ (K)[f]	−1.8	−8[e] −5.4[d]	−8[e] −5.9[d]
μ_{eff} (μ_B)	5.94	5.85[g] 5.93[e]	5.87[e]
J (K)	0.216	0.17[d]	0.2[d]
J' (K)	~0	~0	~0
D (K)	−0.09[h]	0.56[d]	0.97[d]

[a] Ref. [7.58].
[b] Ref. [7.152].
[c] Ref. [7.153].
[d] Ref. [7.144].
[e] Ref. [7.151].
[f] A negative value implies antiferromagnetic coupling, $C_i/(T-\Theta_i)$.
[g] Ref. [7.110].
[h] Ref. [7.154].

7.2.2 Magnetic Hamiltonian for Donor Compounds

The electronic configuration of the Eu^{2+} ion is $^8S_{7/2}$, which arises from a half-filled electronic orbital configuration ($4f^7$), giving rise to a large total spin value of $S = 7/2$ with zero orbital angular momentum. In zero applied magnetic field, the spins lie in the intercalate layer and since the nearest-neighbor exchange is antiferromagnetic, the spins form an antiferromagnetic frustrated triangular spin lattice. This spin system exhibits a variety of magnetic phases brought about by variations of the temperature and applied magnetic field.

The magnetic properties of donor magnetic GICs are distinctly different from those of the magnetic acceptor compounds because of differences relating to the basic magnetic coupling mechanism which occurs both in the pristine magnetic intercalates, which are metals, and in the intercalation compounds, where the dominant metallic behavior is associated with the graphite π bands. The in-plane structure of Eu-metal is not preserved upon intercalation, even though the Eu–Eu distances are similar (Sect. 7.1.3b, Fig. 7.4, and Table 7.2).

For the donors formed from the rare-earth metals, the magnetic Hamiltonian is written for the tightly bound magnetic $4f$ cores interacting with the itinerant conduction electrons. Thus the magnetic interaction of the $4f$ ion is through the polarization of the nearly free conduction electrons, and is described by the RKKY (or indirect exchange) interaction, which has been widely studied for magnetic metals (Sect. 7.2.1) [7.131, 132]. The substitution of the graphite π electrons for the s and p electrons of conventional metals does not change the basic physical mechanism of the RKKY interaction, but does, however, modify the details of the calculation and in particular introduces a large anisotropy in the magnetic properties in the stage-1 Eu GIC.

The only magnetic donor compound whose magnetic properties have been studied extensively to date is the stage-1 Eu GIC, which has the stoichiometry C_6Eu [7.48, 155]. The crystal structure of C_6Eu shown in Fig. 7.4 is commensurate with the adjacent graphite layers and has an in-plane unit cell of $(\sqrt{3} \times \sqrt{3})R30°$, where $a_0 = 4.31$ Å in the GIC as compared with 4.58 Å in the metallic Eu. The structure of C_6Eu is discussed in Sect. 7.1.3b.

The configuration of the Eu^{2+} ion has been determined by Mössbauer spectroscopy studies (Fig. 7.8) of the hyperfine interaction of the Eu nucleus [7.156]. These Mössbauer studies confirm that the configuration of Eu^{2+} in the intercalation compound is $^8S_{7/2}$ (Fig. 7.8). Since the total orbital angular momentum is zero, there is no spin–orbit interaction and the lowest energy level remains nondegenerate, suggesting Heisenberg spin exchange. Mössbauer spectroscopy studies show, however, that the spins lie in the layer planes, consistent with an XY anisotropy term in the magnetic Hamiltonian [7.132]. Also present is a weak hyperfine field ($B_{\mathrm{eff}} = -10.7$ tesla at 4.2 K) due to a small anisotropy caused by dipolar and crystalline electric field effects [7.156]. The temperature dependence of B_{eff} is well fit by a Brillouin function up to a temperature of

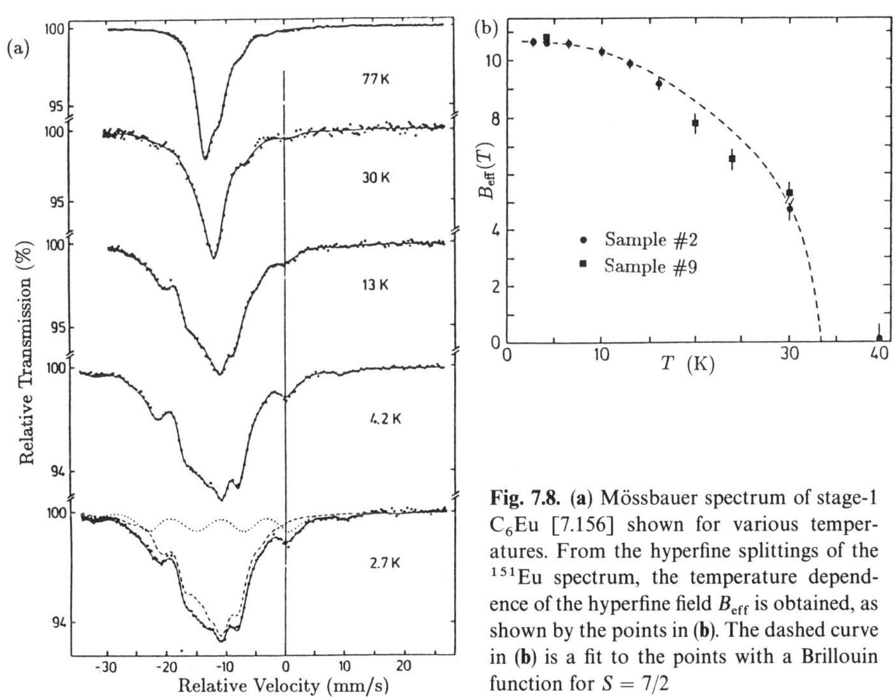

Fig. 7.8. (**a**) Mössbauer spectrum of stage-1 C_6Eu [7.156] shown for various temperatures. From the hyperfine splittings of the ^{151}Eu spectrum, the temperature dependence of the hyperfine field B_{eff} is obtained, as shown by the points in (**b**). The dashed curve in (**b**) is a fit to the points with a Brillouin function for $S = 7/2$

~ 30 K. Departures near T_N are explained in terms of low-energy magnons [7.156].

The compound C_6Eu is of particular interest as a 3D-*XY* spin system insofar as the Eu spins align within the magnetic planes. The unusual features of the observed magnetic behavior of C_6Eu include the large anisotropy in magnetic properties observed for $\boldsymbol{H} \perp \hat{c}$ and $\boldsymbol{H} \parallel \hat{c}$, the difference in magnetic properties of the Eu species resulting from intercalation, the large number of magnetic phases that are observed as a function of magnetic field, and the observation of an unusual ferrimagnetic (also called metamagnetic) phase [7.157]. To account for this ferrimagnetic phase, explicit calculations for the *XY* model [7.132, 158, 159] and Monte Carlo simulations [7.145, 160] both show that in addition to the usual two-spin interaction terms, $J_{ij}\boldsymbol{S}_i \cdot \boldsymbol{S}_j$, higher fourth-order terms in S are required in the Hamiltonian to account for the experimental magnetization results [7.157, 158, 160]. Thus, because of the large value of its spin ($S = 7/2$), C_6Eu provides a physical system where fourth-order magnetic interactions can be studied in detail, thereby enlarging our knowledge of basic magnetic phenomena in solid-state physics. (Although this ferrimagnetic phase has been called metamagnetic in the literature, there are no reports of tricritical points in this system.)

Because the nearest-neighbor exchange is antiferromagnetic (AF), the spins in zero magnetic field form an antiferromagnetic frustrated triangular spin lattice, but lie in the intercalate layer. The large in-plane anisotropy associated with this spin frustration has been estimated experimentally [7.119] and is in agreement with the calculations of *Akera* and *Kamimura*, who introduced a term $-0.008S_z^2$ into their magnetic Hamiltonian [7.132]. This frustrated spin ordering on a triangular lattice can easily be modified by introducing weak interactions, so that a variety of magnetic phases can be realized by variation of the temperature and magnetic field.

The Hamiltonian for the magnetic system that has been widely used to explain a variety of experiments in C_6Eu is [7.158, 159]

$$\begin{aligned}\mathscr{H}_{C_6Eu} = & -J_0 \sum_{nn} \boldsymbol{S}_i \cdot \boldsymbol{S}_j - J_1 \sum_{nnn} \boldsymbol{S}_i \cdot \boldsymbol{S}_j - J' \sum_{nn} \boldsymbol{S}_i \cdot \boldsymbol{S}_k \\ & - B \sum_{nn} (\boldsymbol{S}_i \cdot \boldsymbol{S}_j)^2 - g\mu_B \sum_i \boldsymbol{S}_i \cdot \boldsymbol{H} \\ & + K \sum_{4-\text{spin ring}} \{(\boldsymbol{S}_i \cdot \boldsymbol{S}_j)(\boldsymbol{S}_k \cdot \boldsymbol{S}_l) + (\boldsymbol{S}_i \cdot \boldsymbol{S}_l)(\boldsymbol{S}_j \cdot \boldsymbol{S}_k) - (\boldsymbol{S}_i \cdot \boldsymbol{S}_k)(\boldsymbol{S}_j \cdot \boldsymbol{S}_l)\} ,\end{aligned} \tag{7.10}$$

where the parameters have been evaluated experimentally, as discussed further in Sect. 7.4.2a. The first term in (7.10) represents the antiferromagnetic in-plane nearest-neighbor exchange interaction ($J_0 < 0$). The second term represents the ferromagnetic in-plane next-nearest neighbor exchange interaction ($J_1 > 0$). The third term is the ferromagnetic interplanar nearest-neighbor exchange interaction ($J' > 0$) (Fig. 7.4). The fourth term is the in-plane nearest-neighbor biquadratic exchange interaction. The fifth term is the Zeeman energy from the external magnetic field. The last term is the in-plane four-spin ring cyclic exchange interaction. The biquadratic and four-spin ring terms represent the symmetry-allowed terms in a power-series expansion in powers of the spin angular momentum. The *XY* anisotropy term of *Akera* and *Kamimura* [7.132] does not explicitly appear in (7.10).

The importance of the RKKY interaction in the magnetic coupling of C_6Eu was first discussed theoretically by *Akera* and *Kamimura* [7.132], based on their band-structure calculations. The large hybridization of the graphene π electrons with the $4f$ electrons on the Eu^{2+} ion gives rise to a large 3D RKKY interaction, the only known example of this model in GICs. Later calculations of the magnetization as a function of temperature [7.119, 161] substantially confirmed the earlier work, leaving little doubt that the RKKY interaction is the correct magnetic coupling mechanism for the C_6Eu system.

Because of the similarity of the basic crystal structure of C_6Eu and other magnetic donor compounds in the lanthanide series, we infer that other magnetic donor compounds, when synthesized with sufficiently high quality for detailed magnetic measurements, will also show the RKKY mechanism for magnetic coupling between the magnetic ions. To date only ternary Eu–K GICs of higher-stage donor compounds have been synthesized [7.123].

7.2.3 The 2D-*XY* Model: Theoretical Considerations

In the early history of 2D magnetism it was conjectured that the 2D magnetic models would either not show any phase transition at all, or perhaps would undergo a somewhat different type of phase transition, but of the order/disorder type, as was also the case of the 2D Ising model [7.4]. High temperature series expansions of the susceptibility indicated [7.5, 6, 162] a possible divergence for the 2D Heisenberg model, and a similar indication was noted for the 2D-*XY* model [7.7]. Aware of *Bloch*'s old spin wave argument [7.163] that a nonzero value of the order parameter (the saturation magnetization M) is unstable at nonzero temperature T in these classical continuous-spin models, *Stanley* and *Kaplan* [7.5] argued that a new kind of phase transition was possible: namely one in which the susceptibility diverges at a finite temperature T_c, but M remains zero for $T < T_c$. Such a phase transition is *not* of the order/disorder type. While the possibility existed that Bloch's argument was wrong [7.5], *Mermin* and *Wagner* [7.164] then proved, using rigorous inequalities, that such is not the case: they showed that M had to be zero at all finite T. Thus, given a divergent susceptibility, the *only* possibility for the occurrence of a phase transition was that it would have to be a new type of transition [7.5, 6]. The existence of such a new type of transition in the case of the 2D-*XY* model was later supported by *Wegner* [7.8], *Berezinskii* [7.9, 10], and *Kosterlitz* and *Thouless* [7.11–13]. Discussions of the nature of the low-T phase [7.5, 6] were given [7.8–12], and the mechanism for the breakup of that phase was proposed by *Kosterlitz* and *Thouless* [7.11–13], enabling quantitative calculations to be made of the critical properties. In contrast to results for the 2D-Ising and the *XY* models, later work on the 2D-Heisenberg model [7.165–168] led to a generally accepted view that no phase transition of any sort exists for the 2D-Heisenberg model (but see *Berezinskii* [7.9] for a contradictory view). Even given this absence of a transition, however, a small planar anisotropy may be sufficient to give an *XY* type of behavior in any real experimental system.

The classical 2D-*XY* (planar model) Hamiltonian (7.2) is given by

$$\mathscr{H}_{\text{planar}} = -2JS^2 \sum_{i<j} \cos(\theta_i - \theta_j) , \tag{7.11}$$

where the sum is over nearest-neighbor pairs of spins, and θ_i is the angle of the ith spin with respect to a fixed in-plane direction. Early attempts at understanding the 2D-*XY* model were based on extensions of the spin wave approximation. *Wegner* [7.8] showed that for a small-angle expansion of the cosine, i.e., cos $(\theta_i - \theta_j) \simeq 1 - (\theta_i - \theta_j)^2/2$, then, in calculating thermodynamic quantities, the sums over angles become simple Gaussian integrals that are easily evaluated. From this approximation, Wegner showed that at low temperatures the two-spin correlation function $\Gamma(r)$ depends on the spin–spin separation distance r according to a power law

$$\Gamma(r) = \langle \boldsymbol{S}_0 \cdot \boldsymbol{S}_r \rangle \sim r^{-\eta(T)} , \tag{7.12}$$

where the exponent $\eta(T)$ is given by

$$\eta(T) = k_{\text{B}} T/2\pi J . \tag{7.13}$$

This algebraic decay is much slower than the exponential decay of the spin order found in conventional systems

$$\Gamma(r) = \langle \boldsymbol{S}_0 \cdot \boldsymbol{S}_r \rangle \sim \frac{\mathrm{e}^{-r/\xi}}{r^{d-2+\eta}} , \tag{7.14}$$

where ξ is the correlation length and d is the spatial dimension of the system. For this reason, the power-law decay can be attributed to quasi-long-range order.

Kosterlitz and *Thouless* [7.11, 12] argued that the line of critical points, with a continuously variable exponent $\eta(T)$, must terminate at some finite temperature due to vortex excitations ignored in the spin-wave approximation. A simple argument for determining the order of magnitude for the vortex phase transition is obtained from free-energy arguments. Using the Coulomb gas analogy, positive and negative vortices (Fig. 7.9) interact in much the same way as positive and negative electrical charges confined in two dimensions. We then expect that the force of one vortex on another will vary with distance r according to a $1/r$ dependence. To obtain the potential energy of the vortex E_{vortex}, the $1/r$ function is integrated from the smallest length scale a, the lattice parameter, up to the longest length scale L, the size of the system. We then obtain with appropriate prefactors

$$E_{\text{vortex}} = 2\pi J \ln(L/a) . \tag{7.15}$$

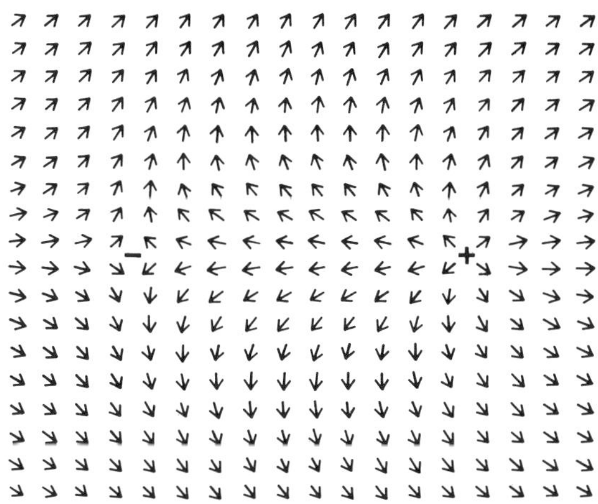

Fig. 7.9. Schematic of a positive and negative spin vortex pair based on a planar square array of spins

The entropy $\mathscr{S}$ of the system is determined by the number of ways a vortex of area $\sim a^2$ can be placed in an area L^2, i.e., $\mathscr{S} = k_B \ln(L^2/a^2)$, which gives an expression for the free energy $\mathscr{F}$ as

$$\mathscr{F} = E_{\text{vortex}} - T\mathscr{S} = 2(\pi J - k_B T)\ln(L/a) . \tag{7.16}$$

Therefore, from (7.16), vortices will spontaneously appear at the temperature $T_{KT} = \pi J/k_B$.

This simplified explanation of vortices ignores the details of vortex unbinding at T_{KT}. It merely shows that vortices will spontaneously appear at T_{KT}, and that the vortices are quite unlike the spin waves that had been previously considered as the dominant spin excitation. Figure 7.9 shows a picture of a vortex/antivortex pair. Above T_{KT} the vortices are only weakly coupled (and are called free or unbound vortices), but as T is reduced below T_{KT}, the vortices bind. The effect of $+/-$pairing (binding of vortices) is to reduce the long-range effects of each vortex, hence lowering their overall energy, and thereby explaining why vortices bind.

In a later paper *Kosterlitz* [7.13] applied renormalization group techniques to obtain $T_{KT} = 1.364\,J/k_B$ and the following functional forms for the susceptibility χ and the correlation length ξ:

$$\chi \sim \begin{cases} \xi^{2-\eta} & T > T_{KT} \\ \infty & T < T_{KT} \end{cases} \tag{7.17}$$

and

$$\xi \sim \begin{cases} \exp(b/\sqrt{t}) & T > T_{KT} \\ \infty & T < T_{KT} \end{cases} . \tag{7.18}$$

Here $t = (T - T_{KT})/T_{KT}$ and $b = 1.5$. The divergence of χ at T_{KT} is shown schematically in Fig. 7.10a. The exponent $\eta(T) = k_B T/2\pi J$ is as determined by

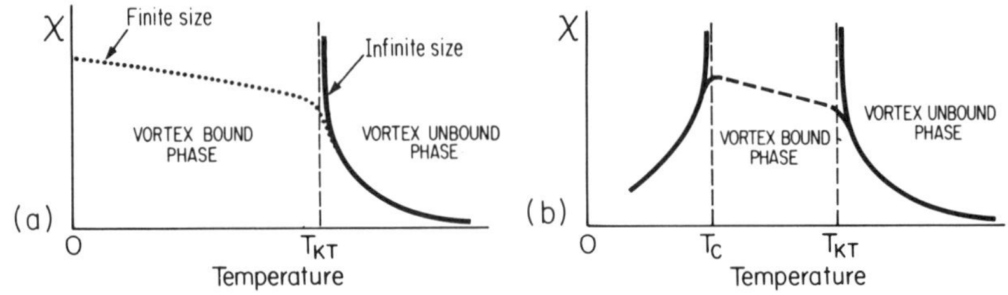

Fig. 7.10. Schematic of the magnetic susceptibility in the Kosterlitz-Thouless model. (**a**) For the model with no in-plane anisotropy, the susceptibility diverges as the Kosterlitz-Thouless transition temperature T_{KT} is approached from above. The inclusion of finite-size effects leads to rounding of the transition and a finite value for the susceptibility below T_{KT} (shown schematically). (**b**) The inclusion of a sixfold anisotropy term H_6 imposes long-range 2D order below T_c. This model then gives two magnetic phase transitions at T_c and T_{KT} [7.169]

the spin-wave approximation for $T < T_{KT}$, and $\eta(T_{KT}) = 1/4$. The value of η at the transition point T_{KT} proved to be an important check between theory and experiment when *Bishop* and *Reppy* [7.30] provided convincing experimental evidence for $\eta = 1/4$ for vortices in superfluid ^{4}He. Other *Kosterlitz* predictions [7.13] were for $\delta = 15$ at $T = T_{KT}$, where the exponent $1/\delta$ governs the magnetization M in a weak external magnetic field according to the relation $M \sim \mathrm{H}^{1/\delta}$. This prediction has not yet been demonstrated. Below T_{KT}, the magnetization is expected to be zero.

Nelson and *Kosterlitz* [7.170] predicted that the spin stiffness of the system for wave vector q_a given by

$$\lim_{\substack{q_a \to 0 \\ T \to T_{KT}}} J(q_a) = \frac{2}{\pi} T_{KT} \tag{7.19}$$

drops discontinuously by $2T_{KT}/\pi$ as the temperature is raised through T_{KT}. This limit can be observed experimentally as a gap in the magnon dispersion relation at $q_a = 0$.

The theoretical interest in the 2D-*XY* model relates to the fact that, within 2D systems, the 2D-*XY* model occupies a position that is on the edge of having a conventional phase transition to a phase with long-range order. The 2D-*XY* model is therefore very sensitive to perturbations, especially a small p-fold in-plane anisotropy field H_p, the presence of any 3D interlayer coupling J', and finite size effects (Fig. 7.10a). *Pokrovskii* and *Uimin* [7.171] showed that a weak p-fold in-plane anisotropy field for $p = 3$, 4, and 6 results in another phase transition at lower temperature (Fig. 7.10b) into a state with long range (2D) order. This result reflects the fact that anisotropy gives certain preferred directions for the spins, whereas the vortex state supports an arbitrary angular distribution of spins.

José et al. [7.169] further developed the *Pokrovskii* and *Uimin* [7.171] result for the in-plane anisotropy, and found that for $p = 4$ the temperature range for stability of the KT state $\to 0$, whereas for $p = 6$ they found a finite temperature range $T_{cl} < T < T_{KT}$ over which the bound vortex KT state is stable (Fig. 7.10). Since the application of an external magnetic field represents a $p = 1$ anisotropy, this result suggests that a weak external magnetic field will suppress the KT state [7.172]. The finite size of the 2D magnetic system is also considered as a possible mechanism for quenching the KT state [7.173]. Further details on theoretical aspects of the 2D-XY model are given in [7.174].

Due to the theoretical complications introduced by the various perturbations to the 2D-XY model, Monte Carlo simulations have confirmed the results of the renormalization techniques and have become a practical avenue of investigation in their own right [7.174]. *Tobochnik* and *Chester* [7.175] established that a system as small as 30×30 spins would undergo a KT transition. Later simulations have taken into account the in-plane sixfold anisotropy field H_6 [7.176], the amount of easy-plane anisotropy [7.177], finite-size scaling effects [7.178], and real magnetic systems using experimentally determined magnetic parameters [7.179]. In the latter investigation, simulations of $CoCl_2$ GICs were carried out including 3D coupling within five layers of 30×30 spins. In general, Monte Carlo simulations provide information on the spin structure and the spin–spin correlation length that can then be compared with experimental values, as are for example obtained from neutron scattering measurements. More recent computer simulations [7.180, 181] have investigated dynamical behavior near T_{KT} due to vortex diffusion.

Experimental studies (e.g., in K_2CuF_4 [7.182]) have been carried out to measure the critical properties at the Kosterlitz-Thouless transition. All experiments, however, are carried out on finite-size samples, which complicates the observation of the critical properties associated with the KT transition. At temperatures close to T_{KT}, observable effects are masked by finite-size effects, whereas at temperatures $T \gg T_{KT}$ where the finite-size effects are less important, the system is no longer in the critical region. However, the qualitative predictions derived from the renormalization group equations agree well with experiments [7.173, 183]. The effect of finite size on the critical properties of GICs is especially important because the magnetic domains are of the order ~ 1000 Å [64]. It is known from Monte Carlo studies [7.175] that systems of this size should show a KT transition. On the other hand, finite-size effects must be taken into account in any quantitative comparison between theory and experiment.

With regard to the effect of perturbations on T_{KT}, the theoretical work of *Hikami* and *Tsuneto* [7.184] predicted that the Kosterlitz-Thouless temperature T_{KT} has only a weak dependence on the amount of XY spin anisotropy. The predicted functional form [7.184]

$$T_{KT} \propto \frac{1}{A - B \ln(J_A/J)}, \tag{7.20}$$

where $A = 1.20$ and $B = 0.08$, was subsequently found to agree with Monte Carlo simulations [7.177]. Equation (7.20) predicts that T_{KT} is relatively insensitive to J_A in the regime $J_A > J/10$, and it is only in the Heisenberg limit ($J_A \to 0$) that $T_{KT} \to 0$ rapidly.

Table 7.7. Summary of types of scattering correlations in solids[a]

	Spin-correlation function	Form of structure factor
Long-range order (LRO)	$\Gamma(r) = \text{const}$	Gaussian (instrumental)
Short-range order (SRO)	$\Gamma(r) \sim r^{-d+2-\eta} e^{-r/\xi}$	$[1 + (\xi q)^2]^{-1+\eta/2}$
2D quasi-LRO[b]	$\Gamma(r) \sim r^{-\eta(T)}$	$\Phi(1 - \eta/2; 1; -q^2 L^2/4\pi)$[c]

[a] The parameters $\Gamma(r)$, ξ, r, and η are as defined in Sect. 7.2.3.
[b] Applies to a system of finite size L.
[c] $\Phi(x, y, z)$ is the Kummer function [7.185].

The primary issue relating to the experimental study of the KT state is the difficulty in identifying an experimental system that is an adequate realization of the 2D-XY model for a magnetic system. Any real system contributes some perturbing terms to the 2D-XY Hamiltonian, and theory suggests that even very small perturbations suppress the formation of a KT bound vortex state. One of the most favorable experimental systems for observation of the KT bound vortex state might be high-stage GICs, where the interlayer exchange interactions are very weak. The experimental characteristics expected of the transition to the KT bound vortex state from higher temperatures is a divergence of the susceptibility χ at T_{KT} (7.17, 18), a vanishing of the magnetization M below T_{KT}, as well as the predicted functional forms of the critical magnetic scattering (Sect. 7.3.4 and Table 7.7), which include a gap in the magnon dispersion relations at $q=0$. This review examines the case for GICs in some detail and shows that for stage index $n \geq 2$, $CoCl_2$ GICs provide evidence for a KT state for temperatures in the vicinity of the magnetic ordering temperatures (Sect. 7.4.1b).

7.3 Experimental Techniques for Studying GICs

In this section, we first briefly review the experimental techniques used to study magnetic GICs and the information provided by each technique, after which in Sect. 7.4 we review our present understanding of several of the more widely studied magnetic GICs.

A variety of experimental techniques are used to study magnetic GICs, each providing complementary data that must then be analyzed within the framework of a specific model describing the magnetism in each of the magnetic GICs. Information on the desired phenomenon or property is often accessible using a combination of techniques. For example, the magnetic phase transitions can be probed using magnetization, magnetic susceptibility, neutron scattering, and transport measurements. The technique of choice for a given study is often dictated by the constraints associated with sample size and perfection. Generally a high degree of crystalline perfection is achieved only in small single crystals (e.g., intercalation compounds based on kish graphite or natural single-crystal graphite flakes). Small samples are adequate for transmission electron microscopy (TEM), X-ray diffraction, susceptibility, magnetization, and transport measurements. However, some of the most sensitive probes of the magnetic properties require large samples. Because of the small magnitude of the magnetic neutron scattering cross sections, studies of the magnetic order using neutron scattering techniques require large samples, and in several cases suitable samples have been prepared by carefully assembling mosaics of small intercalated kish single crystals.

7.3.1 Magnetic Susceptibility

Susceptibility measurements are usually the most convenient magnetic measurement technique to use on newly synthesized magnetic GIC materials. With this technique, measurements on small samples (< 1 mg) can identify the magnetic ordering temperature and provide information about the type of magnetic ordering, e.g., ferromagnetic or antiferromagnetic. The ac susceptibility measurement ($\chi = \partial M/\partial H$) probes the change in the magnetization induced by a change in magnetic field. The magnitude of the modulation field h is kept small ($h \ll 1$ Oe) to avoid perturbation of the magnetic system. For example, in the study of the model 2D-XY system $CoCl_2$ GIC, there has been concern that too large a modulation field might destroy the existence of a KT (Kosterlitz-Thouless) bound vortex state (Sect. 7.2.3) [7.112]. The susceptibility measurements are carried out in an applied magnetic field $\boldsymbol{H}$, which is usually directed parallel to the very much smaller modulating field $\boldsymbol{h}$. Although χ is highly anisotropic, almost all of the susceptibility measurements on GICs have been made with the applied field $\boldsymbol{H}$ and the probing field $\boldsymbol{h}$ parallel to the layer planes to yield the in-plane susceptibility

$$\chi_{aa} = \frac{\partial M_a}{\partial H_a} . \tag{7.21}$$

Some measurements have also been made with $\boldsymbol{H}$ and $\boldsymbol{h}$ along the c axis to yield χ_{cc}. Other geometries that have on occasion been applied to GICs include $\boldsymbol{H} \perp \hat{c}$ and $\boldsymbol{h} \parallel \hat{c}$ [7.141, 186]. Unless otherwise specified, χ' is used to denote the real part of the in-plane susceptibility, which is written as χ_{aa} in (7.21).

As in other magnetic materials, the signature of a magnetic phase transition in GICs is a peak in the magnetic susceptibility. Since peaks in the susceptibility are observed near the magnetic transition temperature both as a function of temperature $\chi'(T)$ and as a function of magnetic field $\chi'(H)$, measurements of the susceptibility as a function of temperature and magnetic field are used to map the magnetic phase diagram. One advantage of an ac susceptibility $\chi'(H)$ measurement compared with a dc magnetization measurement of $M(H)$ at the same temperature is that the presence of first-order field-induced phase transitions is more easily observable in $\chi'(H)$ measurements. For example, peaks in $\chi'(H)$ have been used to determine magnetic phase transitions in stage-1 $CoCl_2$

GICs [7.187] and $NiCl_2$ GICs [7.62] and in stage-2 $NiCl_2$ GICs [7.61]. In several cases, measurements have been made for both $\boldsymbol{H} \perp \hat{\boldsymbol{c}}$ and $\boldsymbol{H} \| \hat{\boldsymbol{c}}$. For example, in the stage-1 $CoCl_2$ GIC, $\chi_{cc}(T)$ was shown to have a peak at 9.77 K, whereas $\chi_{aa}(T)$ was shown to have two peaks at $T_{c1} = 8.22$ K and $T_{c2} = 9.77$ K [7.186]. For this stage-1 compound, the upper transition at T_{c2} was determined by elastic neutron scattering measurements to be the Néel temperature (i.e., $T_{c2} = T_N$) at which 3D antiferromagnetic (AF) ordering occurs [7.103, 118]. The nature of the anomaly at T_{c1} is still the subject of research [7.188].

A measure of the magnitude of the AF interlayer exchange coupling J' for the $CoCl_2$ GICs and $NiCl_2$ GICs can be obtained from an investigation of the low-temperature in-plane susceptibility $\chi'(H)$. Such χ' traces show a field-induced peak at H_t, and this transition can be interpreted as the field value at which the interplanar AF ordering is overcome. The two quantities J' and H_t are related through the relation

$$g_{\perp}\mu_B H_t = 2\tilde{z}\, J'S\,, \tag{7.22}$$

where $\tilde{z}$ is the number of interplanar nearest neighbors. This relation is extensively used in Sects. 7.4.1a, b to obtain an estimate of J' for the $CoCl_2$ GICs and $NiCl_2$ GICs. An alternative measure of J' could also be obtained from the zero-temperature extrapolation of $\chi'(T)$, which should scale inversely with the AF coupling $\chi'(0) \propto 1/J'$.

A general feature of the temperature dependence of the in-plane ac susceptibility $\chi_{aa} \equiv \partial M_a/\partial H_a$ for GICs in magnetically ordered phases is the rapid increase in the magnitude of the susceptibility maximum in $\chi'(T)$ per magnetic ion (denoted by χ_{max}) as the stage index n is increased, corresponding to a decrease in the 3D interplanar coupling. This increase in χ_{max} per magnetic ion has been reported for both $CoCl_2$ GICs [7.189] and $FeCl_3$ GICs [7.56].

The high-temperature susceptibility for magnetic GICs tends to follow either a Curie ($\chi = C/T$) or a Curie-Weiss

$$\chi = \frac{C}{T - \Theta} \tag{7.23}$$

law. In general, different values of C and Θ are obtained for $\boldsymbol{H} \| \hat{\boldsymbol{c}}$ and for $\boldsymbol{H} \perp \hat{\boldsymbol{c}}$, denoted by C_i and Θ_i (where $i = \|, \perp$), and reflecting the characteristic anisotropy of these materials. The Curie-Weiss behavior is usually studied by dc magnetization techniques and we therefore denote the correponding susceptibility by χ_{dc}. The sign of Θ_i in these high-temperature fits provides information on the sign of the dominant intraplanar and interplanar exchange interactions ($\Theta_i > 0$ for $J > 0$ and $\Theta_i < 0$ for $J < 0$) [7.144]. The magnitude of Θ_i provides information on J_i. If the dominant exchange term is the in-plane exchange, then

$$\Theta_{\|} = \frac{2z\, J_{\|} S(S+1)}{3}\,, \tag{7.24}$$

where z is the number of in-plane nearest neighbors, while the Curie-Weiss constant C_i provides a measure of the effective g-factor g_i through

$$C_i = \frac{N_A S(S+1) g_i^2 \mu_B^2}{3k_B}, \tag{7.25}$$

where N_A is Avogadro's number. Thus the anisotropy in the g-factor g_i found from measurements of the in-plane and c-axis g factors can be used to determine the in-plane exchange anisotropy in, for example, the $CoCl_2$ GIC system using the *Lines* theory [7.40]. Measurements of χ over a wide range for stage-1, stage-2, and stage-3 $CoCl_2$ GICs have been carried out for both the $\|$ and $\perp$ geometries [7.190]. The deviations from Curie-Weiss (7.23) behavior as the temperature is lowered have been described by high-temperature series expansions [7.7, 191], and these results have, for example, been used to show that 2D-XY high-temperature series expansions are most appropriate for describing the magnetic behavior in high-stage $CoCl_2$ GICs [7.191].

Most of the susceptibility studies have been confined to measurements of the real part of the susceptibility:

$$\chi(\omega) = \chi'(\omega) - i\chi''(\omega)\,. \tag{7.26}$$

Note that in most papers the authors use χ for the real part of the susceptibility rather than χ'. However, a few studies have also probed the imaginary part χ'' [7.117, 192, 193]. For example, in Fig. 7.11 [7.118] the temperature dependence of the real and imaginary parts of the susceptibility, $\chi'(T)$ and $\chi''(T)$ are plotted for two different GICs. In both cases, $\chi'(T)$ shows a single peak at T_{max} while the much weaker $\chi''(T)$ signal clearly shows well-defined peaks at both T_{cl} and T_{cu}.

Measurements of $\chi''(\omega)$ provide information on energy loss or dissipation, as could for example occur through motion of magnetic domains in magnetic GICs. Just as anomalies in $\chi'(\omega)$ are associated with magnetic phase transition, a similar interpretation is given to anomalies in $\chi''(\omega)$. Though the theory for the KT transition discussed in Sect. 7.2.3 is restricted to the real part of $\chi(\omega)$,

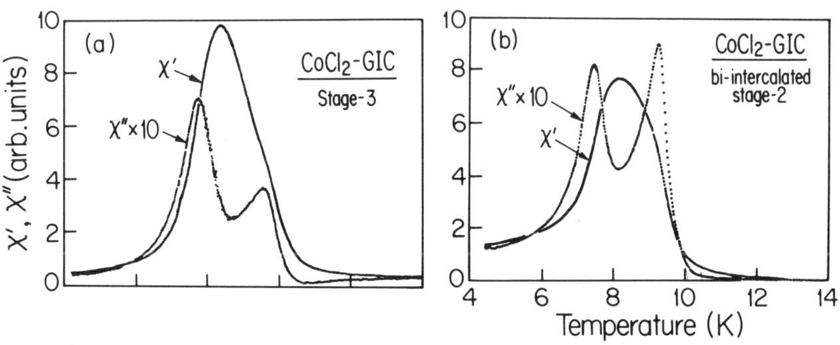

Fig. 7.11. The susceptibilty $\chi(T) = \chi'(T) - i\chi''(T)$ of (**a**) stage-3 $CoCl_2$ GIC, and (**b**) stage-2 $CoCl_2$ GIC bi-intercalated with $AlCl_3$. The real component of the susceptibility exhibits a peak at T_{max}, whereas the imaginary component $\chi'' \times 10$ exhibits peaks at T_{cl} and T_{cu} [7.188]

there are theoretical predictions [7.194] for the temperature and frequency dependence of χ'', which have been used in dynamic studies of superfluid ^{4}He and 2D arrays of superconducting Josephson junctions. It is expected that these theories can also be extended to a magnetic 2D-XY system such as the high-stage $CoCl_2$ GICs. The possibility of using the frequency dependence of the complex susceptibility $\chi(\omega) = \chi'(\omega) - i\chi''(\omega)$ to investigate the Kosterlitz-Thouless transition has, however, not yet been explored in magnetic GICs. By analogy with the superfluid ^{4}He and superconducting Josephson junction arrays, it is inferred that detailed studies of the functional form of $\chi'(\omega)$ and $\chi''(\omega)$ could be used to identify the position of T_{KT}.

In more general magnetic studies, both χ' and χ'' have been measured, as for example in the $FeCl_3$ GICs [7.192] and in the $CoCl_2$ GICs [7.177, 193]. In these studies, measurements of $\chi''(\omega)$ have been used to identify magnetic phase transitions and to study dissipative mechanisms. In the very low frequency regime (< 1 mHz), unusual training effects have been observed in stage-2 $CoCl_2$ GICs, and these effects have been identified with spin-glass behavior [7.193, 195, 196]. Such hysteretic effects in the stage-2 $CoCl_2$ GICs have recently been reviewed by *Suzuki* [7.24].

7.3.2 Magnetization

Magnetization measurements provide a powerful tool for studying the magnetic properties of magnetic GICs. From saturation magnetization measurements in conjunction with accurate weight-uptake measurements, the magnetic moment per magnetic intercalate ion can be determined accurately (within 5%), which in turn gives information about the effective spin and the g values (as an alternative to ESR measurements discussed in Sect. 7.3.7). The anisotropy of the magnetization and of the g factor gives direct information on the anisotropy of the Hamiltonian. Anomalies in the field dependence of the magnetization as a function of temperature are associated with magnetic phase transitions, thereby providing information on the magnetic phase diagram of the system. An especially good example of magnetization measurements to study a magnetic phase diagram in magnetic GICs has been for stage-1 C_6Eu [7.119, 161]. Detailed $M(H, T)$ measurements have also been used to extract the critical exponent δ, near the magnetic phase transition, where δ is defined by $M \sim H^{1/\delta}$ for weak fields. For $T > T_{cu} = 22$ K in stage-2 $NiCl_2$ GICs, it is found experimentally that the parameter $\delta = 1$ [7.139] and that δ increases to $\delta = 2$ at $T_{cl} = 19.5$ K, and remains at $\delta = 2$ as T is further decreased. Magnetization measurements are also useful because they can be made on small intercalated natural graphite flake samples using the high sensitivity of a superconducting quantum interference device (SQUID) magnetometer. Hysteresis and memory effects have also been studied by magnetization measurements [7.117].

The two most widely used instruments to measure the magnetization of GICs are the vibrating sample magnetometer (VSM), which can be used to measure $M(H)$ over a wide range of fields, and the SQUID magnetometer, which is usually used to measure $M(T)$ very accurately for medium field strengths (< 5 tesla) and over a wide temperature range (4–300 K). From measurements of $M(T)$ in a fixed applied field H, the dc susceptibility $\chi_{dc}(T)$ is found over a wide temperature range from the relation $M(T)/H = \chi_{dc}(T)$. Both $\chi_{dc}(T)$ and $M(H, T)$ can be used to estimate some of the magnetic exchange constants, the g values and their anisotropy $g_{\|}$ (for $\boldsymbol{H} \| \hat{c}$) and $g_{\perp}$ (for $\boldsymbol{H} \perp \hat{c}$) [7.144]. For certain magnetic GICs such as the $CoCl_2$ GICs, the g values are highly anisotropic, and for this reason the samples are subjected to large torque effects $\boldsymbol{M} \times \boldsymbol{H}$ when the magnetic field is applied normal to the easy xy plane, where $\boldsymbol{M}$ denotes the total magnetization of the sample. In a vibrating sample magnetometer experiment, the GIC sample should thus be glued firmly to a vibrating glass rod to prevent rotation of the sample.

High-field magnetization measurements in the saturation regime can be used to measure the g factors. An example of such a measurements is given in Fig. 7.12 for a stage-1 $CoCl_2$ GIC at $T = 4.2$ K [7.111, 118, 143, 186] showing highly anisotropic g values: $g_{\perp} = 4.60$ and $g_{\|} = 2.80$. These results are in fair agreement with theoretical predictions ($g_{\perp} = 4.98$ and $g_{\|} = 2.64$) [7.142] and also with SQUID measurements [7.190]. In the case of the stage-1 $CoCl_2$ GICs, the results for $g_{\perp} - g_{\|}$ have been compared with the theory of *Lines* [7.40] to give a value of the anisotropy $J_A/J \simeq 0.5$, which is close to that of pristine $CoCl_2$ [7.1].

Two types of hysteresis effects have been observed in the magnetization of GICs. The first is an unusual high-field (0–20 tesla) hysteretic behavior in the c-axis magnetization $M_{\|}(H)$ observed in the stage-1 $CoCl_2$-GIC samples at 4.2 K (Fig. 7.12) [7.111, 143], but not for the magnetic field orientation $\boldsymbol{H} \perp \hat{c}$. Nor is it observed for stage-2 samples when $\boldsymbol{H} \| \hat{c}$ [7.186]. This hysteresis (Fig. 7.12) may be the result of a strong magnetostriction effect (a length change Δl caused by a change in H), which may be attributed to a strong coupling between the 3D lattice (which has a rhombohedral stacking of the intercalate) and the 3D antiferromagnetically ordered spin sheets below $T_N = T_{c2}$. Incomplete saturation of the c-axis magnetization $M_{\|}(H)$ at 20 tesla is observed in

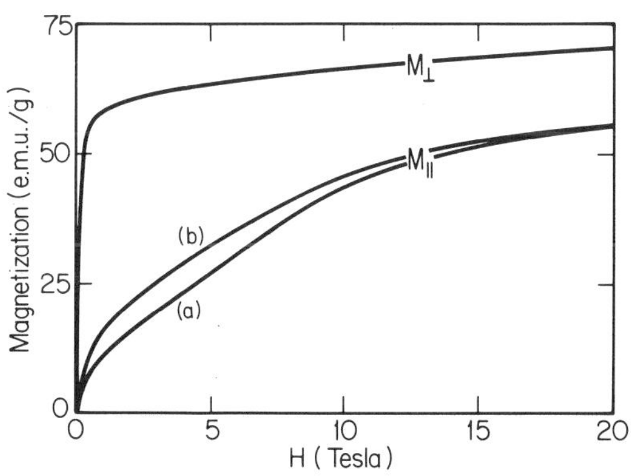

Fig. 7.12. Magnetization of stage-1 $CoCl_2$ GIC at 4.2 K [7.143]. The virgin trace $M_{\|}$ (for $\boldsymbol{H} \perp \hat{c}$) is labeled (**a**), and on reducing the magnetic field, the curve (**b**) is obtained showing considerable hysteresis of the magnetization. The curve for $M_{\perp}$ (for $\boldsymbol{H} \perp \hat{c}$) shows no hysteresis effects

Fig. 7.12, possibly because at 20 tesla the applied magnetic field is not sufficiently large to fully overcome J_A ($\approx$ 15 tesla).

A second type of hysteresis effect has been reported by *Matsuura* et al. [7.117] in stage-2 $CoCl_2$ GICs in terms of changes in the low-field magnetization relaxation times when the samples are cycled through T_{cl} and T_{cu} in zero magnetic field or in a finite magnetic field. These phenomena have recently been reviewed by *Suzuki* [7.24].

7.3.3 Heat Capacity

From thermodynamic considerations, anomalies in the heat capacity of magnetic materials can be identified with magnetic phase transitions. The functional form of the heat capacity $C(T)$ provides an important characterization tool to distinguish between first- and higher-order phase transitions and to classify the dimensionality and other characteristics of specific magnetic phase transitions [7.16]. A discontinuous heat capacity and hysteresis behavior in $C(T)$ at the magnetic ordering temperature and a latent heat associated with the magnetic ordering are characteristic of first-order transitions. In this context, anomalies in the heat capacity associated with magnetic phase transitions have been observed in the magnetic donor GIC C_6Eu [7.119] and in the magnetic acceptor GICs with $CoCl_2$ [7.57, 197–200], $NiCl_2$ [7.57, 201], and $FeCl_3$ [7.56, 57, 200, 201]. Also, the heat-capacity results in $CoCl_2$ GICs and $FeCl_3$ GICs have recently been reviewed [7.200]. In these acceptor compounds a peak in the magnetic contribution to $C(T)$ is observed at T_{cu}, signaling the onset of magnetic order [7.202].

To illustrate this point, we show in Fig. 7.13 early results by *Karimov* et al. [7.19] on the magnetic contribution to the heat capacity $C(T)$ of a nominal stage-2 $CoCl_2$ GIC. In the analysis of the heat-capacity data, the temperature dependence of the magnetic contribution to the heat-capacity $C_M(T)$ is obtained by subtracting the contributions from the nonmagnetic processes (such as the

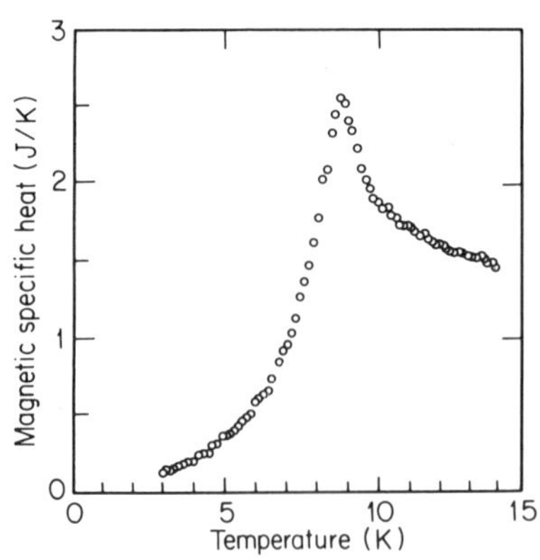

Fig. 7.13. Temperature dependence of the magnetic specific heat for a nominal stage-2 $CoCl_2$-GIC sample [7.203], showing clear evidence for a magnetic phase transition at $T \sim 9$ K

phonon and electron contributions) from the observed heat capacity $C(T)$. A detailed study for stage-1 $CoCl_2$ GICs yielded a T^3 dependence for $C_M(T)$ at low temperature, which was attributed to 3D magnons [7.200]. The phonon and electron contributions are determined by plausible low-temperature extrapolations of measurements made well above the magnetic ordering temperature or by applying a sufficiently high magnetic field to align all the spins. The integral

$$\mathscr{S}_M(T) = \int_0^T \frac{C_M(T')}{T'} dT' \tag{7.27}$$

gives the entropy $\mathscr{S}_M(T)$ associated with the magnetic ordering , which can be compared with the molar entropy contribution $\mathscr{S}'$ expected from the thermodynamic relation $\mathscr{S}' = R\ln(2J+1)$, where J is the total spin. The ratio $\mathscr{S}_M/\mathscr{S}'$ is interpreted in terms of the fraction of the spins that are aligned at a given temperature T.

In general, for 2D systems the entropy change associated with the spin ordering is spread over a wide temperature range and the features are not as pronounced as those that occur for 3D magnetic transitions. For the 2D-XY model a weak feature in $C(T)$ at a temperature somewhat above T_{KT} has been observed in Monte Carlo simulations of small systems [7.175].

Experimental measurements on stage-2 (or higher-stage) $NiCl_2$ GICs show a weak, broad magnetic contribution [7.201, 202], consistent with a transition to a quasi-2D magnetic phase with a relatively small entropy change. Specific-heat measurements of $FeCl_3$ GICs [7.200] show a much sharper anomaly at the magnetic ordering temperature for the stage-1 compound than for stage 2, consistent with the more 3D magnetic interactions in the stage-1 compound. Similarly, the anomaly in $C_{\mathrm{M}}(T)$ at T_{N} for the stage-1 C_6Eu magnetic donor GIC is relatively sharp [7.119].

7.3.4 Neutron Scattering

The primary advantage of neutron scattering measurements over other magnetic techniques is due to the direct interaction between the spin of the neutron and the spin of the magnetic intercalate ions, thereby allowing an explicit determination of the magnetic structure of magnetic GICs through elastic neutron scattering experiments. Inelastic scattering experiments allow study of the low-energy magnetic excitations.

A large scattering volume is usually required to carry out neutron scattering measurements, and for this reason the measurements on GICs are usually restricted to powder samples (where the in-plane anisotropy information is lost) or large mosaic arrays of 40–50 oriented intercalated kish crystals [7.147]. The c-axis mosaic spread of such an array is significant (e.g., $\sim 9°$), and the in-plane properties are powder averaged. The availability of sufficiently large single crystals and high-flux neutron sources would greatly enhance progress in this field. In some species (e.g., Eu), the neutron absorption cross section is so high that neutron scattering experiments cannot readily be carried out.

Elastic neutron scattering can be used to study both crystal structure and magnetic ordering. With regard to crystal structure, elastic neutron scattering is in some cases a preferred technique because of the large penetration depth of the neutron beam relative to electrons or X-rays, thereby allowing more sensitive structural characterization of bulk samples. The elastic neutron scattering intensity can be measured by setting the analyzer for zero energy transfer. Specifically, elastic neutron-nuclear scattering can be used to monitor the bulk staging fidelity throughout the sample. For example, for a mosaic array of intercalated kish graphite flakes, it was discovered, from a *Hendricks-Teller* analysis [7.204] of the Bragg structure peaks, that a sample characterized as a nominal stage-2 $CoCl_2$ GIC array (on the basis of X-ray characterization) was in fact only $\sim 70\%$ stage 2 [7.147].

Despite the value of elastic neutron scattering as a structural characterization tool, most of the elastic neutron scattering experiments have been carried out to determine the magnetic ordering. Elastic neutron scattering thus can identify the magnetic ordering through the observation of superlattice magnetic scattering intensity superimposed on the nuclear scattering associated with the crystal lattice. For example, in stage-2 $CoCl_2$ GICs, elastic neutron scattering measurements confirmed that the spins lie in the xy plane, and that interplanar antiferromagnetic order sets in below T_{c1}, through the appearance of $(0, 0, \frac{l}{2})$ magnetic Bragg peaks [7.103].

In the neutron scattering experiment, the coherent elastic magnetic scattering cross section is proportional to the spatial Fourier transform of the two-spin correlation function $\Gamma(r) = \langle \boldsymbol{S}_0 \cdot \boldsymbol{S}_r \rangle$ for spins separated by a distance r. From the functional form of $\Gamma(r)$, we can then distinguish whether the spins exhibit long-range order (LRO), short-range order (SRO), or 2D quasi-long-range order (Table 7.7). In Table 7.7, ξ is the spin–spin coherence length, d refers to the spatial dimensionality of the system (e.g., $d = 2$ or $d = 3$), and η is a critical exponent. Study of the temperature dependence of $\Gamma(r)$ and of its Fourier transform allows a determination of the spin and spatial dimensionality of the magnetic interactions as well as the ranges of these interactions, thereby yielding important information on the magnetic transitions in zero applied field. For example, study of elastic neutron scattering in the temperature region $T_{c1} < T < T_{cu}$ for the stage-2 $CoCl_2$ GICs has shown short-range magnetic order with $\xi_c(0) \sim 3I_c$ well below T_{c1} through determination of the functional form of the spin-correlation length [7.205]. For the case of the stage-1 $CoCl_2$ GICs, elastic neutron scattering measurements have shown long-range correlations ($\xi_c \gg I_c$) for the antiferromagnetic arrangement of the ferromagnetic spin sheets below T_{c2} [7.103]. In these studies of the magnetic structure, both the magnetic Bragg peaks and the diffuse magnetic background have been investigated as a function of temperature and magnetic field (Sect. 7.4.1b). Study of the diffuse background (quasielastic scattering) is particularly important for investigation of magnetic vortex behavior, should this be present.

Study of the temperature dependence of the in-plane coherence length $\xi_a(T)$ also has been interpreted to provide valuable information, insofar as the lower critical temperature T_{c1} in the stage-1 acceptor compounds (e.g., stage-1 $CoCl_2$ GICs) was identified with the onset of a decrease in $\xi_a(T)$ above T_{c1} [7.206].

Inelastic neutron scattering has traditionally been the method of choice in determining the low-energy collective excitation in solids. From inelastic neutron scattering measurements, the low-energy phonon branches [7.120] and the magnon dispersion relations for GICs can be determined. Measurements of the phonon dispersion relations for the magnetic acceptor $FeCl_3$ GICs have been carried out at room temperature [7.120, 207], far above the magnetic ordering temperature. No measurements have been reported to date on the use of inelastic neutron scattering to probe the effect of magnetic interactions at low temperature on the phonon dispersion relations. Magnon-phonon interactions in $CoCl_2$ GICs were considered by *Wiesler* [7.205], and he concluded that there was no evidence from neutron scattering for such an interaction.

On the other hand, inelastic neutron scattering has provided a powerful tool for studying magnetism in GICs through measurements of the magnon dispersion relations as a function of temperature and magnetic field. For stage-2 $CoCl_2$ GICs at low temperatures [7.146], such studies have been of particular interest insofar as the magnetic structure of pristine $CoCl_2$ is well established [7.1], thereby allowing detailed study of the effect of intercalation on the magnetic ordering and on the low-energy excitations (magnons) of the antiferromagnetic ground state. From these studies, values of the in-plane and interplanar exchange interactions J and J' can in principle be determined. For the case of $CoCl_2$ GICs, the dispersion [7.146] along the c-axis is too weak to obtain any quantitative measure of J' and J'_A. The inelastic neutron scattering technique has also been used in elegant studies of free and bound magnetic vortices in stage-2 $CoCl_2$ GICs, as predicted in the KT theory [7.208] and discussed in Sect. 7.4.1b

In summary, for elastic scattering studies, scans are taken along $(Q_a, 0, 0)$ and $(0, 0, Q_c)$ to probe the in-plane and c-axis ordering where $\boldsymbol{Q}$ denotes the neutron wave vector. Scans along Q_c about $(Q_a, 0, Q_c)$ have been used to study critical scattering between T_{cl} and T_{cu} in stage-2 $CoCl_2$ GICs, the reason for use of nonvanishing Q_a values being the suppression of nonmagnetic background noise. Scans at $(Q_a, 0, \pi/I_c)$ are taken to determine the magnon dispersion relations.

7.3.5 Electrical Resistivity and Magnetoresistance

Electronic transport measurements (electrical resistivity, magnetoresistance, and Hall effect) are usually not the method of first choice in examining the properties of a magnetic material, because the transport coefficients provide only indirect information on the critical exponents. In contrast, susceptibility, magnetization, or specific-heat measurements provide direct information on the critical exponents through their temperature and magnetic field dependences. Nevertheless, information on the magnetic properties of GICs can be obtained through study of the coupling of the magnetic spins on the intercalate layer to the graphitic π electrons, which is sensitively probed by transport measurements in magnetic GICs. The carrier transport in this case occurs primarily through the high mobility π electrons in the graphene layers. These π electrons are scattered by the spins on the magnetic intercalate layers and are thus sensitive to the magnetic order. There is essentially no transport by electrons in the magnetic intercalate layers because of their low mobility in the case of the donors, and because of the ionic nature of the intercalate layer in the case of the acceptor compounds. Transport measurements provide a complementary tool and can be used along with other techniques to study the magnetic phases in magnetic graphite intercalation compounds (GICs), as well as to demonstrate new magnetic phenomena generally, as summarized below.

Magnetic transport phenomena were first studied in the stage-1 magnetic donor compound C_6Eu, where a strong coupling is found between the spins on the magnetic intercalate species and the conduction electrons on the graphite layers because of their relatively small spatial separation and the resulting strong RKKY interaction. In the study of C_6Eu, the magnetic field dependence of the resistivity has been measured over a wide temperature range for various geometrical configurations for the magnetic field: $\boldsymbol{j} \parallel \boldsymbol{H} \perp \hat{c}$; $\boldsymbol{j} \perp \boldsymbol{H} \perp \hat{c}$; and $\boldsymbol{H} \parallel \hat{c} \perp \boldsymbol{j}$. In all cases evidence for magnetic phase transitions was found [7.119, 145, 160, 209]. The most interesting result that emerged from these magnetic transport studies was the identification of a fan spin phase that was not clearly resolved in previous magnetization experiments [7.209] (Sect. 7.4.2a).

Although there are many more examples of magnetic acceptor GICs than donors, few magneto-transport measurements on magnetic acceptor GICs have been carried out until recently. Magnetic effects in the transport properties of the acceptor compounds are small because of the weak coupling between the graphite conduction π electrons and the localized magnetic spins on the intercalate layer, since the magnetic spins are separated from the conduction electrons on the graphite layers by an intervening halogen (usually a chlorine) layer. Despite this weak coupling, changes in the resistivity and magneto-resistance of magnetic acceptor GICs associated with magnetic phase transitions have recently been observed [7.126, 128].

Early work in Japan [7.210] explored the magnetoresistance of several acceptor magnetic GICs and reported a positive magnetoresistance at fields up to ~ 1 tesla, though the evidence for a magnetic phase transition at lower fields was inconclusive. The most interesting result obtained from the transport studies of the magnetic GICs was the identification of an anomalous increase in the in-plane resistivity $\varrho_a(T)$ of stage-1 $CoCl_2$ GICs as the temperature was lowered below T_N. This anomalous behavior was explained by zone-folding effects associated with the antiferromagnetic interplanar coupling, giving rise to a doubled magnetic unit cell, thereby affecting the Fermi surface and electron states on the Fermi surface near the new Brillouin-zone boundary [7.126, 211, 212]. While this model predicts an energy gap opening at the Fermi level due to the antiferromagnetic arrangement of the spin sheets [7.211], no modification to the Fermi surface has been observed thus far in a direct Fermi surface measurement using the Shubnikov-de Haas effect. An anomalous behavior similar to that observed in $\varrho_a(T)$ for stage-1 $CoCl_2$ GICs has also been reported in $\varrho_c(T)$ near T_N [7.188].

In generalizing to other magnetic acceptor GICs, an anomalous increase in resistivity $\varrho_a(T)$ below T_c is also observed in the stage-1 $NiCl_2$ GICs [7.62] and stage-1 $Co_{0.77}Mg_{0.23}Cl_2$ GIC [7.213], which likewise exhibit an antiferromagnetic stacking (along the c axis) of ferromagnetic spin planes. For the stage-2 acceptor $CoCl_2$ GICs, the dominant magnetic scattering mechanism for the π electrons is spin-disorder scattering, which is abruptly reduced as T is decreased below T_{cu} [7.128]. A decrease in $\varrho_a(T)$ was also observed in stage-1 $FeCl_3$ GICs below about 5.5 K, also identified with the suppression of spin disorder scattering, since there is no reported antiferromagnetic stacking of spin sheets in this compound.

Measurements of $\varrho_a(H)$ have been especially useful in confirming the H_t versus T magnetic phase diagram of stage-1 $NiCl_2$ GICs and $CoCl_2$ GICs, as determined by susceptibility measurements, described in Sect. 7.3.1 [7.62, 128]. For the case of the stage-1 $NiCl_2$ GICs [7.62], the value of the transition field $H_t(T)$ is determined from the position of the maximum in $|\partial\varrho_a(H)/\partial H|$. The results obtained in this way from the $\varrho_a(H)$ data were found to agree with H_t as determined from susceptibility measurements.

7.3.6 Thermal Transport

Thermal conductivity is not generally regarded as one of the more sensitive probes of the magnetic properties of materials. Indeed, though magnons are known to contribute to the thermal currents in magnetic materials, the separation of the magnon contribution to thermal transport from that of the electrons and phonons is a delicate problem in 3D systems. Moreover, in electrically conducting samples, since the electrical conductivity and the electronic contribution to the thermal conductivity are related through the Wiedemann-Franz law, the magnetic field used to suppress the effect of the spin waves also modifies the electronic contribution to the thermal conductivity [7.214].

Nevertheless, thermal transport measurements have been carried out on magnetic GICs and magnetic effects have been observed in the thermal transport properties. This work is reviewed in Chap. 6.

Anomalies associated with magnetic phase transitions have been identified in the low-temperature thermal conductivity of $FeCl_3$ GICs [7.215]. Specifically, an enhancement of the thermal conductivity was found in the temperature range where there is an excess specific heat of magnetic origin. This suggested the presence of an additional contribution to the thermal conductivity due to magnons. In contrast, the scattering of phonons or charge carriers by magnons should lead to a decrease in thermal conductivity, which is contrary to what was observed [7.215]. However, the separation of the magnon contribution from other contributions in such magnetic GICs has not yet been attempted.

With regard to the thermoelectric power, it was suggested [7.215] that magnetic GICs may turn out to be the materials of choice for study of the magnon-drag mechanism, where electrons are driven along the temperature gradient by the spin-wave excitation. Indeed, in magnetic GICs, magnetic phase transitions occur at low T (typically below 10 K), whereas phonon-drag occurs at higher temperatures because of the high Debye temperature. It was found experimentally that the thermoelectric power of first-stage magnetic compounds is much larger than that observed in nonmagnetic compounds [7.215, 216]. Similarly, the thermopower for a stage-1 bi-intercalation compound (of C–$CoCl_2$–C–$AlCl_3$ layers) is much lower than for the binary stage-1 magnetic $CoCl_2$ GIC [7.215]. There has, however, not yet been a definitive experiment demonstrating that the magnon-drag mechanism proposed by *Sugihara* [7.212] is responsible for the observed thermoelectric power enhancement in magnetic GICs [7.215]. Perhaps the additional magnetic scattering of the charged carriers may be important [7.214].

7.3.7 Electron-Spin Resonance

Electron-spin resonance (ESR) has provided a key tool for examining the effective g factors in magnetic salts. The technique can in principle determine the Zeeman terms in the magnetic Hamiltonians [7.105] discussed in Sects. 7.2.1, 2.

Electron-spin resonance measures electronic g factors through the Zeeman interaction

$$\hbar\omega = g\mu_B H_0 \langle S \rangle \, , \tag{7.28}$$

where $\hbar\omega$ and H_0 are respectively the resonant photon energy and resonant magnetic field as the conduction electrons are excited from the ground state in a spin-flip transition. Both conduction and localized electrons can be studied by the ESR technique. The experiments are normally carried out in a microwave resonant cavity at constant frequency as the magnetic field is swept through the resonant field H_0, from which the g factor in the direction of $\boldsymbol{H}_0$ is determined using (7.28). The asymmetry of the absorption line shape (Dysonian line shape) is sensitive to the fraction of the microwave power absorbed by the conduction electrons and to the ratio of the spin-diffusion time to the spin–spin relaxation time [7.217].

Of particular interest to magnetic studies in GICs is the anisotropy in the g factor and its temperature dependence, where $g_{\parallel}$ and $g_{\perp}$ respectively denote the g factor for the magnetic field parallel and perpendicular to the crystalline c axis, and where $H_0^{\parallel}$ and $H_0^{\perp}$ respectively denote the corresponding resonant fields. The anisotropy in the g factor provides information about the preferred spin alignment and, in particular, through use of the *Lines* theory [7.40], provides a determination of the in-plane exchange anisotropy J_A/J in the $CoCl_2$ compounds. Also the anisotropy in the g shift, $\Delta g = (g - g_0)$ where $g_0 = 2.0023$ is the free-electron g factor, can be related to the anisotropy in the magnetic susceptibility by the relations [7.218]

$$\Delta g_{\perp}/g_0 = (\chi_{\perp} - \chi_{\|})/2\chi_{\perp} ,$$

$$\Delta g_{\|}/g_0 = -(\chi_{\perp} - \chi_{\|})/\chi_{\|} , \tag{7.29}$$

where $\chi_{\|} = \chi_{cc}$ and $\chi_{\perp} = \chi_{aa}$ are the susceptibilities for $\boldsymbol{H} \| \hat{c}$ and $\boldsymbol{H} \perp \hat{c}$ (Sect. 7.3.1). The ESR experiment is also sensitive to anisotropy in the basal plane, and this has been explored in $CuCl_2$ GICs, which exhibit structural in-plane distortions from a triangular lattice [7.219].

In addition to the angular dependence, study of the temperature dependence of the g factors provides much useful information. The observation of a temperature dependence of the g factor as T is lowered indicates the presence of short-range order, while a divergence of the g factor at yet lower temperatures indicates long-range order. If we define the characteristic field

$$\bar{H} = [H_0^{\|}(H_0^{\perp})^2]^{1/3} , \tag{7.30}$$

then a temperature-independent $\bar{H}$ would be associated with 2D-Heisenberg behavior. However, a rapid increase in $H_0^{\|}$ at low T accompanied by a rapid decrease in $H_0^{\perp}$ is characteristic of 2D-XY behavior. Study of the temperature dependence of the ESR spectrum can also be used to determine structural phase transitions. The detailed temperature dependence of $g_{\|}$ and $g_{\perp}$ is often complicated [7.219] and cannot be explained by simple models.

Also of importance in the ESR measurements is the resonant linewidth ΔH, its anisotropy, and its temperature dependence, because of the relation of ΔH to spin–spin correlations and spin fluctuations. If the spin–spin correlations are dominated by spin fluctuations with wave vectors $q \approx 0$, then an angular dependence $\Delta H(\theta) \sim (3\cos^2\theta - 1)^2$ is expected for 2D behavior. Such behavior in $\Delta H(\theta)$ has been reported for stage-2 $NiCl_2$ GICs [7.134], stage-2 $MnCl_2$ GICs [7.151, 153, 219, 220], and stage-1 and stage-2 $CuCl_2$ GICs [7.219]. The corresponding temperature dependence of $\Delta H(\theta)$ provides information on the nature of the spin–spin interaction as magnetic phase boundaries are traversed. For example, the linewidth ΔH at $\theta = 25°$ in stage-2 $NiCl_2$ GICs [7.134] increases rapidly at low T and diverges at 6 K, characteristic of the expected behavior of ΔH in a 2D ferromagnet.

The ESR technique has been used by *Vaknin* et al. to explore the magnetic properties of OsF_6 GICs [7.221] and MoF_6 GICs [7.222] in the paramagnetic phase. The observed ESR spectra were interpreted as showing evidence of a well-defined crystal field and a well-defined orientation of the intercalate species with the C_3 axis of the OsF_6^- molecular ion aligned parallel to the c axis of the GIC. Complementary susceptibility studies indicated an antiferromagnetic in-plane and c-axis coupling with large exchange constants ($\sim$ 15 K) for the in-plane interactions. However, no long-range order was found down to 2 K, suggesting that this 2D magnetic system is consistent with the *Mermin-Wagner* theorem [7.164, 223]. MoF_6 GICs [7.222] show contributions to the ESR spectra from both conduction electrons and localized moments. The magnitude of the in-plane exchange interaction is estimated by ESR to be $\sim$ 10 K, but the GIC shows no sign of long-range order. Various other transition-metal fluorides such as IrF_6 GICs [7.224] have been studied by the group in Jerusalem [7.106, 221, 222, 225–228] with regard to 2D structural phase transformations, though no magnetic order has been identified in these materials, either by ESR or by other techniques.

7.3.8 Nuclear Magnetic Resonance

Nuclear magnetic resonance (NMR) is another example of a technique that probes the local order about the magnetic ion and can be used to confirm the magnetic Hamiltonians discussed in Sects. 7.2.1, 2. By measuring the shift of the NMR line from that of the free magnetic ion, the crystal-field splitting parameters of the magnetic ion in the GIC can be determined. For example, NMR experiments in pristine $NiCl_2$ [7.229, 230] and in pristine $CoCl_2$ [7.230] were used to give information related to the Hamiltonians given by (7.7, 8). In addition, study of the temperature dependence of the NMR linewidth provides information on the spin–spin and spin-lattice relaxation times.

Nuclear magnetic resonance has been studied in stage-1 $CoCl_2$ GICs by *Tsuda* et al. [7.231]. These workers reported resonance spectra for both the ^{35}Cl, ^{37}Cl and the ^{59}Co nuclei. Two spin-echo signals were observed at 1.4 K: one broad signal between 17 MHz and 29 MHz with a peak at 22.5 MHz associated with the Cl nucleus, and another resonance between 230 MHz and 410 MHz associated with the Co nucleus. The observed broadness of the spectra is consistent with the incommensurate structure of these compounds, whereby specific nuclear species see a distribution of local fields.

In addition to the interactions of the nuclear spin with the local crystal fields, the nuclear spins interact with the electron spins through the hyperfine interaction (Sect. 7.3.9). The strength of the hyperfine field in stage-1 $CoCl_2$ GICs is found to be lower than in pristine $CoCl_2$. This reduced hyperfine field is a reflection of the strength of interaction of the ^{35}Cl nuclear moment with the magnetic moment of the Co^{2+}; a weaker field implies that the Cl^- octahedra are more relaxed in the GIC than in the pristine material. This relaxation is also reflected in the high-field magnetization measurements of J_A/J discussed in Sect. 7.3.2

7.3.9 Mössbauer Spectroscopy

The Mössbauer effect is the recoilless resonance fluorescence of γ-rays emitted from nuclei bound in solids. In typical solids, the Mössbauer line is split by the hyperfine interaction, allowing measurement of hyperfine fields and electric field gradients, thereby yielding information on crystal symmetries, crystal field splittings, and the internal magnetic fields seen at specific sites in the lattice. Study of the anisotropy of the electric field gradient yields the spin orientation, while the temperature dependence of the internal magnetic field B_{eff} gives the magnetic ordering temperature.

The two Mössbauer isotopes that have been studied in magnetic GICs are ^{57}Fe and ^{151}Eu. Mössbauer spectroscopy has also been carried out in pristine $CoCl_2$ [7.232] using ^{57}Co as the source and $Na_2Fe(CN)_6 \cdot 10H_2O$ as the absorber. The *K* capture of the ^{57}Co nucleus gives a ^{57}Fe which then emits a γ-ray. The resulting spectrum is characteristic of an isolated Fe^{2+} ion in a $CoCl_2$ matrix. In addition, conventional Mössbauer experiments have been performed on a system where $\sim 2\%$ ^{57}Fe is doped into pristine $CoCl_2$ and the sample is then used as the absorber [7.233]. This approach should also give information on the effect of an isolated impurity probe (^{57}Fe) in a GIC, but this approach has not been applied to $CoCl_2$ GICs.

Extensive Mössbauer experiments have been carried out on the $FeCl_3$ and $FeCl_2$ GICs [7.234–237] because ^{57}Fe is one of the most convenient Mössbauer nuclei. These experiments have focused on the great difficulty in obtaining a pure Fe^{2+} or a pure Fe^{3+} system. All $FeCl_3$ GICs examined to date show an admixture of Fe^{2+} with the dominant Fe^{3+} species. Since the magnetic moments and magnetic interactions of these two ions are very different, these crystals must be considered as random magnetic alloys. Spin-glass behavior has in fact been reported for nominal $FeCl_3$ GICs [7.238], consistent with a random arrangement of these spin species.

A detailed Mössbauer study has been carried out on the ^{151}Eu nucleus in C_6Eu [7.156]. From study of the isomer shift it was determined that the europium was in an Eu^{2+} state with negligible Eu^{3+} admixture. Through study of the hyperfine interaction shown in Fig. 7.8a, the internal magnetic field B_{eff} was measured to be very small in magnitude, while its temperature dependence (shown in Fig. 7.8b) could be used to determine the Néel temperature by fitting B_{eff} to a Brillouin function for spin 7/2. From measurements of the electric quadrupole interaction, the electric field gradient was determined and from its angular dependence, the axial symmetry of the electric field gradient was established, showing that the Eu spins lie in the *xy* plane. The large negative isomer shift was interpreted in terms of a strong charge transfer from the Eu to the graphene layers and a strong hybridization of the wave functions of the Eu ions and the π electrons.

7.3.10 Other Techniques

Optical techniques have been used to study the electronic structure of magnetic GICs. These experiments, however, have not contributed much to our understanding of the magnetic nature of these compounds. Faraday rotation experiments, which would be sensitive to the magnetic structure, have been made for the pristine $NiCl_2$ [7.140] and $MnCl_2$ [7.38], but have not yet been reported for the magnetic GICs.

Though one-magnon and two-magnon Raman scattering has been reported in pristine transition-metal halides [7.239], these excitations involve very small frequency shifts and require very good optical surfaces. No Raman results due to magnon excitations have yet been reported in magnetic GICs.

Magnetostriction measurements [7.129], which measure anomalies in the change of the sample length in an applied magnetic field, can be used to monitor magnetic phase transitions. Hysteresis behavior in magnetostriction studies can be used to distinguish first-order magnetic transitions from those of higher order [7.240]. Preliminary magnetostriction measurements [7.129] on stage-1 $CoCl_2$ GICs have been carried out showing anomalies in the magnetic field dependence of the magnetostriction. The effects are large but occur at field values that appear to be inconsistent with other magnetic observations on the same compounds.

7.4 Overview of Magnetic GICs

In this section we examine some of the experimentally studied magnetic GICs and summarize the current status of this research. Rather than an exhaustive listing of all the known data, we restrict our report to an in-depth coverage of a few of the more thoroughly researched systems. Relatively more work has been done on magnetic acceptor GICs than on the magnetic donor GICs, and even for the magnetic acceptors most of the studies have been on the $3d$ transition metal chloride GICs. The $4d^n$ series (Y, Zr, Nb, Mo, Tc, Ru, Rh, Pd, Ag) halides and the $5d^n$ series (La, Hf, Ta, W, Re, Os, Ir, Pt, Au) halides are also likely candidates for interesting magnetic acceptor GICs.

In this section we focus on the stage-1 and stage-2 GICs. The magnetic properties of the stage-1 compounds are unique and must be discussed separately. With regard to the higher-stage compounds, there have been few detailed magnetic studies on high-stage samples ($n > 2$). One difficulty with the higher-stage compounds is that as the stage index increases, so do problems with sample homogeneity. In a few cases, some very precise measurements have been made on high-stage compounds ($n > 2$), and in these cases many features of the magnetic properties are similar to those for stage 2. Thus the stage-2 compounds give important information on the behavior expected for the higher-stage compounds.

7.4.1 Overview of Acceptors

One advantage of studying magnetic interactions in magnetic acceptor GICs is that for most of these compounds there is a close similarity between the crystal structure of the magnetic intercalate layer and that of the parent magnetic material (Table 7.3). In such cases, the intraplanar magnetic exchange interactions can be inferred from those of the parent materials, which have been studied extensively in the past because of their magnetic anisotropy. The further interest in the magnetic acceptor GICs relates to the relatively large spatial separation between the magnetic species along the c axis, thereby enhancing the two-dimensionality of these compounds relative to their parent layered materials.

In this context, some of the magnetic graphite intercalation compounds have been proposed as experimental systems that approximate 2D-XY behavior. To optimize the XY approximation, the intercalate is chosen to maximize the magnitude of (J_A/J) for a magnetic system, where J_A and J have the same sign. To achieve 2D spatial behavior, it is necessary to synthesize high-stage compounds. Approximations to 2D magnetic phases have been achieved in specific magnetic GICs over a limited range of temperatures and magnetic fields, as discussed below.

The experimental work on magnetic acceptor GICs exhibiting 2D-XY behavior dates back to the early 1970s when *Karimov* [7.19, 57, 202, 203, 241–244] and his collaborators [7.124, 245] first discovered anomalies in the magnetic susceptibility of a variety of transition-metal chloride GICs [7.203]. In this pioneering work, attention was also given to the identification of temperature ranges where spontaneous magnetization could be found and where anomalies in the specific heat could be correlated with anomalies in the susceptibility. Much of the early work on magnetic acceptors was on the $NiCl_2$ GICs [7.19, 124, 202, 203, 243, 244], the first system we discuss in some detail.

(a) $NiCl_2$ GICs

Of the various magnetic acceptor GICs, the $NiCl_2$ GICs have the highest magnetic ordering temperature (~ 22 K), thereby exhibiting the strongest magnetic interactions. Because of the similarity of the in-plane lattice constants [7.114] in the pristine $NiCl_2$ and in the corresponding GICs, their in-plane magnetic exchange constants should be similar (Table 7.4).

As mentioned in Sect. 7.2.1a, though the anisotropy term in the magnetic Hamiltonian (7.7) is relatively small, the spins in pristine $NiCl_2$ and in the GICs are not isotropically arranged, but rather lie in the xy plane. For this reason, by studying the $NiCl_2$-GIC system, the quasi-2D-XY behavior observed in the $CoCl_2$ GIC system can be investigated for a case where the strength of the anisotropy term J_A/J is very much reduced. It would therefore be of particular interest to explore the magnetic behavior of high-stage $NiCl_2$ GICs. Unfortunately, only well-staged samples of stage-1 and stage-2 $NiCl_2$ GICs have been prepared thus far. The stage-2 samples are somewhat easier to synthesize and have been studied far more extensively, whereas it is only recently that stage-1 samples have been synthesized and studied magnetically [7.62, 135, 136]. However, mixed-phase samples of $NiCl_2$ with $AlCl_3$ have been prepared with stage indices as high as 5 [7.246].

With regard to the magnetic properties of $NiCl_2$ GICs, the magnetic species is the Ni^{2+} ion, with $S = 1$, and the magnetic properties are described by the magnetic Hamiltonian of (7.7). If the Ni^{2+} ion with configuration $3d^8$ were in a ligand field of exactly octahedral symmetry, then the g factor would be isotropic and close to 2. However, if there is trigonal distortion lowering the octahedral (cubic) symmetry, then the three levels for $S = 1$ will no longer be degenerate in zero field, and this will be reflected in a shift in the g factor and hence in the measured ESR resonances. *Suzuki* et al. [7.134] measured the anisotropy of the g factor by ESR at 300 K and found $g_{\perp} = 2.156 \pm 0.002$ and $g_{\parallel} = 2.096 \pm 0.003$ in stage-2 $NiCl_2$ GICs, consistent with similar measurements by *Flandrois* et al. in stage-1 $NiCl_2$ GICs [7.135]. The magnetic properties of the $NiCl_2$ compounds (pristine, stage 1, and stage 2) are summarized in Table 7.4. The similarity of the various g values in Table 7.4 indicates that the value of D for the GICs (7.7) is probably close to that of pristine $NiCl_2$ ($D = 0.4$ K), suggesting that the spins in the magnetic GICs also lie in the xy plane at low temperatures, consistent with direct observations in stage-2 compounds using neutron scattering [7.147]. By careful analysis of the angular and temperature dependence of the ESR resonant fields and linewidths in the stage-2 compound, it was inferred that the system changes from XY-like behavior (below ~ 30 K) to Heisenberg-like spin symmetry (above ~ 30 K) [7.134].

Despite the small size of the anisotropy D, we conclude from a variety of measurements that the spin structures of stage-2 $NiCl_2$ GICs are very similar to those of stage-2 $CoCl_2$ GICs [7.147] where the exchange anisotropy term is large ($J_A/J \simeq 0.5$); i.e., we infer that stage-2 $NiCl_2$ GICs undergo a two-step ordering process at temperatures T_{cu} and T_{cl}. The stage-2 $NiCl_2$ GICs behave quasi-two dimensionally in the temperature range $T_{cl} < T < T_{cu}$, and exhibit weak antiferromagnetic order between the layers for $T < T_{cl}$, associated with a spin-correlation length along the c axis of only $\xi_c(0) \simeq 22$ Å (corresponding to spin correlation only with nearest-neighbor Ni^{2+} planes), as determined from elastic neutron scattering measurements [7.147]. The curves for the temperature dependence of the neutron scattering intensity for the stage-2 $NiCl_2$ GICs and for the corresponding $CoCl_2$ GICs are very similar with regard to both the half-integral $(0, 0, l/2)$ Bragg peaks associated with the antiferromagnetic (AF) interplanar spin stacking (which is quenched at T_{cl}) and the diffuse background which dominates the scattering in the range $T_{cl} < T < T_{cu}$ (Sect. 7.4.1b) [7.147]. One difference between the two magnetic species is the factor of two greater integrated intensity for the AF Bragg peaks in the case of the stage-2 $NiCl_2$ GICs, perhaps associated with some stage-3 admixture in the samples.

As is the case for all the acceptor GICs, measurements of T_{cl} and T_{cu} for the same nominal stage but for different samples (and often by different groups) and/or using different measurement techniques (specific heat [7.57], susceptibility [7.61], and magnetization [7.139]) yield somewhat different T_{cl} and T_{cu} values. (At present no magnetic transport measurements have been reported on stage-2 $NiCl_2$ GICs.) Reasons for these discrepancies in T_{cl} and T_{cu} might be different intercalate filling factors or a small admixture of other stages or of unstaged regions (unique to each sample) and the different sensitivities of the various techniques to these minority phases.

At high temperatures ($T \gg T_{cu}$), $NiCl_2$ GICs exhibit typical paramagnetic behavior with a small amount of spin anisotropy. In this temperature range, the most useful information comes from ESR (summarized above) and dc magnetic-susceptibility measurements, which for the stage-2 compounds show typical Curie-Weiss behavior (Table 7.4) with positive in-plane Θ values, approximately independent of stage, consistent with the strong in-plane ferromagnetic exchange coupling [7.135, 139]. No Curie-Weiss constants for the $NiCl_2$ GICs have been reported for $\boldsymbol{H} \parallel \hat{c}$.

The behavior of the $NiCl_2$ GICs near the magnetic ordering temperature has been studied in some detail by magnetic-susceptibility measurements. The most noticeable feature in the in-plane ac susceptibility $\chi'(T)$ (Fig. 7.14) [7.61] of both the stage-1 [7.62] and stage-2 $NiCl_2$ GICs [7.61] is the maximum of $\chi'(T)$ near 20 K, which is related to the magnetic ordering. The measurements reveal only a single peak close to T_{cu}; clear measurements of T_{cl} and T_{cu} from $\chi''(T)$ have not yet been reported. Assuming that the Hamiltonian of (7.7) is appropriate to the $NiCl_2$ GICs, and that the value of J (21.7 K) is retained upon intercalation, then the observed susceptibility maxima are expected at approximately $JS^2/k_B = 21.7$ K, where $S = 1$. The fact that the susceptibility maxima do not vary much with stage index indicates that the $NiCl_2$ GICs show more quasi-2D magnetic behavior and less 3D magnetic behavior relative to pristine $NiCl_2$ ($T_N = 52.3$ K) [7.2] (Table 7.4).

Several authors [7.61, 139] have attempted to determine the exponent γ in the relation

$$\chi'(T) \sim \left| \frac{T - T_{cu}}{T_{cu}} \right|^{-\gamma}, \tag{7.31}$$

which describes the divergence of the susceptibility at the upper critical temperature T_{cu}. The observed values of $\gamma \simeq 2.05$ [7.61] and 2.02 [7.139] are close to 2. The detailed fits of $\chi'(T)$ near T_{cu} [7.61] to the Kosterlitz form (i.e., $\sim \exp(b/\sqrt{t})$) are poor, and the reason for the poor fits is not well understood. One possible explanation is that the finite domain size of the intercalate limits the spin-correlation length, making the Kosterlitz form invalid. No attempt has been made to fit high-temperature series expansions [7.7] to the

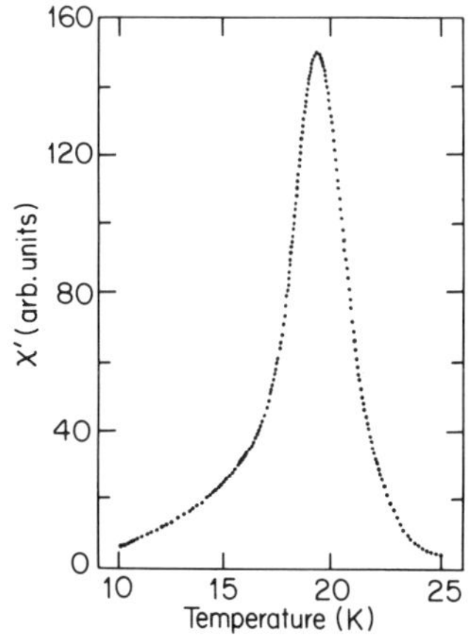

Fig. 7.14. In-plane ac magnetic susceptibility of stage-2 $NiCl_2$ GIC in zero magnetic field. The maximum of $\chi'(T)$ near 20 K is related to the magnetic ordering [7.61]

susceptibility $\chi'(T)$ for the stage-2 $NiCl_2$ GICs, as was, for example, done for stage-3 $CoCl_2$ GICs [7.191].

Figure 7.15 shows in-plane susceptibility $\chi'(H)$ measurements at constant temperature of the stage-2 $NiCl_2$ GICs. At low temperatures, magnetic-field-dependent structure is observed in the region below 50 Oe. As the temperature is raised, this structure disappears at $\sim$19.5 K, which provides one of the most definitive determinations of the 3D magnetic ordering temperature. From the elastic neutron scattering measurements, it is known that the low-temperature magnetic order in these compounds is not long-range; however, the disappearance of the field structure in the low-temperature limit was used by *Suzuki* and *Ikeda* [7.61] to estimate the ratio of the antiferromagnetic interlayer exchange to the intraplanar ferromagnetic exchange to be $|J'/J| \simeq 3\times10^{-4}$. No detailed magnetic phase diagram was reported in this work [7.61].

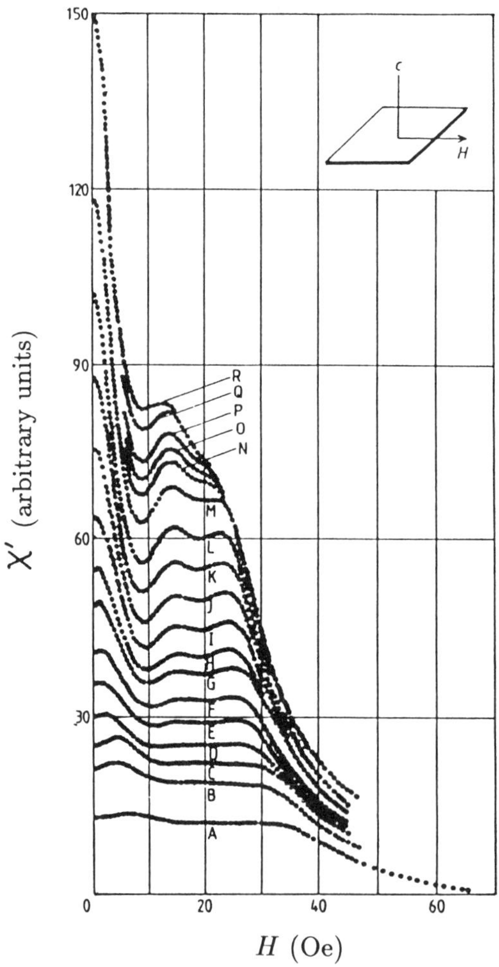

Fig. 7.15. A series of traces showing the field dependence of the magnetic susceptibility $\chi'(H)$ for a stage-2 $NiCl_2$ GIC for various temperatures. As shown in the inset, the magnetic field H is applied in the easy plane ($H \parallel a$-axis). (A) 15.73 K, (B) 15.83 K, (C) 16.26 K, (D) 16.60K, (E) 16.94 K, (F) 17.19 K, (G) 17.46 K, (H) 17.63 K, (I) 17.81 K, (J) 17.99 K, (K) 18.17 K, (L) 18.36 K, (M) 18.54 K, (N) 18.70 K, (O) 18.79 K, (P) 18.89 K, (Q) 19.09 K, and (R) 19.39 K [7.61]. The field dependent structure persists up to 19.39 K and above (not shown). The temperature at which the structure in $\chi'(H)$ disappears provides a good measure of T_{c1}

The ordering at the magnetic phase transition for the stage-2 $NiCl_2$ GICs has also been explored by magnetization studies [7.139, 202, 203]. One objective of these studies was to obtain a value of the exponent $\delta(T)$ in the relation $M \sim H^{1/\delta}$. The experimental values of $\delta(T)$ were found to be temperature independent for $T > T_{cu}$, where $\delta \simeq 1.0$ (as expected for the paramagnetic phase), but for $T < T_{cl}$ a value of $\delta \simeq 2.0$ was found, with a linear increase in δ from T_{cu} to T_{cl} [7.139]. These values of δ are, however, not close to the Kosterlitz estimate of $\delta(T_{KT}) = 15$ for a 2D-XY system. A small residual magnetization was observed below T_{cu} and a finite spontaneous magnetization was observed between T_{cl} and T_{cu} [7.139]. These results were interpreted in terms of a 2D spin phase, later shown to be consistent with neutron scattering studies [7.147].

The magnetic behavior of the stage-1 $NiCl_2$ GICs is intermediate between that of pristine $NiCl_2$ and the stage-2 compound, and bears great similarity to that of the stage-1 $CoCl_2$ GICs. No neutron scattering measurements have been carried out on stage-1 $NiCl_2$ GICs, so the spin arrangement is not known in detail. Preliminary ESR measurements have also been reported on stage-1 $NiCl_2$ GICs and the measured $g = 2.25$ is found to be isotropic above 100 K [7.135]. At lower temperatures anisotropy effects are observed and the anisotropy increases as T is reduced, in a similar fashion to the corresponding observations in stage-2 $NiCl_2$ GICs and pristine $NiCl_2$.

With regard to susceptibility studies in the stage-1 compound, we show in Fig. 7.16 $\chi'(H)$ scans for the stage-1 $NiCl_2$ GICs, where the peak in $\chi'(H)$ is identified with the transition field H_t, defined as the field at which the AF interplanar coupling is overcome by the externally applied field. From the value of H_t in the limit $T \to 0$ K, the value of J' is determined from (7.22) (the relation $H_t = 2zJ'S/g_\perp \mu_B$, where $z = 6$ is the number of neighbors on adjacent layers), yielding $J' = 0.012$ K (Table 7.4). From the disappearance of the magnetic field structure at higher temperature, the Néel temperature $T_N = 22$ K is obtained. Figure 7.17 shows the resulting $H_t - T$ phase diagram for stage-1 $NiCl_2$ GICs. From the similarity of the magnetic properties (i.e., with regard to the $\chi'(H)$,

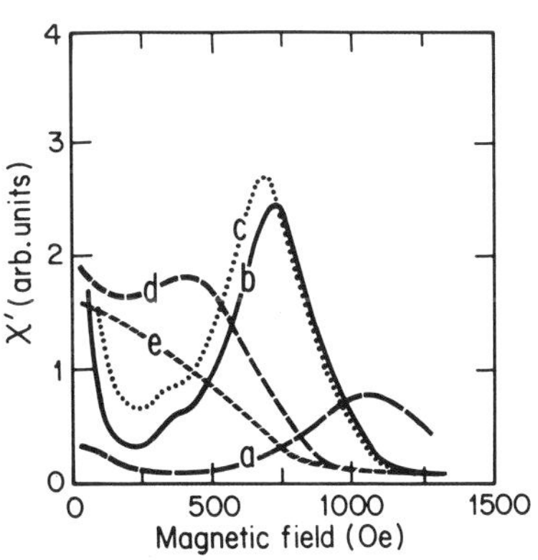

Fig. 7.16. In-plane ac susceptibility $\chi'(H)$ of HOPG-based stage-1 $NiCl_2$-GIC sample for various temperatures: (*a*) 4.2 K, (*b*) 17.4 K, (*c*) 19.0 K, (*d*) 21.8 K, and (*e*) 23.0 K [7.136]

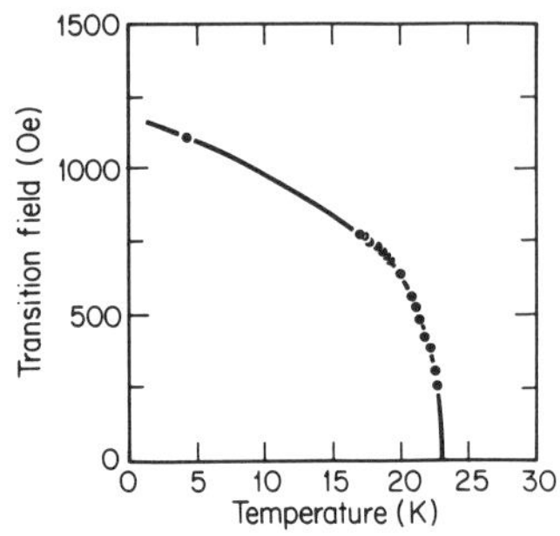

Fig. 7.17. The position of the dominant peak at the transition field H_t for stage-1 $NiCl_2$ GICs is plotted versus temperature T [7.136]. The solid line is a guide to the eye

$\varrho_a(H)$ and $\varrho_a(T)$ experiments) to those of stage-1 $CoCl_2$, it is estimated that J' in stage-1 $NiCl_2$ GICs is reduced by a factor of 60 [7.136] from the value in pristine $NiCl_2$. It is proposed [7.62] that the stage-1 compound in zero magnetic field will undergo only one magnetic transition, at the Néel temperature $T_N \simeq 22$ K, into an antiferromagnetic phase with a somewhat longer AF correlation length along the c axis than its stage-2 counterpart.

(b) $CoCl_2$ GICs

Of all the magnetic acceptor intercalation compounds, the $CoCl_2$ GICs have been most extensively studied, and the results for the stage-2 $CoCl_2$ GICs have been recently reviewed [7.24]. Particular attention has been given to the stage-1 and stage-2 compounds, with only a few measurements having been reported for stage-3 and higher compounds, primarily due to the difficulty in preparing well-staged, homogeneous high-stage compounds. From the point of view of 2D magnetism, stage-2 and higher-stage compounds are of particular interest. However, stage-1 $CoCl_2$ GICs are easier to synthesize, more homogeneous, and have therefore been studied more extensively. In this section we summarize the present understanding of the stage-1 and higher-stage $CoCl_2$ GICs. Although the $CoCl_2$ GICs exhibit much larger spin anisotropy than the $NiCl_2$ GICs discussed in Sect. 7.4.1a, both families of compounds show very similar behavior with regard to the dependence of the magnetic properties on temperature, magnetic field, and stage.

Of the various magnetic GICs that have been synthesized, the $CoCl_2$ GICs exhibit the largest XY anisotropy ($J_A/J \approx 0.5$), and therefore have been selected by several research groups as the best GIC approximation to the 2D-XY behavior. Since it is possible to synthesize $CoCl_2$ GICs for a variety of stages, this system can in principle be used to investigate stage-dependent effects and 2D–3D spatial crossover phenomena. Unfortunately, the intercalate layer does not extend uniformly throughout the sample, but rather breaks up into a network of domains (Sect. 7.1.3a). Thus the interpretation of magnetic studies is complicated by finite-size effects, though the domain sizes are large enough ($\gg 10^4$ spins/domain) to contain a statistically meaningful number of spins (Sect. 7.1.3a).

As a result of the interest in 2D-XY behavior, the $CoCl_2$ GICs have been extensively studied by many techniques, such as susceptibility [7.247, 248], magnetization [7.143, 248], neutron scattering [7.103, 147, 205, 249], transport [7.211], heat capacity [7.198, 200, 201], magnetostriction [7.129], and NMR [7.231]. Many of the above techniques have been used to provide a determination of T_{cl} and T_{cu} for the high-stage compounds ($n \geqslant 2$), as well as the 3D magnetic ordering temperature T_N for the stage-1 compound. Specifically, for stage-2 compounds, $\chi'(H)$ gives T_{cl}, neutron scattering gives both T_{cl} and T_{cu}, transport gives T_{cu}, and heat capacity gives T_{cu}.

SQUID magnetometer measurements in the paramagnetic region (30 K $< T <$ 300 K) provide a general indication of the magnetic properties of the $CoCl_2$ GICs [7.144], yielding values for the Curie-Weiss parameters $C_{\|}$, $C_{\perp}$, $\Theta_{\|}$, and $\Theta_{\perp}$ (Table 7.5). Though qualitatively similar to the results for $NiCl_2$ GICs (Table 7.4), the results for the $CoCl_2$ GICs show much larger spin anisotropies (as can be seen by comparing the $C_{\|}$, $C_{\perp}$, $\Theta_{\|}$, and $\Theta_{\perp}$ values in Table 7.5 with the corresponding values in Table 7.4) [7.145, 147]. This is consistent with the crystal-field arguments discussed in Sect. 7.1.2. NMR [7.231] and magnetization [7.111, 143] measurements show that the ground state can be considered as an effective $S = 1/2$ state with a very large anisotropy of the g factor, as is the case for pristine $CoCl_2$ [7.142] (Table 7.5). Just as for the $NiCl_2$ GICs, the spins in the low-temperature magnetically ordered state for $CoCl_2$ GICs are arranged ferromagnetically in each of the layer planes, but the layers themselves are stacked antiferromagnetically (Fig. 7.3), consistent with neutron scattering experiments [7.144, 147] and high-temperature dc susceptibility measurements [7.144].

Low-temperature susceptibility measurements ($\chi(T) = \chi'(T) - i\chi''(T)$) are especially useful for identification of the magnetic transition temperatures. For stage 1, two peaks are seen (Fig. 7.18) yielding T_{c1} and T_{c2}, identified as the temperature of the lower and upper peaks in $\chi'(T)$ [7.145]. Two peaks (associated with T_{c1} and T_{c2}) are also found in the measurement of $\chi''(T)$. With regard to the stage-2 compound, the peak in $\chi'(T)$ occurs at a temperature T_{max} between T_{cl} and T_{cu}, while the two peaks in $\chi''(T)$ occur at T_{cl} and T_{cu}, respectively [7.75]. Shoulders in the $\chi'(T)$ curves for $T > T_{max}$ have been reported for stage-2 and higher compounds, and in some cases this shoulder has been identified with T_{cu} [7.188].

The real and imaginary components of the zero-field susceptibility of a stage-3 $CoCl_2$ GIC are shown in Fig. 7.11a. The features are similar to the stage-2 results; $\chi'(T)$ has one peak at T_{max}, whereas $\chi''(T)$ exhibits peaks at $T_{cl} = 7.7$ K and $T_{cu} = 9.5$ K. The position T_{cu} of the upper peak in $\chi''(T)$ coincides with a weak shoulder that can be seen in the real part of the susceptibility $\chi'(T)$. In an earlier study [7.250] of a stage-3 $CoCl_2$ GIC, this shoulder in $\chi'(T)$ had been claimed to be the upper transition T_{cu}.

As a function of stage, the intensity of the $\chi'(T)$ peak per Co^{2+} ion increases as the stage index increases (Fig. 7.18) [7.189], consistent with the 2D magnetic properties of the higher-stage $CoCl_2$ GICs. Note that in the 2D limit of infinite interplanar separation (or zero interplanar coupling), $\chi'(T)$ is expected to diverge (7.18) (Fig. 7.10). However, the peak in $\chi'(T)$ remains finite due to perturbations associated with the finite-size effects of the intercalate domains as well as the finite values of the interplanar coupling J' and of the sixfold in-plane anisotropy field H_6. The effect of the anisotropic exchange interaction J_A (7.8) is not expected to significantly shift the peak position in $\chi'(T)$ [7.251].

Strong experimental evidence in support of 2D-XY behavior in the high-stage $CoCl_2$ GICs comes from the high-temperature results of both susceptibility and neutron scattering. One argument in support of the 2D-XY behavior comes from a comparison of the experimental magnetic susceptibility $\chi'(T)$ for a stage-3 compound to the high-temperature series expansion calculation for different spin dimensionalities [7.172]. Using previously determined values of the exchange constant J [7.1] and g values [7.142] for pristine $CoCl_2$, dimensionless plots of $\chi' T/C$ versus $k_B T/J$ were calculated for the 2D-Heisenberg, 2D-XY, and 2D-Ising models, using no adjustable parameters, as shown in Fig. 7.19 [7.191]. A good fit to the experimental data (Fig. 7.18) for stage-3 was obtained for the 2D-XY model, while the fits to the 2D-Heisenberg and 2D-Ising models were poor. Although fits to the high-temperature series expansions give good agreement for the in-plane J values, attempts to fit $\chi'(T)$ to

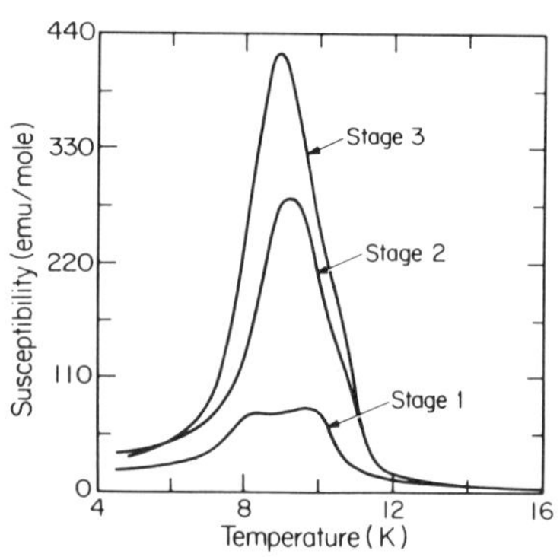

Fig. 7.18. Stage dependence of the low-temperature magnetic susceptibility in zero magnetic field for stage 1, 2, and 3 $CoCl_2$ GICs [7.145, 189]

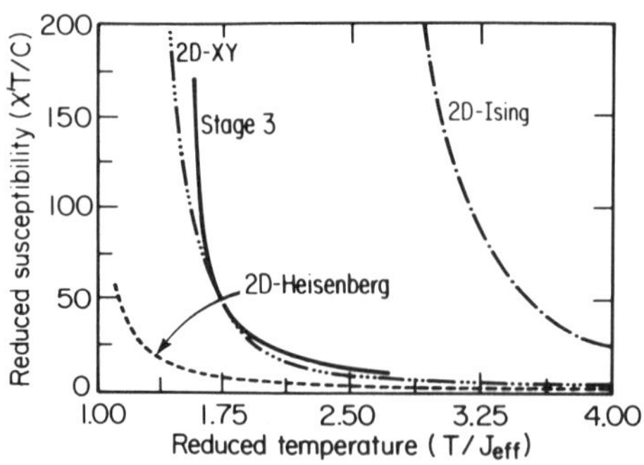

Fig. 7.19. High-temperature series expansion fit of various 2D spin Hamiltonians to the reduced susceptibility ($\chi' T/C$) data (solid curve) of stage-3 $CoCl_2$ GICs. The best fit is obtained for the 2D-XY model [7.191]. The value for the in-plane exchange is taken to be that for pristine $CoCl_2$

the Kosterlitz-Thouless form (7.17) just above T_{KT} have not been successful, as, for example, with the stage-2 $CoCl_2$ GICs [7.247].

Whether or not bound vortices exist *below* a transition temperature T_{KT}, there is strong experimental evidence from neutron scattering in the high-temperature regime *above* T_{cu} for the existence of unbound vortex excitations and vortex diffusion in stage-2 $CoCl_2$ GICs [7.205, 208]. These results from neutron scattering are generally in agreement with the high-temperature series expansion fits [7.191] to the high-temperature susceptibility data discussed above. Fits to the measured line shapes of the central peak in the inelastic neutron scattering spectrum [7.205] were found to be best described by a Lorentzian squared profile, which remains finite with decreasing wave vector, in agreement with the theory of *Mertens* et al. [7.180, 181] for the inelastic neutron scattering intensities $S(\omega, \boldsymbol{q})$ vs ω for the diffusion of unbound vortex excitations. Also, the correlation lengths and the vortex velocity obtained from the measured linewidths of the central peak were found to be in close agreement with theoretical predictions [7.180, 181, 205].

Mertens et al. [7.180, 181] also predicted that correlations between spins arranged in free vortex configurations that were diffusing through the system above T_{KT} would give rise to quasielastic neutron scattering. These authors further evaluated the quasielastic scattering function $S(q)$ and predicted that the functional form of the q dependence was that of a Lorentzian raised to the power 3/2. Figure 7.20 shows the T dependence of the power p to which the q dependent Lorentzian is raised. For $T \geqslant 12$ K (i.e., $T > T_{cu}$), the exponent p in Fig. 7.20 is $p = 1.5$, as predicted [7.180, 181] for vortex diffusion. As T is lowered

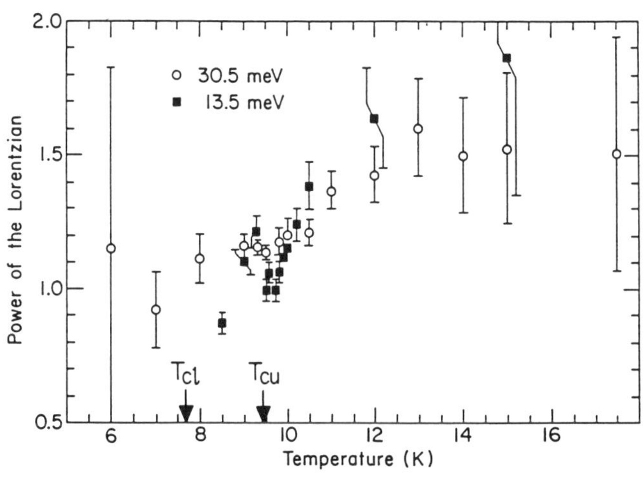

Fig. 7.20. Exponent of the Lorentzian in the quasielastic scattering profile versus temperature [7.205]. The results show a crossover from $p = 1.5$ above T_{cu} to $p = 1$ below T_{cl}. *Mertens* et al. [7.180, 181] predict $p = 1.5$ for spin vortex diffusion (above T_{cu}), whereas the Ornstein-Zernike result $p = 1$ holds for 3D correlations

toward the critical temperature, p crosses over to a value of about 1.25 at T_{cu} and to about 1 for $T < T_{cl}$. Within the large experimental error, good agreement is obtained below T_{cl} with the behavior of 3D systems which typically exhibit the Ornstein-Zernike result $p \approx 1$.

Thus, both inelastic and quasielastic neutron scattering line shapes provide evidence in support of unbound spin vortices for $T > T_{cu}$. However, the evidence for the Kosterlitz-Thouless transition from the neutron scattering line shapes is not equally convincing as $T \to T_{cu}$, nor is the evidence as strong for temperatures in the range $T_{cl} < T < T_{cu}$. However, the diffuse scattering that contributes to the elastic scattering line shape below T_{cu} [7.103] may also be related to a vortex scattering mechanism discussed below.

A critical test of the 2D-XY model comes from measuring the in-plane spin-correlation length by the quasielastic neutron scattering technique, and comparing the results with the theoretical predictions of the spin-correlation function (and its Fourier transform) discussed in 7.3.4, and summarized in Table 7.7. To avoid problems associated with the mosaic spread of their samples, *Wiesler* and *Zabel* [7.64] chose to carry out a series of neutron scattering scans along the $(h, 0, 0.075)$ ridge of stage-2 $CoCl_2$ GICs at various temperatures (Fig. 7.21). Each curve is obtained through a subtraction of data at $T = 20$ K, well above the magnetic ordering temperature. As the sample is cooled below 20 K, the magnetic component of the scattering shows the emergence of a broad diffuse peak. This broad peak grows more rapidly as T falls below T_{cu} (Fig. 7.21). Below T_{cl} a sharper component develops above the diffuse reflection background, and as the sample is further cooled, the sharp peak continues to gain in intensity at the expense of the diffuse component (Fig. 7.21).

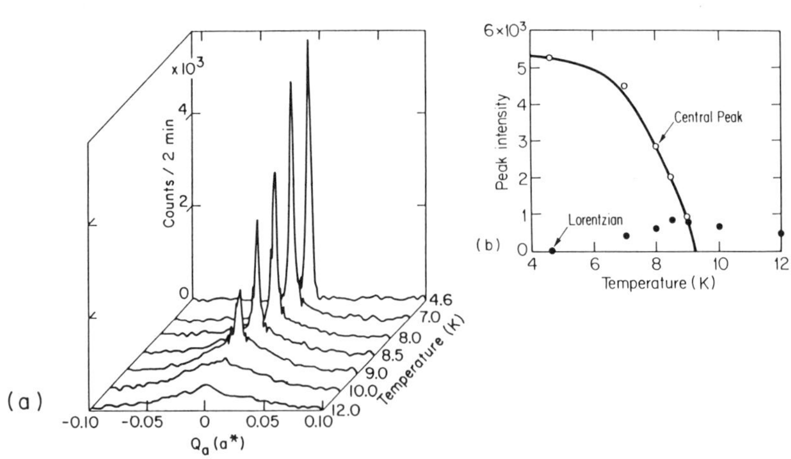

Fig. 7.21. Magnetic scattering component of stage-2 $CoCl_2$ GICs for (**a**) a series of transverse $(h, 0, 0.075)$ neutron scans across the magnetic c^* ridge at $l = 0.075$, taken at various temperatures, and (**b**) the temperature dependence of the sharp central peak and of the broad Lorentzian component of (**a**) [7.64]

For low temperatures the diffuse background and the sharp central peak have been fitted with structure factors describing short-range order (SRO) and long-range order (LRO), respectively. The width of the central peak corresponds to LRO of length $\xi_a \simeq 900 \pm 150$ Å. The intensity of the sharp central peak disappears at $T_{cu} = 9.4$ K, and hence there is strong evidence that the magnetic behavior of these compounds cannot be described by a theory of superparamagnetism, as proposed by *Rancourt* [7.252]. Though the domain size of the stage-1 $CoCl_2$ GICs is consistent with the Rancourt model, the larger in-plane coherence distances that are now reported for the higher-stage compounds are not consistent with this model of superparamagnetism [7.24]. The correlation length and intensity associated with the diffuse SRO peak both exhibit maxima near $T_{cu} = 9.4$ K. The scans at 9 K (in the intermediate temperature regime $T_{cl} < T < T_{cu}$) show diffuse tails in q_a and no sharp cutoff at T_{cu}. However, these scans could not be quantitatively fit with a Kummer function that describes quasi-LRO [7.64].

Evidence that the low-temperature spin structure of the stage-2 $CoCl_2$ GICs consists of planar sheets stacked in an antiferromagnetic arrangement along the c axis comes from the observation of antiferromagnetic $(0, 0, \frac{l}{2})$ Bragg peaks. Figure 7.22 shows the variation of the scattering intensity of the $(0, 0, \frac{1}{2})$ Bragg peak as a function of temperature. As the temperature is raised, the intensity of the $(0, 0, \frac{1}{2})$ Bragg peak falls, and the fall is very rapid near T_{cl}. The smeared behavior above T_{cl} is attributed to a diffuse background contribution which becomes sharply attenuated as $T \to T_{cu}$. This diffuse background may be due to scattering by bound vortex excitations. The small residual tail above T_{cu} may be due to scattering by unbound vortices discussed above. From the width of the antiferromagnetic $(0, 0, \frac{1}{2})$ Bragg peaks, the spin-correlation length along the c axis can be determined; the results obtained from such traces at various

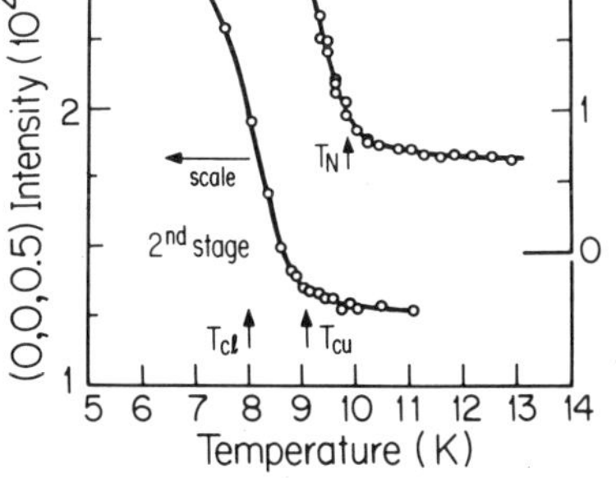

Fig. 7.22. Temperature variation of the neutron scattering intensity at the antiferromagnetic $(0, 0, \frac{1}{2})$ peak position of stage-1 and stage-2 $CoCl_2$ GICs [7.103]. The positions of T_N for stage 1 and T_{cl} and T_{cu} for stage 2 obtained from susceptibility measurements are indicated

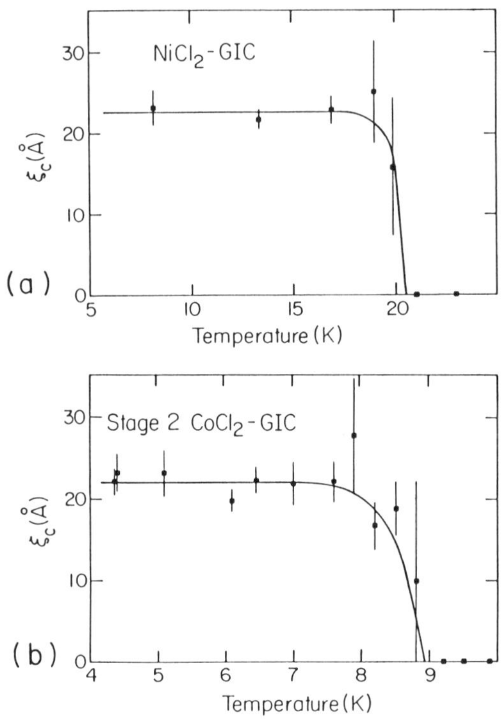

Fig. 7.23. Temperature dependence of the spin-correlation length along the c axis, taken from the width of the antiferromagnetic $(0, 0, \frac{1}{2})$ scattering peaks for **(a)** stage-2 $NiCl_2$ GIC and **(b)** stage-2 $CoCl_2$ GIC. The solid curves are a guide to the eye [7.147]

temperatures are shown in Fig. 7.23. Extrapolation of the data in Fig. 7.23 to $T \to 0$ K gives a value of $\xi_c(0) \sim 22$ Å for both stage-2 $CoCl_2$ GICs and stage-2 $NiCl_2$ GICs, in each case corresponding to a coherence length of approximately three magnetic layers of the stage-2 intercalation compound, a central magnetic layer, and the adjacent nearest-neighbor magnetic layers. Similar measurements of the width of the $(0, 0, \frac{1}{2})$ peak of the stage-2 $CoCl_2$ GICs from the data of *Ikeda* et al. [7.103] give a value of $\xi_c \simeq 70$ Å at 6 K, three times larger than in Fig. 7.23. The origin of this discrepancy is not understood, though a small admixture of stage-1 material into the stage-2 compound could have a significant effect on ξ_c.

To determine explicit values for the exchange interactions, the in-plane magnon (spin-wave) dispersion curve has been measured [7.146] for a stage-2 $CoCl_2$ GIC. The results are shown in Fig. 7.24 along with the same dispersion curve (dashed curve) for pristine $CoCl_2$ [7.1]. Because of the rather large mosaic spread associated with composite sample used in the measurements on the stage-2 GIC, it was not possible to follow the dispersion curve experimentally with any accuracy for $q_a < 0.05$. The zero magnetic field spin excitations at a point q relative to a magnetic Bragg point $\boldsymbol{G}_{\mathrm{m}}$ are given by [7.1]

$$E_{\pm} = S[(zJ - J_q \pm z'J' + J'_q)(zJ - J_q + z'J' \mp J'_q + zJ_{\mathrm{A}})]^{1/2} . \qquad (7.32)$$

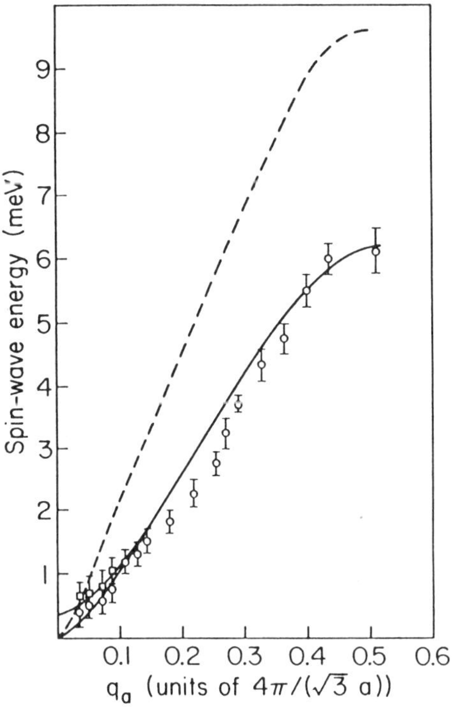

Fig. 7.24. Spin-wave dispersion relations for pristine $CoCl_2$ (dashed) [7.1] and for stage-2 $CoCl_2$ GICs [7.146]. The solid curves are calculated from (7.32)

Here $S = 1/2$ is the effective spin, $z = 6$ is the number of in-plane nearest-neighbor spins that are coupled via J, and $z' = 2$ is the number of nearest-neighbor spins in adjacent planes with coupling J' and $\langle J_q \rangle = 6\,J\,\mathscr{J}_0(q_a r_{||})$, where $\mathscr{J}_0$ is the zeroth-order Bessel function. The parameters of the magnetic Hamiltonian obtained from fitting the curves are $J = 14.6 \pm 2$ K and $J_A = 7.3$ K [7.146]. These are not in good agreement with those deduced from static susceptibility measurements on the same compound or with those for pristine $CoCl_2$ (Table 7.5) [7.144]. The splitting of the dispersion curve near $q_a = 0$ is a direct consequence of the interplanar coupling; however, the J' calculated from this splitting is in strong disagreement with other measurements of J' (Table 7.5).

Elastic neutron scattering experiments have also been carried out on the stage-1 compound where the quenching of the $(0, 0, \frac{1}{2})$ magnetic scattering peak has been identified with the Néel temperature ($T_N = 9.9$ K), as shown in Fig. 7.22 [7.103]. This result for T_N is in good agreement with the transition temperature T_{c2} as identified by the peak in the ac susceptibility $\chi'(T)$. The antiferromagnetic Bragg peak for stage-1 $CoCl_2$ GICs is quite sharp, corresponding to AF order in the c direction with a coherence distance of $\sim$200 Å [7.103]. Therefore, in the stage-1 compounds, there is full 3D antiferromagnetic order at low temperatures, in contrast to the much shorter range interplanar order observed in the stage-2 compound (Fig. 7.23) by *Wiesler* and *Zabel* [7.64] and by *Ikeda* et al. [7.103].

A great deal of information about the magnetic behavior of $CoCl_2$ GICs is also found by studies of the various magnetic properties as a function of magnetic field. For example, measurements of $\chi'(H)$ on stage-1 and stage-2 compounds display significant differences in the magnetic properties as a function of stage. In-plane susceptibility $\chi'(H)$ measurements at constant temperature of the stage-1 $CoCl_2$ GICs are shown in Fig. 7.25. At low temperatures the field-induced peak at H_t occurs at 380 Oe. As the temperature is raised, this peak becomes sharper and stronger in magnitude and then disappears at ~ 10 K, a temperature about 0.3 K above the Néel temperature T_N. The peak at H_t has been identified by neutron scattering experiments to be the transition at which the applied magnetic field overcomes the antiferromagnetic ordering along the c axis [7.118]. Hence, from the extrapolation of H_t to low temperature $H_t(0) = 380$ Oe, we can use (7.22) to obtain an estimate for the interlayer coupling J' (Table 7.5).

For the stage-2 $CoCl_2$ GICs, scans of $\chi'(H)$ show that the magnitude of $\chi'(T)$ is very sensitive to small applied in-plane magnetic fields [7.189] ($H \leqslant 50$ Oe). Furthermore, there is a noticeable trend that $\chi'(T)$ becomes increasingly sensitive to small H fields as the stage number increases, as expected for quasi-2D systems. In the low-temperature scans of $\chi'(H)$ for the stage-2 $CoCl_2$ GICs, a weak shoulder is observed at ~ 10–15 Oe [7.145, 188]. By comparison with the intense structure observed at 380 Oe in similar traces for the stage-1 compound (Fig. 7.25), we conclude that the antiferromagnetic interplanar order in the stage-2 compound is much shorter range, and that J' for stage 2 is an order of magnitude smaller than J' for stage 1 (Table 7.5). As a function of temperature, the weak shoulder at 10–15 Oe is found to disappear near T_{cu}, an unexpected result considering that T_{cl} is the expected 3D ordering temperature. A corres-

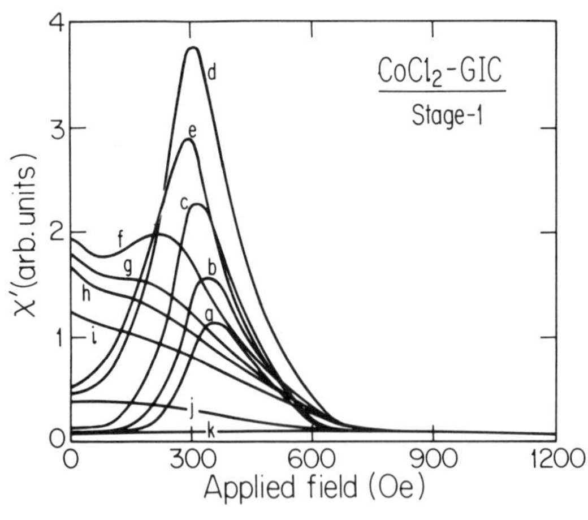

Fig. 7.25. Magnetic field dependence of the susceptibility $\chi'(H)$ for a stage-1 $CoCl_2$ GIC with $\boldsymbol{H} \perp \hat{\boldsymbol{c}}$ for various temperatures: (*a*) 6.8 K, (*b*) 7.41 K, (*c*) 8.02 K, (*d*) 8.70 K, (*e*) 8.90 K, (*f*) 9.72 K, (*g*) 10.00 K, (*h*) 10.14 K, (*i*) 10.41 K, (*j*) 10.93 K, and (*k*) 13.40 K [7.130]

ponding feature has also been observed at 6 K in the scan of the (0, 0, $\frac{1}{2}$) magnetic Bragg peak by neutron scattering [7.141].

The application of hydrostatic pressure on pristine $CoCl_2$ increases the antiferromagnetic interlayer coupling J'. The change in J' with pressure for pristine $CoCl_2$ has been studied by *Lukin* et al. [7.74]. The variation of J'_{pristine} with I_c was modeled by $J'_{\text{pristine}} \sim \exp(-\beta_{\text{pristine}} I_c)$, where the decay constant $\beta_{\text{pristine}} \approx 5\,\text{Å}^{-1}$ was obtained by fitting the experimental results to the exponential model (Fig. 7.6). Recent studies of the pressure dependence of the magnetic properties of $CoCl_2$ GICs [7.75] indicate that the application of pressure increases the interplanar magnetic coupling and shifts T_N to higher temperatures, while having little effect on J, J_A, and H_6. The dependence of J' on pressure is shown in Fig. 7.6, where it is seen that J' changes by a factor of 2 upon application of a pressure of 15 kbar, which causes a direct change in the interplanar layer separation. Finally, the different slopes in Fig. 7.6 indicate that the dominant magnetic interplanar interactions in the pristine $CoCl_2$ and in the stage-1 and stage-2 compounds are all different. Specifically, for stage-1, a π-electron modified superexchange interaction mechanism is suggested, whereas for stage-2, dipolar interactions become important (Sect. 7.2.1).

Transport measurements also provide a method for measuring the magnetic ordering temperature in both stage-1 and stage-2 $CoCl_2$ GICs through measurement of the anomaly in $\varrho_a(T)$. For both the stage-1 and stage-2 compounds, the anomalies observed in the resistivity measurements at the magnetic phase transitions are explained by an interaction based on the π-d electron coupling (Sect. 7.3.5). For the stage-1 compound, $\varrho_a(T)$ increases sharply and anomalously as T is lowered below T_N (Fig. 7.26), while the behavior for $T \gg T_N$ is similar to that for nonmagnetic acceptor GICs [7.126, 211]. This anomalous

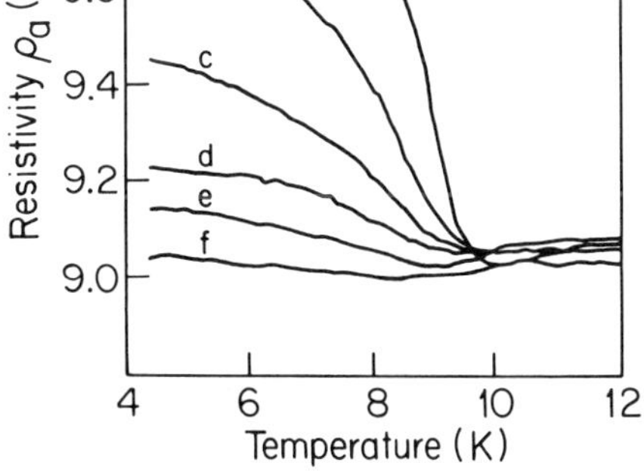

Fig. 7.26. In-plane resistivity versus temperature for a stage-1 $CoCl_2$ GIC for various in-plane applied magnetic fields with the current normal to the applied field: (*a*) 0, (*b*) 450, (*c*) 600, (*d*) 900, (*e*) 1000, and (*f*) 1200 Oe. Above T_N, the resistivity is almost independent of field ($H \leqslant 1200$ Oe) [7.130]

increase in $\varrho_a(T)$ below T_N has been identified with an additional scattering effect, arising from a Fermi surface modified by c-axis zone folding due to the onset of the antiferromagnetic stacking of the ferromagnetically ordered spin planes [7.211, 212, 253]. The long-range coherence of this antiferromagnetic stacking in the stage-1 compound has been established by magnetic neutron scattering [7.118]. The increase in resistivity accompanying the onset of the antiferromagnetic phase is explained by the greater range of the charged-impurity Coulomb potential (smaller effective screening) in the antiferromagnetic phase, thereby giving rise to increased scattering of the conduction electrons on the graphene layers [7.211]. Because of the long antiferromagnetic correlation length in the stage-1 compound, this anomalous effect (Fig. 7.26) makes a much larger contribution than the spin-disorder scattering effect, which is dominant in stage-2 $CoCl_2$ GICs and is typical of metallic magnetic systems. The decrease in $\varrho_a(T)$ in the stage-2 compound as T is lowered below T_{cu} (Fig. 7.27) is associated with the suppression of spin disorder scattering as the magnetic phase is established. Application of a magnetic field parallel to the layer plane suppresses the spin disorder scattering, so for a sufficiently high field (Fig. 7.27), all the spins become aligned along the magnetic field and the spin disorder scattering effect is effectively suppressed [7.126]. Likewise, application of a magnetic field suppresses the anomalous increase in resistivity below T_N associated with zone folding due to the antiferromagnetic layer stacking (Fig. 7.26). The effect of a magnetic field in this case is to line up the spins parallel to the applied field. The contrasts in the T- and H-dependences of the resistivity between the stage-1 and stage-2 compounds are thus attributed to the different correlation lengths in the c-axis antiferromagnetic ordering.

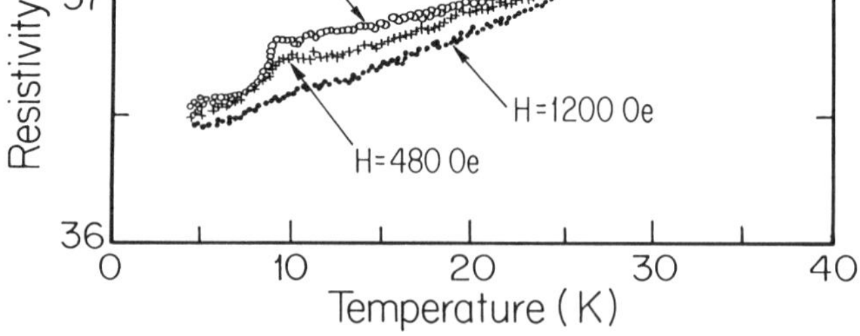

Fig. 7.27. In-plane resistivity versus temperature for a stage-2 $CoCl_2$ GIC for several in-plane applied magnetic fields ($\boldsymbol{H} \perp \hat{c}$): 0, 480, and 1200 Oe [7.128]

(c) $MnCl_2$ GICs

The Mn^{2+} magnetic ions in the $MnCl_2$ GICs exhibit antiferromagnetic in-plane coupling [7.151, 153, 254], consistent with the antiferromagnetic behavior of pristine $MnCl_2$ and the half-filled shell ($3d^5$ state) and spin $S = 5/2$. Thus the $MnCl_2$ GICs are the best example of a 2D antiferromagnet in a magnetic GIC.

Structurally, the Mn^{2+} ions in the GICs are arranged on an incommensurate triangular lattice with essentially the same in-plane lattice constant as the pristine material (Table 7.3), suggesting that the Hamiltonian for pristine $MnCl_2$ (7.9) also applies to the GICs. At low temperature, pristine $MnCl_2$ undergoes two magnetic phase transitions (at $T_N = 1.96$ K and at 1.81 K) to some complex and not fully understood antiferromagnetic structures [7.58]. High-temperature magnetic-susceptibility measurements [7.144, 151], showing Curie-Weiss behavior down to 20 K, have been very useful for the determination of many of the magnetic parameters for the $MnCl_2$ GICs (Table 7.6). These measurements show that the Curie-Weiss parameters C and Θ are only slightly anisotropic, and that the g values are close to 2. From the anisotropy of the Curie-Weiss temperatures, the value of D in (7.9) is determined from the relation

$$D = \frac{10(\Theta_\perp - \Theta_{||})}{4S(S+1) - 3}, \tag{7.33}$$

and the in-plane exchange constant J is found from the relation

$$J = \frac{-(\Theta_{||} + 2\Theta_\perp)}{6zS(S+1)}. \tag{7.34}$$

From the deviations of the low-temperature susceptibility $\chi'(T)$ from the Curie-Weiss law and the positive value of D in (7.9), *Wiesler* et al. [7.144] concluded that the spin ordering is of the XY type, consistent with neutron scattering measurements (Fig. 7.28) that show an in-plane unit cell of $2\sqrt{3} \times 2\sqrt{3}$ relative to the Mn^{2+} in-plane triangular lattice structure [7.254].

ESR measurements on stage-2 $MnCl_2$ GICs above 4.2 K [7.153] provide support for a paramagnet which exhibits 2D antiferromagnetism below the magnetic ordering temperature. The close similarity between the ESR data for the stage-1 and stage-2 compounds [7.151] suggests similar magnetic order for the stage-1 compound. Evidence for the 2D behavior comes from the $(3\cos^2\theta - 1)^2$ angular dependence of the ESR linewidth, ΔH, observed for both stage-1 and stage-2 compounds. The $(3\cos^2\theta - 1)$ angular dependence of the resonant field also supports this conclusion [7.153]. The divergence of the linewidths $\Delta H_{||}$ and $\Delta H_\perp$ at low temperature in both compounds is associated with the divergence of the staggered susceptibility $\chi'_c(T)$ as antiferromagnetic ordering is achieved. The high temperature (> 77 K) ESR results for $g_{||}$ and $g_\perp$ show little anisotropy with values of $g_{||} = 1.912$ and $g_\perp = 1.977$ for the stage-2 compound [7.153]. Anisotropy in the g factor begins to appear below 50 K, and becomes more pronounced below 30 K; this anisotropy is identified with the onset of

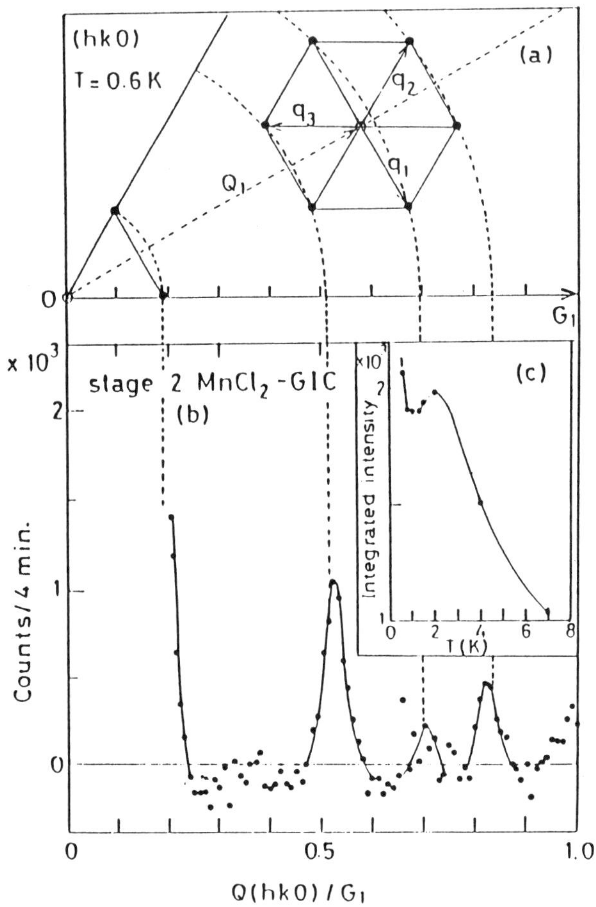

Fig. 7.28. (**a**) The $(h, k, 0)$ reciprocal-lattice plane at 0.6 K for stage-2 $MnCl_2$ GIC. (**b**) Magnetic scattering intensity along $(h, k, 0)$ at 0.6 K. (**c**) Temperature variation of integrated magnetic Bragg intensity at $|\boldsymbol{Q}_i| = |\boldsymbol{Q}_1| = |\boldsymbol{Q}_1 - \boldsymbol{q}_2| = |\boldsymbol{Q}_1 + \boldsymbol{q}_3|$ with $l = 0$ [7.254]

short-range order. The appearance of short-range order is also identified as the cause for the anomaly in the specific heat at 46.7 K [7.153].

Because of the low magnetic ordering temperature in $MnCl_2$ GICs (1.30 K in stage-1 and 1.16 K in stage-2), relatively few direct observations have been made of the magnetically ordered phases in the $MnCl_2$ GICs. Neutron scattering measurements on a stage-2 $MnCl_2$-GIC sample based on kish graphite show magnetic scattering intensity at $|\boldsymbol{Q}_i| = 1.977\ \text{Å}^{-1} = 2\sqrt{3}|\boldsymbol{q}_i|$, where the $\boldsymbol{q}_i$ are the structural unit vectors in $\boldsymbol{k}$ space (Fig. 7.28) [7.254] and the magnetic scattering intensity in the figure is obtained by subtracting the spectrum at 0.6 K from that at 20 K. Thus, the in-plane spin structure has a $(2\sqrt{3} \times 2\sqrt{3})$ periodicity, which is *different* from the $(\sqrt{3} \times \sqrt{3})$ periodicity of the Mn^{2+} spins on three sublattices forming an angle of 120° with respect to each other. For a simple 2D antiferromagnet on a triangular lattice the $(\sqrt{3} \times \sqrt{3})$ frustrated spin structure is expected. The temperature dependence of the integrated magnetic Bragg intensity at

$$|\boldsymbol{Q}_i| = |\boldsymbol{Q}_1| = |\boldsymbol{Q}_1 - \boldsymbol{q}_2| = |\boldsymbol{Q}_1 + \boldsymbol{q}_3| \tag{7.35}$$

presented in the inset (Fig. 7.28) shows a minimum at ~ 1.2 K (consistent with the peak in $\chi'(T)$) and sharply decreases at higher temperatures. The formation of the $(2\sqrt{3} \times 2\sqrt{3})$ superlattice may be due to a small interplanar exchange coupling J' or due to next-nearest-neighbor in-plane interactions.

Detailed information about the magnetically ordered phase is also provided by ac susceptibility and static magnetization measurements of *Kimishima* et al. [7.59, 152], where the ordering temperature is identified as a peak in $\chi'(T)$, yielding 1.30 K for stage 1 and 1.16 K for stage 2. Similar behavior in $\chi'(T)$ is observed for both the stage-1 and stage-2 compounds, suggesting that the $\chi'(T)$ is only weakly affected by interplanar interactions J'. Application of a magnetic field reduces the magnitude of $\chi'(T_p)$ and reduces the susceptibility peak temperature T_p. These observations are inconsistent with the predictions of a frustrated triangular 120° antiferromagnetic three sublattice spin structure [7.255].

The field cooling remanent magnetization studies on the stage-1 $MnCl_2$ GICs [7.152] also indicate more complex magnetic interactions than simply the nearest-neighbor antiferromagnetic interactions, which would give rise to the 120° frustrated spin lattice. No detailed picture of the spin structure has thus far been proposed for the magnetically ordered $(2\sqrt{3} \times 2\sqrt{3})$ spin superlattice, nor have the $\chi'(T)$ and magnetization experiments been interpreted in terms of this more complex spin structure. Interestingly, a complex spin structure also occurs in pristine $MnCl_2$ [7.58].

(d) $FeCl_3$ GICs

The $FeCl_3$ GICs have been studied in some detail by a variety of techniques and have been considered for various practical applications because of the ability to synthesize the $FeCl_3$ GICs for a range of stages and because of their relatively good chemical stability. Structural studies show that for high-stage compounds (e.g., stage 4) the magnetic layers become uncorrelated structurally, resulting in a random stacking of unit cells of length I_c [7.256]. Electronically, the free Fe^{3+} ion has a half-filled d shell ($3d^5$) implying a ^6S ground state and spin $S = 5/2$. Thus one might expect high-stage $FeCl_3$ GICs to be candidates for 2D-Heisenberg behavior.

The magnetic properties of $FeCl_3$ GICs have been described by the following Hamiltonian [7.257]:

$$\mathscr{H}_{\mathrm{FeCl_3}} = -2J \sum_{i<j} \boldsymbol{S}_j \cdot \boldsymbol{S}_j + D \sum_i (S_i^z)^2 \, . \tag{7.36}$$

This is similar (except for the sign of J) to that for $NiCl_2$ (7.7) (Sect. 7.2.1a).

Interestingly, the magnetic properties of the $FeCl_3$ GICs show relatively little anisotropy, and this small anisotropy is attributed to crystal-field effects [7.235]. From the measurements of the anisotropy in the magnetic susceptibility, values of $D = 0.13$ K and 0.23 K were obtained for stage-1 and stage-2 $FeCl_3$ GICs, respectively [7.235]. The corresponding anisotropy fields of $|H_A| = 1.4$ kOe and 3.6 kOe were obtained for the stage-1 and stage-2 compounds, far greater than the interlayer exchange fields for these compounds [7.257].

Experimental magnetic studies on the $FeCl_3$ GICs were first carried out by *Karimov* et al. [7.19] on polycrystalline samples, and they reported a peak in the magnetic susceptibility for the stage-1 compound at 3.6 K and for the stage-2 compound at 8.5 K. They identified these peaks as the onset of antiferromagnetic ordering. Evidence for magnetic phase transitions was also provided by specific-heat measurements (Sect.7.3.3). [7.197, 199–201]. *Ohhashi* and *Tsujikawa* [7.257] found long-range order at finite temperatures in the $FeCl_3$GICs using Mössbauer spectroscopy [7.235], which they followed up with a detailed susceptibility study [7.257] on oriented samples, so that they could obtain information on the magnetic anisotropy of these compounds. Their analysis gave antiferromagnetic in-plane coupling, with $J/k_B = -0.41$ K for stage 1 and -0.35 K for stage 2, while $T_N = 3.9$ K and 3.6 K were reported for the stage-1 and stage-2 compounds, respectively. This contrasts with the ferromagnetic coupling reported by Karimov for the stage-2 compounds. Intercalation was found to give rise to a trigonal distortion of the Cl^- ions from ideal octahedral symmetry around the Fe^{3+} ions [7.235]. In addition to the phase transitions reported at higher temperatures, neutron scattering measurements on a stage-2 compound revealed a phase transition at 1.7 K, where a doubling of the in-plane magnetic unit cell was observed, and was identified with antiferromagnetic order [7.258].

Above ~ 25 K, good fits to Curie-Weiss behavior were found, so that the Curie-Weiss parameters could be determined (Table 7.8). Relatively little anisotropy was observed, and the g values were found to be close to 2. Direct comparison to the magnetic behavior of pristine anhydrous $FeCl_3$ could not be made because of the helical-type antiferromagnetic behavior of pristine $FeCl_3$ [7.55, 259, 260].

Another reason for the broad interest in the magnetic $FeCl_3$ GICs is the presence of the ^{57}Fe nucleus, the prototype nucleus for Mössbauer studies. Thus many groups have studied $FeCl_3$ GICs extensively by the Mössbauer spectroscopy technique [7.235–238, 261]. The most significant conclusion from the Mössbauer studies is that the $FeCl_3$ intercalate layer in GICs (basically a Heisenberg antiferromagnet) always contains significant amounts of $FeCl_2$ (an Ising antiferromagnet). The random nature of these mixed-valence $(FeCl_3)_x(FeCl_2)_y$ GICs suggests that the Hamiltonian will become much more complicated. For the case $y \ll x$, the leading term in (7.36) will have random Ising character with probability y/x, and will have Heisenberg character otherwise. This mixture of the trivalent and divalent magnetic species was found to lead to inhomogeneous samples, with magnetic properties dependent on the

Table 7.8. Comparison between the magnetic properties of stage-1 and stage-2 $FeCl_3$ GICs and pristine $FeCl_3$

Property	Pristine $FeCl_3$	Stage-1 $FeCl_3$ GIC	Stage-2 $FeCl_3$ GIC
T_N (K)	8.83[a]	3.9[b,c] 1.7[d]	3.6[b] 8.5[c] 1.7[d,e,f]
$g_{\|\|}$	—	1.99 ± 0.02[b]	1.99 ± 0.02[b]
$g_{\perp}$	—	2.03 ± 0.02[b]	2.03 ± 0.02[b]
$C_{\|\|}$ (emu K/mole)	4.0[g]	4.31 ± 0.10[b]	4.30 ± 0.10[b]
$C_{\perp}$ (emu K/mole)	4.0[g]	4.47 ± 0.07[b]	4.46 ± 0.10[b]
$\Theta_{\|\|}$ (K)	−14[g]	-11.4 ± 1.8[b]	-9.0 ± 2[b]
$\Theta_{\perp}$ (K)	−14[g]	-8.2 ± 0.8[b]	-6.0 ± 1[b]
J (K)	—	−0.41[b,h]	−0.35[b,h]
D (K)	—	0.13[i]	0.23[i]

[a] Ref. [7.55].
[b] Ref. [7.258].
[c] Ref. [7.19].
[d] Ref. [7.56].
[e] Ref. [7.259].
[f] Ref. [7.201].
[g] Powder measurements [7.259].
[h] Using high-temperature series expansion. The stage-1 and stage-2 samples showed magnetic ordering at $T_N = 3.9$ and 3.6 K.
[i] See (7.9) [7.257].

nature of the graphite host material [7.261], and variability in the magnetic properties of samples with the same nominal stage synthesized by different groups. Since the magnetic exchange interactions of Fe^{2+} and Fe^{3+} are quite different (Sect. 7.4.1e), the mixed-valence compound is thought to cause random fields. Several investigators have suggested that $FeCl_3$ GICs form a spin-glass phase [7.56, 262].

More recent susceptibility (both χ' and χ'') studies on $FeCl_3$ GICs have shown susceptibility peaks near 1.7 K, which have been investigated as a function of temperature T, magnetic field H, and stage n [7.56, 192]. Since the $FeCl_3$ GICs can be prepared for a range of stages, this system has been especially useful in showing that the peak susceptibility per magnetic ion increases significantly as the stage index increases [7.56], as expected from the increased quasi-2D behavior. A long spin-lattice relaxation time has been identified with the susceptibility anomaly near 1.7 K [7.192]. No quantitative modeling of this effect has yet been carried out [7.183]. At present, the relation between the susceptibility peaks in $\chi'(T)$ at 1.7 K [7.56, 192] and those near 3.6 K [7.19, 235] remains unclear, but may possibly be attributed to the formation of another phase when the GIC is exposed to water in the air. On the other hand, the random mixture of Fe^{2+} and Fe^{3+} ions may provide an excellent vehicle for study of the experimental properties of a magnetic spin glass, though a full experimental study of the expected frequency dependence of the cusp in $\chi'(T)$ has not yet been reported.

Magnetoresistance measurements on stage-1 $FeCl_3$ GICs have been reported by *Issi* [7.214, 253], who found a peak at 5.5 K that was not consistent with any Fermi-surface gaps.

(e) $FeCl_2$ GICs

The magnetic properties of pristine anhydrous $FeCl_2$ have been extensively studied experimentally [7.3, 39, 263–265] because they are so different from the other transition-metal chlorides. The magnetic ordering in anhydrous $FeCl_2$ consists of planes of spins that are arranged ferromagnetically within each plane of Fe ions, but neighboring planes of spins are stacked antiferromagnetically below $T_N = 23.55$ K, as also occurs in the case of pristine $NiCl_2$ and $CoCl_2$. However, in contrast to the other transition-metal chlorides, the spins in anhydrous $FeCl_2$ are oriented normal to the planes (Ising behavior), rather than in the planes (XY behavior), as in $NiCl_2$ and $CoCl_2$. Furthermore, in the presence of an applied magnetic field, metamagnetic behavior is observed, with a tricritical point at 11 kOe and 20.5 K [7.3]. The very weak interplanar coupling in the pristine compound (i.e., a field of 11 kOe is sufficient to overcome the antiferromagnetic interplanar coupling) suggests that the $FeCl_2$ GICs would be good candidates for the study of 2D-Ising behavior.

To date, very few studies have been carried out on the $FeCl_2$ GICs, and then only for stage-1 and stage-2 compounds. As is the case for many of the other transition-metal compounds, the in-plane structure of $FeCl_2$ GICs is closely related to that of pristine $FeCl_2$ with the same in-plane triangular lattice and lattice constants [7.266] (Table 7.2). The $FeCl_2$ GICs, however, exhibit a rotational locking angle of $\theta_{\text{epitaxy}} = 0°$ between the a axes for the graphene and the intercalate layers.

The Fe^{2+} ion is in a $3d^6$ state in the free ion with spin $S = 1$, and in a cubic crystal field the Fe^{2+} ion assumes a $^5T_{2g}$ ground state. Mössbauer experiments show that the Fe^{2+} ion is found on two sites [7.261, 266, 267], with two thirds of the sites exhibiting distortions from octahedral symmetry similar to pristine anhydrous $FeCl_2$. One third of the sites exhibit a different kind of distortion from octahedral symmetry. The crystal-field splittings of Fe^{2+} in $FeCl_2$ GICs have been modeled theoretically for these two types of sites [7.266] with some success in explaining the complex temperature-dependent Mössbauer spectra.

The magnetic properties of the $FeCl_2$ GICs have been investigated by susceptibility, magnetization [7.54, 266], and Mössbauer spectroscopy [7.266, 267] studies. In both the paramagnetic and magnetically ordered phases, surprisingly little anisotropy is observed in the g factors (3.85), which differ significantly from 2, and in the Curie-Weiss parameters ($\Theta = 16.6$ K), and the saturation magnetization (5.5 μ_B). Relatively little hysteresis is observed in magnetization loop experiments (coercive force of ≈ 1 kOe) [7.266]. Nevertheless, the similarity in T_N and other magnetic parameters between stage 1 and stage 2 suggests very weak interplanar exchange coupling in the $FeCl_2$ GICs, consistent also with the similarity between the magnetic ordering temperature ($T_N = 15.5$ K) and the Curie-Weiss temperature parameter ($\Theta = 16.6$ K). This indicates a dominance of the strong ferromagnetic in-plane interaction over the weak antiferromagnetic interplanar interaction. The magnetic properties of $FeCl_2$ GICs are thus representative of spins with 2D spatial dimensionality and 1D spin dimensionality (the 2D-Ising model).

(f) $CuCl_2$ GICs

The magnetic properties of $CuCl_2$ GICs have been of special interest for several reasons. The synthesis conditions for $CuCl_2$ GICs are relatively favorable in terms of intercalation temperature, intercalation time, and the ability to prepare well-staged samples for several stages. Secondly, with a spin $S = 1/2$ for the Cu^{2+} ion, and an antiferromagnetic nearest-neighbor Heisenberg interaction for the pristine $CuCl_2$, the $CuCl_2$ GICs would be expected to have interesting magnetic properties. Since $CuCl_2$ GICs and $FeCl_3$ GICs have similar Hamiltonians, similarities in magnetic properties are expected. However, some differences in magnetic behavior for the $CuCl_2$ GICs are also expected because of their lower crystal symmetry, which can be described as a distorted triangular in-plane structure with nearest-neighbor Cu–Cu distances of 3.30 Å along the *b* axis (or chain axis) and next-nearest-neighbor Cu–Cu distances of 3.80 Å between two chains in the *bc* plane (which becomes the intercalate layer in the GIC) [7.104]. The Cu–Cu distance linking chains outside the layer plane is much larger, so that the magnitudes of the exchange interactions corresponding to third and fourth neighbor distances in the pristine $CuCl_2$ are much reduced. The chain-like structure of the Cu^{2+} ions along the *b* axis may stabilize an antiferromagnetic interaction in this pseudohexagonal (or pseudotriangular) structure. Although the in-plane structures for pristine $CuCl_2$ and its GICs differ in detail, both are distorted triangular structures with Cu–Cu nearest-neighbor and next-nearest-neighbor distances that differ by less than 1.3% [7.91, 104, 268]. X-ray diffraction and high-resolution TEM studies show a significant density of defects and staging imperfections in $CuCl_2$ GICs [7.268].

With regard to pristine $CuCl_2$, the ordering to an antiferromagnetic state below $T_N = 23.9$ K is observed by both specific-heat [7.52, 53] and magnetization and susceptibility measurements [7.269]. Fits of the susceptibility measurements $\chi'(T)$ to high-temperature series expansions yield a value for the antiferromagnetic exchange interaction for pristine $CuCl_2$ of $J = 37$ K [7.269]. Of special interest regarding the magnetic properties of pristine $CuCl_2$ is a spin-flop transition observed at 4.2 K in the magnetization $M(H)$ with $\boldsymbol{H} \parallel \boldsymbol{b}$ axis at a spin-flop field of $H_{SF} = 40.6$ kOe (but not for H along either of the two mutually orthogonal directions) [7.270] and corresponding to a jump in magnetization of 65 emu/mole at H_{SF}. The anisotropy of the spin-flop effect suggests spin alignment along the *b* axis, and yields values of the very small anisotropy field $H_A \sim 840$ Oe, and of the saturation magnetization $M_S \sim 3.2$ kOe, based on the measured exchange field $\sim 10^6$ Oe (or $J \sim 37$ K) [7.270].

Magnetic measurements of $CuCl_2$ GICs have been mainly limited to SQUID magnetometer magnetization and dc susceptibility measurements on stage-2 $CuCl_2$ GICs [7.104, 271]. The results obtained by the two groups are generally consistent and complementary. Both groups have reported a temperature dependence of the magnetic susceptibility of the form

$$\chi_{dc}(T) = \frac{C}{T - \Theta} + \chi_0(T) , \qquad (7.37)$$

where the first term is dominant at low temperature (Θ is very small, $\Theta < 1$ K) and $\chi_0(T)$ is dominant at higher temperatures ($T \geqslant 30$ K). The first term has been attributed to finite-size spin chain effects associated with odd numbers of spins [7.104]. After subtraction of the $C/(T - \Theta)$ term from $\chi_{dc}(T)$, the bulk spin contribution $\chi_0(T)$ has been obtained [7.104, 271]. A remarkable similarity of $\chi_{dc}(T)$ for the GICs to $\chi_{dc}(T)$ for pristine $CuCl_2$ [7.269] was observed for all samples studied [7.104, 271], with $\chi_{dc}(T)$ exhibiting a broad peak near $T_{max} \simeq 75$ K, associated with short-range spin order along the chains. A decrease in the absolute value of $\chi_0(T)$ was found with increasing stage, consistent with the smaller concentration of the magnetic species [7.271]. Anisotropy effects were reported in the stage-2 compound both for T_{max} (with $T_{max\,\|} = 66$ K and $T_{max\,\perp} = 72.1$ K) and for the coefficient C in (7.37) with $[(C_\perp - C_\|)/(C_\perp + C_\|)] = 0.085$ [7.104]. Here the subscripts $\perp$ and $\|$ refer, respectively, to the angle between H and the normal to the intercalate layers. Interestingly, the anisotropy effect of $\chi_0(T)$ in the stage-2 $CuCl_2$ GICs was found to change sign as a function of temperature, with $(\chi_{0\perp} - \chi_{0\|}) < 0$ at low temperature so that the easy axis is in the intercalate plane. At higher temperatures (>16 K), a sign change in $(\chi_{0\perp} - \chi_{0\|})$ was observed so that the easy axis switches to a direction normal to the layer planes [7.104].

While low-temperature anomalies in $\chi_{dc}(T)$ were reported for both stage-1 $CuCl_2$ GICs [7.271] and the stage-2 compound [7.104], no claims were made for a magnetically ordered state at low temperature, largely because of the uncertainties in the subtraction process of the $C/(T - \Theta)$ term in (7.37) and because of artifacts introduced by finite spin chain effects [7.104]. *Rancourt* et al. [7.104] further concluded that "islands" of ~ 100 Å in diameter (Sect. 7.1.3a) with a distribution of finite quasi-Heisenberg antiferromagnetic chains of Cu^{2+} ions inhibited 3D order (down to 1.9 K) because of spin frustration. No spin-flop magnetic transitions (as observed in pristine $CuCl_2$) were reported for the $CuCl_2$ GICs.

Although no definitive reports of magnetic ordering in $CuCl_2$ GICs have yet been made, the system merits further magnetic study, especially by techniques such as neutron scattering, which could provide more definitive information on the spin structure, both in pristine $CuCl_2$ and in the associated graphite intercalation compounds.

(g) $CrCl_3$ GICs

Because of the difficulty in preparing $CrCl_3$ GICs, relatively little work has been done with this compound. Structural studies show that the $CrCl_3$ GICs can be prepared as a stage-3 compound [7.94, 135] and as a mixed stage-2/stage-3 material [7.94]. The in-plane structure is very similar to that of the $FeCl_3$ GICs (Sect. 7.1.3a), and the c axis intercalate stacking sequence has not yet been determined unambiguously [7.94]. Table 7.3 summarizes the structural parameters of the $CrCl_3$ GICs.

The magnetic properties of pristine $CrCl_3$ determined by neutron diffraction techniques [7.272] described by the Hamiltonian (7.9) are similar to those of pristine $CoCl_2$ and $NiCl_2$, insofar as the spins lie in the layer planes (XY model), where they couple ferromagnetically, while the layers of spins are stacked antiferromagnetically, exhibiting a magnetic ordering temperature of 17 K. Values for the parameters in (7.9) are $zJ = 13.56$ K and $z'J' = -0.065$ K, while the anisotropy is given by $DS/g\mu_B = -2000$ Oe [7.50]. The ability to prepare stage-3 $CrCl_3$ GICs is therefore of considerable interest for the study of 2D-XY magnetic behavior.

Only a few magnetic measurements have been made on the $CrCl_3$ GICs, and these are generally consistent with the magnetic behavior of the pristine $CrCl_3$. For example, high-temperature susceptibility measurements on the intercalation compounds [7.135] show Curie-Weiss behavior with appropriately reduced values of the Curie-Weiss parameters C and Θ as compared with pristine $CrCl_3$ [7.273] (Table 7.9). The value for the magnetic moment of the Cr^{3+} ion in the intercalation compound is 3.60 μ_B, as compared with 3.85 μ_B in the free ion. Electron-spin-resonance measurements show a Dysonian line with an A/B ratio of ~ 1.5 as compared to ~ 6 in pristine graphite; this is also consistent with the more metallic behavior of the $CrCl_3$ GICs. The ESR measurements yield g values that are close to 2 and are nearly isotropic down to 50 K [7.135], but below ~ 30 K, $g_\perp$ increases rapidly while $g_\parallel$ decreases rapidly, similar to the behavior for $NiCl_2$ GICs. The rapid increase in the ESR linewidths $\Delta H_\parallel$ and $\Delta H_\perp$ below ~ 30 K in the $CrCl_3$ GICs is also similar to that for the $NiCl_2$ GICs [7.135]. The overall similarity in the observed magnetic properties of the $CrCl_3$ GICs to those of the $NiCl_2$ GICs suggests a similar form for the magnetic Hamiltonian (7.7). Magnetization measurements for both $\boldsymbol{H} \parallel \hat{c}$ and $\boldsymbol{H} \perp \hat{c}$ show a phase transition to a magnetically ordered state below ~ 10 K. Large hysteresis effects are observed upon cooling in a small magnetic field (5 Oe), suggesting spin-glass behavior below ~ 10 K [7.135]. Whether this spin-glass behavior is intrinsic or due to structural sample inhomogeneities awaits further detailed study.

(h) $MoCl_5$ GICs

The $MoCl_5$ GICs have the advantage that they can be prepared for a variety of stages [7.60, 277], though their hexagonal in-plane structure is complex, displaying the basic dimer structure of the pristine compound [7.98]. The free Mo^{5+} ion has a $(4d^1)$ configuration giving rise to $S = 1/2$. High-temperature susceptibility measurements show ferromagnetic intraplanar interaction with a Curie-Weiss temperature of $\Theta \sim 15$ K and $J \sim 5$ K, and the interplanar interaction was identified as weak and antiferromagnetic [7.60, 277]. ESR [7.60]

Table 7.9. Magnetic parameters for some other layered compounds

GIC	stage[a]	Θ (K)[b]	g[b]	T_N[c] (K)	J (K)
MnF_2	0	− 113[d]	—	67.3[e]	− 1.76[e]
FeF_2	0	− 117[d]	—	79[d]	− 2.69[e]
CoF_2	0	− 52.7[d]	—	37.7[d]	—
NiF_2	0	− 116[d]	—	73.2[d]	—
OsF_6	0	~ 0[f]	—[f]	—	—
OsF_6	1–5	∥ − 40, ⊥ − 10[g]	∥ 1.65, ⊥ 3.25[g]	—	—
TcF_6	1, 2	~ 0[h]	—[h]	—	—
$MoCl_5$	0	15[i]	1.98	22[i]	5.0
$MoCl_5$	2–5	15[i]	∥ 1.987, ⊥ 1.979[i]	T_{cu} = 1.63[i]	5.0
$CrCl_2$[d]	0	− 149[d]	—	40	—
$CrCl_3$[j]	0	− 31	—	16.8	5.25[e]
$CrCl_3$	2–3	24[k]	—	~ 10	—
$CuCl_2$	0	− 109[d]	—	23.9[d]	4.52[l]
$CuCl_2$	1, 2	< 1[m]	—[m]	—	—
$FeCl_2$	0	48[d]	4.1[e]	23.6[q]	3.9[e]
$FeCl_2$	1, 2	∥ 16, ⊥ 14[n]	∥ 3.85, ⊥ 3.85[n]	15.5[o]	7.9[n]
$CoBr_2$	0	− 20[p]	∥ 3.06, ⊥ 4.91[q]	19[p]	—
$CuBr_2$	0	− 246[d]	—	193[d]	—
$CrBr_3$[r]	0	37	—	T_c = 32.7	8.25[e]
CoI_2[s]	0	—	∥ 3.3, ⊥ 4.69[q]	8[s]	—
CrI_3[r]	0	70	—	T_c = 16.8	13.5[e]

[a] Stage "0" denotes pristine material.
[b] Measurements on powder samples unless noted.
[c] Antiferromagnetic Néel temperature except as noted.
[d] Ref. [7.274]. [e] Ref. [7.31]. [f] Ref. [7.228].
[g] HOPG host material [7.221].
[h] Ref. [7.226].
[i] Ref. [7.60, 70]. T_{cl} is stage dependent and is 0.69, 0.50, and 0.40 for stages 3, 4, and 5 respectively.
[j] Ref. [7.94]. [k] Ref. [7.51]. [l] Ref. [7.50]. [m] Ref. [7.271]. [n] Ref. [7.266].
[o] Ref. [7.54]. [p] Ref. [7.138]. [q] Ref. [7.142]. [r] Ref. [7.275]. [s] Ref. [7.276].

measurements indicated only a small anisotropy of the g factor ($g_\perp = 1.979$ and $g_\| = 1.987$), suggesting a Heisenberg Hamiltonian with a small XY anisotropy (7.7) [7.60].

Early susceptibility and magnetization measurements by *Zvarykina* et al. [7.277] using powder samples indicated that the stage-1 compound exhibited a high T_c value of 15 K, consistent with the high T_c value for the pristine compound ($T_c \sim 22$ K) [7.277]. More detailed magnetic-susceptibility measurements on kish-based compounds show that the higher-stage ($3 \leqslant n \leqslant 5$) compounds exhibit a much lower magnetic ordering temperature [7.60]. For the stage-3, stage-4, and stage-5 compounds, two peaks in $\chi'(T)$ were observed and identified with T_{cl} and T_{cu} with $T_{cu} = 1.60$ K for each of these compounds, while $T_{cl} = 0.69$ K for stage 3, decreasing to $T_{cl} = 0.40$ K for stage 5 extrapolating to $T_{cl} \sim 0$ K at stage 8. The small interplanar coupling was interpreted to imply the formation of a 2D-XY phase for the high-stage compounds [7.60]. For the stage-3 compound a finite magnetization was reported in the intermediate temperature region $T_{cl} < T < T_{cu}$, in contrast to ideal 2D-XY behavior [7.60]. This observation was attributed to finite-size effects.

Very recently bi-intercalation compounds based on stage-3 and stage-4 $MoCl_5$ GICs have been prepared by the insertion of layers of alkali metal K (Sect. 7.4.1k). Raman spectral shifts indicate that the graphite layers adjacent to the $MoCl_5$ intercalate layers are charged positively, while those adjacent to the potassium layers are charged negatively [7.278].

(i) Fluoride Compounds

The magnetic properties of the pristine transition-metal halides have been studied in some depth [7.279, 280]. The structures and Néel temperatures of a number of these compounds are listed in Table 7.10. The magnetic properties of the chlorides, bromides, and iodides of a given cation tend to be similar. The

Table 7.10. Néel temperatures of some pristine halides of transition metals

Compound	Structure	T_N (K)
CoF_2 [a]	rutile	37.7
CoF_3 [b]	perovskite	460[b]
$CoCl_2$ [c]	$CdCl_2$	24.7
$CoBr_2$ [c]	CdI_2	19[d]
CoI_2 [c]	CdI_2	12
CrF_2 [a]	rutile	53
$CrCl_2$ [a]	rutile	20
FeF_2 [a]	rutile	79[e]
FeF_3 [b]	perovskite	394[b]
$FeCl_2$ [c]	$CdCl_2$	23.6[f]
$FeBr_2$ [c]	CdI_2	11
FeI_2 [c]	CdI_2	10
MnF_2 [a]	rutile	72[e]
MnF_3 [b]	perovskite	43[b]
$MnCl_2$ [c]	$CdCl_2$	1.96
$MnBr_2$ [c]	CdI_2	2.16
MnI_2 [c]	CdI_2	3.40
NiF_2 [a]	rutile	73.2[e]
$NiCl_2$ [g]	$CdCl_2$	52
$NiBr_2$ [g]	CdI_2	54
NiI_2 [c]	$CdCl_2$	75

[a] Ref. [7.279].
[b] Ref. [7.281].
[c] Ref. [7.280].
[d] Ref. [7.138].
[e] Ref. [7.274].
[f] Ref. [7.3].
[g] Ref. [7.31].

fluoride compounds are also magnetic but show somewhat different crystal structures (mostly rutile, which is a body-centered tetragonal structure) and hence the magnetic couplings are quite different from the hexagonal structures of the other halides. The effect of intercalation would also be expected to be very different because the 2D square lattice is not compatible with the graphene unit cell.

The 4*d* and 5*d* halides generally form GICs, and many of these have been studied, at least from a preparative and structural standpoint [7.105]. To date, little work has been done on the magnetic properties of these fluorine-based GICs. The magnetic properties of the transition-metal hexafluoride compounds have, however, been reported by *Selig* [7.106], showing strong paramagnetic behavior, but no transitions to magnetically ordered states have been reported.

Some of the magnetic results and references are summarized in Table 7.9 for pristine-fluorine-based compounds, and in a few cases for their corresponding intercalation compounds.

(j) Bromide Compounds

Some of the pristine transition-metal bromides have interesting magnetic phases and transition temperatures comparable to those of the corresponding transition-metal chloride compounds. Many transition-metal bromide GICs have been synthesized [7.90], but have not yet been studied for their magnetic properties. Some of these are expected to be of interest with regard to low-dimensional magnetism. For example, pristine $CoBr_2$ can be described by the Lines Hamiltonian (7.8) [7.282], and the $CoBr_2$ GICs could very well be another system for observing 2D-XY behavior.

(k) Bi-Intercalation Compounds

A number of bi-intercalation compounds containing magnetic species have been synthesized (Table 7.11) [7.71]. Bi-intercalation compounds are generally prepared by intercalating the second intercalate species into a stage-2 (or higher-stage) compound of the species having the higher intercalation temperature. The attraction of the bi-intercalation compounds is the possibility of inserting a nonmagnetic species N between two magnetic intercalate layers M, thereby greatly expanding their separation in a typical sequence . . . CMCNC . . . where C is the graphene layer. In most cases, the two intercalate species are both acceptors or both donors, but in some cases alternate donor and acceptor intercalate layers have been prepared [7.71]. Of particular interest here is the bi-intercalation compound C–$CoCl_2$–C–Na–C, in which the Na layers are part of a low-stage compound, in contrast to the binary Na GICs, which can be prepared only as high-stage compounds ($n \geq 6$) [7.287]. (A binary GIC consists of a single intercalate species, in contrast to ternary compounds consisting of two distinct intercalate species.) In one case, the two intercalate species are different magnetic layers $CoCl_2$ and $FeCl_3$ [7.70]. In all cases, the repeat distance of the bi-intercalation compound is found by adding the layer thicknesses of the

Table 7.11. Graphite magnetic acceptor bi-intercalation compounds

Intercalate	I_c[a] (Å)	Reference
$YCl_3/FeCl_3$	—	[7.283]
$CoCl_2/AlCl_3$	19.05	[7.68, 188, 284]
$CoCl_2/GaCl_3$	19.14	[7.284]
$CoCl_2/FeCl_3$	18.60	[7.70]
$NiCl_2/FeCl_3$	19.00	—[b]
$FeCl_3/AlCl_3$	18.88	[7.284]
$FeCl_3/GaCl_3$	18.93	[7.284]
$InCl_3/FeCl_3$	18.94	[7.68]
$CoCl_2/Na$	~23.0[c]	[7.285]
$CoCl_2/K$	~14.7	[7.69, 285]
$MoCl_5/K$	21.35[d]	[7.278]
$MoCl_5/K$	23.32[e]	[7.278]
$CuCl_2/ICl$	16.52[f]	[7.286]
$CuCl_2/ICl$	23.64[g]	[7.286]

[a] Unless specified, the compounds are stage-1 with a sequence C*M*C*N*C, where C denotes a graphene layer, *M* the magnetic species, and *N* the nonmagnetic species.
[b] Mentioned in Ref. [7.284].
[c] Corresponds to a stage-3 compound with a sequence C*M*C*N*C*N*C*N*C.
[d] Corresponds to a sequence C*M*C*N*CCC.
[e] Corresponds to a sequence C*M*C*N*CC*N*C.
[f] Corresponds to a formula $C_{10}CuCl_2(ICl)_{0.6}$ in a sequence CMC*N*C.
[g] Corresponds to a formula $C_{15}CuCl_2(ICl)_{1.2}$ in a sequence CMC*N*C*N*C.

constituents. It is further found that the second intercalated species does not occupy the unoccupied sites of the first intercalated species in cases where the filling of the intercalate layers is incomplete. Thus bi-intercalation gives rise to layered compounds where the two intercalate layers are truly distinct.

Of particular significance is the possibility of preparing higher-stage bi-intercalation compounds, such as have been prepared with alternate $MoCl_5$ and potassium intercalate layers from stage-3 and stage-4 $MoCl_5$ GICs [7.278], following the sequence C*M*C*N*CC for the stage-3, and C*M*C*N*CCC and C*M*C*N*CC*N*C for the stage-4 compounds. According to $(0, 0, l)$ X-ray diffraction scans and Raman scattering experiments, the $MoCl_5/K$ bi-intercalation compounds show good staging fidelity. Since these structures have graphene planes surrounded by either magnetic or nonmagnetic intercalate layers, Raman spectroscopy could be used to identify whether excess holes or electrons have been transferred to the graphene layers (Sect. 7.4.1h).

Few magnetic measurements have thus far been made on magnetic bi-intercalation compounds. Preliminary experiments on a $CoCl_2/AlCl_3$ bi-intercalation compound suggest that both T_{cl} and T_{cu} are lowered from 8.0 K and

9.1 K, respectively, in the stage-2 $CoCl_2$ GIC, to 5 K and 6 K in the $CoCl_2/AlCl_3$ bi-intercalation compound [7.284]. However, later work (Fig. 7.11b) showed little shift of these critical temperatures ($T_{cl} = 7.4$ K and $T_{cu} = 9.2$ K) [7.188]. In addition to a lowering of T_{cl} and T_{cu}, the magnitude of the susceptibility per unit magnetic species was found to increase in the bi-intercalation compound relative to the stage-2 compound, consistent with enhanced two-dimensionality through bi-intercalation [7.188]. For the case of the $CoCl_2/FeCl_3$ bi-intercalation compound, T_{cl} and T_{cu} were found to be almost the same as for the stage-2 $CoCl_2$ GIC (with $T_{cl} = 8.0$ K and $T_{cu} = 8.8$ K) [7.70], even though the interplanar distance between the Co^{2+} ions was increased from 12.65 Å to 18.60 Å. Likewise, the rapid suppression of the peak in $\chi'(T)$ by small magnetic fields is also observed in the bi-intercalation compound, though more complicated behavior attributed to the presence of the $FeCl_3$ layer is observed in the $\chi'(H)$ curves in the vicinity of 150 Oe for various constant-temperature sweeps [7.70]. Study of a bi-intercalation compound formed by the intercalation of $FeCl_3$ into a stage-2 $NiCl_2$ GIC showed the $NiCl_2$ layers to order magnetically at 20.5 K, similar to the behavior of the binary $NiCl_2$ GICs, while the $FeCl_3$ layers were not found to be ordered down to 4.2 K [7.288].

(l) Magnetic Alloys and Dilution Compounds

In prior sections, each magnetic layer nominally consisted of a single magnetic species. In this section, we discuss two generalizations of the magnetic layer. The first generalization is to a dilution compound where a nonmagnetic species is mixed substitutionally into the intercalate layer; an example of such a system is stage-1 $Co_xMg_{1-x}Cl_2$ GICs. The second generalization is to a magnetic alloy where two magnetic species are present in each magnetic layer; an example of such a compound is stage-2 $Co_xNi_{1-x}Cl_2$ GICs.

Of the two types of substitutional alloys, the stage-1 dilution compounds (2D random magnetic sites) are simpler to understand. Here we would expect that the substitution of nonmagnetic ions for magnetic ions would lower T_N and would lower the field H_t required to overcome the interplanar antiferromagnetic coupling, and these effects are indeed observed [7.213]. Due to the similarity of the structure of pristine $MgCl_2$ and $CoCl_2$ and the similarity of the ionic size of Mg^{2+} and Co^{2+}, it is expected that the crystal field experienced by the Co^{2+} ions in stage-1 $Co_xMg_{1-x}Cl_2$ GICs will be the same as that experienced in the pristine $CoCl_2$ GICs. Therefore, the form of the Hamiltonian (7.8) and the exchange constant $J_{(Co-Co)}$ between in-plane nearest-neighbor Co^{2+} ions will be the same in the dilute compound, though there will be a decrease in the average number of nearest-neighbor Co^{2+} ions as the dilution is increased. The disappearance of long-range magnetic order is identified with a percolation threshold for the magnetic Co^{2+} ion network.

The critical percolation concentration x_p has been interpreted in a manner that sheds light on the microstructure of the intercalate lattice in transition-metal chloride acceptor GICs, in the following way. In all studies of magnetic acceptor GICs, the distribution of the voids (estimated to be 15% of the volume of the intercalate gallery that is not filled with intercalate) has been controversial (Sect. 7.4.1d). With this in mind, recent studies of a magnetic/nonmagnetic random-alloy GIC, stage-1 $Co_xMg_{1-x}Cl_2$ GICs, have revealed percolation behavior consistent with the structural results (Sect. 7.1.3a), indicating the presence of a connected network of intercalate.

An example where a nonmagnetic species has been introduced to promote the intercalation of magnetic metal chlorides is the replacement intercalation of $CoCl_2$ into $AlCl_3$ GICs [7.289]. In this case, the replacement was found to be incomplete, giving rise to segregated magnetic $CoCl_2$ islands together with nonmagnetic $AlCl_3$ regions [7.198, 247], thereby greatly complicating the analysis of the magnetic properties of $CoCl_2$ GICs.

With regard to the magnetic substitutional alloys where a magnetic dopant species is substituted for the magnetic species, two cases have been considered. In the first, Ni was substituted for Co in stage-2 $Co_xNi_{1-x}Cl_2$ GICs, where each of the constituents has magnetic exchange interactions of the same sign but of different magnitudes. This provided an example of a 2D random magnetic alloy [7.290]. With regard to this magnetic alloy, both the pure stage-2 $NiCl_2$ GICs and stage-2 $CoCl_2$ GICs exhibit ferromagnetic *XY* behavior, and the experimental Curie-Weiss parameters of stage-2 $Co_xNi_{1-x}Cl_2$ GICs agree with those derived from a molecular-field approximation, assuming a random magnetic mixture of two magnetic ions. The ferromagnetic intraplanar coupling J_{Co-Ni} between the Co^{2+} and Ni^{2+} ions is determined to be

$$J_{Co-Ni} \simeq 1.28\sqrt{J_{Co-Co}J_{Ni-Ni}}\,, \tag{7.38}$$

where J_{Ni-Ni} and J_{Co-Co} are the J values found from analysis of dc susceptibility measurements of the pure stage-2 compounds, and the results are given in Tables 7.4, 5.

In the second case of a magnetic substitutional alloy, the two magnetic species have different signs for their exchange interactions. For example, in the stage-2 random-alloy $Co_xMn_{1-x}Cl_2$ GICs, the nearest-neighbor in-plane coupling constant between two nearest-neighbor Mn^{2+} ions is opposite in sign to that between two Co^{2+} ions, and interesting magnetic effects are observed in this case [7.291]. For example, as a function of x the Curie-Weiss temperature Θ changes sign (positive for large x and negative for small x) [7.291].

Another example of a magnetic substitutional dopant is the magnetic alloy stage-2 $Co_xFe_{1-x}Cl_2$ GICs, where the magnetic behavior is dominated by the stronger Co–Co exchange interaction [7.292]. A fourth example of a magnetic substitutional alloy is the $(FeCl_3)_x(FeCl_2)_y$ -GIC system that is unintentionally formed in the preparation of $FeCl_3$ GICs (Sect. 7.4.1d). In such cases, spin-glass behavior is expected and has been reported [7.56, 262].

7.4.2 Overview of Donors

The magnetic graphite donor intercalation compounds are generally better understood than the acceptor compounds because the magnetic interactions are stronger and therefore easier to measure, and because the experiments in donors have thus far only been carried out on anisotropic 3D systems, where much of the physics had been previously studied in other anisotropic magnetic metals. Almost all of the published work on magnetic donor GICs relates to the first-stage Eu GIC with composition C_6Eu (Sect. 7.4.2a). There has been special experimental theoretical interest in C_6Eu arising from the large spin ($S = 7/2$) of Eu^{2+} and the consequent strong magnetic interactions, giving rise to higher-order magnetic phenomena that have thus far been observed only in this compound (Sect. 7.2.2).

(a) The Magnetic Donor C_6Eu

Because of their relatively strong interlayer coupling, the stage-1 Eu GICs exhibit 3D-XY magnetic behavior. Since the Eu atoms in C_6Eu have a commensurate $(\sqrt{3} \times \sqrt{3})R30°$ in-plane structure, the nearest-neighbor and next-nearest-neighbor Eu–Eu distances are very different in the stage-1 Eu GIC as compared with metallic Eu. Therefore, the in-plane and c-axis exchange interactions are also quite different. Thus the determinations of the exchange interactions in the magnetic donor compound C_6Eu requires a different approach from that for stage-1 $CoCl_2$ GIC, where the in-plane Co–Co distances remain essentially unchanged upon intercalation. It is known from Mössbauer experiments [7.156] (Sect. 7.3.9) that the europium in the GIC is the Eu^{2+} ion with a half-filled f shell ($S = 7/2$) and that below the Néel temperature T_N the in-plane Eu spins order antiferromagnetically.

From the relatively large negative Mössbauer isomer shift of Eu^{2+} in C_6Eu it is concluded that most of the valence charge on the Eu is transferred to the graphite, and that the Eu and the carbon π conduction bands are strongly hybridized [7.156]. The large magnitude of V_{zz} in the electric-field-gradient tensor is consistent with a large structural anisotropy relative to other compounds containing Eu^{2+} ions. Since the electronic states near the Fermi level on the Eu layer are itinerant, the dominant mechanism for the exchange interaction in C_6Eu is of the RKKY form.

Spin disorder scattering plays a major role in the transport properties at room temperature and below, so that the room-temperature in-plane resistivity of C_6Eu is found to be 50 $\mu\Omega$ cm [7.119], which is large compared with other first-stage nonmagnetic GICs with similar in-plane crystal structures [7.65]. As the temperature is lowered (Fig. 7.29), the resistivity decreases, showing an anomaly at T_N, followed by a large decrease in resistivity [7.119], so that at low temperatures where the spins become antiferromagnetically ordered (Fig. 7.30a) the resistivity is comparable to that of other first-stage donor GICs (~ 3 $\mu\Omega$ cm at 4.2 K) [7.65].

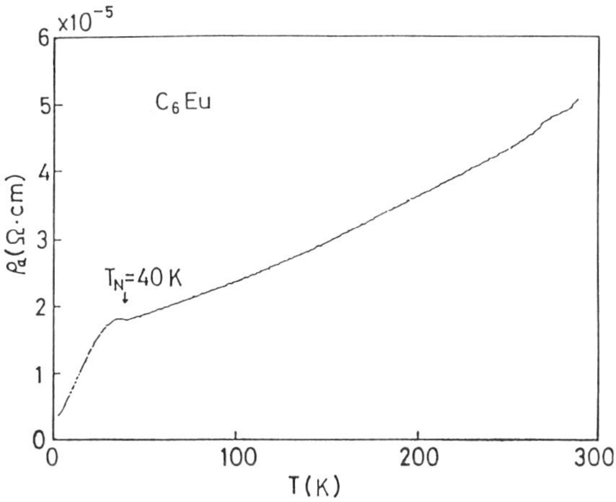

Fig. 7.29. Plot of the in-plane resistivity $\varrho_a(T)$ for C_6Eu over a wide temperature range showing the quenching of the spin disorder scattering below T_N [7.119]

Near room temperature C_6Eu exhibits paramagnetic behavior with anisotropic Curie-Weiss constants (Table 7.12 [7.119]). As the temperature is lowered in zero magnetic field, the susceptibility of C_6Eu shows only weak structure at the magnetic ordering temperature of 40 K, characteristic of antiferromagnetic systems, while the specific heat shows a strong anomaly at 40.0 ± 1.0 K. The occurrence of a magnetic phase transition at 40 K is also confirmed by direct magnetization [7.119] and transport [7.157, 160] measurements.

Electron-spin-resonance measurements [7.218] have been performed in the temperature range 10–300 K for $\boldsymbol{H} \perp \hat{c}$ and $\boldsymbol{H} \parallel \hat{c}$, and these experiments reveal highly anisotropic behavior at low temperature. For $T > T_N$, the resonance field H_0 for $\boldsymbol{H} \perp \hat{c}$ shifts to lower fields with decreasing temperature and to higher fields for $\boldsymbol{H} \parallel \hat{c}$. For $T < T_N$ three different resonance branches are observed (i.e., the g value increases for $\boldsymbol{H} \perp \hat{c}$ and decreases for $\boldsymbol{H} \parallel \hat{c}$ as T decreases). This anisotropic behavior in the g factor can be understood qualitatively in terms of the anisotropic susceptibility given by (7.29), confirming the highly anisotropic nature of the spins in C_6Eu. The ESR spectra below T_N are very complex, and it has not yet been possible to explain these spectra in detail (Sect. 7.3.7).

Table 7.12. Magnetic parameters for C_6Eu [7.158]

	$\boldsymbol{H} \perp \hat{c}$	$\boldsymbol{H} \parallel \hat{c}$
Néel temperature T_N (K)	40.0 ± 1.0	40.0 ± 1.0
Curie-Weiss constant Θ (K)	1.0 ± 5.0	3.0 ± 5.0
Molar Curie constant C (emu K/mole)	$7.2 + 0.1$	5.6 ± 0.1
Effective paramagnetic moment (μ_B)	7.6	6.7
g factor at 150 K (from ESR)	1.94	1.92
Saturation moment at $H = 30$ T (μ_B)	6.5	6.0

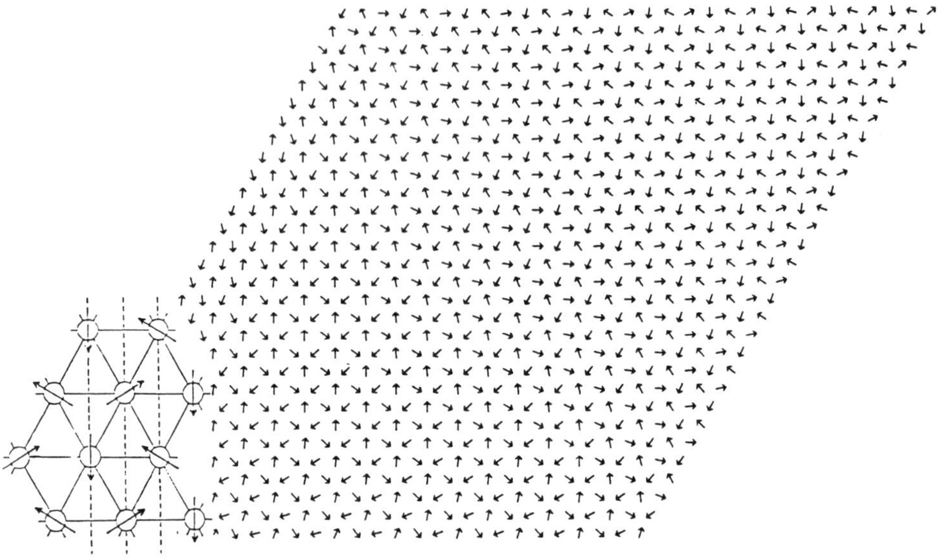

(a)

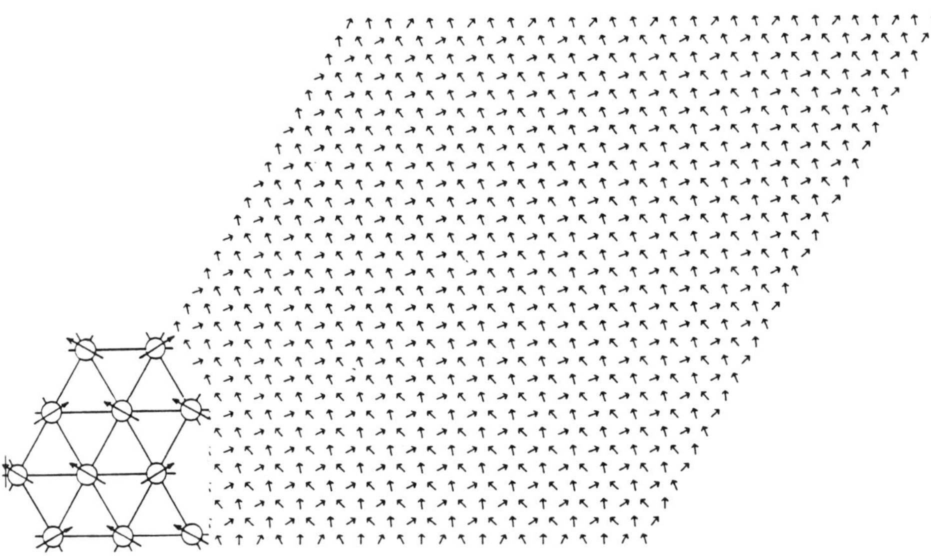

(c)

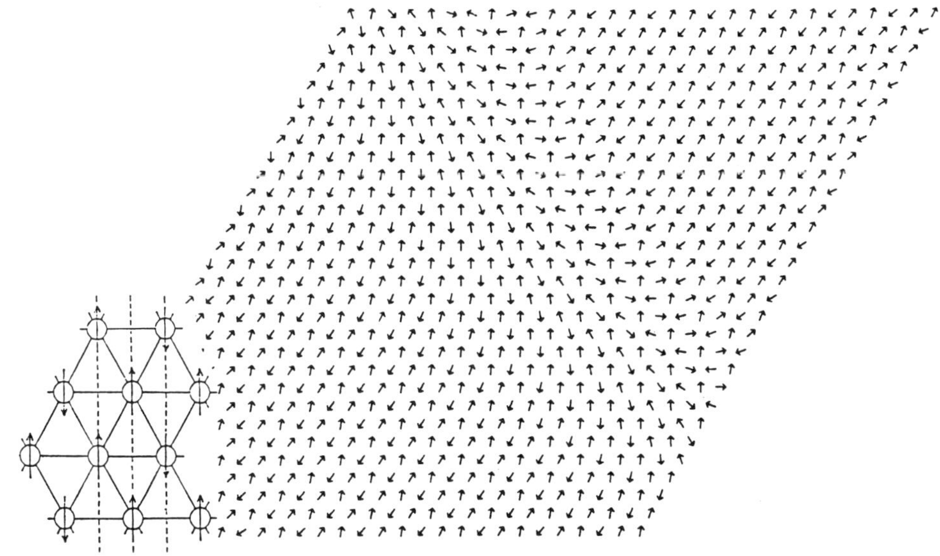

(b)

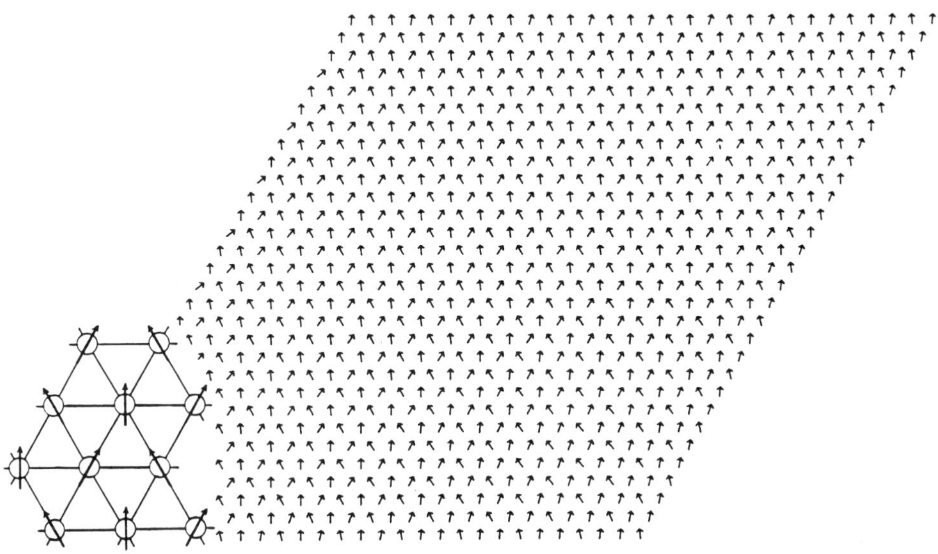

(d)

Fig. 7.30. Spin arrangements of a plane of 30 × 30 spins in C_6Eu for the various magnetic phases obtained from a Monte Carlo simulation: (**a**) triangular spin frustrated phase (Δ state) at $H = 0$, (**b**) ferrimagnetic (three-sublattice) phase at $H = 3T$, (**c**) canted fan 120° phase at $H = 13T$, (**d**) canted fan 60° phase at $H = 17$ T. The spin aligned paramagnetic phase where all the spins are aligned along the magnetic field is not shown [7.160]

The most remarkable behavior of C_6Eu is found in the magnetic-field-dependent magnetization measurements. The observation of four distinct magnetic phases [7.161] (Fig. 7.31) in the low-temperature magnetization curves taken for various temperatures below T_N in fields up to 40 tesla ($\boldsymbol{H} \perp \hat{\boldsymbol{c}}$) led to a great deal of interest in this material. At low magnetic fields ($\boldsymbol{H} \perp \hat{\boldsymbol{c}}$) and below T_N, a large susceptibility is observed in the magnetization measurements, which was readily identified with a triangular spin ordering in the xy plane (Fig. 7.30a) [7.161]. The most curious region in the magnetic phase diagram for C_6Eu (Fig. 7.32) occurs in the applied magnetic field range between $2.2 \leqslant H_0 \leqslant 8.2$ tesla, where the magnetization (Fig. 7.31) remains roughly constant at a value of about one third of the moment of a Eu^{2+} ion (2.2 to 2.7 μ_B per Eu ion). This value of the magnetization is consistent with a ferrimagnetic spin ordering with two spin-up sublattices and one spin-down planar sublattice (Fig. 7.30b).

At fields above 8.2 tesla, a third magnetic phase is found, exhibiting a linear dependence of the magnetization on field ($8.2 \leqslant H \leqslant 20.5$ tesla), and this magnetic phase is identified with a canted (120°) spin arrangement (Fig. 7.30c). Finally, above 20.5 tesla, the magnetization saturates at a value (6.2 to 6.7 μ_B),

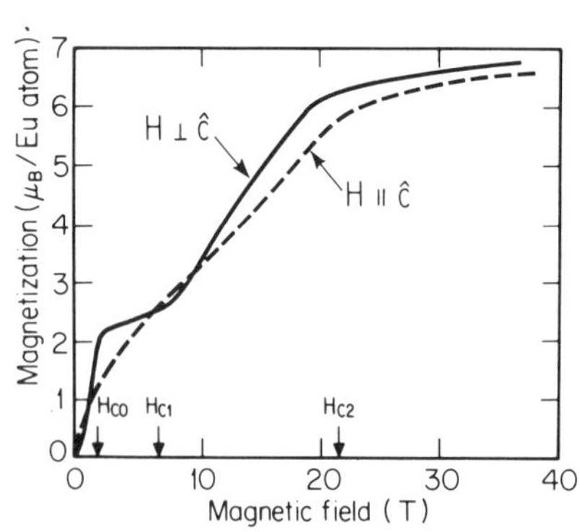

Fig. 7.31. Magnetic field dependence of the magnetization for C_6Eu at 4.2 K [7.119, 145]. Phase transitions occur at H_{c0} (from the triangular spin frustrated phase to the three-sublattice ferrimagnetic phase), at H_{c1} to a canted phase, and finally at H_{c2} to a spin aligned paramagnetic phase. No structure is seen in the $M(H)$ plot at H_{c12} denoting the transition from a canted fan 120° phase to a 60° phase. Results for the magnetic field parallel and perpendicular to the c axis are presented

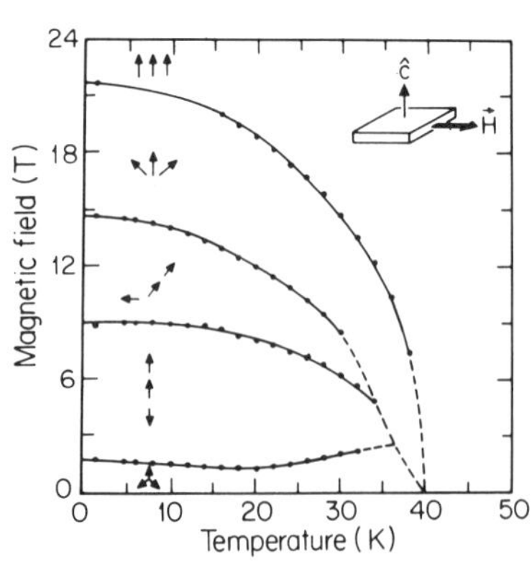

Fig. 7.32. Magnetic phase diagram of C_6Eu from magnetoresistance data for $\boldsymbol{H} \perp \hat{\boldsymbol{c}}$ [7.160]

close to the full moment (7 μ_B) of the Eu^{+2} ion, thereby indicating aligned spins at high magnetic fields. The identification of the spin arrangement for each of the magnetic phases was initially made through a calculation of the magnetization [7.158] and later confirmed by a Monte Carlo simulation [7.145, 160].

Transport measurements provide information complementary to magnetization measurements with regard to the magnetic phase diagram shown in Fig. 7.32 [7.119, 160]. Resistivity measurements in high magnetic fields ($\boldsymbol{H} \perp \hat{c}$) [7.157, 160] (Fig. 7.33) have confirmed all the magnetic phase transitions reported from the magnetization measurements and have also indicated an additional first-order transition to a new phase at fields $H_{c12} \sim 15$ tesla, and associated with the spin fan phase (Fig. 7.30d). The magnetic phase transitions observed in the transverse $\varrho_t(H_\perp)$ and longitudinal $\varrho_l(H_\perp)$ resistivities (Fig. 7.33) are more distinct than those observed in the magnetization, over the whole temperature range of Fig. 7.33 and especially at high magnetic fields. The additional phase transition at $H_{c12} = 15$ T to a 60° spin fan phase is found experimentally in both $\varrho_l(H_\perp)$ and $\varrho_t(H_\perp)$.

Monte Carlo simulations [7.160] show that the phase transition at H_{c12} is from a 120° spin spiral configuration to a 60° fan configuration (Fig. 7.30d).

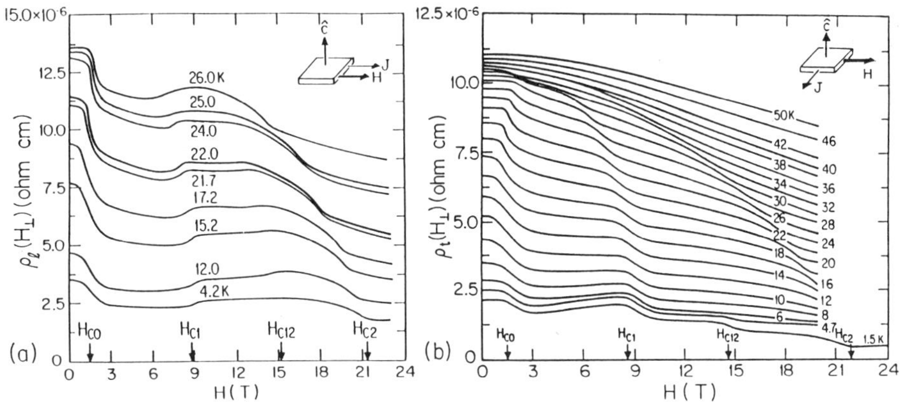

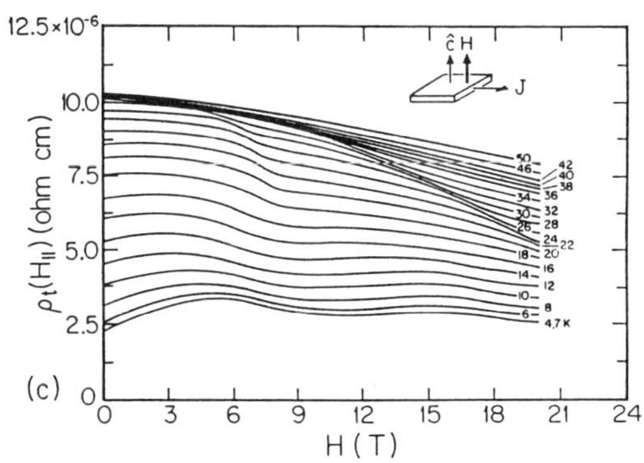

Fig. 7.33. The magnetoresistance for C_6Eu at various temperatures [7.160]: (**a**) The longitudinal magnetoresistance for $\boldsymbol{H} \perp \hat{c}$ and $\boldsymbol{j} \parallel \boldsymbol{H}$ [7.157]. (**b**) The transverse magnetoresistivity for $\boldsymbol{H} \perp \hat{c}$ and $\boldsymbol{j} \perp \boldsymbol{H}$. (**c**) The transverse magnetoresistance for $\boldsymbol{H} \parallel \hat{c} \perp \boldsymbol{j}$ [7.145]. The transition fields are indicated in (**a**) and (**b**) for $\boldsymbol{H} \perp \hat{c}$

These Monte Carlo simulations [7.160] also showed very smooth behavior in the magnetization versus magnetic field curve between H_{c1} and H_{c2}, thus accounting for the fact that the transition at H_{c12} was not reported in the magnetization measurements [7.119]. The magnetization curve between H_{c0} and H_{c1} calculated from the Monte Carlo simulation [7.160] shows a magnetic moment equal to one third of the saturation high-field value above H_{c2}, which is in qualitative agreement with the experimental data.

The temperature and field dependence of the magnetization and resistivity have been used to determine the magnetic phase diagram experimentally (Fig. 7.32) and to evaluate the parameters of the magnetic Hamiltonian (7.10) [7.158]. The values of these parameters resulting from the analysis of the magnetization [7.158] and magneto-transport [7.145, 160] results are given in Table 7.13. Although the parameters B and K are very small compared with J_0, the large S value ($S = \frac{7}{2}$) makes the contribution from these terms important to (7.10). Neutron scattering experiments, which normally provide the most sensitive identification of the spin configuration, have not been carried out on C_6Eu, primarily because of the very small size of the GIC samples and the high neutron absorption cross section for the Eu nucleus.

The Monte Carlo simulation also allows exploration of the effect of variation of the various magnetic parameters on the magnetic phase diagram [7.145, 160]. Thus, if we increase the value of KS^2, the Monte Carlo simulation shows that H_{c1} will increase and H_{c0} will decrease. With a sufficiently large KS^2 value (for example, 0.1 K), the critical field H_{c0} will be reduced to zero and the ferrimagnetic state (Fig. 7.30b) will become the ground state, while the Δ state (Fig. 7.30a) will no longer exist [7.145].

The C_6Eu system has also been studied for $\boldsymbol{H} \parallel \hat{c}$, but in less detail. For this geometry, we expect the magnetic field to tip the spins out of the xy plane. Figure 7.33c shows the magnetic field dependence for $\varrho_t(H_{\|})$, where anomalies in $\varrho_t(H_{\|})$ can be identified with magnetic phase transitions. Inspection of Fig. 7.33c reveals two phase boundaries, though the phase transitions and associated spin states have not yet been identified. High-field ($H_{\|}$) magnetization measurements $M(H)$ at low temperatures do not show well-defined magnetic transitions, but $M(H)$ does show a saturation behavior above $\sim$24 tesla, yielding 6 μ_B per Eu^{2+} ion in high magnetic fields [7.218].

Table 7.13. Parameters for magnetic interactions in C_6Eu

	J_0(K)	J_1(K)	J'(K)	BS^2(K)	KS^2(K)
Sakakibara and Date's Model[a]	−0.5	0.4	0.1	0.02	0.05
Monte Carlo simulation[b]	−0.656	0.525	0.131	0.0262	0.0656

[a] Model developed to explain dependence of magnetization on magnetic field [7.158].
[b] Developed to identify magnetic phases and the magnetic phase diagram determined from resistivity measurements [7.145, 160].

To date, stage-1 Eu GIC is the only stage that has been synthesized in any significant quantity and quality. It would be particularly interesting to synthesize higher-stage Eu GICs to study the effect of the reduction of J' on the Eu-GIC system. Rather different behavior from the magnetic acceptor compounds would be expected, since the exchange interaction in the stage-1 Eu GICs is the RKKY interaction, while that for the stage-1 magnetic acceptor compounds is dominated by a modified superexchange mechanism (Sect. 7.2.1). Since the RKKY interaction is short range and falls off rapidly with increasing stage index, the dipole–dipole interaction is expected to dominate the magnetic behavior for high-stage Eu GICs.

(b) Other Donors

In view of the very interesting magnetic properties of C_6Eu, the other rare-earth GICs are also expected to exhibit interesting magnetic properties. Other rare-earth GICs (such as with Sm, Tm, and Yb) have been synthesized [7.48], but have not yet been studied with regard to their magnetic properties.

It was also reported that the mixed-crystal $Sm_{1-x}K_x$ GIC can be synthesized in stages 2, 3, and 4 [7.123]. The stage-3 $Eu_{0.75}K_{0.25}$ GIC shows two anomalies in the magnetization curve at 30 K and 80 K. The magnetic behavior resembles that of stage-1 C_6Eu [7.123].

Studies on these and other magnetic donor compounds await progress in the synthesis of suitable samples for a variety of stages, with good staging fidelity and of large enough physical size for the various measurements.

7.5 Summary

Magnetic graphite intercalation compounds represent an attractive system for a detailed study of 2D magnetism. We have reviewed the present state of this research and, in several cases, identified new promising research directions. A status report has been given for both acceptor and donor magnetic GICs, including a summary of progress on stage-1 magnetic GICs, which generally exhibit highly anisotropic 3D magnetic behavior. The stage-1 compounds are of interest in their own right as examples of new magnetic phenomena, as for example the anomalous resistivity near T_N for the stage-1 $CoCl_2$ GIC and the unusual ferrimagnetic phase observed in the magnetization and transport properties of the stage-1 Eu GIC.

The most serious impediment to further progress in the study of 2D magnetism in magnetic GICs is the difficulty in synthesizing of well-staged, homogeneous, magnetic GICs of the appropriate intercalates and (high) stages. In addition, further theoretical developments will be needed to treat a variety of perturbations that affect 3D–2D crossover and are present in magnetic GICs, including multiple symmetry-breaking fields and finite-size effects. Although these problems are very difficult, significant progress has been made for the case of the 2D-XY model with ferromagnetic nearest-neighbor interactions. Many unanswered questions remain concerning the phase diagrams of magnetic GICs.

Acknowledgements. We wish to thank Drs. J. Speck, N.-C. Yeh, S.T. Chen, K.Y. Szeto, D.G. Wiesler, H.J. Zeiger, Prof. T.A. Kaplan, Prof. H. Zabel, Prof. K. Sugihara, Prof. M. Suzuki, Prof. J.-P. Issi, and Prof. Y. Iye for valuable discussions regarding this work. This work was supported by AFOSR Contract Grant #F49620-83-C-0011 in its initial stages and by NSF Grant #DMR 88-19896 thereafter.

References

7.1 M.T. Hutchings: J. Phys. C **6**, 3143 (1973)
7.2 P.A. Lindgard, R.J. Birgeneau, J. Als-Nielsen, H.J. Guggenheim: J. Phys. C **8**, 1059 (1975)
7.3 R.J. Birgeneau, W.B. Yelon, E. Cohen, J. Makovsky: Phys. Rev., B**5**, 2607 (1972)
7.4 L. Onsager: Phys. Rev. **65**, 117 (1944)
7.5 H.E. Stanley, T.A. Kaplan: Phys. Rev. Lett. **17**, 913 (1966)
7.6 H.E. Stanley, T.A. Kaplan: J. Appl. Phys. **38**, 975 (1967)
7.7 H.E. Stanley: Phys. Rev. Lett. **20**, 150 (1968)
7.8 F. Wegner: Z. Phys. **206**, 465 (1967)
7.9 V.L. Berezinskii: Zh. Eskp. Teor. Fiz. **59**, 907 (1970) [Sov. Phys. JETP **32**, 493 (1971)]
7.10 V.L. Berezinskii: Zh. Eskp. Teor. Fiz. **61**, 1144 (1971) [Sov. Phys. JETP **34**, 610 (1972)]
7.11 J.M. Kosterlitz, D.J. Thouless: J. Phys. C **5**, L124 (1972)
7.12 J.M. Kosterlitz, D.J. Thouless: J. Phys. C **6**, 1181 (1973)
7.13 J.M. Kosterlitz: J. Phys. C **7**, 1046 (1974)
7.14 H.E. Stanley: *Introduction to Phase Transitions and Critical Phenomena* (Oxford University Press, London 1971)
7.15 C. Domb, M.S. Green: *Phase Transitions and Critical Phenomena* (Academic, New York 1972) Vols. 1–12
7.16 P. Pfeuty, G. Toulouse: *Introduction to the Renormalization Group and to Critical Phenomena* (Wiley, New York 1977)
7.17 K.G. Wilson, J. Kogut: Phys. Rep. C**12**, 75 (1974)
7.18 K. Binder, D. Stauffer: In Applications of the Monte Carlo Method in Statistical Physics, ed. by K. Binder, Topics Curr. Phys. Vol. 36 (Springer, Berlin, Heidelberg 1984)
7.19 Yu. S. Karimov, M.E. Vol'pin, Yu. N. Novikov: Zh. Eksp. Teor. Fiz. Pis. Red. **14**, 217 (1971) [JETP Lett. **14**, 142 (1971)]
7.20 M.S. Dresselhaus: Festokörper Probleme: Advances in Solid State Physics, **XXV**, 21 (1985)
7.21 M.S. Dresselhaus: In *Two-Dimensional Magnetism in Graphite Intercalation Compounds*, Springer Proc. Phys. Vol. 14 (Springer, Berlin, Heidelberg 1986) p. 172
7.22 G. Dresselhaus, M.S. Dresselhaus: In *Intercalation in Layered Materials*, ed. by M.S. Dresselhaus, NATO ASI Series B: Physics (Plenum, New York 1987) Vol. 148, p. 407
7.23 M.S. Dresselhaus: Bull. Mater. Res. Soc. **12**, 24, (1987)
7.24 M. Suzuki: CRC Rev. in Solid State and Materials Science Vol. 16, 237 (1990)
7.25 R.J. Birgeneau, G. Shirane: In *Physical Properties of High Temperature Superconductors* (World Scientific, Singapore 1989) p. 151
7.26 Y. Iye: In Extended Abstracts on the Symposium on Graphite Intercalation Compounds (MRS, 1988) p. 49
7.27 S.G.J. Mochrie, A.R. Kortan, R.J. Birgeneau, P.M. Horn: Phys. Rev. Lett. **53**, 985 (1984)
7.28 C.J. Lobb: Physica B**126**, 319 (1984)
7.29 H.W. Jiang, A.J. Dahn: Phys. Rev. Lett. **62**, 1396 (1989)

7.30 D.J. Bishop, J.D. Reppy: Phys. Rev. **B22**, 5171 (1980)
7.31 L.J. de Jongh, A.R. Miedema: Adv. Phys. **23**, 1 (1974)
7.32 J. Kwo, D.B. McWhan, M. Hong, E.M. Gyorgy, L.C. Feldman, J.E. Cunningham: In Proceedings of the Materials Research Society: Layered structures, epitaxy and interfaces (MRS, Pittsburgh 1985) Vol. 37, p. 509.
7.33 M.B. Salamon, S. Sinha, J.J. Rhyne, J.E. Cunningham, R.W. Erwin, J. Borchers, C.P. Flynn: Phys. Rev. Lett. **56**, 259 (1986)
7.34 S.O. Demokritov, N.M. Kreines, V.I. Kudinov, S.V. Petrov: Zh. Eksp. Theor. Fiz. **94**, 283 (1988) [Sov. Phys. JETP **67**, 2552 (1988)]
7.35 M. Pomerantz, F.H. Dacol, A. Segmüller: Phys. Rev. Lett. **40**, 246 (1978)
7.36 J.B. Goodenough: *Magnetism and the Chemical Bond* (Wiley, New York 1963)
7.37 J.S. Griffiths: *The Theory of Transition Metal Ions* (Cambridge University Press, New York 1961)
7.38 M. Regis, Y. Farge: J. de Phys. **37**, 627 (1976)
7.39 K. Ôno, A. Ito, T. Fujita: J. Phys. Soc. Jpn. **19**, 2119 (1964)
7.40 M.E. Lines: Phys. Rev. **131**, 546 (1963)
7.41 C.E. Moore: Atomic energy levels, Circular 467, National Bureau of Standards (1952)
7.42 S. Sinha, J. Cunningham, R. Du, M.B. Salamon, C.P. Flynn: J. Magn. Magn. Mater. **54–57**, 773 (1986)
7.43 J. Kwo, E.M. Gyorgy, D.B. McWhan, M. Hong, F.J. DiSalvo, C. Vettier, J.E. Bower: Phys. Rev. Lett. **55**, 1402 (1985)
7.44 J. Kwo, E.M. Gyorgy, F.J. DiSalvo, M. Hong, Y. Yafet, D.B. McWhan: J. Magn. Magn. Mater. **54–57**, 771 (1986)
7.45 L.L. Chang, B.C. Giessen (eds.): *Synthetic Modulated Structures* (Academic, Orlando 1985)
7.46 G. Timp, M.S. Dresselhaus: J. Phys. C. **17**, 2641 (1984)
7.47 R.W.G. Wyckoff: *Crystal Structures*, Vol. 1 (Wiley, New York 1964)
7.48 M. El Makrini, D. Guérard, P. Lagrange, A. Hérold: Physica **99B**, 481 (1980)
7.49 E.P. Wohlfarth: *Ferromagnetic Materials* (North Holland, Amsterdam 1980) Vol. 1
7.50 A. Narath: Phys. Rev. **131**, 1929 (1963)
7.51 S. Flandrois, P. Biensan, J. Amiell, B. Agricole: In Extended Abstracts of the 19th Biennial Carbon Conference (State College, PA) (American Carbon Society, University Park, PA 1989) p. 522
7.52 J.W. Stout, R.C. Chisholm: J. Chem. Phys. **36**, 979 (1962)
7.53 D. Billerey, C. Terrier, R. Mainard, N. Ciret, A.J. Pointon: J. Magn. Magn. Mater. **30**, 55 (1982)
7.54 Yu. S. Karimov, A.V. Zvarykina, Yu. N. Novikov: Fiz. Tverd. Tela **13**, 2836 (1971) [*Sov. Phys.–Solid State* **13**, 2388 (1972)]
7.55 P.B. Johnson, S.A. Friedberg, J.A. Rayne: J. Appl. Phys. **52**, 1932 (1981)
7.56 A.K. Ibrahim, G.O. Zimmerman: Phys. Rev. **B35**, 1860 (1987)
7.57 Yu. S. Karimov: Zh. Eksp. Teor. Fiz. **66**, 1121 (1974) [JETP Lett. **39**, 547 (1974)]
7.58 R.B. Murray: Phys. Rev. **128**, 1570 (1962)
7.59 Y. Kimishima, A. Furukawa, M. Suzuki, H. Nagano: J. Phys. C **19**, L43 (1986)
7.60 M. Suzuki, A. Furukawa, H. Ikeda, H. Nagano: J. Phys. C **16**, L1211 (1983)
7.61 M. Suzuki, H. Ikeda: J. Phys. C **14**, L923 (1981)
7.62 J.T. Nicholls, J.S. Speck, G. Dresselhaus: Phys. Rev. **B39**, 10047 (1989)
7.63 S.G.J. Mochrie, A.R. Kortan, P.M. Horn, R.J. Birgeneau: Phys. Rev. Lett. **58**, 690 (1987)
7.64 D.G. Wiesler, H. Zabel: Phys. Rev. **B36**, 7303 (1987)
7.65 M.S. Dresselhaus, G. Dresselhaus: Adv. Phys. **30**, 139 (1981)
7.66 S.A. Solin: Adv. Chem. Phys. **49**, 455 (1982)
7.67 A. Hérold: In *Physics and Chemistry of Materials with Layered Structures*, ed. by F. Lévy (Reidel, Dordrecht 1979) p. 323, Vol. 6
7.68 G. Furdin, L. Hachim, N.E. Nadi, L. Lelaurain, R. Vangelisti, A. Hérold: C.R. Acad. Sci. Paris, Vol. **301 II**, 87 (1985)

7.69 M. Suzuki, P.C. Chow, H. Zabel: Phys. Rev. **B32**, 6800 (1985)
7.70 M. Suzuki, I. Oguro, Y. Jinzaki: J. Phys. C**17**, L575 (1984)
7.71 A. Hérold, G. Furdin, D. Guérard, L. Hachim, N.E. Nadi, R. Vangelisti: Ann. de Phys. **11**, 3 (1986) suppl. 2
7.72 M. Suzuki, A. Furukawa, H. Ikeda, H. Nagano: In *Graphite Intercalation Compounds* ed. by S. Tanuma and H. Kamimura. Summary report of the Special Distinguished Research Project supported by the Ministry of Education, Japan (1984)
7.73 I. Rosenman, F. Batallan, Ch. Simon, L. Hachim: J. de Phys. **47**, 1221 (1986)
7.74 S.N. Lukin, P.V. Vodolazskii, S.M. Ryabchenko: Fiz. Nizk. Temp. **3**, 1465 (1977) [Sov. J. Low Temp. Phys. **3**, 705 (1977)]
7.75 J.T. Nicholls, C. Murayama, H. Takahashi, N. Mōri, T. Tamegai, Y. Iye, G. Dresselhaus: Phys. Rev. **B41**, 4953 (1990)
7.76 L.A. Kolodziejski, R.L. Gunshor, T.C. Bonsett, R. Venkatasubramanian, S. Datta, R.B. Bylsma, W.M. Becker, N. Otsuka: Appl. Phys. Lett. **47**, 169 (1985)
7.77 G.C. Osbourn: J. Appl. Phys. **53**, 1586 (1982)
7.78 G.C. Osbourn, R.M. Biefeld, P.L. Gourley: Appl. Phys. Lett. **41**, 172 (1982)
7.79 P. Laszlo: Science **235**, 1473 (1987)
7.80 S.A. Solin: In *Intercalation in Layered Materials*, ed. by M.S. Dresselhaus NATO ASI Series B: Physics, (Plenum, New York 1987) Vol. 148, p. 145
7.81 P.A. Politowicz, G.H. Weiss, J.J. Kozak: Chem. Phys. Lett. **120**, 388 (1985)
7.82 N. Wada, M. Suzuki, D.R.Hines, K. Koga, H. Nishihara: J. Mater. Res. **2**, 864 (1987)
7.83 W.Y. Liang: In *Intercalation in Layered Materials* ed. by M. S. Dresselhaus NATO ASI Series B: Physics (Plenum, New York 1987) Vol. 148, p. 31
7.84 S.S.P. Parkin, R.H. Friend: Physica **99B**, 219 (1980)
7.85 S.S.P. Parkin, R.H. Friend: Phil. Mag. **B41**, 65 (1980)
7.86 S.S.P. Parkin, R.H. Friend: Phil. Mag. **B41**, 95 (1980)
7.87 C.M. Pereira, W.Y. Liang: J. Phys. C **18**, 6075 (1985)
7.88 F.R. Gamble, T.H. Geballe: *Treatise on Solid State Chemistry*, (Plenum, New York 1976) Vol. **3**, p. 89
7.89 M.B. Dines: Science **188**, 1210 (1975)
7.90 E. Stumpp: Mater. Sci. Eng. **31**, 53 (1977)
7.91 J.S. Speck, J.T. Nicholls, B.J. Wuensch, J.M. Delgado, M.S. Dresselhaus, H. Miyazaki: Phil. Mag. B**64**, 181 (1991)
7.92 S. Flandrois, J.M. Masson, J.C. Rouillon, J. Gaultier, C. Hauw: Synth. Met. **3**, 1 (1981)
7.93 J.S. Speck: *Structural correlations in graphite and its layer compounds*, PhD thesis, Massachusetts Institute of Technology (1989)
7.94 R. Vangelisti, A. Hérold: Carbon **14**, 333 (1976)
7.95 A.F. Wells: J. Chem. Soc. p. 1670 (1947)
7.96 W. Rüdorff, H. Schulz: Z. Anorg. Allg. Chem **207**, 340 (1932)
7.97 J.M. Cowley, J.A. Ibers: Acta Crystallogr. **9**, 421 (1956)
7.98 A.W. Syme-Johnson: Acta Crystallogr. **23**, 770 (1967)
7.99 P. Behrens, W. Metz: Synth. Met. **23**, 81 (1988)
7.100 J.S. Speck, M.S. Dresselhaus: Synth. Met. **34**, 211 (1989)
7.101 P. Behrens, W. Metz: In *Intercalation in layered materials* ed. by M.S. Dresselhaus, NATO ASI Series B: Physics (Plenum, New York 1987) Series B: Physics Vol. 148, p. 229
7.102 P. Behrens, W. Metz: Synth. Met. **34**, 223 (1989)
7.103 H. Ikeda, Y. Endoh, S. Mitsuda: J. Phys. Soc. Jpn. **54**, 3232 (1985)
7.104 D.G. Rancourt, C. Meschi, S. Flandrois: Phys. Rev. B**33**, 347 (1986)
7.105 D. Davidov, H. Selig: In *Intercalation in Layered Materials*. ed. by M.S. Dresselhaus NATO ASI Series B: Physics (Plenum, New York 1987) Vol. 148, p. 433
7.106 H. Selig: In *Inorganic Solid Fluorides* ed. by P. Hagenmuller (Academic, New York 1985) p. 381
7.107 E. Stumpp, G. Nietfeld: Z. Anorg. Allg. Chem. **456**, 261 (1979)
7.108 R.C. Croft: Aust. J. Chem. **9**, 184 (1956)

7.109 H. Stahl: Z. Anorg. Allg. Chem. **428**, 269 (1977)
7.110 F. Baron, S. Flandrois, C. Hauw, J. Gaultier: Solid State Commun. **42**, 759 (1982)
7.111 G. Chouteau, R. Yazami: Europhys. Lett. **3**, 229 (1987)
7.112 M. Elahy: *Phase Transitions in Magnetic Graphite Intercalation Compounds*, PhD thesis, Massachusetts Institute of Technology (1983)
7.113 G.K. Wertheim: Solid State Commun. **38**, 633 (1981)
7.114 S. Flandrois, A.W. Hewat, C. Hauw, R.H. Bragg: Synth. Met. **7**, 305 (1983)
7.115 J. Gaultier, C. Hauw, J.M. Masson, J. C. Rouillon, S. Flandrois: C.R. Acad. Sc. Paris **289C**, 45 (1979)
7.116 M. Elahy, M. Shayegan, K.Y. Szeto, G. Dresselhaus: Synth. Met. **8**, 35 (1983)
7.117 M. Matsuura, Y. Murakami, K. Takeda, H. Ikeda, M. Suzuki: Synth. Met. **12**, 427 (1985)
7.118 G. Chouteau, J. Schweizer, F. Tasset, R. Yazami: Synth. Met. **23**, 249 (1988)
7.119 H. Suematsu, K. Ohmatsu, T. Sakakibara, M. Date, M. Suzuki: Synth. Met. **8**, 23 (1983)
7.120 H. Zabel: Lattice Dynamics: I Neutron Studies In *Graphite intercalation compounds I*, Springer Ser. Mater. Sci. Vol. 14 (Springer, Berlin, Heidelberg 1989)
7.121 M.S. Dresselhaus, G. Dresselhaus: *Light Scattering in Solids III*, ed. by M. Cardona, G. Güntherodt, Topics Appl. Phys. Vol. 51 (Springer, Berlin Heidelberg 1983) p. 3
7.122 D.M. Hwang, D. Guérard: Solid State Commun. **40**, 759 (1981)
7.123 T. Takamoto, H. Suematsu, Y. Murakami: Synth. Met. **34**, 53 (1989)
7.124 M.E. Vol'pin, Y.N. Novikov, N.D.Lapkina, V.I. Kasatochkin, Y.T. Struchkov, M.E. Kazakov, R.A. Stukan, V.A. Poritskij, Y.S. Karimov, A.V. Zvarykina: J. Am. Chem. Soc. **97**, 3366 (1975)
7.125 J. Kanamori: J. Phys. Chem. Solids **10**, 87 (1959)
7.126 N.-C. Yeh: *Electronic and Magnetic Properties of Graphite Intercalation Compounds*, PhD thesis, Massachusetts Institute of Technology (1988)
7.127 P.W. Anderson: In *Solid State Physics*, Vol. 14, ed. by F. Seitz, D. Turnbull. (Academic, New York 1963) pp. 99–214
7.128 N.C. Yeh, K. Sugihara, M.S. Dresselhaus, G. Dresselhaus: Phys. Rev. B**40** 622 (1989)
7.129 K. Miura, Y. Iye, J.T. Nicholls, G. Dresselhaus: In Extended Abstracts of the Symposium on Graphite Intercalation Compounds (MRS, Pittsburgh 1988), p. 69
7.130 J.T. Nicholls: *Magnetic Studies of Graphite Intercalation Compounds*, PhD thesis, Massachusetts Institute of Technology (1990)
7.131 H.J. Zeiger, G.W. Pratt: *Magnetic Interactions in Solids* (Clarendon, Oxford 1973)
7.132 H. Akera, H. Kamimura: Solid State Commun. **48**, 467 (1983)
7.133 G.D. Jones, C.W. Tomblin: Phys. Rev. **B18**, 5990 (1978)
7.134 M. Suzuki, K. Koga, Y. Jinzaki: J. Phys. Soc. Jpn. **53**, 2745 (1984)
7.135 S. Flandrois, J. Amiell, B. Agricole, E. Stumpp, C. Ehrhardt, P. Schubert: Synth. Met. **34**, 531 (1989)
7.136 J.T. Nicholls, G. Dresselhaus: Synth. Met. **34**, 519 (1989)
7.137 D. Billerey, C. Terrier, A.J. Pointon, J.P. Redoules: J. Magn. Magn. Mater. **21**, 187 (1980)
7.138 H. Bizette, C. Terrier, B. Tsai: Compt. Rendue **243**, 1295 (1956)
7.139 H. Suematsu, R. Nishitani, R. Yoshizaki, M. Suzuki, H. Ikeda: J. Phys. Soc. Jpn. **52**, 3874 (1983)
7.140 J. de Gunzbourg, S. Papassimacopoulos, A. Miedan-Gros, Y. Allain: J. de Phys. **32**, C I-125 (1971)
7.141 M. Suzuki, H. Ikeda, Y. Endoh: Synth. Met. **8**, 43 (1983)
7.142 G. Mischler, D.J. Lockwood, A. Zwick: J. Phys. C **20**, 299 (1987)
7.143 J.T. Nicholls, Y. Shapira, E.J. McNiff, Jr., G. Dresselhaus: Synth. Met. **23**, 231 (1988)
7.144 D.G. Wiesler, M. Suzuki, P.C. Chow, H. Zabel: Phys. Rev. **B34**, 7951 (1986)
7.145 S.T. Chen: *Magnetic Properties of Graphite Intercalation Compounds*, PhD thesis, Massachusetts Institute of Technology (1985)
7.146 H. Zabel, S.M. Shapiro: Phys. Rev. **B36**, 7292 (1987)
7.147 D.G. Wiesler, M. Suzuki, H. Zabel: Phys. Rev. **B36**, 7051 (1987)
7.148 I.S. Jacobs, S.D. Silverstein: Phys. Rev. Lett. **13**, 272 (1964)
7.149 K. Yosida: Prog. Theor. Phys. **6**, 691 (1951)

7.150 M.H.L. Pryce: Phys. Rev. **80**, 1107 (1950)
7.151 O. Gonzalez, S. Flandrois, A. Maaroufi, J. Amiell: Solid State Commun. **51**, 499 (1984)
7.152 Y. Kimishima, A. Furukawa, H. Nagano, P.C. Chow, D.G. Wiesler, H. Zabel, M. Suzuki: Synth. Met. **12**, 445 (1985)
7.153 K. Koga, M. Suzuki: J. Phys. Soc. Jpn. **53**, 786 (1984)
7.154 R.B. Murray, L.D. Roberts: Phys. Rev. **100**, 1067 (1955)
7.155 M. El Makrini, G. Furdin, P. Lagrange, J.F. Marêché, E. McRae, A. Hérold: Synth. Met. **2**, 197 (1980)
7.156 G. Kaindl, J. Feldhaus, U. Ladewig, K.H. Frank: Phys. Rev. Lett. **50**, 123 (1983)
7.157 H. Suematsu, H. Minemoto, K. Ohmatsu, Y. Yosida, S.T. Chen, G. Dresselhaus, M.S. Dresselhaus: Synth. Met. **12**, 377 (1985)
7.158 T. Sakakibara, M. Date: J. Phys. Soc. Jpn. **53**, 3599 (1984)
7.159 T. Sakakibara: J. Phys. Soc. Jpn. **53**, 3607 (1984)
7.160 S.T. Chen, M.S. Dresselhaus, G. Dresselhaus, H. Suematsu, H. Minemoto, K. Ohmatsu, Y. Yosida: Phys. Rev. B**34**, 423 (1986)
7.161 M. Date, T. Sakakibara, K. Sugiyama, H. Suematsu: In *High Field Magnetism*, ed. by M. Date (North-Holland, Amsterdam 1983) p. 41
7.162 G.S. Rushbrooke, P.J. Wood: Molec. Phys. **1**, 257 (1958)
7.163 F. Bloch: Z. Physik **61**, 206 (1930)
7.164 N.D. Mermin, H. Wagner: Phys. Rev. Lett. **17**, 1133 (1966)
7.165 A.M. Polyakov: Phys. Lett. **59B**, 79 (1975)
7.166 A.A. Migdal: Zh. Eksp. Theor. Fiz. **69**, 1457 (1975) [Sov. Phys. JETP **42**, 743 (1975)]
7.167 C.J. Hamer, J.B. Kogut, L. Susskind: Phys. Rev. D**19**, 309 (1979)
7.168 S.H. Shenker, J. Tobochnik: Phys. Rev. B**22**, 4462 (1980)
7.169 J.V. José, L.P. Kadanoff, S. Kirkpatrick, D.R. Nelson: Phys. Rev. B**16**, 1217 (1977)
7.170 D.R. Nelson, J.M. Kosterlitz: Phys. Rev. Lett. **39**, 1201 (1977)
7.171 V.L. Pokrovskii, G.V. Uimin: Zh. Eksp. Theor. Fiz. **65**, 1691 (1973) [Sov. Phys. JETP **38**, 847 (1974)]
7.172 K.Y. Szeto, G. Dresselhaus: Phys. Rev. B**32**, 3186 (1985)
7.173 K.Y. Szeto, G. Dresselhaus: Phys. Rev. B**32**, 3142 (1985)
7.174 P. Minnhagen: Rev. Mod. Phys. **59**, 1001 (1987)
7.175 J. Tobochnik, G.V. Chester: Phys. Rev. B**20**, 3761 (1979)
7.176 M.S.S. Challa, D.P. Landau: Phys. Rev. B**33**, 437 (1986)
7.177 C. Kawabata, M. Takeuchi, A.R. Bishop: J. Magn. Magn. Mater. **54–57**, 871 (1986)
7.178 J.F. Fernández, M.F. Ferreira, J. Stankiewicz: Phys. Rev. B**34**, 292 (1986)
7.179 C. Kawabata, A.R. Bishop: Z. Phys. B **65**, 225 (1986)
7.180 F.G. Mertens, A.R. Bishop, G.M. Wysin, C. Kawabata: Phys. Rev. Lett. **59**, 117 (1987)
7.181 F.G. Mertens, A.R. Bishop, G.M. Wysin, C. Kawabata: Phys. Rev. B**39**, 591 (1989)
7.182 K. Hirakawa, H. Yoshizawa, K. Ubukoshi: J. Phys. Soc. Jpn. **51**, 2151 (1982)
7.183 K.Y. Szeto: *Two-Dimensional XY Model and Its Application to Graphite Intercalation Compounds*, PhD thesis, Massachusetts Institute of Technology (1985)
7.184 S. Hikami, T. Tsuneto: Prog. Theor. Phys. **63**, 387 (1980)
7.185 P. Dutta, S.K. Sinha: Phys. Rev. Lett. **47**, 50 (1981)
7.186 R. Yazami, G. Chouteau: Synth. Met. **18**, 543 (1987)
7.187 K.Y. Szeto, S.T. Chen, G. Dresselhaus: Phys. Rev. B**33**, 3453 (1986)
7.188 J.T. Nicholls, G. Dresselhaus: J. Phys.: Condensed Matter **2**, 8391 (1990)
7.189 S.T. Chen, K.Y. Szeto, M. Elahy, G. Dresselhaus: J. de Chim Phys. **81**, 863 (1984)
7.190 D.G. Wiesler, M. Suzuki, H. Zabel, S.M. Shapiro, R.M. Nicklow: Physica **136B**, 22 (1986)
7.191 K.Y. Szeto, S.T. Chen, G. Dresselhaus: Phys. Rev. B**32**, 4628 (1985)
7.192 A.K. Ibrahim, G.O. Zimmerman: Phys. Rev. B**34**, 4224 (1986)
7.193 M. Matsuura, Y. Endoh, T. Kataoka, Y. Murakami: J. Phys. Soc. Jpn. **56**, 2233 (1987)
7.194 V. Ambegaokar, B.I. Halperin, D.R. Nelson, E.D. Siggia: Phys. Rev. B**21**, 1806 (1980)
7.195 Y. Murakami, M. Matsuura, T. Kataoka: Synth. Met. 12, 443 (1985)

7.196 M. Matsuura, N. Tanaka, Y. Karaki, Y. Murakami: Jpn. J. Appl. Phys. **26**, 797 (1987)
7.197 M. Shayegan, L. Salamanca-Riba, J. Heremans, G. Dresselhaus, J.P. Issi: In Proc. of the Symposium on Intercalated Graphite, Vol. 20 (North-Holland, New York 1983) p. 213
7.198 M. Shayegan, M.S. Dresselhaus, L. Salamanca-Riba, G. Dresselhaus, J. Heremans, J.P. Issi: Phys. Rev. B**28**, 4799 (1983)
7.199 M. Shayegan: *High Magnetic Field Studies on the Graphite Intercalation Compounds*, PhD thesis, Massachusetts Institute of Technology (1983)
7.200 Ch. Simon, F. Batallan, I. Rosenman, C. Ayache, E. Bonjour: Phys. Rev. B**35**, 5816 (1987)
7.201 D.G. Onn, M.G. Alexander, J.J. Ritsko, S. Flandrois: J. Appl. Phys. **53**, 2751 (1982)
7.202 Yu. S. Karimov, Yu. N. Novikov: Zh. Eksp. Teor. Fiz. Pis. Red. **19**, 268 (1974) [JETP Lett. **19**, 159 (1974)]
7.203 Yu. S. Karimov: Zh. Eksp. Teor. Fiz. **68**, 1539 (1975) [Sov. Phys.-JETP **41**, 772 (1976)]
7.204 S. Hendricks, E. Teller: J. Chem. Phys. **10**, 147 (1942)
7.205 D.G. Wiesler: *Magnetic Properties of Transition Metal Dichloride-Graphite Intercalation Compounds*, PhD thesis, University of Illinois (1989)
7.206 Ch. Simon: Private communication.
7.207 G. Dresselhaus, R. Al-Jishi, J.D. Axe, C.F. Majkrzak, L. Passel, S.K. Satija: Solid State Commun. **40**, 229 (1981)
7.208 D.G. Wiesler, H. Zabel, S.M. Shapiro: Synth. Met. **34**, 505 (1989)
7.209 H. Suematsu, H. Minemoto, K. Ohmatsu, Y. Yosida, S.T. Chen, G. Dresselhaus, M.S. Dresselhaus: Synth. Met. **12**, 377 (1985)
7.210 I. Shiozaki, T. Fukuhara, M. Suzuki, I. Oguro: Synth. Met. **20**, 57 (1987)
7.211 K. Sugihara, N.C. Yeh, M.S. Dresselhaus, G. Dresselhaus: Phys. Rev. B**39**, 4577 (1989)
7.212 K. Sugihara, I. Shiozaki, S.M. Sampere, M. Suzuki, J.T. Nicholls, G. Dresselhaus: Synth. Met. **34**, 543 (1989)
7.213 J.T. Nicholls, G. Dresselhaus: Phys. Rev. B**41**, 9744 (1990)
7.214 J.-P. Issi: In *Intercalation in Layered Materials*, ed. by M.S. Dresselhaus, NATO ASI Series B: Physics (Plenum, New York 1987) Vol. 148, p. 347
7.215 M. Kinany-Alaoui, L. Piraux, J.P. Issi, P. Pernot, R. Vangelisti: Solid State Commun. **68**, 1065 (1988)
7.216 L. Piraux, M. Kinany-Alaoui, J.P. Issi, A. Perignon, P. Pernot, R. Vangelisti: Phys. Rev. B**38**, 4329 (1988)
7.217 F.J. Dyson: Phys. Rev. 98, 349 (1955)
7.218 H. Suematsu, K. Ohmatsu, T. Sakakibara, M. Date, M. Suzuki: Graphite Intercalation Compounds, Summary report of the Special Distinguished Research Project supported by the Ministry of Education, Japan: ed. by S. Tanuma and H. Kamimura (1984) p. 111
7.219 K. Koga, M. Suzuki, H. Yasuoka: Synth. Met. **12**, 467 (1985)
7.220 G. Ceotto, S. Rolla, G.E. Barberis, C. Rettori: in Extended Abstracts of the Symposium on Graphite Intercalation Compounds (MRS, Pittsburgh 1986) p. 187
7.221 D. Vaknin, D. Davidov, H. Selig, V. Zevin, I. Felner, Y. Yeshurun: Phys. Rev. B**31**, 3212 (1985)
7.222 D. Vaknin, D. Davidov, H. Selig, Y. Yeshurun: J. Chem. Phys. **83**, 3859 (1985)
7.223 N.D. Mermin: Phys. Rev. **176**, 250 (1968)
7.224 I. Bleish, D. Vaknin, D. Davidov, H. Selig, J.K. Kjems: Synth. Met. **23**, 351 (1987)
7.225 J.K. Kjems, D. Vaknin, D. Davidov, H. Selig, Y. Yeshurun: Synth. Met. **23**, 113 (1987)
7.226 H. Selig, D. Vaknin, D. Davidov, Y. Yeshurun: Synth. Met. **12**, 479 (1985)
7.227 I. Ohana, D. Vaknin, H. Selig, Y. Yacoby, D. Davidov: Phys. Rev. B**35**, 4522 (1987)
7.228 D. Vaknin, D Davidov, V. Zevin, H. Selig: Phys. Rev. B**35**, 6423 (1987)
7.229 F.V. Bragin, S.M. Ryabchenko: Fiz. Tverd. Tela **12**, 3579 (1970) [Sov. Phys. Solid State **12**, 2905 (1971)]
7.230 F.V. Bragin, S.M. Ryabchenko: Fiz. Tverd. Tela **15**, 1050 (1973) [Sov. Phys. Solid State **15**, 721 (1973)]
7.231 T. Tsuda, H. Yasuoka, M. Suzuki: Synth. Met. **12**, 461 (1985)
7.232 J.F. Cavanagh: Phys. Status Solidi **36**, 657 (1969)

7.233 T. Fujita, A. Ito, K. Ôno: J. Phys. Soc. Jpn. **27**, 1143 (1969)
7.234 B.V. Liengme, M.W. Bartlett, J.G. Hooley, J.R. Sams: Phys. Lett. **25**, 127 (1967)
7.235 K. Ohhashi, I. Tsujikawa: J. Phys. Soc. Jpn. **36**, 422 (1974)
7.236 S.E. Millman, M.R. Corson, G.R. Hoy: Phys. Rev. B**25**, 6595 (1982)
7.237 M.R. Corson, S.E. Millman, G.R. Hoy, H. Mazurek: Solid State Commun. **42**, 667 (1982)
7.238 S.E. Millman, G.B. Zimmerman: J. Phys. C **16**, L89 (1983)
7.239 D.J. Lockwood: In *Light Scattering in Solids III*, ed. by M. Cardona, G. Güntherodt, Topics Appl. Phys. Vol. 51 (Springer, Berlin, Heidelberg 1982) p. 59
7.240 Y. Shapira, R.D. Yacovitch, C.C. Becerra, S. Foner, E.J. McNiff, Jr., D.R. Nelson, L. Gunther: Phys. Rev. B**14**, 3007 (1976)
7.241 Yu. S. Karimov, A.V. Zvarykina, Yu. N. Novikov: Fiz. Tverd. Tela **13**, 28 (1971) [Sov. Phys.-Solid State **13**, 21 (1972)].
7.242 Yu. S. Karimov: Zh. Eksp. Teor. Fiz. Pis. Red. **15**, 332 (1972) [JETP Lett. **15**, 235 (1972)]
7.243 Yu. S. Karimov: Zh. Eksp. Teor. Fiz. **65**, 261 (1973) [Sov. Phys. JETP **38**, 129 (1974)]
7.244 Yu. S. Karimov: Zh. Eksp. Teor. Fiz. **66**, 1121 (1974) [JETP Lett. **39**, 547 (1974)]
7.245 M.E. Vol'pin, Y.N. Novikov, Y.T. Struchkov, V.A. Semion: Izv. Akad. Nauk. SSSR, Ser. Khim **11**, 2608 (1970)
7.246 M. Elahy, C. Nicolini, G. Dresselhaus, G. Zimmerman: Solid State Commun. **41**, 289 (1982)
7.247 M. Elahy, G. Dresselhaus: Phys. Rev. **B30**, 7225 (1984)
7.248 Y. Murakami, M. Matsuura, M. Suzuki, H. Ikeda: J. Magn. Magn. Mater. **31–34**, 1171 (1983)
7.249 D.G. Wiesler, H. Zabel, S.M. Shapiro: in Extended Abstracts of the Symposium on Graphite Intercalation Compounds (MRS, Pittsburgh 1988) p. 73
7.250 Y. Karaki, N. Tanaka, M. Matsuura, Y. Murakami, M. Suzuki: in Extended Abstracts of the Symposium on Graphite Intercalation Compounds (MRS, Pittsburgh 1986) p. 32
7.251 C. Kawabata, A.R. Bishop: Solid State Commun. **60**, 169 (1986)
7.252 D.G. Rancourt: J. Magn. Magn. Mater. **51**, 133 (1985)
7.253 M. Kinany-Alaoui, L. Piraux, V. Bayot, J.P. Issi, P. Pernot, R. Vangelisti: Synth. Met. **34**, 537 (1989)
7.254 M. Suzuki, D.G. Wiesler, P.C. Chow, H. Zabel: J. Magn. Magn. Mater. **54–57**, 1275 (1986)
7.255 D.H. Lee, J.D. Joannopoulos, J.W. Negele: Phys. Rev. Lett. **52**, 433 (1984)
7.256 S. Hashimoto, K. Forster, S.C. Moss: Acta Crystallogr. A**44**, 897 (1988)
7.257 K. Ohhashi, I. Tsujikawa: J. Phys. Soc. Jpn. **36**, 980 (1974)
7.258 Ch. Simon, F. Batallan, I. Rosenman, G. Furdin, R. Vangelisti, H. Lauter, J. Schweitzer, C. Ayache, G. Pepy: Ann. de Phys. **11**, 143 (1986)
7.259 J.P. Stampfel, W.T. Oosterhuis, B. Window, F. deS. Barros: Phys. Rev. B**8**, 4371 (1973)
7.260 J.W. Cable, M.K. Wilkinson, E.O. Wollan, W.C. Koehler: Phys. Rev. **127**, 714 (1962)
7.261 J.G. Hooley, R.N. Soniassy: Carbon **8**, 191 (1970)
7.262 I. Rosenman, F. Batallan, Ch. Simon, C. Ayache, J. Schweitzer, H. Lauter, R. Vangelisti: Synth. Met. **12**, 439 (1985)
7.263 I.S. Jacobs, P.E. Lawrence: Phys. Rev. **164**, 866 (1967)
7.264 J. Kanimori: Prog. Theor. Phys. **20**, 890 (1958)
7.265 R. Alben: J. Phys. Soc. Jpn. **26**, 261 (1969)
7.266 K. Ohhashi, I. Tsujikawa: J. Phys. Soc. Jpn. **37**, 63 (1974)
7.267 J.G. Hooley, M.W. Bartlett, B.V. Liengme, J.R. Sams: Carbon **6**, 681 (1968)
7.268 C. Hauw, J. Gaultier, S. Flandrois, O. Gonzalez. D. Dorignac, R. Jagut: Synth. Met. **7**, 313 (1983)
7.269 D. Billerey, C. Terrier, R. Mainard, M. Perrin, J. Hubsch: Phys. Lett. **68A**, 275 (1978)
7.270 D. Billerey, C. Terrier: Phys. Lett. **68A**, 278 (1978)
7.271 H. Nishihara, I. Oguro, M. Suzuki, K. Koga, H. Yasuoka: Synth. Met. **12**, 473 (1985)
7.272 J.W. Cable, M.K. Wilkinson, E.O. Wollan: J. Phys. Chem. Solids **19**, 29 (1961)
7.273 H. Bizette, C. Terrier, A. Adam: Compt. Rendue **252**, 1571 (1961)
7.274 T. Nagamiya, K. Yosida, R. Kubo: Adv. Phys. **4**, 1 (1955)
7.275 J.F. Dillon, Jr., H. Kamimura, J.P. Remeika: J. Phys. Chem. Solids **27**, 1531 (1966)
7.276 S.R. Kundersma, J.P. Sanchez, C. Haas: Physica B**111**, 231 (1981)

7.277 A.V. Zvarykina, Yu, S. Karimov, M.E. Vol'pin, Yu. N. Novikov: Fiz. Tverd. Tela **13**, 28 (1971) [Sov. Phys.-Solid State **13**, 21 (1971)]
7.278 Y. Murakami, T. Kishimoto, H. Suematsu, R. Nishitani, Y. Sasaki, Y. Nishina: Synth. Met. **34**, 205 (1989)
7.279 R.S. Tebble, D.J. Craik: *Magnetic Materials* (Wiley, London 1969)
7.280 L.G. Van Uitert, H.J. Sherwood, J.J. Rubin: J. Appl. Phys. **36**, 1029 (1965)
7.281 E.O. Wollan, H.R. Child, W.C. Koehler, M.K. Wilkinson: Phys. Rev. **112**, 1132 (1958)
7.282 H. Yoshizawa, K. Ubukoshi, K. Hirakawa: J. Phys. Soc. Jpn. **48**, 42 (1980)
7.283 E. Stumpp: Physica **105B**, 9 (1981)
7.284 A. Hérold, G. Furdin, D. Guérard, L. Hachim, M. Lelaurin, N.-E. Nadi, R. Vangelisti: Synth. Met. **12**, 11 (1985)
7.285 G. Furdin, L. Hachim, D. Guérard, A. Hérold: C.R. Acad. Sc. Paris Vol **301 II**, 579 (1985)
7.286 V.V. Avdeev, V.Y. Akim, N.B. Brandt, V.N. Davydov, V.A. Kulbachinskii, S.G. Ionov: Zh. Eksp. Teor. Fiz. **94**, 188 (1988) [Sov. Phys. JETP **67**, 2496 (1988)]
7.287 A. Metrot, D. Guérard, D. Billaud, A. Hérold: Synth. Met. **1**, 363 (1980)
7.288 D.G. Rancourt, B. Hun, S. Flandrois: Can. J. Phys. **66**, 776, (1988)
7.289 H. Schäfer: Angew. Chem **88**, 775 (1976)
7.290 M. Yeh, M. Suzuki, C.R. Burr: Phys. Rev. B**40**, 1422 (1989)
7.291 M. Suzuki, L.J. Santodonato, M. Yeh, S.M. Sampere, A.V. Smith, C.R. Burr: J. Mater. Res. **5**, 422 (1990)
7.292 S.M. Sampere, L.J. Santodonato, M. Suzuki, C.R. Burr: In Extended abstracts of the 1988 MRS symposium on graphite intercalation compounds (MRS, Pittsburgh 1988) p. 81

8. Intercalation of Graphite Fibers

By Mildred S. Dresselhaus and Morinobu Endo

With 45 Figures

Intercalated carbon fibers are of interest for both applications and basic science. The availability of intercalation compounds in fibrous form opens up a number of scientific opportunities, many of which exploit the favorable geometry of fibrous samples. For example, the large aspect ratio (length/diameter) greatly improves the sensitivity of transport measurements such as resistivity, magnetoresistance, thermopower, and thermal conductivity, thereby allowing systematic investigations of these properties to be made in graphite intercalation compounds (GICs). For this reason, there is a strong connection between the study of transport properties in GICs (Chap. 6) and the present review of intercalated carbon fibers. For example, with fibrous intercalation compounds it has been possible to separate the lattice and electronic contributions to the thermal conductivity of GICs by measuring the thermal and electrical conductivities on the same fiber sample and invoking the Wiedemann–Franz law [8.1]. The small fiber diameters are of particular utility for structural studies using transmission electron microscopy, where observations can be made directly without thinning of the specimens. The small fiber diameters also allow preparation of compounds not readily synthesized in bulk form, such as stage-1 $NiCl_2$ GICs.

Among the examples of new science that has been explored through the availability of fibrous host materials are precision measurements of the unusual functional form of the temperature dependence of the resistivity of GICs in the low-temperature limit, exploiting the favorable geometry of fibers for high-resolution studies. Closely related are the weak localization studies of transport properties of GICs, exploiting the two-dimensionality of fibrous acceptor compounds, and the ability to control the degree of disorder in the sample by control of the heat treatment temperature and the general fiber microstructure. Studies of the "metal-insulator" transition in ammoniated alkali-metal GICs have also exploited the favorable fiber geometry for transport measurements [8.1].

From an applications standpoint, many of the applications of intercalated carbon fibers exploit the high specific conductivity of GICs, which can be expressed as a figure of merit in terms of the conductivity σ divided by the mass density ϱ_m, which for a good conductor like copper is $\sim 6 \times 10^{-2}$ cm^2/gμΩ. Composite wires based on graphite intercalation compounds can be formed by packing GIC powder into a Cu or Ag tube, which is cold-worked down to a thin thread of suitable diameter [8.2]. However, then the light weight performance is severely reduced by using the metal cladding. The high conductivity of individual small flakes of crystalline graphite is not reflected in the final conductivity

Springer Series in Materials Science, vol. 18
H. Zabel · S.A. Solin (eds.): Graphite Intercalation Compounds II

of the composite wire because of the high contact resistance of the particles and because of the difficulty in preparing a wire out of properly oriented crystallites in the drawing process. Since GICs are difficult to melt without compositional changes, it is not practical to form a wire in a spinning process using a nozzle, as is done for metals or for pristine mesophase pitch fibers. For these reasons, it is advantageous to use fibrous graphite host materials to form a GIC yarn. Therefore, GICs based on fibrous host materials are of particular interest for applications of GICs. Furthermore, intercalated fibers are also of interest as a filler for conducting polymers with high mechanical performance requirements, such as for electromagnetic interference (EMI) shielding applications. Another area of practical interest is for conducting paint applications for printed circuit boards on polymer substrates. Through intercalation, the use of carbon fibers can perhaps be expanded beyond the conventional structural applications to new and broader functional applications areas.

Thus the intercalation of carbon fibers offers important scientific and technological opportunities. In this chapter, both of these aspects are emphasized.

8.1 Precursor Graphite Fibers

Although several different types of carbon fibers have been prepared (Fig. 8.1), it is those fibers possessing the highest degree of graphitic structural order that can best be intercalated [8.3]. We therefore briefly review the main types of available

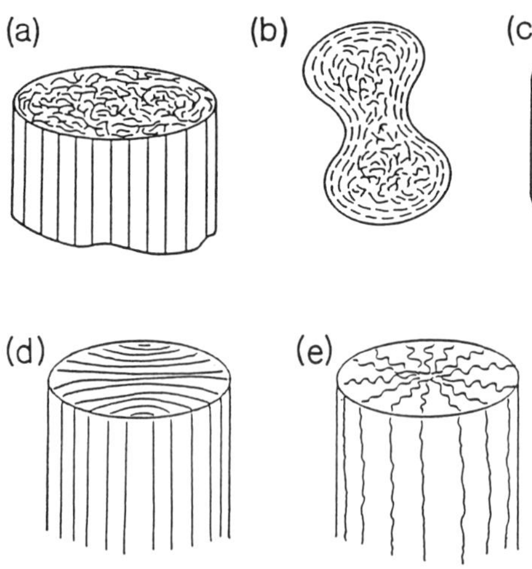

Fig. 8.1. Observed ex-polymer fiber cross sections: (**a**) circular, showing a disordered arrangement of ribbons; (**b**) "dog-bone" section observed in partially graphitic ex-PAN fibers, with a circumferential arrangement of ribbons in the sheath region, and random in the core; (**c**) ex-pitch (mesophase pitch) "PAC-man" section showing a radial arrangement of the straight ribbons; (**d**) ex-pitch (mesophase pitch) oriented core or "PAN-AM" section showing a transverse alignment with nearly parallel graphene planes; and (**e**) wavy radial structure, which adds strength to pitch-based fibers

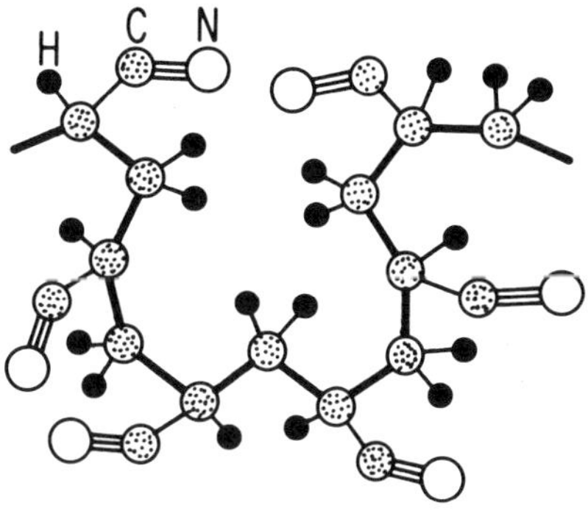

Fig. 8.2. Schematic diagram for the chemical structure of a typical polymer precursor for carbon fibers, polyacrylonitrile (PAN). The carbon and nitrogen atoms, as well as the small solid circles representing hydrogen, are labeled explicitly

carbon fibers, giving particular emphasis to the types of fibers that can be intercalated.

Most commercial carbon fibers are synthesized from polymer precursors. Ex-polymer fibers are manufactured by spinning a polymer through a nozzle into a continuous filament, after stabilization treatment at about 200 °C to 350 °C in air, then heat-treating the filaments to temperatures of about 1300 °C to carbonize the filaments by the removal of H, N, . . . etc. (Figure 8.2 shows the chemical structure of a typical polymer precursor, polyacrylonitrile (PAN). Another common polymer precursor for graphite fibers is mesophase pitch, a highly oriented anisotropic liquid-crystal phase, which results from heating petroleum asphaltine or coal tar at temperatures of ~ 350 °C.) Further heat treatment to elevated temperatures (T_{HT}), typically between 2500 and 3000 °C, is referred to as a "graphitization" step. The fibers with lower heat-treatment temperatures T_{HT} are often classified as *carbon-fibers* and for T_{HT} higher than ~ 2500 °C as *graphite-fibers.* The most important precursors today are PAN and pitch, though rayon precursor materials were important until very recently. It is convenient to denote a polymer-based carbon fiber by its polymer precursor, as for example an ex-PAN or an ex-pitch carbon fiber.

The most obvious feature of a carbon fiber is its small diameter, which is typically in the range 7–15 μm, with PAN fibers being in the thinner range and pitch-based fibers thicker. These carbon fibers are small compared with human hairs (diameter ~ 70 μm). Single carbon fibers can barely be seen by eye, and appear as featureless black filaments under optical magnification. Under closer (100 Å) examination using a scanning electron microscope, a powerful macroscopic characterization technique, the cross sections are often found to be circular (Fig. 8.1a). Sometimes, particularly in the case of ex-PAN fibers, the cross section is double-lobed (Fig. 8.1b). Such double-lobed or kidney-shaped cross sections often have a "dog-bone" shape. Ex-pitch fibers sometimes have a roughly circular cross section, in which a segment has been removed, sometimes called a "PAC-man" section (Fig. 8.1c). Recently, mesophase pitch fibers with fully circular cross sections – with either a straight (Fig. 8.1d) or wavy (Fig. 8.1e) arrangement of layer planes – have become important commercially due to their optimal mechanical properties [8.4, 5].

Under even closer (atomic) scrutiny using X-ray diffraction and transmission electron microscopy, it is found that the basic structural unit of carbon fibers is the graphene plane, the planar monolayer structure occurring in graphite, in which the carbon atoms lie in hexagonal honeycomb arrays. Characteristically, the planar structure of carbon fibers has many defects. Nevertheless, the planar graphene arrays lie roughly aligned so that the direction of the fiber axis is in a graphene layer. This alignment of the graphene planes is responsible for the exceptional mechanical properties of carbon fibers. The high strength and modulus of carbon fibers exploit the fact that the C–C bond in graphite (and also in diamond) is the strongest bond occurring in nature.

In a carbon fiber, these graphene planes are stacked on top of each other to form a ribbonlike structure, having typical widths of several tens of angstroms. Figure 8.3 illustrates the typical arrangement of these planes in an ex-PAN fiber [8.6]. For specific applications, the microstructure can be controlled to yield fibers with high modulus (see Fig. 8.3a, where a high degree of structural ordering is seen) or with high strength (see Fig. 8.3b, where the graphene planar ribbons are heavily intertwined). A classification of the various types of high-strength and high-modulus commercial fibers is given in the plot of tensile strength versus Young's modulus (Fig. 8.4).

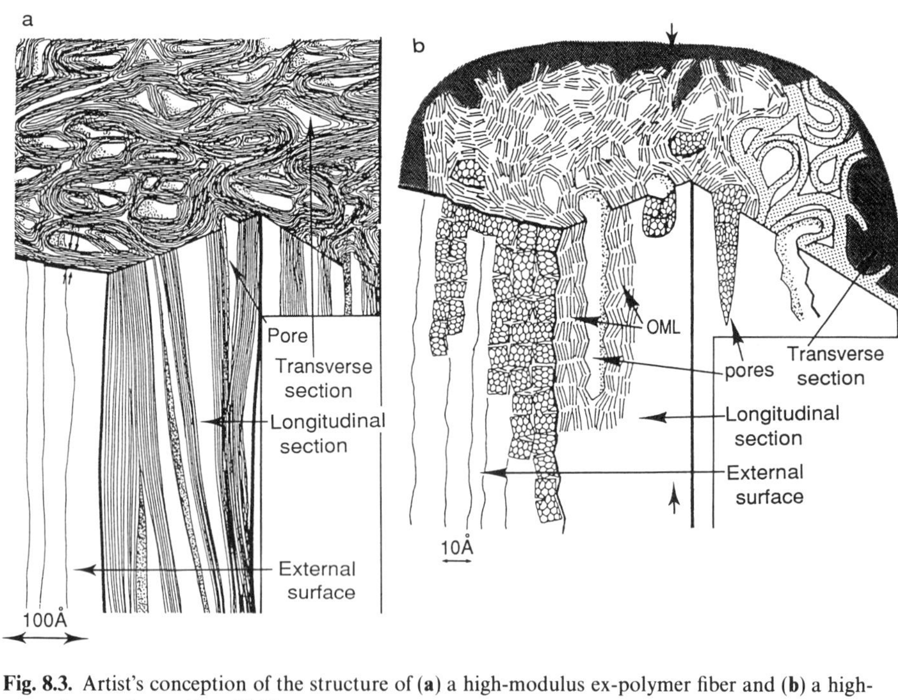

Fig. 8.3. Artist's conception of the structure of (**a**) a high-modulus ex-polymer fiber and (**b**) a high-strength ex-polymer fiber [8.6]

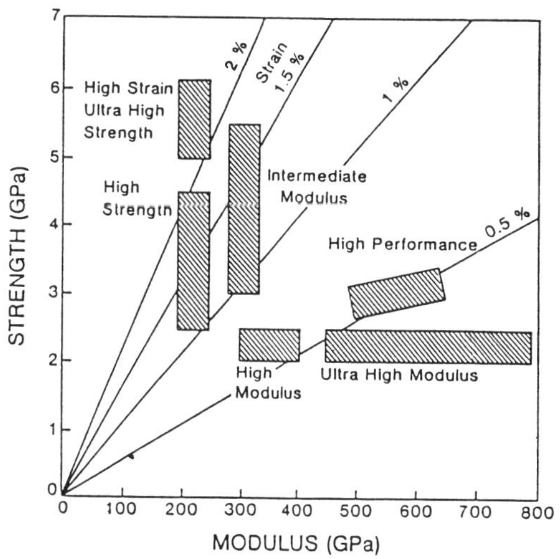

Fig. 8.4. The strength of various types of carbon fibers plotted as a function of Young's modulus. Lines of constant strain can be used to estimate the strain to failure [8.5]

If the fiber cross section is viewed along the fiber axis, the ribbons (stacks of graphene planes) are seen to undulate and separate, but most stay roughly aligned along the fiber axis. Fibers typically have needle-shaped voids (Fig. 8.5a), which reduce the mass density below the value expected for a planar hexagonal array of carbon atoms. It is useful to characterize the crystallinity of carbon fibers by their "crystallite" dimensions $L_{a\|}$ and L_c, representing the extent of relatively straight portions of the lattice fringes along their length, and the stacking height of the graphene planes in the ribbons, respectively, as shown in Fig. 8.5a. Correspondingly, the crystallite dimension $L_{a\perp}$ provides a measure of crystallite dimension in the graphene basal plane and normal to $L_{a\|}$. The structural arrangement of the graphene ribbons in Fig. 8.5a corresponds to high-modulus ex-polymer fibers, while the basket-weave twisted-ribbon structure with smaller microstructural units of Fig. 8.5b is characteristic of high-strength fibers [8.7, 8].

The cross section normal to the fiber axis (Fig. 8.3) indicates a much higher state of disorder than in the longitudinal section (Fig. 8.5). For example, the cross section in Fig. 8.3 shows that the planes are twisted over short distances compared with the ribbon widths, and that there is a random orientation of planes over the diameter of the fibers. These twists in the planar arrays prevent them from stacking in a regular fashion, as in hexagonal 3D graphite.

When the carbon layers are disordered, as in a stack of sheets of papers that lie flat with respect to their neighbors but are randomly rotated, the carbon is said to be *turbostratic*. Such turbostratic fibers cannot be homogeneously intercalated and do not form staged intercalation compounds [8.9, 10]. Nevertheless, the intercalation of turbostratic fibers can result in significant property modifications. Intercalated turbostratic fibers may be very important for practical applications because of their excellent mechanical properties.

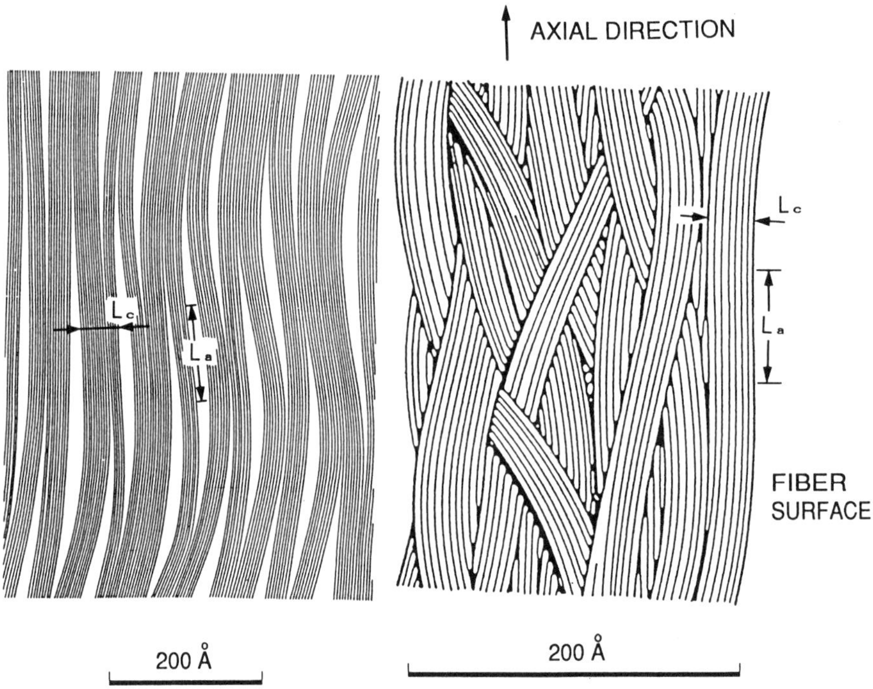

Fig. 8.5. Sketches of the cross section of ex-PAN fibers along the axis direction. (**a**) The in-plane and c axis crystallite sizes $L_{a\|}$ and L_c for a high-modulus fiber [8.8]. (**b**) The corresponding schematic cross section for a high-strength fiber, showing a basket-weave arrangement of ribbons [8.7, 8]

Increasing the heat-treatment temperature (T_{HT}) allows the graphene planes to grow in size. High-resolution electron micrographs of the surface layers of an ex-PAN fiber [8.11] show that when carbonized at 1000 °C, the graphene planes are less than 10 nm in length and are preferentially oriented along the fiber axis. At 1500 °C the graphene planes become more extensive and better aligned. Further improvements in the structural ordering of the ribbons both along and perpendicular to the fiber axis occur on heat treatment of the fibers to $T_{HT} \geqslant 2500$ °C, and further alignment is achieved by stretching the fibers during the graphitization process. It is only after heat treatment to temperatures above 2500 °C that the structural order necessary for systematic intercalation is achieved. Even after high-temperature heat treatment, some crystallites in the carbon fibers remain poorly aligned.

Heat treatment of carbon fibers to $T_{HT} \sim 3000$ °C allows development of interlayer correlations [8.3], as demonstrated by the appearance of weak (hkl) X-ray diffraction peaks when the interlayer spacing becomes less than 0.34 nm. The fiber is then described as partially graphitic. Because of their more fully developed and better oriented graphene layers, graphite fibers generally have higher Young's moduli of elasticity than carbon fibers with lower T_{HT}, though the breaking strengths of the graphite fibers are often observed to be lower than for high-strength carbon fibers. The high modulus ex-PAN fibers can be intercalated and show some tendency toward staging. Ex-mesophase pitch fibers generally have higher Young's modulus and greater structural ordering, and consequently show more of a tendency toward staging. In general, fibers with a turbostratic structure do not show well-developed staging in their intercalation compounds, which directly relates to the need for interlayer correlation to establish preferred in-plane sites for the intercalated atoms. These considerations are fully consistent with the intercalation science for bulk host materials [8.12–14].

The graphene planar network in carbon fibers contains defects, many in the form of multiple vacancies or voids. These voids are formed as the polymer molecules fuse together to form larger graphene ribbons. Impurities, such as H and N, will preferentially attach to either the dangling bonds at these vacancies or the edges of the ribbons. Furthermore, defects exist which produce changes in the bond directions, so that all orientations of the hexagons will be found over the length of a ribbon. This, coupled with stacking disorder, ensures that all orientations of the in-plane axes will occur in the small regions of the fiber probed by X-ray or electron diffraction.

Carbon whiskers or carbon filaments are of particular interest because their physical properties approximate "ideal" graphite values. Carbon whiskers were first produced by striking a dc arc to a carbon electrode under an argon gas pressure of $\sim$ 92 atm [8.15]. The structure of arc-grown carbon whiskers [8.15] is much more regular than that in ex-polymer fibers or as-grown vapor-grown filaments (described below). As Fig. 8.6 shows, the carbon planes of carbon whiskers are arranged in a circular fashion around the whisker axis. X-ray diffraction patterns show single-crystal spots, indicating a highly correlated layer-stacking sequence. Based on this and other observations, it was proposed that the layers are scrolled, and that whisker growth occurs by adding atoms to the scroll edges. Along the fiber axis, the whiskers exhibit properties similar to single-crystal graphite, even in the transport properties, which are sensitive to defects [8.15]. Though no intercalation studies of carbon whiskers have been carried out, one would expect these materials to provide satisfactory hosts for intercalation.

Fig. 8.6. Sketch illustrating the scroll structure of a carbon whisker. The graphene planes are indicated schematically on an exaggerated scale [8.15]

Vapor grown carbon fibers (VGCF) are prepared by the decomposition of gas-phase hydrocarbons onto a catalyzed surface, typically at 1100 °C. After fiber synthesis, heat treatment of the filaments to above $\sim$ 3000 °C results in a structural reorganization, yielding large crystallite dimensions ($L_a > 100$ nm and $L_c > 100$ nm) [8.16]. Normally the vapor-grown carbon fibers (Fig. 8.7) are prepared in short lengths. However, when the CVD growth is carried out on a continuous PAN-based fiber (Fig. 8.8), a continuous VGCF can effectively be synthesized [8.17, 18].

The VGCF filaments have certain similarities to graphite whiskers in that the layers are arranged in a circular fashion about the core (Figs. 8.6, 7a). However, as-deposited filaments ($T_{\mathrm{HT}} \sim 1100$ °C typically) exhibit poor interlayer correlations, and the graphene planes are relatively straight only over short distances (Fig. 8.7a), corresponding approximately to the L_a and L_c values found in ex-PAN carbon fibers after heat treatment. Heat treatment of the VGCF filaments to $T_{\mathrm{HT}} \sim 3000$ °C results in almost complete graphitization, as the crystallite size increases to $L_a \sim 1$ µm (Fig. 8.7b). This tendency for VGCFs to graphitize more readily than ex-polymer fibers is related to the much greater structural organization occurring in as-grown VGCF filaments in comparison with ex-polymer fibers carbonized at the same temperature [8.3]. The circular arrangement of the carbon planes in the VGCF filaments is as a tree ring (Fig. 8.7a), rather than a scroll (Fig. 8.6), and the tree-ring morphology is related to the growth process for the VGCF filaments, which occurs along the fiber axis. After heat treatment above about 2600 °C, faceting of the VGCF filaments is observed (Figs. 8.7b, 8b), and fibers with polyhedral cross sections are formed [8.19]. Because of their very high degree of structural order, these filaments provide very favorable host materials for GICs, and the intercalated VGCF filaments show well-developed staging structure, suitable for scientific investigations of graphite intercalation compounds.

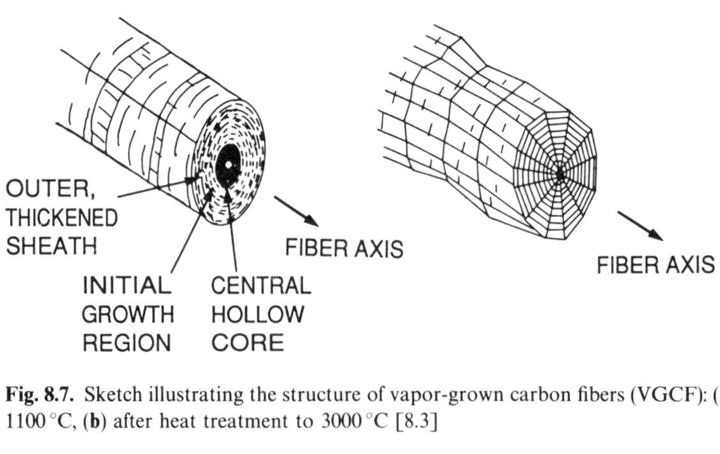

Fig. 8.7. Sketch illustrating the structure of vapor-grown carbon fibers (VGCF): (**a**) as deposited at 1100 °C, (**b**) after heat treatment to 3000 °C [8.3]

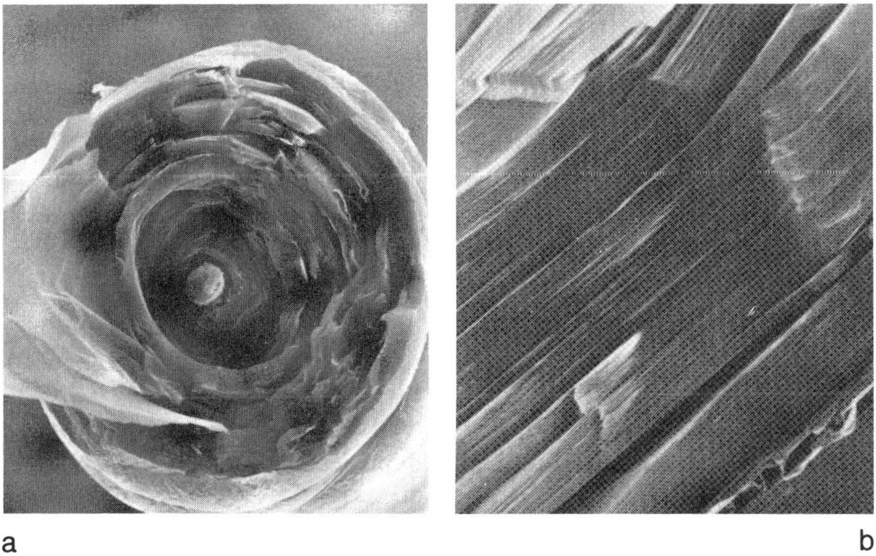

Fig. 8.8. (a) Cross section of a fiber prepared by the chemical vapor deposition process of Fig. 8.44 on a PAN-based continuous core, after heat treatment at 3000 °C. (b) An enlarged (~ 13 ×) view of (a) showing faceting [8.17, 18]

8.2 Intercalation

Intercalation is facilitated by a high degree of structural order of the graphite host material and is retarded by defects, structural imperfections, and disorder. Good staging fidelity is achieved only in hosts with a high degree of structural perfection. Thus, basic scientific studies on intercalated graphite fibers have focused on highly ordered graphite fiber host materials (especially VGCF fibers). In contrast, the requirements of specific commercial applications of intercalated graphite fibers may involve rather different issues, such as the amount of intercalate uptake, quality control of a commercial intercalation process, long-term stability of the intercalated fibers under a variety of environmental conditions, and the availability of continuous intercalated fiber lengths or a continuous intercalation process. Furthermore, for an onion-skin or tree-ring fiber configuration, the presence of some defect sites may facilitate the initiation of the intercalation [8.20].

Graphite fibers are normally intercalated by techniques similar to those used for the corresponding HOPG-based GICs, though the specific intercalation conditions may be different with regard to intercalation temperature, time, and other conditions. For example, the intercalation of a typical chemical species (such as $CuCl_2$) into fiber hosts is successful over a smaller temperature range than for a bulk HOPG host material [8.21, 22]. Because of the small size of fibrous host materials, intercalation times tend to be much shorter. With regard

Table 8.1. Weight uptake and staging for $CuCl_2$ intercalation [8.21]

Graphite host	T [°C]	t [days]	Composition[a]	Stage
Pyrographite (HOPG)	535	7	$C_{10}CuCl_2$	2
Ex-meso pitch fiber $T_{HT} = 3200\,°C$	520	7	$C_{12}CuCl_2$	2, 3, 4
Ex-benzene BDF $T_{HT} = 2900\,°C$	500	2	$C_{11}CuCl_2$	2

[a] From measurements of susceptibility, X-ray diffraction and weight uptake.

to the kinetics, the intercalation of VGCFs is initiated at the free ends of the fibers and then proceeds along the fiber length [8.23]. Fibers prepared from polymeric precursors can be intercalated in a radial direction [8.24].

Table 8.1 compares intercalation conditions for HOPG and two fiber hosts [8.21]. The more graphitic VGCF (ex-benzene) filaments intercalate more similarly to HOPG than to ex-mesophase pitch fibers annealed at comparable heat-treatment temperatures T_{HT} [8.25]. When heat treated to high temperature (e.g., $T_{HT} \sim 2800\,°C$), the VGCF filaments show better staging and a larger intercalant uptake than the ex-mesophase pitch fibers.

Studies of the relative intercalation rates in fibrous host materials are conveniently carried out by in situ resistance measurements during the intercalation process [8.26]. However, the resistivity decrease is not a quantitative measure of the intercalate uptake. The effect of the structure of the fibrous host material on the intercalation rate and on the degree of intercalation is shown in Fig. 8.9. In this figure the resistance change during potassium intercalation is plotted as a function of intercalation time for various hosts [8.26]. Measurements were made with a four-terminal method, and by varying the position of the fiber sample in the two-zone furnace, the temperature difference between the graphite and potassium intercalate could be varied (Fig. 8.9a). The results for the less graphitic fibers are shown in Fig. 8.9b, while those for the graphitic VGCF ($T_{HT} = 2800\,°C$), for mesophase pitch (UC 4104B), and for PAN GY70 fibrous host materials are shown in Fig. 8.9c.

Figure 8.9b shows the surprising result that potassium can to some degree be intercalated into quite disordered carbon fibers prepared at the low growth temperature of $\sim 1000\,°C$, for isotropic pitch, mesophase pitch, or as-grown vapor-grown fibers. In contrast, it is not possible to intercalate AsF_5 [8.23] or Br_2 [8.28] into these disordered fibers. In this figure, the initial decrease in resistance is attributed to the initiation of intercalation, and the subsequent increase in resistance is associated with the degradation on the fiber structure and the breaking off of pieces of fiber near the fiber surface due to local intercalation-induced strains [8.29]. The amount of intercalate uptake is least for the isotropic pitch fibers, and hence the smaller the mechanical degradation, while the degradation effect is most pronounced for the as-grown vapor-grown filaments [8.26]. Though the resistance curves in this figure are plotted on a

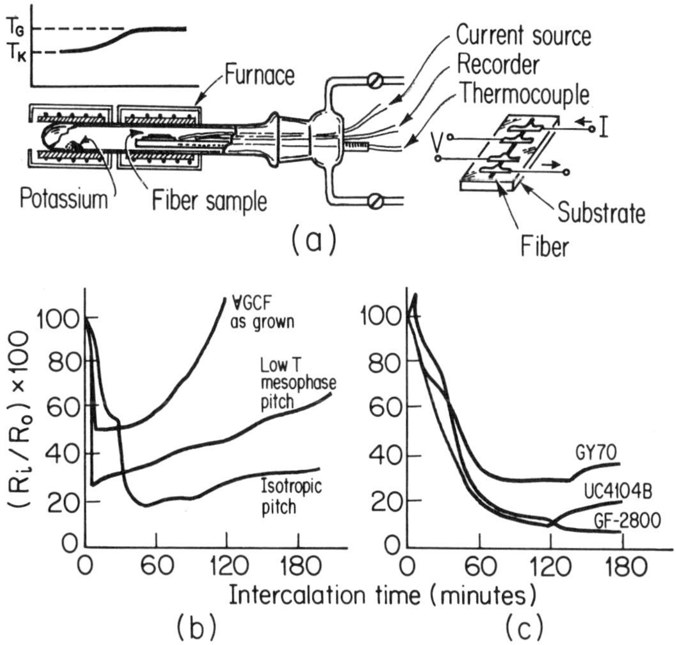

Fig. 8.9. In situ monitoring by resistance measurements of the intercalation process for the donor intercalate potassium into different host materials: (**a**) Schematic illustration of the experimental setup used for in situ resistance measurements as the potassium intercalation proceeds. The furnace arrangement allows for a temperature difference between the fiber sample T_G and the potassium to be intercalated T_K. The four-terminal electrode arrangement for resistance measurements is also shown. (**b**) Plots of the resistance during intercalation R_i relative to R_0 prior to intercalation for various poorly ordered host fibers. The discontinuity in the curves for short times is attributed to mechanical breakup of the fiber due to intercalation-induced stress in the poorly ordered fibers. (**c**) Plots of resistance change during potassium intercalation into PAN (GY70), mesophase pitch (UC4104B), and VGCF (GF-2800) filaments. After about two hours of intercalation, the furnace was turned off, stopping further intercalation [8.27]

normalized scale, the actual resistivities of the isotropic pitch fibers are two orders of magnitude greater than for the as-grown vapor-grown carbon fibers.

The curves of Fig. 8.9c show that intercalation with potassium results in a greater decrease in resistance in the graphitized vapor-grown and mesophase pitch fibers than for the GY70 host material. After about two hours of intercalation with potassium (Fig. 8.9c), the intercalation was stopped and the fibers were cooled [8.26]. The decrease in resistance observed for the intercalated VGCF filament on cooling is typical of metals and of intercalated HOPG bulk hosts, while the intercalated mesophase pitch and PAN fibers show an *increase* in resistance on cooling, characteristic of disordered carbons. The intercalation-induced decrease in the room-temperature resistance is more pronounced for the intercalated UC4104B fibers than for the GY70 fibers, again consistent with the greater structural order of the mesophase pitch fibers, thereby permitting more effective intercalation than for PAN fibers. The anomalous increase in resistance for the GF-2800 fiber at very small intercalation times may be due to the randomness of the initial in-plane intercalate density, which gives rise to a larger degradation in carrier mobility than an increase in carrier density.

For a given fibrous host material, the choice of intercalate species affects both the intercalation rate and the magnitude of the decrease in resistance that is observed as a result of intercalation. In this context, we show in Fig. 8.10a in situ resistance measurements for the HNO_3 acceptor intercalate for the most highly ordered fibrous host material, VGCF with $T_{HT} > 2900\,^{\circ}C$. An apparatus useful for in situ resistance measurements on acceptor compounds based on liquid intercalate species is shown in Fig. 8.10b for the case of liquid HNO_3 intercalation. In situ resistivity measurements using reactors based on vapor-phase intercalation (Fig. 8.10c) are also commonly used [8.32], yielding results similar to those obtained with reactors based on liquid-phase intercalation. The results

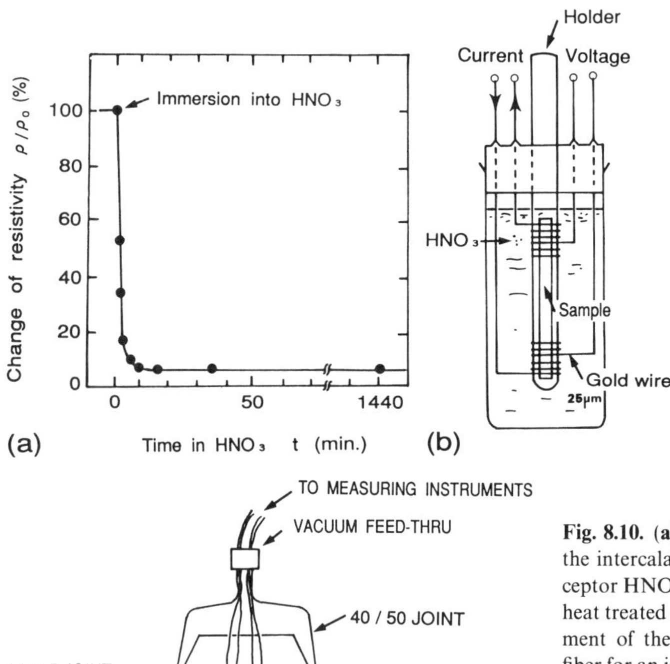

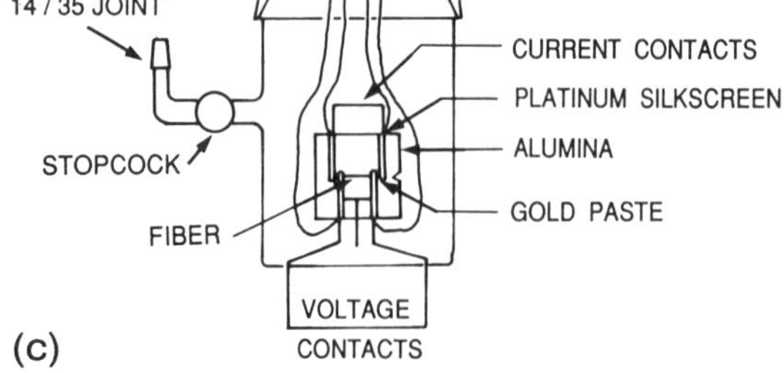

Fig. 8.10. (**a**) In situ monitoring of the intercalation process of the acceptor HNO_3 into VGCF filaments heat treated to 2900 °C by measurement of the resistance of a single fiber for an intercalation time t normalized to the resistance of the fiber before intercalation [8.30]. (**b**) An apparatus used for in situ resistivity measurements of a single fiber during liquid-phase HNO_3 intercalation [8.31]. (**c**) A reaction chamber used for in situ resistivity measurements of a single fiber during gas-phase intercalation [8.32]

of Fig. 8.10a show that rapid intercalation occurs with HNO_3. Rapid (though somewhat slower) intercalation is found with Br_2, and the eventual decrease in resistance is less for Br_2 than for HNO_3. Of all intercalates that have been studied thus far, the largest decrease in resistivity occurs for AsF_5, though the intercalation time is significantly longer than for the intercalates Br_2 and HNO_3.

The presence of defects in fibers implies that one needs a larger driving force to initiate intercalation [8.33]. Nevertheless, the small dimensions of the fiber cross sections can lead to a striking enhancement of the rate of intercalation relative to bulk graphite. For example, the preparation of stage-1 or stage-2 AsF_5 intercalation compounds in HOPG typically takes one or more days. Fibers, however, can be fully intercalated in less than one hour. This is because the fibers are small and thin, and because most commercially produced fibers have a sufficient density of exposed basal plane edges so that intercalation occurs fairly uniformly along the fiber length. Thus, intercalation of a 180-cm-long yarn of ex-PAN fibers [8.24] was observed to occur at roughly the same rate as a 2 cm single filament [8.34]. Enhanced intercalate uptake (at least near the fiber surface) is observed for various intercalates so that, for example, stage-1 $NiCl_2$ and $FeCl_3$ graphite intercalation compounds are more easily prepared in graphite fiber hosts than in bulk graphite [8.35].

For practical applications, it is important that intercalation times are short and that the process is convenient and efficient, for appropriate scale-up and reduction of processing costs. Thus intercalation by immersion of the fibers in liquid bromine is attractive from the point of view of convenience and commercial scale-up [8.19]. One attractive design for the preparation of larger quantities of bromine intercalated carbon fibers is shown in Fig. 8.11 [8.19, 28]. An advantage of this intercalation apparatus is the ability to carry out a quasi-continuous intercalation cycle and then to reuse the liquid-bromine intercalate

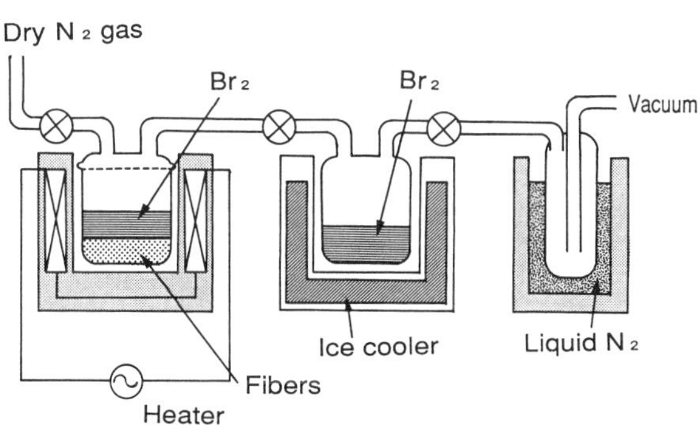

Fig. 8.11. Experimental setup for the bromination of carbon fibers, using a continuous cycle: (**a**) reaction vessel with heater, (**b**) Br_2 reservoir to receive bromine from the reaction vessel of (**a**) after bromination, and (**c**) liquid N_2 trap [8.19, 28]

for subsequent intercalation steps. In this apparatus, bromine at room temperature is added to the bromine reservoir chamber, and the fibers are subsequently added to the fiber reaction chamber Fig. 8.11a. Bromine is then transferred from the bromine reservoir chamber (Fig. 8.11b) to the fiber sample chamber by slightly heating the bromine. The fibers are kept in the reaction vessel at $\sim 25\,^{\circ}$C in the liquid bromine while the intercalation takes place. After the intercalation is completed, the bromine can be transferred from the reaction vessel to the bromine reservoir chamber using flowing dry N_2 gas [8.19, 28].

In most cases, the higher density of structural defects in the fibers tends to inhibit intercalation relative to bulk HOPG [8.28, 29]. The threshold pressure for intercalation is found to depend upon whether the fiber has been previously intercalated and on the species that had been intercalated [8.36]. If a fiber is deintercalated and then subsequently reintercalated, the threshold pressure is significantly reduced. Furthermore, fibers that could not be intercalated with one halogen (e.g., bromine) because of an excessively high threshold pressure, could be intercalated after initial intercalation with another halogen (e.g., first intercalation with ICI and then subsequent intercalation with bromine) [8.36].

In addition to the fibrous ionic GICs discussed above, the synthesis of covalent GICs in fibrous form is of interest for special applications. For example, carbon fibers intercalated with fluorine to achieve $(CF)_n$ and $(C_2F)_n$ (Fig. 8.12) are most important from an applications standpoint [8.26, 37]. In

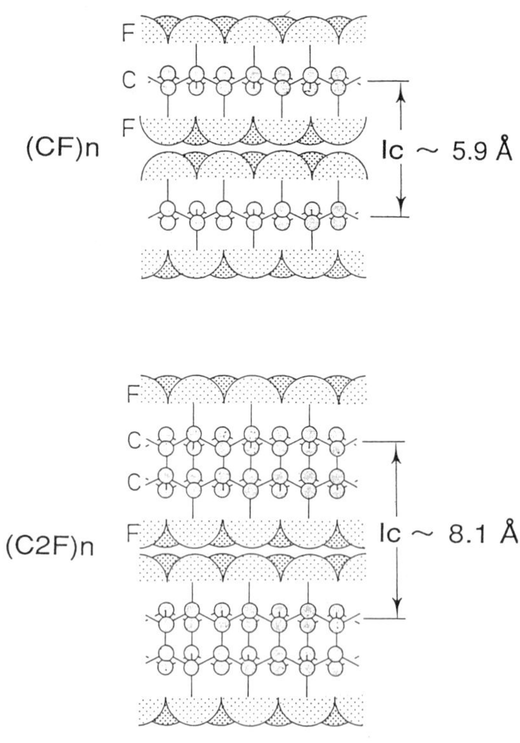

Fig. 8.12. Crystal structures of the covalent $(CF)_n$ and $(C_2F)_n$ compounds. The puckered graphene layers are represented by the small open circles and the fluorine intercalated layers by the large speckled circles. The I_c values are ~ 5.9 Å for stage-1 $(CF)_n$ and ~ 8.1 Å for stage-2 $(C_2F)_n$ [8.37, 38]

these covalent graphite intercalation compounds, the carbon atoms in the graphene planes are displaced into a nonplanar boat-like structure [8.37–39], as indicated by the zigzag lines in Fig. 8.12. The white stage-1 $(CF)_n$ compound is prepared by reaction of turbostratic carbon with fluorine at 500 °C, while the grayish or brown stage-2 compound is prepared between 300° and 400 °C. Their most important application is for very high energy density batteries, while the high chemical stability and low surface energy of $(CF)_n$ and $(C_2F)_n$ make these compounds attractive for use as lubricants [8.28]. As discussed in Sect. 8.8, improved battery performance is achieved by use of fibrous host materials for intercalation, especially for the case of the $(C_2F)_n$ compound, which can only be obtained using highly graphitized vapor-grown fibers. Interestingly, fluorine can also be intercalated into carbon fibers to form ionic GICs with enhanced conductivities [8.39, 40], while the covalent compounds have very low conductivities (are almost insulating) [8.37, 41].

Closely related to the fluoride compounds are the covalent graphite oxide compounds shown schematically in Fig. 8.13a [8.42]. A major difference between the graphite oxide compounds and other compounds is the reactivity of these compounds with water. Graphite oxide absorbs water molecules between the layer planes, and the I_c value of the intercalation compound changes with water content. Graphite oxide is prepared from a well graphitized host material,

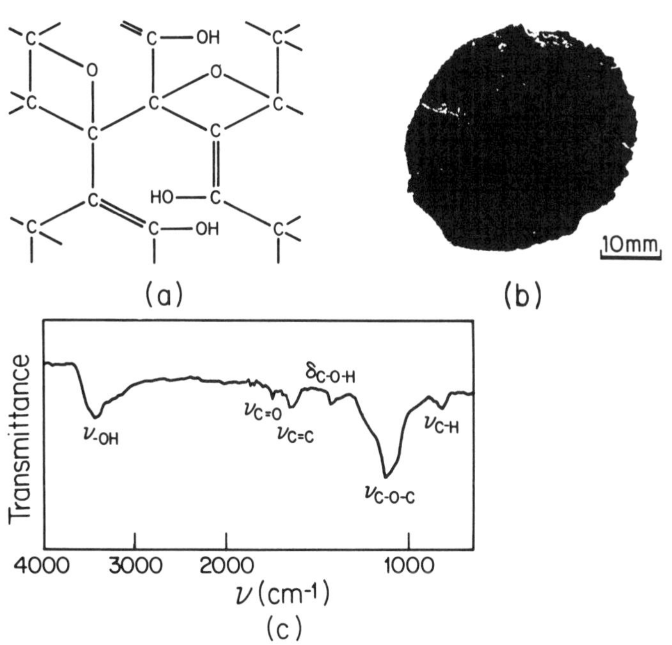

Fig. 8.13. (**a**) Schematic diagram of the structure of the covalent graphite oxide compound. (**b**) Photograph of a graphite oxide sheet after one day of oxidation. (**c**) Optical transmittance spectrum of a typical graphite oxide fiber sample [8.42]

similar to the case of $(C_2F)_n$. To form a fibrous graphite oxide material, well-graphitized vapor-grown fibers provide a suitable host. Figure 8.13b shows a sheet of graphite oxide obtained from heat-treated vapor-grown carbon fibers and having the chemical composition $C_8O_2(OH)_2$. It has been postulated that in the covalent graphite oxide, the oxygen atoms lie in layers adjacent to the boatlike carbon layers, while the OH groups lie between the oxygen layers [8.43]. Also of interest is the fact that the infrared transmission spectra for graphite oxide bulk and fibrous samples (Fig. 8.13c) are the same, showing evidence for C–O, C–OH, and C–H bonds. The connection between these bonds and the proposed layer stacking [8.40] is unclear. It has been shown that the fibrous graphite oxide is a high-performance cathode material in a Li primary battery, similar to the case of $(CF)_n$. This is further discussed in Sect. 8.8.

Of particular significance for both scientific studies and practical applications is the higher stability of intercalated VGCF fibers with regard to deintercalation under ambient conditions [8.19, 21, 23, 44, 45] and when heated to elevated temperatures. These issues are, further discussed in Sect. 8.5.2.

8.3 Structure and Staging

Since a relatively high degree of structural order is required for intercalation of the host fibers, a well-established graphite structure provides a framework for most of the fiber-based intercalation compounds. The dominant structure of intercalated graphite fibers is the staging periodicity along the *c* axis, which is normal to the fiber axis. Just like bulk GICs, intercalated fibers under ideal conditions may exhibit in-plane ordering which can be either commensurate or incommensurate with respect to the adjacent graphite layers [8.12, 46]. For a given intercalate species, the ideal *c* axis and in-plane ordering of the intercalation compound is the same in fibrous graphite hosts as for HOPG [8.46], though the degree of ordering and the degree of staging that is achieved in the actual samples tends to be significantly lower in the fibrous host materials. For example, the density of the fibers tends to be lower than that for HOPG, both before and after intercalation [8.21], consistent with the presence of voids in the pristine and intercalated fibers. The larger variation in density after intercalation of the fibers is consistent with lower homogeneity of the intercalation process in fiber hosts (Table 8.2).

Fibers provide a convenient host material for studying *c* axis structural phenomena, because the lattice imaging method can be applied generally to thin fibers (diameter $\leqslant 1\ \mu m$), without further thinning. In contrast, *c* axis lattice images are more difficult to observe in bulk graphite hosts, since observation of *c* axis lattice fringes requires that the sample edges be turned up to achieve the proper geometry for this observation [8.47, 48].

To carry out properties measurements, it is important to characterize the intercalated carbon fibers with regard to the degree of structural order and

Table 8.2. Fiber densities before and after intercalation [8.21]

Fibers	Density before intercalation [g/cm^3]	Density after intercalation [g/cm^3]
Ex-benzene (vapor deposited)	2.08 ± 0.05	2.40 ± 0.1
Ex-mesophase pitch[a] ($T_{HT} = 2500\,^\circ C$)	2.12 ± 0.05	2.36 ± 0.1
Ex-mesophase pitch[a] ($T_{HT} = 3200\,^\circ C$)	2.12 ± 0.05	2.45 ± 0.1
Graphite (calculated values)	2.26	2.56 (2nd stage)

[a] From Union Carbide Corp.

staging. Such fiber characterization is typically carried out using X-ray diffraction and electron microscopy techniques. The more ordered regions of a carbon fiber intercalate better than the disordered regions. For this reason, intercalation compounds based on fibrous host materials tend to be inhomogeneous, with relatively high internal stresses, due to the swelling of the intercalated regions relative to the unintercalated regions. For disordered fibers, structural studies focus on a determination of the amount of intercalation that has occurred, the distribution of the intercalate over the fiber cross section and along the fiber length, and the degree of staging in the intercalated regions. While similar characterization measurements are made on the more ordered intercalated fibers, the focus of structural studies of more highly ordered fibers is more often directed to a comparison between the structure and staging of the intercalated fibers and that of HOPG intercalated with the same species to the same nominal stage.

To compare the properties of GICs based on fibers and on bulk graphite hosts, knowledge of the stage of the GIC is important. Therefore, considerable effort has been directed to the characterization of intercalated graphite fibers with regard to both stage index and staging fidelity (both longitudinal and transverse to the c axis). The intercalated fibers can be characterized for stage by a number of techniques. For example, the staging of a bunch of intercalated graphite fibers can be conveniently measured using the Debye-Scherrer X-ray diffraction technique (Fig. 8.14a). This technique has been applied to characterize the staging in intercalated ex-PAN, ex-mesophase pitch, and vapor-grown graphite fibers. An example illustrating the degree to which staging can be achieved in a vapor-grown graphite fiber is given in Fig. 8.14b, which shows a Debye-Scherrer pattern for a nominal stage-2 $FeCl_3$-GIC fiber sample [8.49]. The $(00l)$ reflections are indexed, and the densitometer trace for this diffraction pattern shows predominantly stage-2 diffraction peaks with some small admixture of stage-3 regions.

While Debye-Scherrer patterns provide average structural information of a statistical nature, they provide little information about the variation of staging

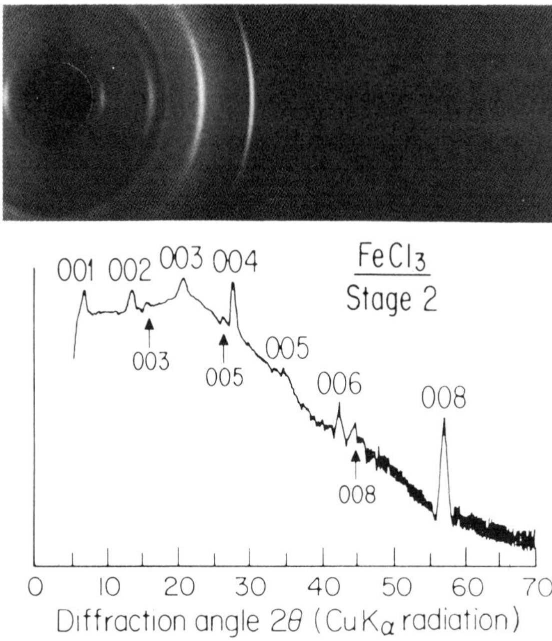

Fig. 8.14. Debye-Scherrer X-ray diffraction pattern taken with Cu $K\alpha$ radiation and the corresponding microdensitometer trace for a stage-2 $FeCl_3$-intercalated VGCF fiber. The $(00l)$ reflections are indexed, with labels above the trace for the stage-2 reflections, and below the trace for the small admixture of stage-3 regions [8.49]

from one fiber to another or along the length of a single fiber. To identify the staging in a single fiber and to determine the staging variation along the length of the fiber, characterization has more generally been carried out using either lattice imaging techniques with a high-resolution transmission electron microscope (TEM) [8.50] or by conventional Raman spectroscopy [8.51] or with a Raman microprobe [8.52, 53]. The field emission scanning transmission electron microscope (FESTEM) has also been used to determine the compositional variation along the fiber length, based on the X-ray fluorescence technique [8.54, 55].

Because of their relatively high degree of structural order, VGCF graphite fibers heat treated to temperatures $T_{HT} \geqslant 2800\,°C$ have been extensively characterized by TEM techniques. High-resolution TEM studies [8.50] show that the degree of ordering in the intercalated fibers increases with increasing order of the pristine fiber host. For example, lattice fringe images taken on VGCF filaments with $T_{HT} \geqslant 2800\,°C$ show defect-free parallel lattice planes for distances extending at least 1000 Å [8.50]. The effect of intercalation is illustrated in Fig. 8.15, which shows lattice fringe patterns for $CuCl_2$, K, and Br_2 for VGCF host filaments previously heat treated to $T_{HT} = 2900\,°C$ [8.9, 56]. Of significance is the good staging fidelity that can be achieved in VGCF filaments under optimal

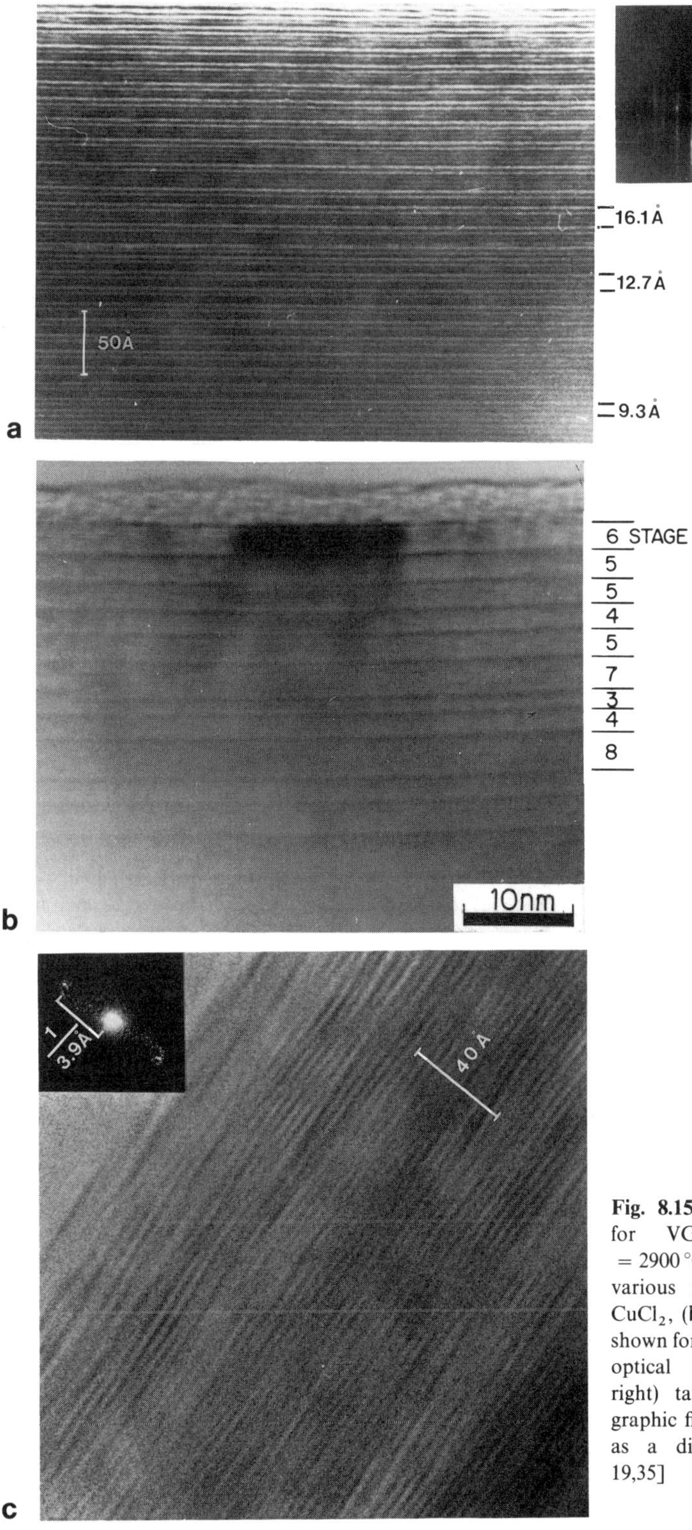

Fig. 8.15. Lattice fringe images for VGCF filaments (T_{HT} = 2900 °C) intercalated with various intercalate species: (**a**) $CuCl_2$, (**b**) Br_2, and (**c**) K. Also shown for the case of $CuCl_2$ is the optical diffractogram (upper right) taken using the photographic film of the lattice fringes as a diffraction grating [8.9, 19,35]

conditions and the variety of defect structures observed for the various intercalate species. For example, the $CuCl_2$ intercalate characteristically shows very straight fringes. In general, acceptor compounds tend to show much straighter fringe patterns than donor GICs. The photograph of Fig. 8.15a shows layer sequences for a dominantly stage $n = 2$ $CuCl_2$-intercalated VGCF fiber, though some $n = 1$ and $n = 3$ staging sequences are easily identified in the photograph where some of the repeat distances I_c are specified. This identification of the lattice planes is made possible by the different atomic numbers of the various chemical species and hence the different in-plane electron densities in the intercalate layers relative to the graphite layers, thereby showing contrast between electrons scattered by the intercalate and by the graphite layers. For the case of $CuCl_2$, the structural coherence distance in the plane (L_a) is much greater than normal to the plane (L_c). These characteristic patterns vary from one intercalate species to another, thereby giving rise to characteristic visual patterns for each intercalate. For example, the potassium intercalate layer (Fig. 8.15c) shows much shorter straight regions and more curved fringes generally than for the case of $CuCl_2$, and the bromine intercalate also shows shorter straight regions than for $CuCl_2$, but longer than for K. It is significant that the extent of the long, defect-free regions of graphite intercalation compounds is strongly dependent on the perfection of the host crystal, the characteristics of the intercalated species, and the conditions of intercalation. Commensurate GICs based on the KHg and Br_2 intercalate species can exhibit long structural coherence distances in bulk GICs [8.57].

It should be emphasized that the high-resolution lattice fringe imaging technique provides microscopic information on the type and density of defects found in intercalated graphite fibers. Statistical information is found by scanning the length of many similarly intercalated fibers. The most common types of defects in relatively well-staged fibers are staging infidelities whereby a stage $n - 1$ or $n + 1$ sandwich is observed directly on the micrograph of a nominally stage n compound (Fig. 8.15).

Debye-Scherrer patterns for intercalated ex-PAN and ex-mesophase pitch fibers show a much lower degree of staging, with the ex-mesophase pitch fibers showing more evidence for staging than the ex-PAN fibers [8.3, 21, 58]. This is consistent with the relative degree of structural perfection found in these three fibrous host materials. The degree of staging fidelity that is achieved in a given fibrous host material also depends on the intercalate species and the intercalation conditions. For example, PAN fibers intercalated with AsF_5 were found to exhibit significant evidence for staging [8.59], though even under the most favorable conditions, single-stage fibers have not been realized in PAN host materials.

X-ray diffraction patterns from intercalated PAN and mesophase pitch fibers do not generally show strong interlayer correlations, and therefore these imperfectly ordered fibers provide suitable host materials for the observation of 2D phenomena (Sect. 8.5.4).

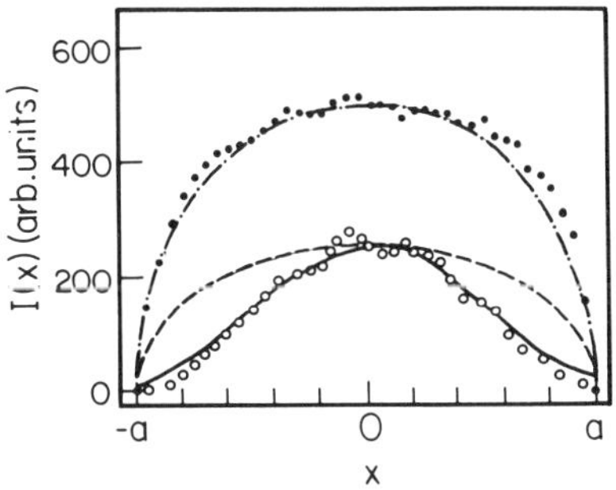

Fig. 8.16. A plot of Br $K\alpha$ X-ray intensity versus the lateral position (x) of the incident beam. The dots denote the experimental data for the Br-rich region, and the dashed line is the calculated intensity from a radially uniform Br distribution. The open circles represent the experimental data points in the Br-poor region, and the solid line is the intensity calculated from a Gaussian Br distribution [8.54, 55]

The field emission scanning transmission electron microscope (FESTEM) has also been used to monitor the homogeneity of the intercalate concentration along the fiber length. Specific studies have been made for a partially desorbed Br_2 intercalated pitch-based fiber by mapping the intensity of the Br $K\alpha_1$ X-ray line, while scanning the fiber length [8.54, 55]. The FESTEM technique has also been applied to study the lateral homogeneity of the bromine distribution across the fiber cross section, showing that the bromine concentration is approximately uniform along the cross section for bromine-rich regions (Fig. 8.16) and follows a Gaussian distribution over the bromine-poor regions. This microscopic information was then used by the authors to model the macroscopic residual resistivity of the fiber [8.54, 55]. Energy dispersive X-ray scans along the fiber diameter of a brominated P100 fiber [8.60] also showed the intercalate distribution to be nonuniform, consistent with the FESTEM observations.

8.4 Raman Characterization

Raman scattering provides a powerful tool for the characterization of carbon fibers both before and after intercalation. Prior to intercalation, Raman spectroscopy is used to determine the degree of structural disorder nondestructively [8.61, 62] by measuring the intensity of the disorder-induced D line at $\sim 1360\ \mathrm{cm}^{-1}$ relative to that of the E_{2g_2} Raman-allowed G line at $1582\ \mathrm{cm}^{-1}$.

When disorder is introduced into the graphite, the $1582\ \mathrm{cm}^{-1}$ G line is broadened and a new disorder-induced line near $1360\ \mathrm{cm}^{-1}$ appears [8.61, 62]. Occasionally, a second disorder-induced line near $1620\ \mathrm{cm}^{-1}$ is observed. Because of the small wave vectors for visible light (typically having a wavelength $\lambda \sim 4880$ Å), the Raman-allowed lines correspond to the lattice modes at the center of the Brillouin zone ($\boldsymbol{q} = 0$). The introduction of disorder relaxes the $\boldsymbol{q}$ conservation selection rule and allows contributions to the Raman spectra from phonons throughout the Brillouin zone, so that the disorder-induced lines correspond to maxima in the phonon density of states. The extent of the structural disorder is therefore characterized by the intensities of the disorder-induced lines relative to the Raman-allowed lines [8.3, 61–64]. That is, the in-plane crystallite dimension (L_a) is proportional to the intensity ratio [8.61] such that

$$\frac{I(1360)}{I(1582)} = \frac{C}{L_a}, \tag{8.1}$$

where $I(1360)$ and $I(1582)$ represent respectively the integrated intensities of the disorder-induced and Raman-allowed peaks, C is a proportionality constant at a given laser wavelength, and L_a is the in-plane crystallite dimension.

For the intercalated fibers, Raman scattering is used to provide staging information for a bunch of fibers using a conventional Raman scattering system or for single fibers using a Raman microprobe. Modifications to the Raman spectra due to intercalation are essentially identical for GICs based on HOPG or on the most graphitic fibrous host materials [8.49]. We therefore can exploit the broad knowledge base that has been developed for bulk GICs [8.3, 12, 62 64] to characterize intercalated fibers.

Studies of the Raman spectra show that the perturbation due to the intercalate layer is largely confined to the graphite bounding layers, which effectively screen the graphite interior layers from the charged intercalate layer. The modes associated with the graphite bounding layers $\hat{E}_{2g_2}$ are upshifted in frequency relative to the interior layer modes $E^0_{2g_2}$ by $\sim 20\,\mathrm{cm}^{-1}$ (Fig. 8.17), almost independent of intercalate species and concentration. The intensity of the upshifted modes grows with increasing intercalate concentration as the relative number of graphite bounding layers increases. The lattice mode structure of the intercalated layer depends in detail on the intercalate species and is related to the corresponding modes in the parent crystalline solid [8.64].

Since the stage dependence of Raman-active frequencies for both the graphite interior layers $\omega(E^0_{2g_2})$ and graphite bounding layers $\omega(\hat{E}_{2g_2})$ is the same for HOPG-based GICs and graphite-fiber-based GICs, the measured Raman frequencies can be used to yield the stage of a given donor or acceptor GIC fiber [8.9, 64]. The relative intensities of the Raman lines associated with the graphite bounding layers and the graphite interior layers (Fig. 8.17c) can also provide information on the stage of the graphite fibers. As the intercalate uptake increases, and the number of graphite bounding layers relative to the number of graphite interior layers increases, so does the intensity of the bounding layer mode $\hat{E}_{2g_2}$ relative to that for the interior layer $E^0_{2g_2}$ mode [8.13]. We note that for state-1 and stage-2 compounds, all the graphite layers are adjacent to intercalate layers, so that the intensity for the $E^0_{2g_2}$ mode due to the graphite interior layers vanishes [8.64].

For acceptor compounds, the frequencies are upshifted as a function of reciprocal stage (Fig. 8.17a), reflecting the decrease in the intraplanar bond lengths as electron charge is transferred from the graphite to the intercalate layers. Of significance are the similar upshifts in frequency observed for a large

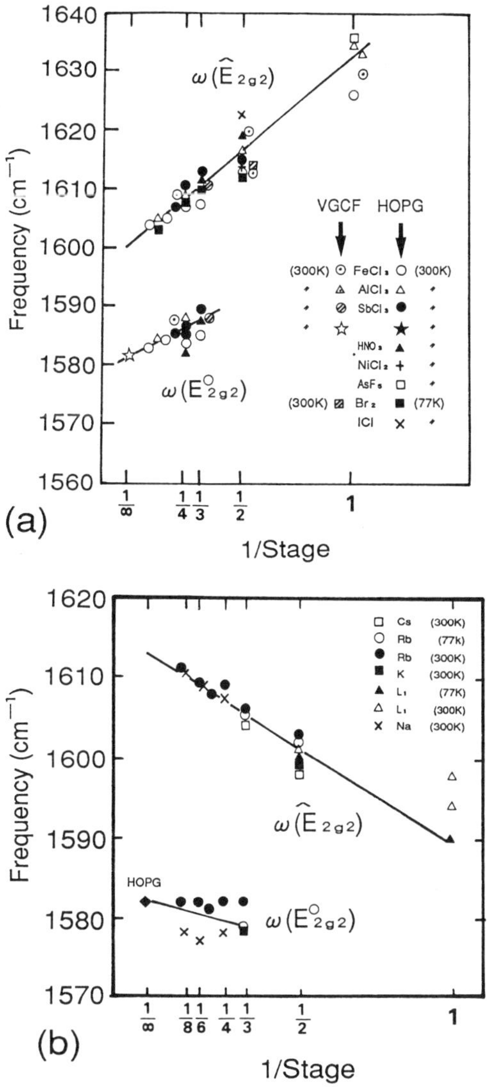

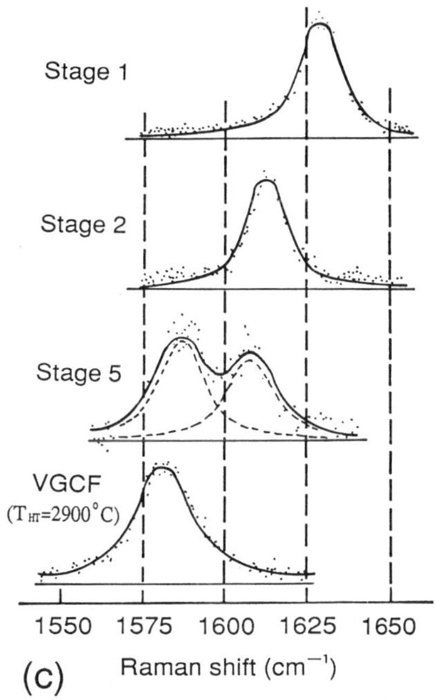

Fig. 8.17. Raman frequencies $\omega(\hat{E}_{2g_2})$ and $\omega(E^0_{2g_2})$ versus reciprocal stage for (**a**) acceptor- and (**b**) donor-intercalated VGCF filaments and bulk HOPG [8.64]. The stage of the intercalated graphite fibers can be found by comparison with the measured Raman frequencies. (**c**) First-order Raman spectra for several stages of $FeCl_3$-intercalated VGCF filaments ($T_{HT} = 2900\,°C$). The solid lines are fits to the experimental points and determine the central frequency and linewidth of each Raman line. The spectrum for the E_{2g_2} mode of pristine graphite fibers (VGCF) is also shown for comparison [8.49]

number of acceptor intercalates and the similarity in behavior found for graphite fibers and HOPG host materials intercalated to the same stage [8.49]. Thus, using the results of Fig. 8.17a, the stage of acceptor-intercalated graphite fibers can be found from the measured Raman frequencies $\omega(\hat{E}_{2g_2})$ and $\omega(E^0_{2g_2})$. In contrast to the acceptor GICs, the donor GICs exhibit a characteristic downshift in frequency, since the bond lengths increase as electrons are transferred to the graphite from the donor intercalate. Figure 8.17b shows the characteristic downshift in frequency $\omega(\hat{E}_{2g_2})$ for a variety of donor compounds, and the results of this figure are used to determine the stage of donor intercalated graphite fibers. Fewer Raman spectra are available for donor-intercalated fibers than for the acceptors.

Narrow Raman lines occur only for well-staged samples. Thus, use of the small shifts in the Raman mode frequency to provide stage identification for graphite fibers requires the fibers to be well staged. In the case of intercalated ex-PAN and mesophase pitch fibers, the Raman spectra for the $\hat{E}_{2g_2}$ and $E^0_{2g_2}$ lines are often very broad [8.51], and only qualitative information on the staging can be obtained from the peak frequencies or the relative intensities of the $\hat{E}_{2g_2}$ and $E^0_{2g_2}$ Raman lines [8.51]. For VGCF graphite fibers heat treated to $T_{HT} \geq 2900\,°C$, the staging fidelity of the intercalated fibers is sufficient to show Raman linewidths comparable to those for HOPG intercalated with the same intercalate species and to the same stage [8.49], thereby allowing definitive stage characterization from the Raman frequencies for the graphite bounding layer mode. This feature is used in the application of the Raman microprobe for the stage characterization of single graphite fibers [8.52], as described below.

A Raman microprobe (beam size of $\sim 2\,\mu m$) can be used to determine the stage index of single fibers and the staging fidelity along the length of a fiber. Using the special focusing optics of the Raman microprobe, stage determination is achieved by measurement of both the mode frequencies and the mode intensities for the graphite bounding layers relative to those for the graphite interior layers. The stage determination of *single* intercalated graphite fibers by a nondestructive technique is important because it allows quantitative and reproducible measurements to be made of the intercalation-induced modifications to the properties of graphite fibers. To illustrate typical variations in the staging of a fiber, we show in Fig. 8.18 Raman microprobe spectra taken along the length of a nominally stage-3 $FeCl_3$-intercalated VGCF graphite fiber ($T_{HT} = 2900\,°C$). The spectra show variations from one region to another, indicative of some admixture of stage-4 regions [8.52]. The variation in the spectra along the length of an intercalated fiber (as demonstrated by the microprobe scans) tends to be significantly larger than for similar scans taken along an HOPG-based sample with the same intercalate species and nominal stage [8.52]. The same

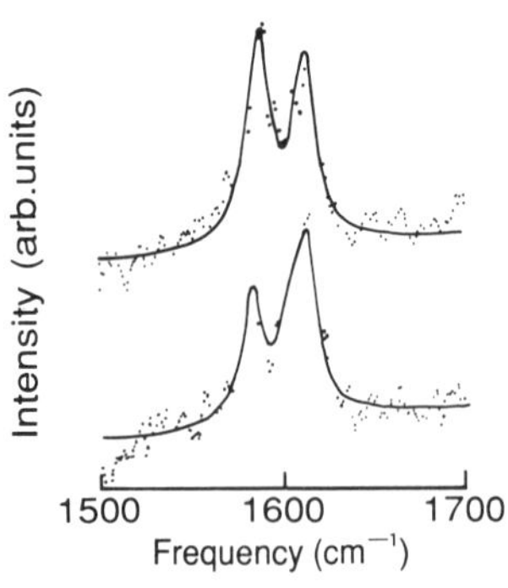

Fig. 8.18. Raman microprobe spectra of a nominally stage-3 $FeCl_3$-intercalated vapor-grown graphite fiber, showing adjacent regions with different mixtures of stages 3 and 4. The two spectra were taken from regions separated by 7.5 μm along the length of the fiber [8.52]

conclusions are reached by lattice fringing studies, which show individual staging defects (e.g., one sequence of stage 3 followed by several stage-2 sequences, as shown in Fig. 8.15). In contrast, the Raman microprobe provides staging information averaged over a macroscopic (2 μm) distance, which contains many unit cells. While the Raman microprobe provides a convenient tool for rapid, nondestructive characterization of single fibers, the lattice fringing technique provides complementary microscopic information on individual defects. However, the lattice fringing method cannot be used for rapid, nondestructive characterization of fibers which are to be subsequently used for other experiments [8.35].

The Raman microprobe has also been applied to a comparative characterization of staging in ex-benzene VGCF and ex-mesophase pitch intercalated fibers, by comparing Raman microprobe spectra taken with light incident along the fiber axis and perpendicular to it [8.21]. The results show that whereas the ex-benzene fibers are approximately homogeneously staged up to the fiber surface, the ex-mesophase pitch fibers show no evidence that there are any intercalate species close to the surface, but rather that the intercalate is present deeper into the fiber. This intercalate distribution has been used to account for the air stability of the intercalated ex-mesophase pitch fibers [8.21]. In contrast, the relative absence of edge sites in the VGCF fiber structure is likely responsible for the air stability of intercalated VGCF filaments and their more uniform intercalate distribution among fibrous GICs.

8.5 Transport Properties

In the past few years, attention has been focused on exploitation of the order of magnitude intercalation-induced enhancement of the electrical conductivity of graphite fibers for the fabrication of practical high-conductivity, lightweight conductors [8.21, 24, 65–70]. Furthermore, the fiber geometry offers advantages relative to HOPG for measurement of several of the transport properties of GICs and for increasing the compositional stability of GICs under ambient conditions and also at elevated temperatures [8.31, 50]. Since the fiber geometry provides the method of choice for various transport measurements in GICs, further information on the transport properties of fibrous and HOPG-based GICs is found in Chap. 6.

As in the case of bulk GICs, intercalation increases the density of carriers by injection of electrons into the graphite planes in the case of donor guest species, and by injection of holes for the case of acceptors. The intercalation-induced decrease in the mobility that results from the increased scattering and the increased effective mass is outweighed by the larger increase in the carrier density, resulting in a large conductivity enhancement. The carriers are localized in the graphene planes, and for high-stage compounds ($n \geqslant 2$), the carrier density falls off rapidly with distance from the graphite bounding layer due to the screening of the charged intercalate layer by the surrounding graphite bounding layers.

Transport measurements on bulk graphite and its intercalation compounds have in the past presented a number of difficulties due to the large anisotropy of the conductivity [8.71]. The large magnitude of the aspect ratio, which is defined as the fiber length over the fiber diameter (L/d), is highly favorable for reliable and accurate *a* axis electrical conductivity, thermal conductivity, thermopower, and magnetoresistance measurements [8.1]. Especially interesting have been the measurements on single fibers and the new physics that these measurements have provided.

8.5.1 Electrical Conductivity

Though both graphite fibers and HOPG experience roughly an order of magnitude enhancement in the in-plane electrical conductivity σ_a (Figs. 8.9c, 10a), the actual conductivities achieved by the intercalation of graphite fibers are generally lower than for the same intercalate and stage in an HOPG host (Table 8.3). This is due to the greater structural disorder in graphite fibers, consistent with the lower conductivities of the fiber hosts prior to intercalation.

Figure 8.19 shows correlated resistivity results for many PAN, mesophase pitch, and vapor-grown carbon fibers, prepared under a variety of heat treatment conditions. In this figure the resistivity of the fibers is plotted before (ϱ_g) and after (ϱ_f) intercalation with bromine [8.28] (Table 8.3). These results show that for the most disordered of these fibers, intercalation has little effect on ϱ_g (upper right-hand corner of the figure). As the crystalline order of the fiber is increased, the ratio (ϱ_g/ϱ_f) for both the mesophase pitch (MP) and vapor-grown (VG) fibers increases, reaching conductivity enhancements approximately equal to that of HOPG, intercalated with the same intercalate species (bromine). Similar results have been reported for other intercalate species [8.3, 21, 77].

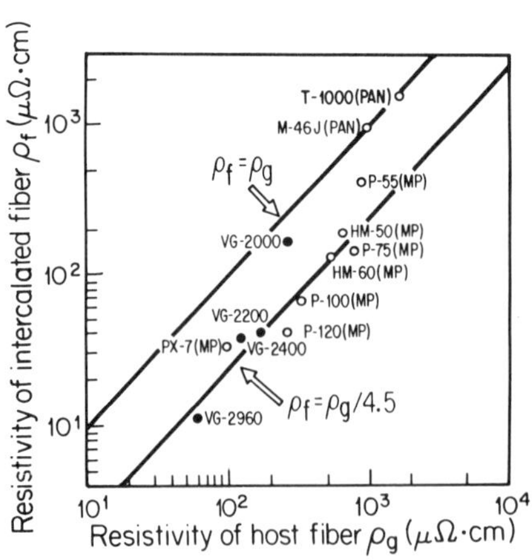

Fig. 8.19. Comparison of the room-temperature resistivity before (ρ_g) and after (ρ_f) bromination for a variety of high-strength and high-modulus PAN-based and mesophase (MP) pitch-based fibers. The vapor-grown (VG) fibers are heat treated at various temperatures in the range $\approx 2000 < T_{HT} < 2960\,°C$. The resistivity of the brominated fibers was measured after five days exposure to air [8.28]

Table 8.3. Enhancement of room-temperature conductivity of carbon fibers by intercalation[a]

Pristine Fiber [b]	Intercalate	Stage	ϱ_g [μΩm]	ϱ_f [μΩm]	Enhancement factor [ϱ_f/ϱ_g]
Thornel 50[c]	HNO_3	–	7.5×10^4	7.0×10^3	9.3
CCVD[d] (2900 °C)	HNO_3	–	8.0×10^3	6.0×10^2	13
CCVD[d] (2900 °C)	Br_2	–	8.0×10^3	1.5×10^3	5.3
CCVD[e] (3300 °C)	AsF_5	–	5.5×10^3	1.1×10^2	50
P-100[f] (3000 °C)	$CuCl_2$	2	1.9×10^4	$(1.3–2.0) \times 10^3$	~ 11
P-100[f] (3200 °C)	$CuCl_2$	2	2.0×10^4	$(2.0–2.2) \times 10^3$	~ 10
P-100[f] (2500 °C)	$CuCl_2$	–	$(3.1–3.4) \times 10^4$	$(3.5–4.0) \times 10^3$	~ 9
CCVD[f] (2900 °C)	$CuCl_2$	2, 3	1.0×10^4	2.0×10^3	5
CCVD[f] (2900 °C)	$CuCl_2$	1	1.0×10^4	8.0×10^2	12
HMPVA3[g] (2800 °C)	C_xF	–	8.3×10^4	1.4×10^4	5.9
VSB-32[g] (2800 °C)	C_xF	–	4.0×10^4	5.0×10^3	8.0
CCVD[h] (2900 °C)	C_xF	–	7.7×10^3	9.1×10^2	8.5
CCVD[h] (2900 °C)	$CuCl_2$	2	7.0×10^3	9.0×10^2	7.7
CCVD[h] (2900 °C)	$FeCl_3$	1	7.0×10^3	7.0×10^2	10
CCVD[h] (2900 °C)	$CoCl_2$	–	7.0×10^3	8.5×10^2	8.2
P-100[i]	$NiCl_2$	–	2.5×10^4	1.9×10^3	13
P-100[i]	$CuCl_2$	–	2.5×10^4	2.5×10^3	10
GY70[i]	$NiCl_2$	–	3.6×10^4	2.05×10^3	18
GY70[i]	$CuCl_2$	2, 3	3.6×10^4	2.5×10^3	14
TP4104B[j]	$CdCl_2$	–	2.0×10^4	$(1.5–2.0) \times 10^3$	~ 11

[a] Denote the resistivity of pristine fibers by ϱ_g, and after intercalation by ϱ_f, so that the intercalation-induced enhancement of the conductivity is given by (ϱ_f/ϱ_g).
[b] Estimated T_{HT} values are given in parenthesis. All CCVD fibers in the table are benzene derived. The Hercules HMPVA3 and Celanese GY70 are PAN fibers, while the Union Carbide VSB-32, P-100, and TP4104B are mesophase pitch fibers.
[c] Ref [8.73]
[d] Ref [8.31]
[e] Ref [8.23]
[f] Ref [8.21]
[g] Ref [8.74]
[h] Ref [8.49]
[i] Ref [8.75]
[j] Ref [8.76]

To achieve the highest electrical conductivity in a fibrous host material, the fiber host must be selected for the lowest residual resistivity ϱ_r (temperature independent) and for the lowest temperature-dependent contribution to the resistivity ($\varrho_i \equiv$ ideal resistivity) [8.78]. Fibers with the lowest residual resistivity before intercalation also have the lowest residual resistivity after intercalation. With regard to the choice of host fibers ϱ_r (shown on the left scale), we see in

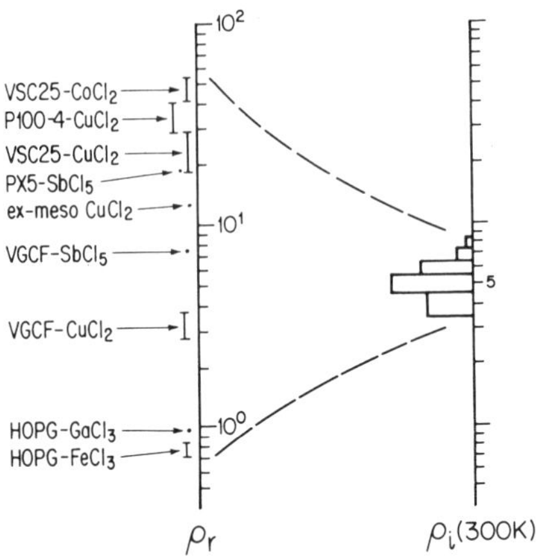

Fig. 8.20. Comparison between the residual resistivity (ρ_r left side) and the room-temperature ideal resistivity (ρ_i right side) of stage-2 acceptor GICs. Resistivity values for a number of fiber hosts and intercalate species are given in μΩ cm [8.78]

Fig. 8.20 that the low-temperature residual resistivity of a host fiber can easily vary over a factor of 20 or more, with a much smaller range of values available for the ideal resistivity ϱ_i (shown on the right scale at 300 K) in the case of metal chloride acceptor intercalation to stage 2 [8.78].

The most extensive effort to produce a high-conductivity graphite fiber has been carried out with the intercalate species AsF_5 [8.23]. The choice of AsF_5 for the intercalate species was motivated by previous measurements on HOPG-based GICs [8.65] showing that the intercalate AsF_5 yields the highest room-temperature electrical conductivity, presumably because of an unusually low ideal resistivity for this intercalate species, coupled with a relatively small intercalation-induced degradation of the carrier mobility; this discussion is consistent with the straight lattice fringe TEM images that are seen for the AsF_5-intercalated fibers [8.65]. A maximum room-temperature conductivity ($9 \times 10^7 \Omega^{-1} m^{-1}$) for AsF_5-intercalated fibers has been reported for a VGCF host material heat treated to 3300 °C [8.65], the conductivity value being significantly above that for copper ($5.9 \times 10^7 \Omega^{-1} m^{-1}$) and above previous measurements on AsF_5-intercalated fibers [8.79]. These high values for σ_a were obtained while the fiber was surrounded by AsF_5 gas in the intercalation ampoule. This high value of the conductivity must at this time be regarded as open to further confirmation, since contact potentials, thermal effects, and changes in the fiber cross section may have contributed to a lowering of the measured potential.

Intercalation-induced increases (by factors of 5 to 12) in the electrical conductivity relative to that of the pristine fibers are common for many kinds of fibers and intercalate species [8.21, 80]. For the most highly ordered fibers, good staging occurs. Nevertheless, the residual resistivity in fibers is large compared with that in intercalated HOPG. Therefore, it cannot be assumed that the dependence of the conductivity σ on intercalate concentration (or reciprocal stage $1/n$) is similar to that of HOPG-based GICs. These show an approximate linear increase in σ with $1/n$ for low intercalate concentrations (high stage), and eventually reach a maximum conductivity σ_{max} at some stage index, followed by a decrease in σ upon further increase in intercalate concentration [8.1, 12]. No systematic study of the stage dependence of σ_a has yet been carried out for intercalated fibers, though related work on in situ measurements of σ_a versus intercalate uptake have been made as discussed below.

As in the case of GICs based on bulk host materials, intercalated fibers do not generally exhibit their maximum conductivity for the stage-1 compounds, but rather at a lower intercalate concentration ranging from about a stage-2 compound for typical metal chloride acceptor intercalates to about a stage-5 or stage-6 compound for the bromine intercalate [8.19, 81]. Figure 8.21 shows a plot of in situ measurements of the electrical conductivity for ionic fluorine GICs versus fluorine uptake in PAN, mesophase pitch, and vapor-grown fibrous host materials. These plots show that the conductance versus intercalate uptake for the most graphitic fibers is closest to that of HOPG, with smaller increases in conductance occurring in the more disordered fibers, which also start from a lower conductance value prior to intercalation [8.28]. Thus the intercalation-induced increase in conductance [8.74] is greatest for VGCF filaments, less for ex-pitch fibers, and still less for ex-PAN fibers (Table 8.3). The general shapes of the curves in Fig. 8.21 are similar to the plots of conductivity versus reciprocal stage for HOPG discussed above, with the maximum conductivity occurring for a stage-3 compound [8.40]. For higher intercalate concentrations, increased carrier scattering occurs and a lower carrier mobility is observed. Furthermore, the intercalation of F is unique among intercalates, resulting in the reduction of the conductivity by over an order of magnitude relative to the pristine fiber, upon increase of the F concentration to achieve about $C_{2.9}F$ stoichiometry [8.82].

For disordered fibers, the misalignment of the crystallite axes with respect to the fiber axis must be considered. The simplest model assumes crystallites of a

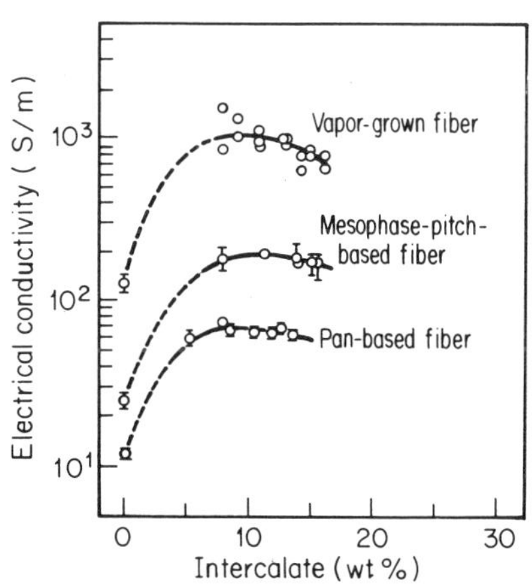

Fig. 8.21. Electrical conductivity of ionic GICs formed from fluorine-intercalated carbon fibers as a function of intercalate concentration for vapor-grown, mesophase-pitch, and PAN fiber host materials [8.28]

given size and relates the conductivity along the fiber axis $\sigma_{\|}$ to that along the a and c axes (σ_a and σ_c) by

$$\sigma_{\|} = A\langle \sin^2 \phi \rangle \sigma_c + (1 - A\langle \cos^2 \phi \rangle)\sigma_a , \tag{8.2}$$

assuming uniform intercalation. However, intercalation favors the ordered regions, and this effect should be taken into account in a more quantitative treatment. To our knowledge, misalignment effects have not been considered in treating the resistivity of intercalated disordered carbon fibers. On the other hand, (σ_a/σ_c) can be greater than 10^6 for selected acceptor GICs, so that in this case only the term in σ_a contributes to $\sigma_{\|}$. If the term in σ_c is neglected, then $\sigma_{\|}$ is proportional to σ_a and the major effect of misalignment is an effective elongation of the fibers.

Of importance for applications of intercalated fibers as practical lightweight high-conductivity filaments is the increase in conductivity that is observed even when the fibers are too disordered structurally to exhibit a well-established staging periodicity [8.19]. For such practical applications, there are advantages in using intercalate species where the conductivity maximum σ_{max} is observed at relatively high stages. Because of the lower amount of intercalate that is required to reach σ_{max}, the amount of intercalate-induced stress is consequently smaller as the small crystallites in the intercalated regions expand along the c axis. Because of the disorder in the fibers, the intercalation-induced maximum conductivity is lower in the fibrous host materials and the sample-to-sample variation is much larger than for the corresponding intercalation into HOPG [8.21, 80].

Studies of the temperature dependence of the conductivity are of particular interest with regard to identification of the conduction mechanism. In contrast to the behavior in many pristine graphite fibers, low-stage intercalated graphite fibers typically show a metallic temperature dependence of the conductivity σ (Fig. 8.22), as is found for intercalated bulk host graphite materials [8.49, 72].

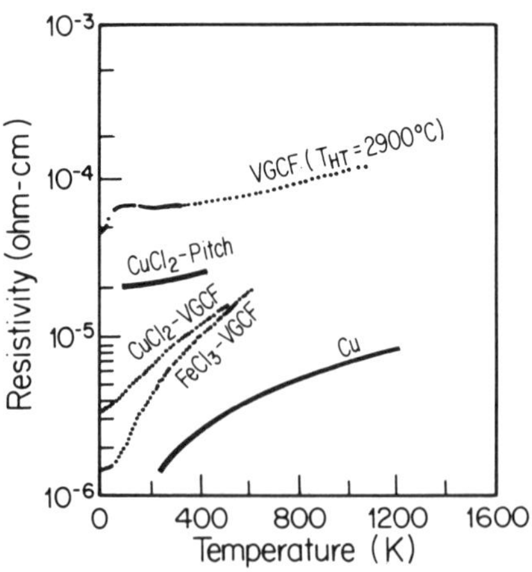

Fig. 8.22. Temperature dependence of resistivity for air-stable acceptor compounds ($FeCl_3$, $CuCl_2$) intercalated into vapor-grown (VGCF) and mesophase-pitch fibers [8.77] over a wide temperature range. The $\rho(T)$ curves for a pristine VGCF filament and for metallic Cu are shown for comparison [8.28]

That is, for low-stage intercalated VGCF and mesophase pitch filaments, the resistivity decreases as the temperature decreases, but the functional dependence of $\varrho(T)$ is quite different from that prior to intercalation (Fig. 8.22). Values of the residual resistance ratio (RRR $= R_{300K}/R_{4.2K}$) provide a figure of merit for characterizing the defect concentration of the fibers. Intercalation generally causes the RRR to increase, consistent with the large increase in carrier concentration and smaller relative decrease in carrier mobility. In this context, we note that the RRRs for intercalated fibers in Fig. 8.22 are higher than for the corresponding pristine fibers. The larger RRR values for HOPG-based GICs relative to fibers intercalated with the same intercalate to the same stage are consistent with the higher defect density of the fibers. Thus the intercalation-induced increase in conductivity [8.74] is greatest for VGCF filaments, less for ex-pitch fibers, and still less for ex-PAN fibers (Fig. 8.9, Table 8.3). Much smaller temperature-dependent changes in resistivity are found for pitch-based fibers ($T_{HT} \simeq 3000\,°C$) [8.77] compared with VGCF filaments intercalated with the same species ($CuCl_2$), as is shown in Fig. 8.22. For more disordered VGCF filaments than are shown in Fig. 8.22, $\varrho(T)$ *decreases* with increasing temperature, as is also the case for these fibers prior to intercalation (Fig. 8.23) [8.19]. In Fig. 8.23 we see an unintercalated VGCF filament (heat treated to 2960 °C) showing an increase in resistivity with decreasing temperature [8.19]. This negative slope in $\varrho(T)$ persists upon intercalation with bromine for a short time ($\sim$ 1 h). But after about six hours of intercalation, the $\varrho(T)$ curves in Fig. 8.23 show a positive slope, which increases in magnitude as the intercalation time increases up to about 48 hours, after which essentially no change in conductivity is observed. Thus, intercalation tends to make the fiber conductivity more metallic. The saturation behavior seen in Fig. 8.23 for long intercalation times reflects in part the relative insensitivity of the conductivity to stage index n near σ_{max} in the $\sigma(1/n)$ curve.

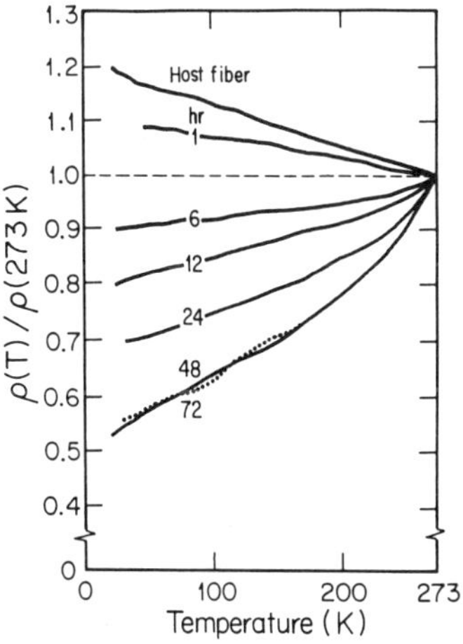

Fig. 8.23. Temperature dependence of the resistivity, normalized to the resistivity at 273 K, for bromine-intercalated vapor-grown carbon fibers heat treated at 2960 °C (fiber diameter 7–13 μm) [8.19]

Study of the low-temperature electrical conductivity of intercalated graphite fibers has also revealed new physics. In general, 3D metals at low temperature follow the Bloch-Grüneisen law for the temperature dependence of the resistivity

$$\varrho_{BG}(T) = A_{BG} + B_{BG} T^n \quad (8.3)$$

where $n \sim 5$. The low-temperature T^5 regime is applicable only when carrier scattering is dominated by low-energy phonons. At high temperatures where all phonons can scatter electrons, a linear temperature dependence for $\varrho(T)$ is found [8.83].

Now for both intercalated graphite fibers and for intercalated HOPG, the functional form of the temperature dependence of the resistivity has been found by many authors to be

$$\varrho(T) = A_\varrho + B_\varrho T + C_\varrho T^2 , \quad (8.4)$$

for many intercalate species, where A_ϱ, B_ϱ, and C_ϱ are constants [8.68]. The room-temperature resistivity is predominantly due to carrier-phonon scattering, while defect and boundary scattering play a significant role at low temperatures in determining the residual resistivity $\varrho_r = A_\varrho$.

The functional form of (8.4) for the intercalated fibers is very different from that for the pristine fibers (Fig. 8.22), which is also different from that for pristine HOPG [8.9, 84]. Both the magnitude of the room-temperature resistivity for fibrous GICs and the temperature dependence $\varrho(T)$ depend strongly on T_{HT} for a given precursor material [8.68], consistent with the strong dependence of ϱ_r on the morphology of the precursor material and on T_{HT}. Size-dependent resistivity measurements of the pristine fibers also show that the residual resistivity A_ϱ in (8.4) increases as the fiber diameter decreases [8.85], consistent with a higher fraction of the fiber being disordered as the diameter decreases.

Recently, high-resolution resistivity measurements [8.86], made possible by the favorable geometry of intercalated graphite fibers, have been carried out at low temperatures to investigate whether a Bloch-Grüneisen effect exists for GICs at low temperatures. High-resolution temperature-dependent resistivity measurements on $CuCl_2$-intercalated VGCF fibers down to 1.5 K confirm the validity of (8.4), showing that the usual low-temperature 3D Bloch-Grüneisen T^n behavior with $n \sim 5$ is not applicable to GICs [8.87]. One explanation for the unusual temperature dependence of $\varrho(T)$ for intercalated fibers at low temperature (down to 40 K) may lie in the dominance of carrier-carrier scattering or of carrier scattering by out-of-plane phonons. These out-of-plane phonons give rise to a temperature dependence $\varrho(T) \sim T^\alpha$, where α is in the range $1.1 < \alpha < 1.5$, because of the low energy and weak dispersion of these phonons [8.88, 89]. However, this theory does not explain the temperature dependence of $\varrho(T)$ below 40K. As discussed below (Sect. 8.5.4), weak localization and carrier-carrier scattering effects become important for intercalated carbon fibers at very low temperatures, thereby modifying the functional form of $\varrho(T)$.

In terms of fundamental science, the favorable fiber geometry also facilitates accurate measurements of charge transfer phenomena in GICs. An example of such an application is the study of the effect of ammoniation on the in-plane resistivity of the ternary $C_{24}K(NH_3)_x$-GIC system as a function of NH_3 concentration x [8.90]. As x was increased, the resistance of a fiber intercalated with $K(NH_3)_x$ was found to remain at the same value as for the stage-2 compound for $x = 0$, namely $C_{24}K$, until a value of $x = 1$ was reached. At the value $x = 1$ (where the concentration of NH_3 was equal to that of the K), the resistance increased by more than a factor of 4, as the transition to a stage-1 compound took place. This dramatic increase in resistivity was explained by a back-donation of electrons to the NH_3, thereby significantly reducing the carrier density and conductance in the graphene layers. As x was further increased to ~ 4.3, an additional 10% decrease in resistance was observed [8.90] and was identified with charge transport in the intercalate layer as the intercalate molecules are squeezed tighter together.

8.5.2 Stability Issues

For the practical use of intercalated carbon fibers, various stability issues can become critical, such as stability of the fibers under ambient conditions, their chemical compatibility with various matrix materials (e.g., polymers or epoxies), thermal stability under wide temperature excursions, and thermal stability at elevated temperatures. We start with a review of these stability issues by considering intercalate desorption under ambient conditions.

Relative to bulk graphite, fibrous host materials generally show less intercalate desorption under comparable conditions. This is of course an attractive feature of the fibrous host materials for practical applications. For the more disordered host materials, intercalate pinning by defects is the dominant desorption retardation mechanism. On the other hand, for the VGCF filaments, the structural organization of the fibers inhibits intercalate desorption, both in vacuum and under ambient laboratory conditions. The physical basis for this effect in the VGCF filaments is the great reduction of surface area where exposed intercalate edge planes could lead to intercalate desorption. Long-term stability (hundreds of days) was achieved for residue compounds of AsF_5 intercalated fibers [8.23]. These residue fibers have conductivities in excess of $10^7 \Omega^{-1} m^{-1}$, high enough for useful applications as a lightweight conductor. A similar result was obtained upon bromine intercalation of similar fibers, as shown in Fig. 8.24a. Long-term (i.e., four month) stability studies on the desorption of C_xF [8.74] show that desorption occurs more slowly in PAN fiber hosts than in VGCF filaments, though both types of fibers have circumferential preferred orientation of the graphene planes. The higher stability of the intercalated PAN fibers is consistent with increased intercalate pinning in the more disordered fibers.

Intercalate desorption was also found to be inhibited in the mesophase pitch fibers [8.44], even though the surface area for exposed edge planes is large.

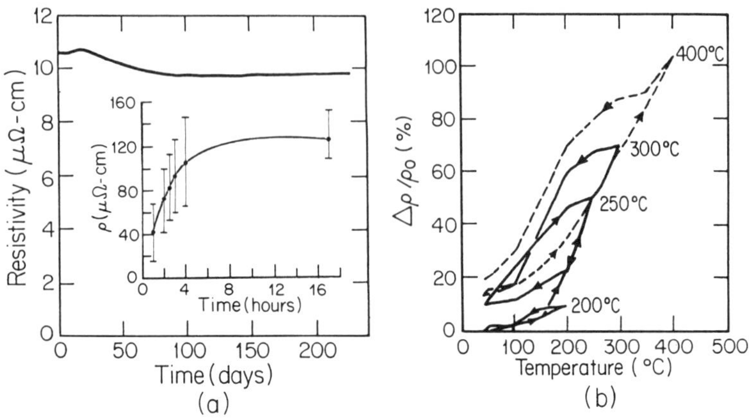

Fig. 8.24. (**a**) Variation of the resistivity of thick (10 μm) fibers (heat treated at 2960 °C and intercalated for 72 hours) over long times under ambient conditions. The results show a relatively long-term stability for the brominated fibers. The decrease in resistivity after about 20 days is due to the lower ambient laboratory temperature during the winter season in Japan. The inset shows data by *Dominguez* et al. [8.91] for the increase in resistivity for AsF_5 intercalated into mesophase-pitch TP4104B fibers at short times. (**b**) Thermal stability of thick (~ 10 μm) bromine-intercalated fibers. The resistivity change $\Delta\rho$ relative to the resistivity of the as-prepared Br_2-intercalated fibers is measured during the indicated heating cycles under ambient conditions (in air) and keeping the fibers at the maximum temperature of the heating cycle for ten minutes [8.19]

Raman microprobe studies of the intercalate distribution in the mesophase pitch fibers did, however, show that intercalate desorption occurred in the near-surface region (< 1 μm), but significant intercalate retention was found in the fiber bulk [8.21]. Studies of intercalate desorption of milled fibers indicate that desorption also occurs at the open fiber ends [8.25, 60]. On a short time scale, *Dominguez* et al. [8.91] showed a rapid increase in resistivity by a factor of ~ 3 for AsF_5-intercalated mesophase pitch fibers exposed to air under ambient conditions for four hours.

Desirable long-term (nine month) chemical stability characteristics have been exhibited by P100 pitch-based fibers intercalated with bromine [8.60], with regard to subsequent exposure to water and organic solvents, such as ethanol, propanol, chloroform, and butyl acetate. Less than a 5% increase in resistivity was observed over a nine-month period upon exposure of the brominated P100 fibers to methanol and acetone, for which some chemical reaction was observed. A rapid increase in resistivity by a factor of ~ 4 in a three-day period was observed for $CuCl_2$-intercalated P100 fibers exposed to 90% humidity at $T = 70\,°C$ [8.22]. Stability studies of fluorine-intercalated pitch-based fibers showed excellent stability characteristics over a 200-day period [8.40], consistent with the results shown in Fig. 8.24a for bromine intercalation. In the case of fluorine compounds [8.40, 43], X-ray diffraction results show the appearance of the graphite (002) line, consistent with a near-surface reaction of the fluorine with moist air, but preventing subsequent decomposition of the intercalation compound.

For practical use of intercalated fibers as lightweight conductors, thermal stability is important. Long-term (3-month) thermal stability has been demonstrated to temperatures above 100 °C [8.92] and to even higher temperatures of 170 °C [8.77] in $CuCl_2$-intercalated ex-pitch P100 fibers, and for shorter times to even higher temperatures for intercalated VGCF filaments [8.35, 50].

Figure 8.24b shows results for thermal stability studies of brominated fibers under various heating cycles on VGCF filaments intercalated with Br_2 for 72 hours. The results show an increase in resistivity upon heating (in air) as expected from the temperature-dependence measurements [8.19], but with significant hysteresis on the heating and cooling cycles. Of particular interest is the very small hysteresis in the resistivity measurements for heating to 200 °C for ten minutes so that almost no increase in resistivity is found after thermal cycling in this case (similar results have also been observed in brominated P100 fibers [8.60]). For thermal cycles to 250 °C, the hysteresis effects become more pronounced, and an increase in resistivity of $\sim$ 10% is observed upon return to room temperature (Fig. 8.24b). For thermal cycles to 300 °C and 400 °C, only a 20% increase in resistivity is found after cooling to room temperature. This increase in ϱ is very small indeed when compared with the 700% decrease in resistivity resulting from the original intercalation. These results are important for the practical use of intercalated filaments for high-conductivity applications. Very thin ($\sim$ 1 μm) intercalated fibers also show high thermal stability below 200 °C for short heating times [8.19]. Long-term heating studies were carried out by measuring the resistance of $CuCl_2$ intercalated P100 fibers, showing no change after three months at 100 °C, but showing an increase in resistance by a factor of $\sim$ 4 after two months at 150 °C, and by a factor of $\sim$ 6 after two months at 165 °C [8.22]. These same intercalated fibers show good short-term stability under heating cycles in agreement with the results of Fig. 8.24b, though significant desorption occurs after long-term heating [8.22].

Also important for practical applications is the high current density (4×10^4 A/cm^2) that intercalated fibers (e.g., P100 fibers intercalated with $CuCl_2$) can withstand without burnout; less than a 5% increase in resistance was observed after 20 minutes at these high current densities [8.22]. The current carrying capacity of the Br_2-intercalated fibers is also comparable to that of Cu wire. However, after prolonged passage of high current density, the fibers will fail mechanically. The failure mode of the intercalated fibers upon heating is also of interest; SEM photographs [8.31] show that intercalated VGCF filaments rupture through a local exfoliation phenomenon, presumably accompanied by the release of the intercalate in gaseous form.

8.5.3 Magnetoresistance

Magnetoresistance measurements on graphite fibers are of particular interest because of the favorable geometry of the fibers for such measurements. To the extent that the transverse magnetoresistance of intercalated graphite fibers yields values for an average carrier mobility μ through the relation

$$\Delta\varrho/\varrho(0) = \frac{1}{2}(\mu B)^2 , \tag{8.5}$$

where B is the magnetic field and the factor of 1/2 arises from the angular average, measurements of $\Delta\varrho/\varrho(0) \equiv [\varrho(B) - \varrho(0)]/\varrho(0)$ can be used with zero-field resistivity $\varrho(0)$ measurements to estimate the temperature dependence of the carrier density and the carrier mobility for the intercalated fibers.

The simple single-band formula $\Delta\varrho/\varrho(0) = (\mu B)^2$ must be used with caution for intercalation compounds, since there are often several types of carrier pockets contributing to the conduction process. For the case of intercalated fibers, defect scattering plays an important role, so that the magnetoresistance tends to be much smaller than for the unintercalated fibers. For disordered fibers, the magnetoresistance is found to be negative and assumes a different functional form than the simple B^2 dependence. A test of the appropriateness of the simple formula $\Delta\varrho/\varrho(0) = (\mu B)^2$ for finding the carrier mobility for specific intercalated fibers can be made by directly fitting the magnetic field dependence of the experimental magnetoresistance measurements.

A variety of different detailed behaviors are observed for the magnetoresistance of intercalated carbon fibers, depending on the ordering of the pristine fiber prior to intercalation and on the intercalation parameters, such as intercalate species and stage and the intercalation temperature and time. For example, the magnetoresistance of the most highly ordered VGCF filaments ($T_{HT} \geqslant 2800\,^{\circ}$C) remains positive after intercalation and shows a quadratic field dependence at low fields [8.31]. Disordered intercalated fibers, however, generally show negative magnetoresistance behavior.

A typical example of positive magnetoresistance behavior in intercalated carbon fibers is shown in Fig. 8.25, where the magnetoresistance at 77 K of

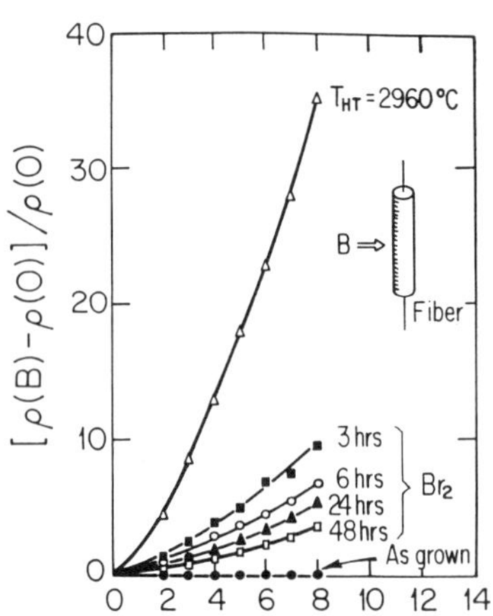

Fig. 8.25. Transverse magnetoresistance of the ~ 10 μm fibers at 77 K after reaction with liquid bromine for various lengths of time. The geometry of the magnetic field arrangement is shown in the inset. Also shown are results for the pristine fibers before heat treatment (as grown) and after heat treatment at 2960 °C [8.19]

bromine-intercalated VGCF filaments ($T_{HT} = 2960\,^{\circ}C$) is plotted as a function of magnetic field. The intercalation in Fig. 8.25 was done in situ by immersion of the same fiber in liquid bromine for different lengths of time [8.19]. Under these intercalation conditions, the maximum conductivity σ_{max} is achieved in 48 h. The magnetoresistance measurements show that bromine intercalation causes a large decrease in the magnitude of the magnetoresistance, and this decrease in $\Delta\varrho/\varrho_0$ continues beyond the reaction time at which σ_{max} is reached. By measuring both the zero-field resistivity and the magnetoresistance, use of the simple formula permits determination of the carrier density and mobility [8.19]. Such results characteristically show a higher mobility for acceptor compounds relative to donor compounds, consistent with the lower intercalation-induced defect density for acceptor GICs, as confirmed by TEM lattice fringe image studies (Fig. 8.15). Mobility values as high as 170 cm^2/Vs have been reported for pitch-based ($T_{HT} \simeq 3000\,^{\circ}C$) fibers intercalated with $CuCl_2$ [8.77].

Intercalation reduces the carrier mobility by increasing the size of the Fermi surface and hence increasing the carrier effective masses, while at the same time introducing additional intercalation-induced scattering centers. A linear field dependence for $\Delta\varrho/\varrho_0$ versus B is often obtained at field values above about 5 kG for intercalated VGCF filaments (Fig. 8.25) [8.93–95]. This effect is also observed in highly graphitized pristine fibers [8.96].

The angular dependence of the magnetoresistance of fibers is complicated for two reasons. First, the fiber axis of actual carbon fibers is not exactly aligned along an in-plane a axis, but rather the fibers show a spread of preferred orientations [8.3]. This misorientation effect is less important in the case of intercalated carbon fibers relative to pristine fibers, because fibers with large orientational misalignments do not intercalate. The second complicating effect concerns the alignment of the applied field relative to the crystallographic directions of the GIC. While an external magnetic field can be aligned along the a axis (assuming no orientational misalignment of the fiber axis), it is not possible to align the magnetic field along $\hat{c}$ for any of the fiber types (Figs. 8.1, 7). In addition, the magnetoresistance of GICs is highly anisotropic, with much larger $\Delta\varrho/\varrho(0)$ values observed for $\boldsymbol{B}\parallel\hat{c}$ than for $\boldsymbol{B}\perp\hat{c}$. Therefore, alignment of the magnetic field in the transverse direction (as in Fig. 8.27) emphasizes contributions to $\Delta\varrho/\varrho(0)$ from $\boldsymbol{B}\parallel\hat{c}$. However, because of the various complications discussed above, a lengthy analysis of the transverse magnetoresistance measurements on fibers is necessary to yield data that can be directly compared with the corresponding measurements on bulk samples [8.93–95, 98].

For fibers exhibiting cross sections lacking rotational symmetry, an angular dependence of the magnetoresistance can be observed as the magnetic field is rotated around the fiber axis in a plane normal to the fiber axis. Such an effect has been observed for a $CuCl_2$-intercalated oriented core pitch fiber (also called the "PAN-AM" cross section, as shown in Fig. 8.1d). Where an approximate $\pm 20\%$ variation in $\varrho(B, T)$ is observed as a function of angle θ [8.99, 100]. This angular anisotropy has been taken into account in interpreting transverse negative magnetoresistance data for $CuCl_2$-intercalated pitch fibers by averaging $[\Delta\varrho(B\cos\theta, T)/\varrho(0, T)]$ over an angle normal to the fiber axis [8.101, 102]. Angular anisotropy due to faceting has also been observed.

A complex magnetic field dependence is observed for intercalated fibers based on less graphitic hosts (Figs. 8.26, 27). For example, P100 pitch-based fibers intercalated with $CuCl_2$ and $PdCl_2$ show a negative magnetoresistance at low fields (< 1 T), followed by a positive magnetoresistance at higher fields when the measurements are made at low temperatures (4 K and 78 K) [8.93, 95]. Similar phenomena are observed for the corresponding pristine fibers, though the magnetoresistance has a much smaller magnitude for the intercalated fibers. The magnitude of the magnetoresistance increases significantly upon lowering the temperature, both for the positive and negative magnetoresistance effects. In this connection, Fig. 8.28 shows the enhancement of the negative magnetoresistance effect by lowering the temperature. The negative magnetoresistance effect [8.97] in intercalated fibers is further discussed in Sect. 8.5.4 in terms of weak localization effects.

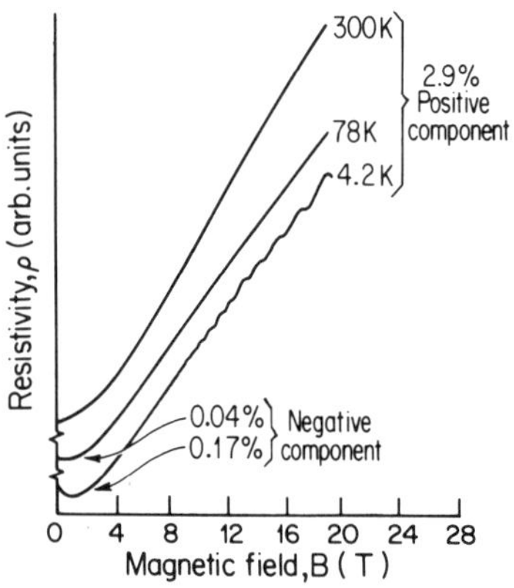

Fig. 8.26. Electrical resistivity versus magnetic field for a $CuCl_2$-intercalated P100-4 fiber at the three temperatures indicated. Note the observation of Shubnikov-de Haas oscillations at low temperatures and high magnetic fields [8.93, 94]

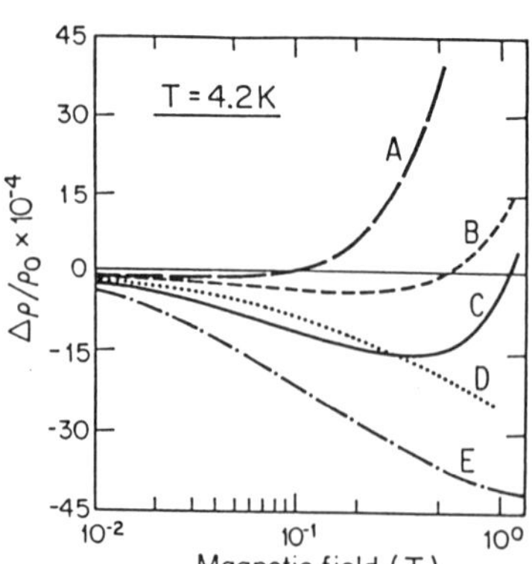

Fig. 8.27. Transverse magnetoresistance as a function of magnetic field for a set of five fibers at 4.2 K: (*A*) BDF-$SbCl_5$, (*B*) PX5-$CuCl_2$, (*C*) PX5-$SbCl_5$, (*D*) VSC25-$CuCl_2$, and (*E*) P100-4-$CuCl_2$ [8.97]

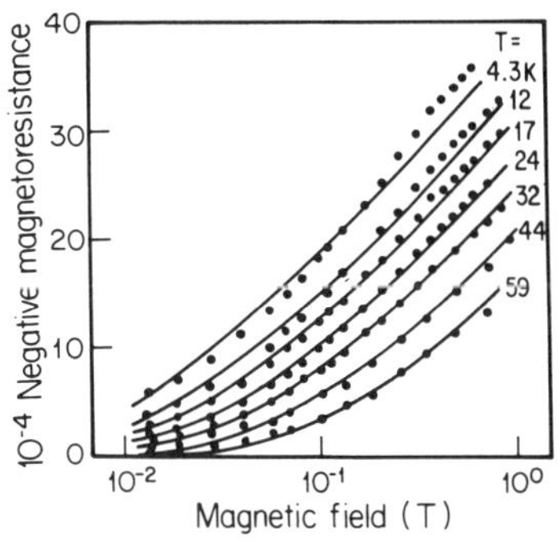

Fig. 8.28. Transverse negative magnetoresistance versus magnetic field at various temperatures for P100-4 fibers intercalated with $CuCl_2$. The solid lines are calculated from a best fit using weak localization theory [8.102, 103]

Even after intercalation, many well-ordered graphite fibers exhibit sufficiently high carrier mobilities to satisfy the condition $\omega_c \tau \gg 1$ for the observation of quantum oscillatory effects. Several reports have already been made on the observation of the Shubnikov–de Haas effect in intercalated graphite fibers, since intercalation enhances the observation of quantum oscillatory effects [8.49, 66, 77]. Shubnikov–de Haas quantum oscillations have also been observed in intercalated fibers that exhibit a negative magnetoresistance at low temperatures and low magnetic fields (see Fig. 8.26, specifically for P100 ex-pitch fibers intercalated with $CuCl_2$). The magnitude of the Shubnikov–de Haas oscillations is found to decrease as the magnitude of the negative magnetoresistance effect increases, consistent with the identification of the negative magnetoresistance effect in intercalated carbon fibers with disorder [8.94, 95]. Consistent with this result is the report of scattering times between a factor of 5–10 shorter in the $CuCl_2$-intercalated fibers compared with the HOPG host material [8.77]. Because of the cylindrical nature of the π-electron Fermi surfaces, the most interesting field direction is along the c axis, which unfortunately is not along a single spatial direction in a fiber. Therefore, the fiber geometry is unfavorable for quantitative interpretation of the observed Shubnikov–de Haas periodicities in terms of band-structure models.

8.5.4 Weak Localization Effects

Intercalation with acceptors increases the resistivity anisotropy ratio ϱ_c/ϱ_a by two or three orders of magnitude, making the transport behavior in GICs more two-dimensional than in graphite itself or in alkali-metal donor GICs. Because of their quasi-2D behavior, partly disordered graphite fibers intercalated with acceptors are good candidates for the study of 2D weak localization [8.104] and carrier-carrier interaction [8.105] effects in the transport properties. Both 2D weak localization and carrier-carrier interaction effects produce a logarithmic decrease in the resistivity $\Delta\sigma(T)$ at low temperatures. These effects are expected to induce a correction in all transport properties of the fibers at low temperatures, including the electrical conductivity, electronic contribution to the thermal conductivity, Hall effect, and the thermopower [8.106]. Both weak localization and electron-electron interaction effects appear in weakly disordered materials, with the disorder characterized by the parameter $k_F l$, where k_F is the Fermi wave vector and l is the mean free path. Both of these effects are observed in the range $1 < k_F l < 10$, whereas a transition to strong disorder (variable-range hopping regime) is observed for $k_F l < 1$, while $k_F l \to \infty$ denotes the perfect metal limit.

The weak localization effect is due to the constructive quantum interference between phase-coherent elastically backscattered electron waves [8.107]. In contrast, the carrier-carrier Coulomb interaction effect results from the poor screening in a disordered electron (or hole) gas, if compared with a perfect metal. In zero magnetic field both weak localization and carrier-carrier interaction effects predict a logarithmic correction $\Delta\sigma$ to the constant conductivity σ expected for a perfect metal at low temperature:

$$\Delta\sigma(T) = \frac{e^2}{2\pi^2\hbar}(ap + \gamma)\ln(T/T_0) \tag{8.6}$$

where $\Delta\sigma(T)$ is the change in the 2D sheet conductivity at temperature T and σ is related to the 3D residual resistivity ϱ_r by $\sigma = I_c/\varrho_r$, where I_c is the c axis repeat distance. The temperature T_0 is a temperature below the resistance minimum (Fig. 8.29). The numerical constant p in (8.6) is related to the temperature dependence of the inelastic scattering rate $1/\tau_{in}$ by T^p. The value of p was found to be ~ 1 by *Piraux* et al. [8.97, 99, 100, 108–110] and has been tentatively attributed to carrier-phonon scattering. The two numerical constants α and γ depend on the parameters of the system and may also include an inhomogeneity parameter. Weak localization is represented in (8.6) by the factor ap, while the carrier-carrier interaction effect is given by the factor γ. Thus, both weak localization and carrier-carrier scattering contribute to a logarithmic decrease in the 2D conductivity with decreasing temperature at low temperature.

These 2D phenomena, which are of current interest in condensed-matter physics, can be quantitatively studied in disordered intercalated carbon fibers,

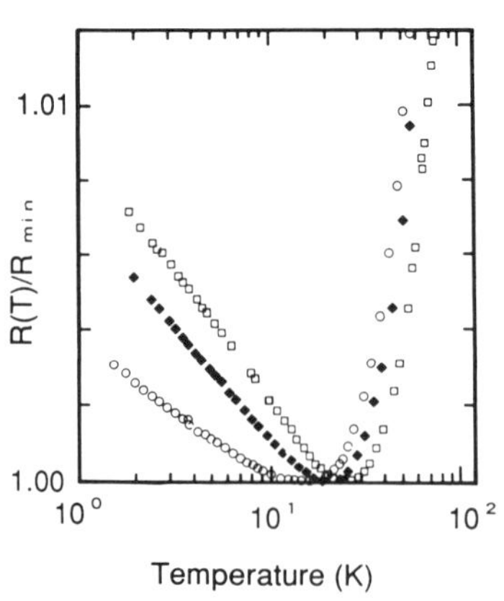

Fig. 8.29. Temperature dependence of the normalized resistance $R(T)/R_{min}$ for three intercalated fiber samples: VSC25-$CuCl_2$ (○), P100-4-$CuCl_2$ (■), and VSC25-$CoCl_2$ (□). These data were extended to higher temperatures by *Piraux* et al. [8.97]

though they have also been observed in bulk intercalated GICs [8.21]. Thus, fibrous intercalation compounds have made possible the quantitative study of new physical phenomena. In this context, early results on the change in the low-temperature resistivity $\Delta\varrho(T)$ – corresponding to $\Delta\sigma(T)$ in (8.6) – for metal chloride intercalated carbon fibers are shown in Fig. 8.29. Each of the fibers in Fig. 8.29 first shows a decrease in resistance with decreasing temperature, characteristic of the metallic regime, followed by a resistance minimum at a temperature, T_{min}, below which the resistance starts to increase logarithmically [8.97, 100, 102] as expected from the weak localization and carrier-carrier interaction effects described above. The experimental data of Fig. 8.29 are well fit by the functional form of (8.6). Values of the parameters used to achieve the fits are given in Table 8.4 [8.97]. By choosing different types of host fibers (mesophase pitch, PAN, or VGCF), by varying the heat treatment temperature T_{HT}, and by choosing different intercalate species and nominal stages, the various parameters of (8.6) can be varied, though in all cases, the experimental data follow the prescribed functional form of (8.6). Recently, large magnitudes for the weak localization effect have been found in fluorine-intercalated VGCF [8.109], thereby permitting a more quantitative study of this effect. Because of the large increase in the resistivity of the fluorine intercalated VGCFs with increasing fluorine concentration (below $x \sim 6$ in C_xF), it is not only possible to study the functional form of the transport properties in the weak localization regime, but also to cross into the strong coupling behavior regime, and to study changes in the structure and properties for these high-concentration ($x \leqslant 3.5$) fluorine-intercalated carbon fibers.

The magnitude of the slope of the negative $\Delta R(T)/R(T_0)$ and the magnitude of the temperature T_{min} (Fig. 8.29) are both strongly correlated with the residual resistivity ϱ_r, as discussed by *Issi* et al. [8.78]. The more disordered the fiber (and consequently the larger ϱ_r), the higher T_{min} and the greater the slope of

Table 8.4. Typical parameters[a] for various types of fibers after intercalation. P55, P100-4, and VSC25 are pitch-based carbon fibers [8.86]

Sample	Intercalate	ϱ_0	Stage	T_{min}	RRR	$ap + \gamma$
P100-4 # 1	$CuCl_2$	35	1 and 2	22	1.14	0.89
P100-4 # 2	$CuCl_2$	31	1 and 2	20	1.16	0.71
VSC25 # 1	$CuCl_2$	22	1 and 2	15	1.23	0.62
VSC25 # 2	$CuCl_2$	19	1 and 2	15	1.20	0.66
VSC25	$CoCl_2$	50	2	27	1.12	0.61

[a] Here ϱ_0 denotes the residual resistance in μΩ/m, T_{min} denotes the temperature below which the resistance increases with decreasing temperature and is given in K, RRR denotes the residual resistance ratio $R(4.2\text{K})/R(300\text{K})$, and p is the exponent in T^p that gives the temperature dependence of the inverse inelastic lifetime. The dimensionless parameter γ depends on the hole-hole screening and the dimensionless parameter a describes the magnitude of the weak localization effect [8.101].

$\Delta\varrho(T)/\varrho(T_0)$. Thus large values of ϱ_r favor the observation of weak localization effects.

Weak localization phenomena also explain the negative magnetoresistance found in acceptor GICs as well as the negative magnetoresistance phenomena previously observed in disordered nonintercalated graphite fibers. It has recently been found that the corresponding phenomena in the nonintercalated fibers can also be explained quantitatively by the same basic mechanism as described above [8.111], rather than the *Bright* mechanism [8.112], which had been previously applied with some success to interpret negative magnetoresistance effects in disordered graphite.

In 2D systems, the application of a magnetic field perpendicular to the 2D carrier system introduces a phase shift between the interacting partial waves and destroys the constructive interference in the backward direction, which is observed in zero magnetic field. The destruction of the weak localization phenomenon destroys the associated increase in resistivity, and is thus observed as a negative magnetoresistance, given by [8.107]

$$\Delta\sigma(B) = \frac{e^2\lambda}{2\pi^2\hbar}\left[\frac{3}{2}\psi\left(\frac{1}{2}+\frac{B_1(T)}{B}\right) - \frac{1}{2}\psi\left(\frac{1}{2}+\frac{B_2(T)}{B}\right) + \ln B + \frac{1}{2}\ln B_2(T) - \frac{3}{2}\ln B_1(T)\right], \tag{8.7}$$

where $\psi(x)$ is the digamma function for dimensionless argument x. The terms $B_1(T)$ and $B_2(T)$ in (8.7) are given by

$$B_1(T) = B_i(T) + \tfrac{4}{3}B_{s.o.} + \tfrac{2}{3}B_s\,,$$
$$B_2(T) = B_i(T) + 2B_s\,. \tag{8.8}$$

Here B_i is the characteristic inelastic scattering field, $B_{s.o.}$ is the characteristic spin-orbit field, and B_s is the characteristic magnetic scattering field [8.107]. For acceptor GICs, $B_{s.o.}$ is very small and can be neglected.

It is expected that B_s is small and can be neglected in the fits to the magnetoresistance data, although it has been used as a fitting parameter to the magnetoresistance data by some authors [8.108, 109]. Two other features of the fit deserve further mention: First, a homogeneity parameter λ is often included (8.7) to take into account the fact that, although the analytical form has been found to fit the data very well, there is a discrepancy between the magnitudes of the observed and theoretical magnetoresistance. This homogeneity parameter λ has been found to decrease from 0.2–0.3 in weakly disordered fibers to 0.03 in more strongly disordered fibers. No satisfactory explanation has been proposed until now to account for the magnitude of λ. Second, B in (8.7) stands for the component of the magnetic field perpendicular to the 2D system. A parallel magnetic field has no effect on the weak localization phenomenon. The onion-ring structure of the fibers is taken into account when fitting the magnetoresistance data by performing an average over the angle between the applied magnetic field and the graphene layers.

The temperature dependence of the inelastic scattering rate is often fitted to a power law:

$$\frac{1}{\tau_{in}} = A_{in}T + B_{in}T^2 . \tag{8.9}$$

Here the linear terms in the temperature have been tentatively attributed to electron-phonon inelastic scattering and carrier-carrier inelastic scattering, respectively, in the clean limit by *Piraux* et al. [8.101, 102]. The difficulty in identifying A_{in} and B_{in} with specific mechanisms comes from the existence of different analytical forms for both carrier-phonon scattering and carrier-carrier scattering, depending on whether one is in the clean metal limit or in the dirty metal limit [8.107].

It is also possible to disentangle the respective contributions of weak localization and carrier-carrier interactions in the logarithmic increase of (8.6) by applying a magnetic field (Fig. 8.30). Indeed, the application of a magnetic field largely destroys the weak localization contribution, but has no effect on the carrier-carrier interaction until high fields are reached. Such high-field measurements [8.113] show that the weak localization and carrier-carrier interaction effects are of the same order of magnitude, with the carrier-carrier interaction slightly dominant in fibers.

8.5.5 Thermal Transport Properties

Because of the greatly enhanced carrier density in GICs relative to graphite, both phonons and electrons contribute to the thermal conductivity κ of GICs (Chap. 6):

$$\kappa = \kappa_e + \kappa_l . \tag{8.10}$$

Here the lattice contribution κ_l is well approximated by the Debye relation

$$\kappa_l = \frac{1}{3} C_v v L_\phi \tag{8.11}$$

in which C_v, v, and L_ϕ are, respectively, the heat capacity, phonon velocity, and phonon mean free path, and the electronic contribution κ_e is approximated by the Wiedemann–Franz law

$$\kappa_e = LT\sigma . \tag{8.12}$$

In the residual resistivity regime where the electrical conductivity σ is independent of temperature, then L is well approximated by the free-electron Lorenz number $L_0 = 2.44 \times 10^{-8}\ V^2 K^{-2}$. The ability to measure σ accurately in fibrous GICs thus allows separation of the electronic and lattice contributions [8.86, 99] without the need for application of high magnetic fields [8.114]. Further discussion of the mechanisms involved in thermal transport in GICs is found in Chap. 6.

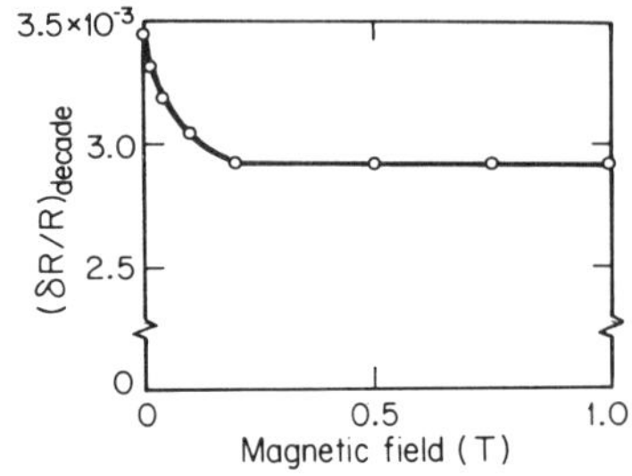

Fig. 8.30. Field dependence of $\delta R_\square(T)/R_\square(T_0)$ in (8.6) over a decade of temperature for a P100-4 $CuCl_2$ intercalated fiber. The points are experimental and the solid line is a guide to the eye [8.101–103]

The maximum total thermal conductivity of typical GICs is on the order of a few hundred $Wm^{-1}K^{-1}$ and is characterized by an approximately linear T dependence at low temperature, a small hump at intermediate temperature, followed by a broad maximum at higher temperature (Fig. 8.31). In some samples, the maximum thermal conductivity occurs above room temperature. Thermal-conductivity measurements can be made on thick single fibers and pertain to the fiber axis direction [8.86]. If the misorientation angle is very small (as for VGCF filaments heat treated above 2500 °C), then the measurements yield the in-plane conductivity κ_a.

A more detailed analysis of the thermal conductivity of GICs can be made by invoking several scattering mechanisms, so that the total scattering probability per unit time is then given by summing over the dominant scattering processes:

$$\frac{1}{\tau} = \frac{1}{\tau_N} + \frac{1}{\tau_U} + \frac{1}{\tau_B} + \frac{1}{\tau_D} , \tag{8.13}$$

where τ_N and τ_U refer to relaxation times associated with normal electron-phonon and umklapp processes, respectively. Boundary scattering gives rise to a relaxation time

$$\tau_B = L_B/v , \tag{8.14}$$

which is independent of phonon frequency ω and where L_B is the boundary scattering length. Scattering from *Daumas-Hérold* domain walls [8.115] could also contribute to this term. The relaxation time τ_D associated with point defect scattering is highly dependent on the phonon frequency ($\tau_D \sim \omega^r$ where $r \sim 4$ for 3D solids and $r \sim 3$ for the 2D case).

Typical results for the thermal conductivity of fibrous acceptor GICs ($CuCl_2$ intercalated VGCF filaments) are shown in Fig. 8.31a. An approximately linear temperature dependence of κ_a at low temperatures ($T < 5$ K) is typically observed for intercalated fiber samples, consistent with a large electronic contribution to κ_a at low T. In this low-temperature limit, contributions from the intercalate phonons may also be significant. Further support for the electronic contribution to κ_a comes from the electrical conductivity measurements made on the same fibers, [8.97, 99] allowing direct computation of κ_e by the Wiedemann–Franz relation. With the VGCF filaments, it is possible to achieve

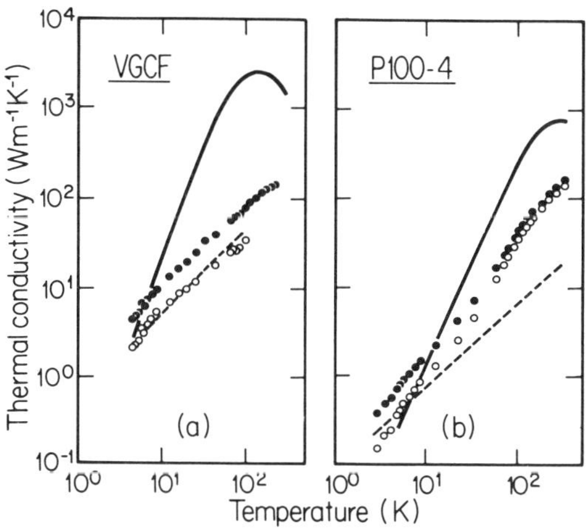

Fig. 8.31. (**a**) The temperature variation of the thermal conductivity of a heat treated ($T_{HT} = 3000\,°C$) VGCF filament intercalated with $CuCl_2$ to stage 2. The solid curve is for the pristine fiber and the solid points are for the intercalated fiber. The dashed curve represents the electronic thermal conductivities κ_e computed from the electrical resistivities measured on the same fiber sample. The open circles represent the lattice thermal conductivity, where $\kappa_l = \kappa - \kappa_e$ [8.86, 99]. (**b**) The same plots as in (**a**) for a P100-4 mesophase pitch fiber intercalated with $CuCl_2$ to stage 2, but containing some stage-1 regions [8.116]

sufficiently large L/A (fiber length/area) values so that four-probe measurements can be carried out quantitatively on a single fiber. Thus a direct determination of κ_l can be made by the subtraction $\kappa_l = \kappa - \kappa_e$, allowing separation of κ into κ_e and κ_l. At higher (intermediate) temperatures ($5 < T < 70$ K), the intercalate phonon modes may continue to contribute to thermal transport. At yet higher temperatures (up to room temperature) κ_a is dominated by the graphitic phonon term κ_{lC}, which is the dominant term overall.

Just as for bulk GICs, intercalation of the graphite fibers results in a decrease in the room-temperature thermal conductivity, where the lattice contribution κ_{lC} dominates (Fig. 8.31a). The decrease in κ_{lC} is due in part to intercalation-induced defects. The smaller values for the thermal conductivity of fibers in both their pristine and intercalated forms relative to the values in the corresponding bulk graphite hosts are also associated with the higher defect concentration in the fibers. This effect is clearly seen by comparison of the thermal conductivity data for GICs based on the VGCF ($T_{HT} = 3000\,°C$) shown in Fig. 8.31a with that for the more disordered P100-4-based intercalated fibers shown in Fig. 8.31b (also intercalated with $CuCl_2$). These curves show that after intercalation, the electronic contribution is more important for the VGCF than for the P100-4 host materials, as expected.

8.5.6 Thermopower

Measurement of the thermoelectric power (TEP) or Seebeck coefficient S provides important information about the sign of the carrier type dominating the transport properties (Chap. 6). The fiber geometry provides an important advantage for carrying out thermopower measurements. A typical apparatus for such measurements in a fiber is shown in Fig. 8.32. As expected, donor GICs exhibit negative thermopower while positive values are found for acceptors.

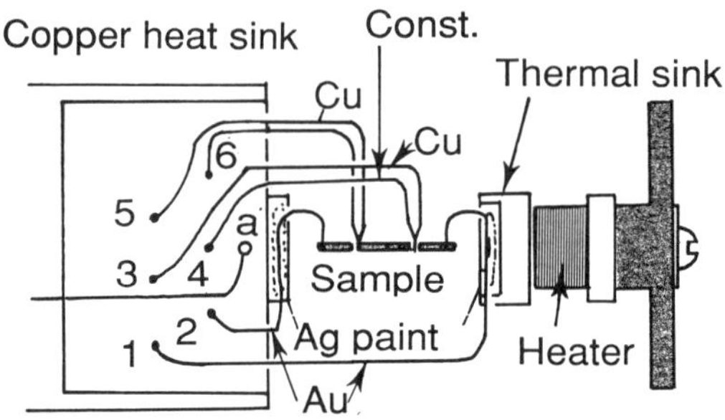

Fig. 8.32. Apparatus for thermoelectric power measurements in GICs [8.117]. Contacts 1–6 denote heat sunk binding posts for the leads. An Au (0.07% Fe) versus chromel thermocouple is attached to contact *a*

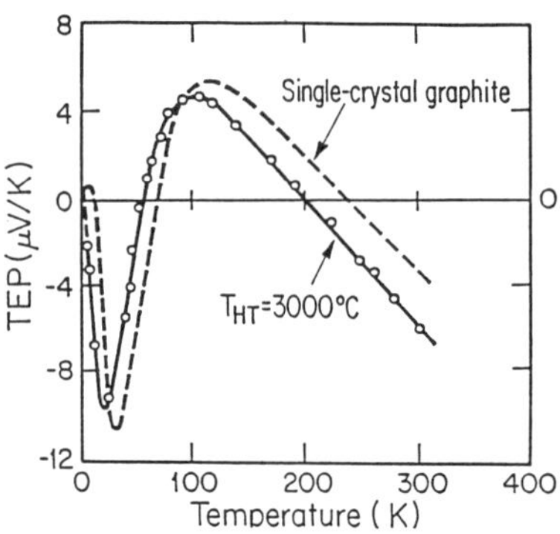

Fig. 8.33. Temperature dependence of the thermopower for unintercalated vapor-grown carbon fibers heat treated at 3000 °C [8.122]. The dashed line is for the corresponding curve for kish single-crystal graphite [8.123, 124]

Intercalation induces a drastic change in the thermopower (Chap 6). In contrast to other transport properties, the thermopower of intercalated fibers is not very sensitive to defects in the fibers prior to intercalation. For carbon fibers, the main interest is in the in-plane thermopower $S_a(T)$, which shows the following features [8.10, 118–121]. First, the $S_a(T)$ curves for GICs are qualitatively different in functional form from those of pristine graphite, as shown in Fig. 8.33, where the thermopower for a VGCF filament heat treated to 3000 °C is compared with that for single-crystal graphite. Of interest is the similarity between the two curves in Fig. 8.33. In contrast, the thermopower results for a similar fiber intercalated with the acceptor species $CuCl_2$ are shown in Fig. 8.34a. At low temperature ($\leqslant$ 100 K), the thermopower for the intercalated fiber follows a linear temperature dependence, while saturation behavior is found above 150 K. The measured $S_a(T)$ curves for intercalated fibers are almost the same both in functional form and magnitude as for bulk HOPG intercalated with the same species to the same stage (Chap. 6). Since the anisotropy in $S(T)$ is relatively small compared with other transport properties [8.3], misalignments in the crystallites are not as important for thermopower measurements on fibers, compared with other transport or mechanical properties [8.87].

Not only is the thermopower of an acceptor compound not sensitive to the choice of the fibrous host material (or HOPG), but also the stage dependence of the thermopower is not strong [8.125]. This is shown in Fig. 8.34b, where a log-

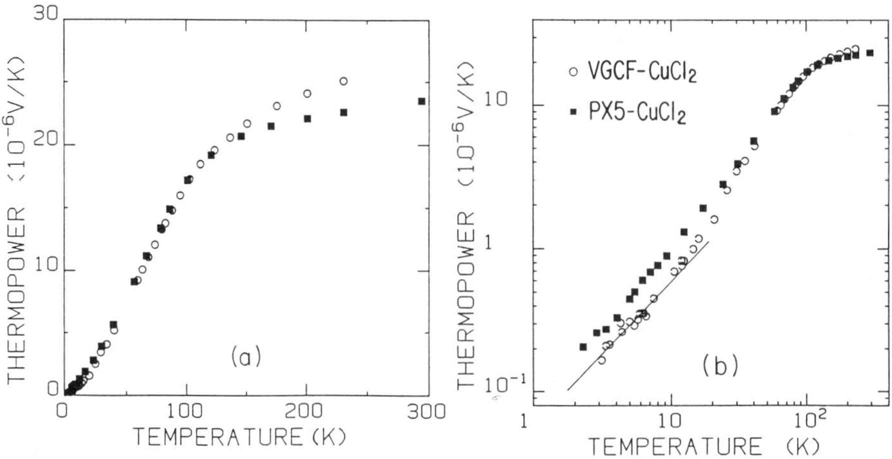

Fig. 8.34. (**a**) Linear plot of the temperature dependence of the thermopower for a benzene-derived VGCF filament and a PX5 pitch-based fiber intercalated with $CuCl_2$ to stage 2 (○) and stage 4 (■), respectively. (**b**) The corresponding log-log plot of the temperature dependence of the thermopower. Note the linear temperature dependence in the low-temperature range [8.99]

log plot of the thermopower is presented for the acceptor $CuCl_2$ intercalated into a benzene-derived VGCF filament (stage 2) and into a pitch-based (PX5) fiber (stage 4). In particular, the saturation effect above ~ 50 K appears to be relatively insensitive to stage index and choice of fibrous host material [8.99]. One exception to this general trend is the anomalous behavior of stage-1 acceptor GICs (observed so far only in bulk samples), which show an anomalous decrease in the thermopower to very small values below ~ 5 K [8.126].

In the low-temperature region below about 100 K, the diffusion contribution to the thermopower (depending on kT/E_F) is dominant, thereby accounting for the approximately linear temperature dependence of the thermopower at low temperature and the difference in behavior between stage-2 and stage-4 compounds. Similar results were also obtained in the HNO_3-intercalated CVD carbon fibers, prepared by vapor growth on a continuous PAN fiber (P75) and subsequent thickening and heat treatment to achieve a highly ordered fibrous host material [8.117]. Above ~ 100 K, saturation occurs in $S_a(T)$, and near ~ 250 K an anomaly in $S_a(T)$ is observed, in good agreement with transport anomalies observed in bulk HNO_3 GICs. In the saturation regime, *Sugihara* has shown [8.121, 127–129] that the phonon-drag contribution becomes important, particularly due to phonon scattering by the strain field introduced by the intercalate. For this mechanism, Sugihara shows a weak temperature dependence

$$S_a = A_s + B_s \ln(C_s/k_B T) \tag{8.15}$$

leading to saturation behavior, where the constants A_s, B_s, and C_s are independent of temperature. By introducing intercalate layers into pristine graphite, many point defects are formed. Owing to the functional dependence of the strain-field scattering rate $1/\tau_I(q) \propto q^3$ on the phonon wave vector, strain-field scattering becomes the predominant scattering process at high temperatures in GICs that have a large cross-sectional diameter of the Fermi surface ($k_F \sim 10^7\ \mathrm{cm}^{-1}$). In pristine graphite this process is unimportant since the Fermi-surface cross section is much smaller: $k_F \sim 10^6\ \mathrm{cm}^{-1}$. The lower thermopower for donors relative to acceptors can be explained by the higher carrier density in the alkali metal donor GICs.

8.6 Mechanical Properties

Strong, high-modulus fibers with high electrical conductivity can be prepared by the intercalation of carbon fibers. Specifically, an order of magnitude increase in electrical conductivity can be obtained with a much smaller decrease in the tensile strength σ_T and in Young's modulus E_Y. Qualitatively similar results are obtained for several fibrous host materials (e.g., mesophase pitch, vapor-grown carbon fibers, rayon-based fibers) and for several intercalates (e.g., alkali metals, HNO_3, Br_2, $CuCl_2$, $NiCl_2$, $FeCl_3$, AsF_5). Because of the unusually high values of the tensile strength and Young's modulus prior to intercalation, intercalated fibers exhibit excellent mechanical properties relative to other materials.

Early work by *Herinckx* et al. [8.130] showed relatively small changes in mechanical properties as a result of potassium intercalation of rayon-based fibers, previously heat treated to 2800 °C. In this work, the mechanical properties of individual filaments of a stage-1 compound several cm in length were measured under 10^{-6} mm Hg vacuum conditions. The force F was applied perpendicular to the fiber axis at the center of the fiber, keeping both ends fixed by means of a calibrated stainless steel spring. The displacement of the center of the filament H and the elongation of the spring were measured. The Young's modulus E_Y of the fibers was obtained from the relation

$$F/H = 4E_Y S \Delta L/L^2\ , \qquad (8.16)$$

where S is the cross-sectional area, ΔL the elongation, and L the length of the filament. Figure 8.35 shows the results for a first-stage K-intercalated rayon fiber in comparison with the corresponding result for the pristine host fiber. The nonlinear behavior in this figure at low F/H is due to the straightening of the filaments. By comparing the two curves, it is seen that the Young's modulus of the intercalated fiber is slightly lower (1.86×10^{12} dynes/cm^2) than that of the host fiber (1.99×10^{12} dynes/cm^2), neglecting the change in the fiber cross section S with potassium intercalation. In this study the tensile strength of the fibers was also estimated roughly, and it was found that the tensile strength decreased by approximately a factor of 4 upon intercalation to stage 1 (from 12×10^9 dynes/cm^2 to $\sim 3.2 \times 10^9$ dynes/cm^2). From these results *Herinckx* et al. [8.130] concluded that in the direction of the fiber axis, it is the carbon layer

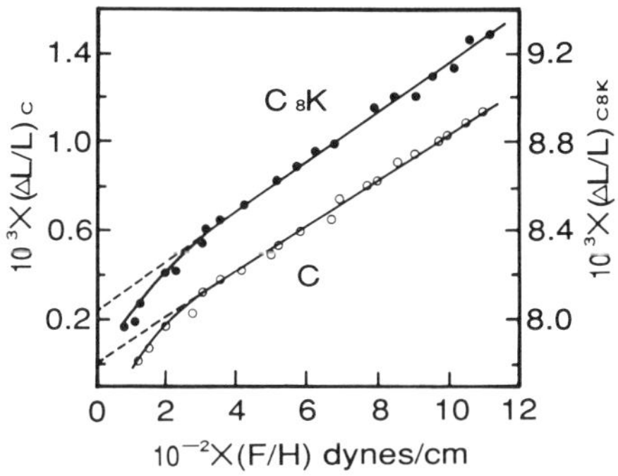

Fig. 8.35. Fractional elongation of a rayon-based fiber before (labeled "C") and after (labeled "C_8K") intercalation with potassium to stage 1 as a function of F/H, where F is the force normal to the fiber axis and H is the displacement of the fiber center [8.130]

structure that determines the mechanical properties of the intercalated fiber, and this layer structure does not significantly change with intercalation, as is indicated in Fig. 8.36.

Systematic studies of the degradation of the tensile strength and the Young's modulus as a function of intercalate uptake are shown in Figs 8.37a, b respectively, for the case of $CuCl_2$ intercalated into the mesophase pitch fiber P120X

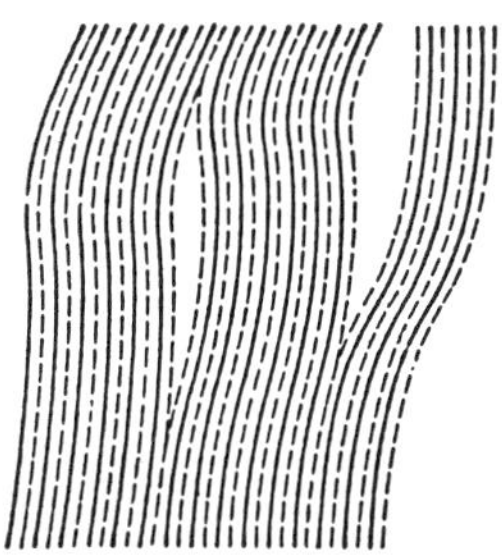

Fig. 8.36. Schematic representation of the insertion of intercalate layers (solid lines) in the ribbon structure (dashed lines) of a stage-1 intercalated fiber [8.130]

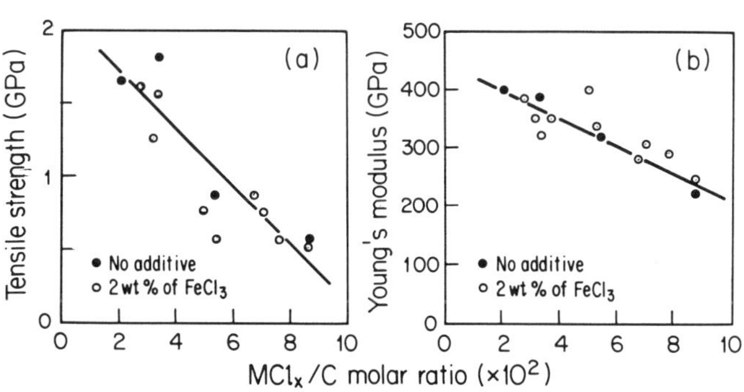

Fig. 8.37. (**a**) Tensile strength σ_T and (**b**) Young's modulus versus $CuCl_2$ uptake. Data are presented both for the intercalation of $CuCl_2$ into a P120X pitch-based fiber and for $CuCl_2$ mixed with 2% by weight of $FeCl_3$ [8.131]

host material [8.131]. As expected, the results show a much larger degradation in the tensile strength σ_T relative to the Young's modulus E_Y for a given amount of intercalate uptake. The larger expected effect on σ_T is due to the increase in local stress resulting from an expansion of the well-graphitized regions of the fiber that intercalate, and almost no expansion of the highly disordered regions that do not intercalate. The lower tensile strength of carbon fibers relative to single-crystal graphite is likely due to the introduction of defects or flaws along the length of the fibers. For both pristine and intercalated fibers, short lengths of fiber are less likely than long lengths to have flaws and are therefore more likely to exhibit better mechanical properties. It is of interest that the mechanical properties seem to be essentially unaltered by the use of a small amount (~ 2%) of $FeCl_3$, which serves to enhance the speed of the intercalation by a factor of ~ 6–8 [8.131].

Of particular importance to applications is the relation between the drop in resistance upon intercalation and the degradation in the tensile strength of the intercalated fibers. Figure 8.38 shows that the resistivity can be decreased by a factor of 3–4 with almost no loss in strength. However, for greater decreases in resistivity, a substantial loss in strength is observed [8.131], as the intercalate-induced flaws and defects grow more numerous and start forming a network of flaws that eventually results in failure. These trade-offs must be taken into account when considering practical electronic applications of intercalated fibers.

The most broad-gauged study of the mechanical properties of intercalated carbon fibers was reported by *Touzain* and *Bagga* [8.132], who studied the effect of intercalation on both the tensile strength σ_T and Young's modulus E_Y for various fiber hosts (PAN, mesophase pitch, and VGCF) and for several acceptor intercalate species ($CuCl_2$, $NiCl_2$, $FeCl_3$). For each precursor, Figs. 8.39a, b show results for σ_T and E_Y before and after intercalation with several intercalates, including some staged compounds. The results show that the fibers with the highest Young's modulus before intercalation tend to have the highest modulus after intercalation (Fig. 8.39a). The mesophase pitch high-modulus P100 fibers showed a decrease of ~ 20% in E_Y, and a similar percentage decrease was found for the P55 fibers intercalated with $CuCl_2$. A larger decrease in modulus (~45%) was found for the VGCF filaments, with no measurable dependence of E_Y on stage index (for intercalation with $CuCl_2$).

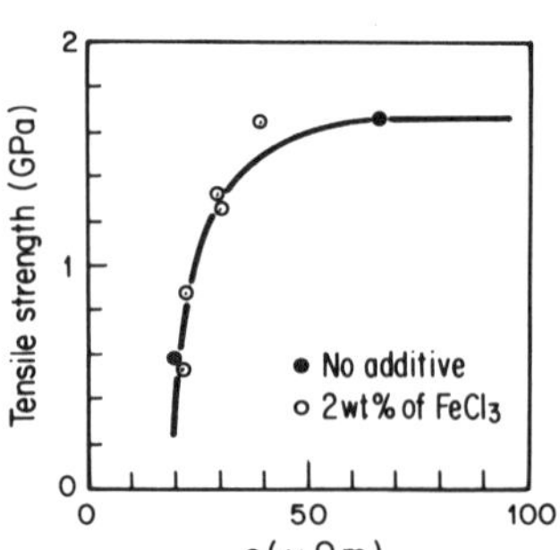

Fig. 8.38. Tensile strength versus resistivity of a P120X pitch-based fiber intercalated with $CuCl_2$ and with $CuCl_2$ mixed with 2% $FeCl_3$ [8.131]

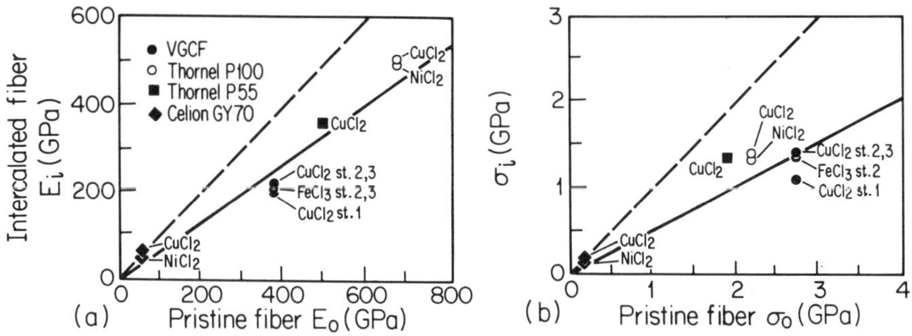

Fig. 8.39. Correlation between (**a**) the Young's modulus (E_Y) and (**b**) the tensile strength (σ) for several fiber types and acceptor intercalate species before (subscript "o") and after (subscript i) intercalation [8.132]. The dashed lines denote no change upon intercalation, while the solid lines denote the average observed change

With regard to the tensile strength σ_T, *Touzain* and *Bagga* [8.132] found a correlation between σ_T in the intercalated and pristine fibers (Fig. 8.39b), with a decrease of ~30% for the Thornel P55 fibers resulting from $CuCl_2$ intercalation, to a decrease of ~40% for the Thornel P100 fibers (after $CuCl_2$ or $NiCl_2$ intercalation), to a decrease of over 50% for the VGCF intercalated with $CuCl_2$ or $FeCl_3$. A larger decrease (~ 60%) in σ_T was found for the VGCF intercalated to stage 1 with $CuCl_2$ relative to that for stages 2 and 3 (Fig. 8.39b). In all cases, the larger decreases in tensile strength can be correlated with correspondingly larger amounts of intercalate uptake. It is of interest, however, to note that alkali metal donor intercalation degrades the host fiber more than acceptor intercalation, keeping all adjustable parameters (such as stage index) fixed. This result is consistent with the high-resolution structural studies carried out with the TEM (Fig. 8.15), showing larger L_a and L_c values in acceptor intercalated fibers than for alkali metal donor intercalated fibers [8.9, 35, 56].

Figure 8.40 shows results for the correlation between the Young's modulus and tensile strength for intercalated P100 mesophase pitch fibers, and for intercalated methane-derived and benzene-derived VGCF filaments [8.21]. The results show that residue compounds exhibit the smallest degradation in mechanical properties, consistent with their lower intercalate uptake.

This behavior suggests that defects introduced during intercalation pin the basal planes together and inhibit shear. With regard to the inhibition of shear, an 18% intercalation-induced increase in interlamellar shear strength has been reported for P100 mesophase pitch fibers intercalated with bromine [8.60]. On the other hand, PAN fibers with $T_{HT} \sim 1600\,°C$, which would be much less ordered, showed essentially no change in the interlamellar shear strength after immersion in a 0.2 M bromine in CCl_4 solution [8.133].

Studies on VGCF intercalated with HNO_3 and Br_2 show that the stress-strain curves for these intercalated fibers are similar to those for the fibers prior to intercalation and that the tensile strength is approximately inversely propor-

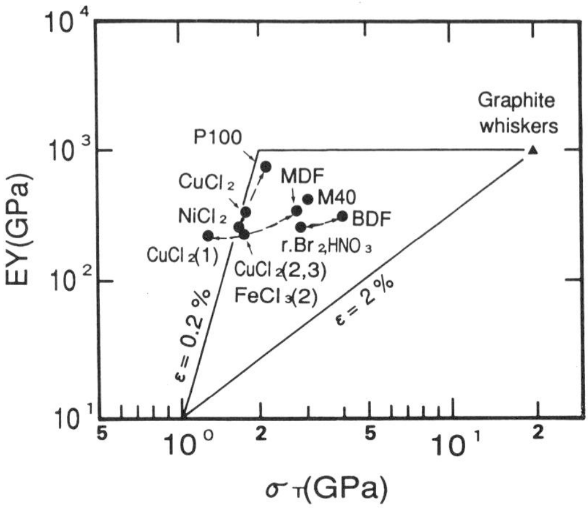

Fig. 8.40. A log-log plot of the Young's modulus E_Y versus tensile strength (σ_T) at room temperature for a series of graphite fibers before and after intercalation (connected by a dashed line). BDF and MDF refer to benzene- and methane-derived fibers (VGCF), and *r* refers to a residue compound. Lines for the normalized elongation ($\Delta L/L$) = 0.2% and 2% are indicated [8.21]

tional to the fiber diameter [8.31]. The triangle in Fig. 8.40 gives reference values for fiber elongations to the break point $\varepsilon = \Delta L/L$ of magnitude 0.2% and 2%. This figure confirms the findings of other workers that intercalation tends to decrease the fiber strength more than the modulus [8.31, 69, 76, 77, 134]. Intercalation generally makes the fibers more brittle (more likely to fracture upon bending), though this effect seems to be less pronounced with smaller diameter fibers [8.69]. On the other hand, exposure to bromine has been reported to plasticize PAN fibers [8.133]. This plasticization effect may result from the penetration of bromine between short and imperfect graphene layers, allowing structural relaxation and increased structural order for PAN fibers subjected to medium heat treatment temperatures ($T_{HT} \sim 1600\,°C$).

Also reported in the literature is an *increase* in modulus (by about 69%) together with a decrease (by about 21%) in tensile strength after the intercalation of Thornel P75 fibers with red fuming nitric acid to stage 2 [8.73]. An intercalation-induced *increase* in modulus is unusual. The tensile strength of carbon fibers is limited by the propagation resistance of cracks as well as the density of defects. The introduction of a foreign species either near the surface or into the bulk of the fibers is likely to form a void, which could reduce the tensile strength in this region. In contrast, the modulus, which depends on the crystal perfection and the parallel arrangement of the layers, may be increased by the passage of a partial dislocation through the lattice. Thus, as a result of inserting a sheet of nitric acid between the hexagonal graphene planes, *Vogel* has inferred [8.73] that the graphene layers become more parallel to one another, thereby increasing the overall preferred orientation of the planes. This effect has been confirmed by other workers [8.133], who showed that the bromination of PAN fibers ($T_{HT} \sim 1600\,°C$) during stress annealing appears to increase both the modulus and strength of these fibers, suggesting that in some cases, chemical treatment could be used instead of higher-temperature heat treatments to enhance mechanical properties [8.133].

Table 8.5. Properties of deintercalated GICs

	GICs-4D[a]	GICs-5D[b]	GICs	P120X
Heat treated temperature	400	500	–	–
MCl_x molar ratio $\times 10^2$	4.16	1.01	8.54	–
Stage number [c,d]	2 + G	(2) + G	2 + (G)	–
Resistivity [μ Ωcm]	68.3	155	20.7	278
Tensile strength [GPa]	0.75	2.0	0.57	2.6
Young's modulus [GPa]	310	480	290	560
Elongation [%]	0.24	0.42	0.19	0.45

[a] Conditions of deintercalation are $T_{HT} = 400\,°C$ in vacuum for 1 h.
[b] Conditions of deintercalation are $T_{HT} = 500\,°C$ in vacuum for 1 h.
[c] Parentheses denote a minority component.
[d] G denotes unreacted graphite.

The measurements of the mechanical properties of the intercalated fibers generally indicate that small cracks both on the fiber surface and inside the fiber more significantly degrade the fiber strength than the modulus. In addition, some defects and microcracks result from the intercalation process itself, and also contribute to the degradation of the mechanical properties. But, as shown in Table 8.5, after deintercalation through heating the $CuCl_2$-intercalated fibers to 400 °C and 500 °C, the mechanical properties of the pristine fibers are almost completely recovered. This important practical result indicates that, for appropriate intercalate species and intercalation conditions, the intercalation-induced defects do not seriously degrade the mechanical properties. In Table 8.5, characterization information is given for various fibers, including intercalate uptake, stage number, resistivity, σ_T, E_Y, and elongation. To reduce this intercalation-induced degradation of the mechanical properties, it is very important to select slow reaction conditions and suitable intercalates that introduce less lattice strain.

Of particular importance is the conclusion that intercalated graphite fibers have superior mechanical strength relative to other metallic conductors, due in large measure to the excellent mechanical properties of the fibers prior to intercalation and to the relatively minor degradation of the mechanical properties resulting from intercalation.

8.7 Thermal Expansion

The anisotropic thermal expansion of intercalated carbon fibers presents serious problems, especially for their application in composite materials, because of the differential thermal expansion between the fiber and matrix material upon thermal cycling. Although there are no reports on the thermal expansion of intercalated carbon fibers, we expect that the general features of the thermal expansion behavior can be predicted from the results on intercalated bulk graphites and pristine carbon fibers, taking into account the microstructure of the host fiber materials. Specifically, little thermal expansion along the fiber length is expected. Also, the macroscopic intercalation-induced expansion along the fiber length is expected to be much less than along the fiber diameter. Because of the inhomogeneous intercalate distribution in a fiber, the anisotropic thermal expansion coefficients (α_a and α_c) for both the intercalated and unintercalated regions of the fibers, and the differential expansion coefficients of the intercalated and unintercalated regions of the fiber, a great deal of internal stress is expected to be introduced into thermally cycled intercalated carbon fibers.

8.8 Applications of Intercalated Carbon Fibers

Several practical applications of intercalated carbon fibers have been implemented or proposed, the most important from a commercial standpoint being battery applications. Other important applications areas exploit the intercalation-enhanced conductivity and other property modifications induced by intercalation.

The unique ability of graphite to electrochemically intercalate both negative and positive ions is central to the use of intercalated fibers in battery applications [8.135–137]. A primary battery based on fluorine and lithium is theoretically the best system [8.74, 138, 139] for a high-energy-density power source. Such a battery has been realized commercially using $(CF)_n$ as a cathode material, Li as an anode material, and 1M $LiClO_4$-propylene carbonate (PC) as the electrolyte [8.140].

Figure 8.41 shows the open circuit voltage of an Li – $(CF)_n$ fiber and an Li – $(C_2F)_n$ fiber cell versus percent discharge, demonstrating the broad stability range for this battery [8.37]. Typical discharge current densities are in the

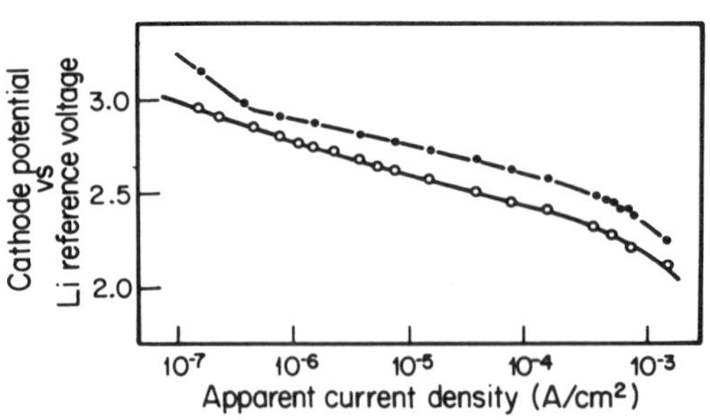

Fig. 8.41. Cathode potential of the CF electrode relative to the Li reference electrode versus apparent current density at 25% discharge. Improved performance is found for the fluorine intercalated graphite fiber electrode ($CF_{0.68}$) (●) relative to the graphite fluoride electrode $CF_{0.64}$ (○) prepared from natural graphite flakes [8.142]

0.02–0.2 mA/cm^2 range [8.37]. The discharge reactions are summarized as follows:

anode reaction: $Li + zPC \rightarrow Li^+ \cdot zPC + e^-$
cathode reaction: $CF + Li^+ \cdot zPC \rightarrow C\text{-}F\text{-}Li \cdot zPC$
total reaction: $(CF)_n + mLi + mzPC \rightarrow (CF)_{n-m} + m(C\text{-}F\text{-}Li \cdot zPC)$
$\rightarrow (CF)_{n-m} + mC + mLiF + mzPC$

Here z is the solvation number. Physically, the Li^+ diffuses between the layered structures of the CF cathode and the discharge reaction proceeds according to a disproportionation reaction forming an intermediate phase. As the discharge reaction continues (over long times), the intermediate product eventually decomposes to graphite and LiF by decomposition and desolvation. The factors of interest for high-performance battery operation include a high open circuit voltage (OCV), a low overpotential, high stability of the discharge potential, and a high percentage utilization of the cathode material in the energy production.

With regard to the cathode material, note that there are two classes of graphite intercalation compounds based on fluorine: *ionic* (metallic) compounds, where charge transfer occurs between the fluorine and the planar graphene planes, and *covalent* (highly insulating) compounds, where there is no charge transfer and the carbon atoms are arranged in puckered layers (cyclohexane rings). In these covalent compounds the A and B carbon atoms are displaced in opposite directions relative to the graphene plane. It should be emphasized that the cathode material used for batteries is the covalent compound, and being insulating, carbon fibers or carbon powder is added to make the cathode electrode conducting. With regard to the covalent compounds, there are two main types [8.26]: $(CF)_n$, which is normally prepared at $\sim 500\,°C$, forming a stage-1 compound, and $(C_2F)_n$, which is prepared at lower temperatures (~ 350 to $\sim 400\,°C$) and forms a stage-2 compound (Fig. 8.12).

Though one of the most important commercial batteries today is based on $(CF)_n$ compounds, the $(C_2F)_n$ compounds have been shown to have a somewhat higher open circuit voltage (by ~ 0.2 V) and a lower overpotential (by ~ 0.1 V), in addition to other advantages cited below. Even better performance is achieved when the $(C_2F)_n$ material is prepared from VGCF filaments [8.141] rather than from graphite flakes, thereby achieving about a 0.5 V increase in the discharge potential at a current density of 0.5 mA/cm^2. This improved performance of $(C_2F)_n$ is in part explained by the higher open circuit voltage attributed to the higher activity of F^- in $(C_2F)_n$ relative to $(CF)_n$ (Fig. 8.41). The smaller overpotential (and consequently higher discharge voltage) for fiber-based cathodes is attributed to improved diffusion of the lithium ions into the inner cathode material because of the presence of defective microdomain structures in the fibers [8.142]. For both $(C_2F)_n$ and $(CF)_n$ cathode materials, the overpotential is constant through the discharge, thereby providing a stable discharge voltage under load. The fiber host materials have additional advantages insofar as the intercalation can be done at lower temperatures and for shorter times [8.38]. Also, higher F/C ratios can be achieved in the stage-2 compound, thereby allowing more energy storage (discharge capacity) in the battery [8.38]. The use of $(C_2F)_n$ fibers rather than the $(CF)_n$ fibers also offers additional advantages with regard to being less brittle, providing a more constant battery discharge voltage and higher cathode utilization ($\sim$80%) [8.142]. Further improvements in performance (discharge potential and cathode utilization) have been achieved with ultrathin (< 1 μm) VGCF filaments [8.143].

The potential of graphite oxide fibers for primary, high-energy-density, and low-cost battery applications has recently been investigated. The relatively flat discharge curves (Fig. 8.42) obtained when graphite oxide fibers are used as a cathode material may be due to the formation of elemental carbon in a fibrous form during the discharge reaction [8.42]. The fibrous graphite oxide cathode material yields the very high energy density of 1770 Wh/kg, which is the same as that of the fibrous $(CF)_n$ material discussed above and $\sim$10% higher than that obtained with similarly intercalated flaky graphite host materials [8.42]. This good performance is especially promising in view of the much lower cost of the graphite oxide material compared with $(CF)_n$. In the case of the fibrous graphite oxide, however, the discharge potential is lower (2.45 V), compared with 3.2 V for the fibrous $(CF)_n$ cathode material [8.26, 42].

Highly ordered vapor-grown graphite fibers have been used with Li as the counter-electrode in a secondary battery with lithium perchlorate-propylene carbonate as an electrolyte [8.144]. The suggested mechanism for the battery operation is

$$C_{96} + ClO_4^- \rightleftharpoons C_{96} \cdot ClO_4 + e^- \ .$$

During the charging, ClO_4^- ions are introduced into the graphite fibers to form a GIC, and during the discharging the intercalate species are released into the electrolyte. A well-developed graphite structure with a high level of preferred orientation in the VGCF has been shown to be necessary for the high performance of this secondary battery [8.144]. The discharge capacity increases as

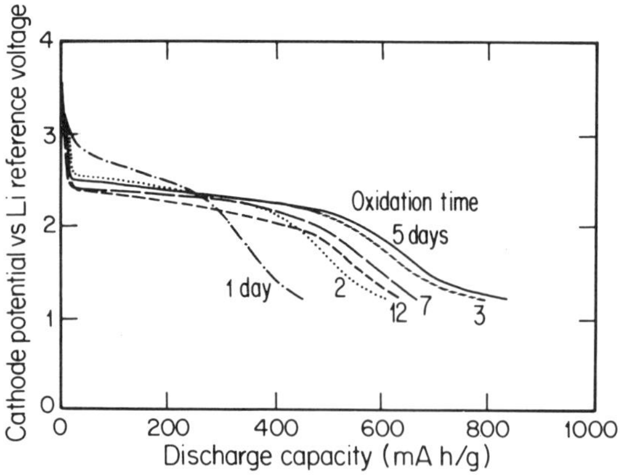

Fig. 8.42. Galvanostatic discharge curves (at 0.1 mA/cm^2) for various fibrous graphite oxide cathodes obtained from oxidation of heat treated vapor-grown carbon fibers for various lengths of time [8.42]

a function of the degree of graphitization of the fibers, and a discharge capacity as high as 270 Coulomb/g has been observed for fibers heat treated at 2700 °C. These secondary batteries exhibit a large positive potential of 4.5 V versus Li/Li^+, which is relatively constant during the discharge time. More than 400 cycles of charging-discharging have been demonstrated with this secondary battery.

Another battery application that uses the intercalation capability of graphite fibers is in concentration cells designed to convert thermal energy directly into electricity. *Lalancette* and *Roussel* [8.145] described such a cell based on bromine intercalation of graphite, whereby the amount of bromine intercalated at the two electrodes would be different due to the dependence of bromine activity on temperature. *Endo* et al. [8.146–148] later showed that the energy conversion efficiency of such a cell could be improved if VGCF filaments were used at the electrodes. *Maeda* et al. [8.149] demonstrated distinct electrochemical advantages of using VGCF filaments (instead of graphite powder) in a similar concentration cell based on nitric acid.

The intercalation-induced increase in electrical conductivity of carbon fibers has also found a number of applications. A conducting paint containing chopped intercalated highly graphitized carbon fibers has been discussed for printed circuit board (and related) applications. Such a paint could be made more conducting than materials now used for this application. The addition of low-intercalate (e.g., bromine) concentration, chopped intercalated carbon fibers into the injection molding of a thermoplastic gives sufficient conductivity to be of interest for a number of aerospace applications which utilize the intercalation enhancement of the fiber conductivity: construction of strong, lightweight electromagnetic interference (EMI) shielded electronic containers; construction of electrically grounded space structures with high stiffness and low thermal expansion; de-icing of aircraft surfaces by direct ohmic heating; and increasing the tolerance of aircraft surfaces to lightning strikes [8.60]. A combination of electrical, optical, and lubrication properties have found numerous applications for intercalated carbon fibers in such areas as toner inks, electroconductive polymers, preparations for photosensitive imaging materials, and electrochromic displays [8.150].

As an example of the enhanced electromagnetic shielding provided by intercalation, we show in Fig. 8.43 that bromine intercalation can enhance the performance of a fiber composite EMI shield [8.60]. By adding chopped low-density brominated P100 pitch-based carbon fibers to an injection-molded polystyrene thermoplastic matrix, more than a 10 dB increase in attenuation is obtained relative to a similarly prepared composite using the same pristine fibers. The largest intercalation-induced rf attenuation is achieved for laminates with fiber layers at 45° with respect to each other [8.3, 60]. To meet government regulations in Europe and the United States, 40 dB of rf attenuation is necessary.

As indicated above, carbon fibers are likely the most suitable host material for the preparation of highly conductive synthetic wire based on GICs. One of

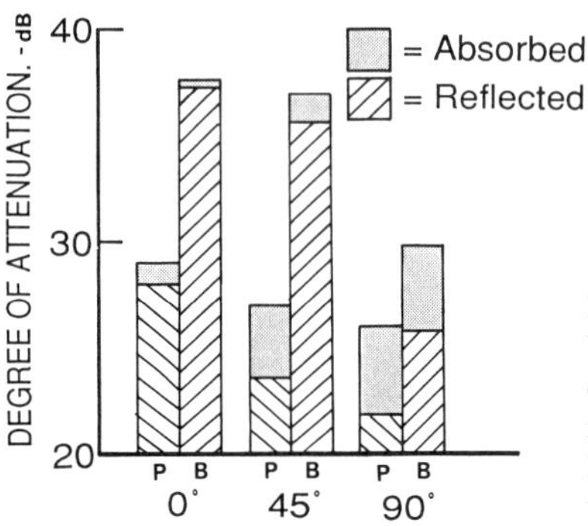

Fig. 8.43. Application of (bromine) intercalation to carbon fiber composites to enhance EMI shielding. The degree of attenuation before (*P*) and after bromine intercalation (*B*) is presented for fiber thermoplastic laminates with three different relative orientations of the fiber directions [8.60]

the most difficult problems for such applications is the low degree of graphitizability of common carbon fibers, especially those made from PAN precursors. Even for mesophase pitch-based carbon fibers exhibiting a relatively high degree of graphitizability, the crystal perfection of graphite is still low in comparison with that of HOPG, and so also is the electrical conductivity. For high conductivity, highly graphitized VGCF filaments are the most suitable host fibers for intercalation, but these fibers are discontinuous. For many applications, it is necessary to make composites with long continuous strands of fibers. These problems can be solved by combining the continuity of PAN-based carbon fibers and the high graphitizability of fibers synthesized by the chemical vapor deposition (CVD) process. Specifically, well-graphitized thick layers have been successfully formed on PAN-based carbon fibers by pyrolysis of cyanoacetylene using a continuous process shown in Fig. 8.44 [8.17, 18, 151–153]. By this process, several meters of long continuous filaments with diameters of about 100 μm can be prepared as a host fiber for intercalation. These host fibers have an annular structure built around the thin core consisting of a PAN-based fiber (~ 10 μm in diameter). The structure of the outer 90 μm of these composite fibers is similar to that of VGCF, as indicated in Fig. 8.8. After heat treatment at 3250 °C, the lattice constant $\tilde{c}$ along the c axis is 6.712 Å

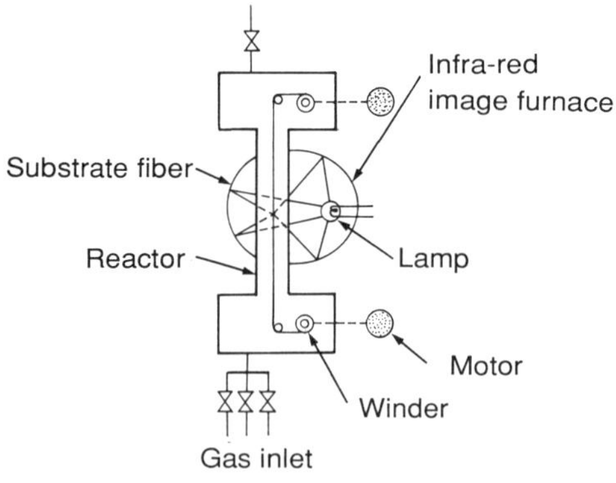

Fig. 8.44. Schematic diagram of the synthesis of continuous graphite fibers by chemical vapor deposition, carried out by means of an image furnace. The as-grown fibers are then graphitized by heat treatment to ~ 3000 °C [8.153]

(essentially the same as that of HOPG) and the electrical conductivity is 1.8×10^4 S/cm, which is similar to that of HOPG. The fibers have enough flexibility to be wound on a spool of 4 cm diameter. By intercalation of the thick fibers with fuming nitric acid, the conductivity increases up to 1.3×10^5 S/cm [8.18]. By exposing the nitrated fibers to air, the conductivity at room temperature decreases slightly at first, but then the value is stable for a long time (> 90 days) [8.17, 18, 151–153]. The small decrease in conductivity of the nitrated fibers was correlated with a structural phase transition from an α-stage-3 compound ($I_c = 14.50$ Å) to a β-stage-3 compound ($I_c = 13.25$ Å) [8.151, 152]. Similar experiments carried out on similarly intercalated and exposed benzene-derived VGCF showed that the CVD-coated continuous fibers had significantly better long-term stability in air than the HNO_3-intercalated benzene-derived fibers [8.152]. Furthermore, by intercalating KHg into the fibers to stage 1, superconducting fibers have been obtained as shown in Fig. 8.45 [8.153]. By winding the superconducting wire in a spool, a strong, lightweight superconducting magnet could perhaps be fabricated. These long and flexible highly conductive and superconductive GIC fibers could have widespread applications in the high-technology area.

The very different gas absorption capacities of K GICs for hydrogen, deuterium, and tritium make alkali-metal GICs an attractive system for isotope separation. A particularly interesting application of these properties would be enrichment of tritium by passage of the ^{1}H, ^{2}H, and ^{3}H mixtures through C_{24} K powder [8.154]. Using fibrous K GICs or by mixing potassium powder with the fibers, the gas can flow more easily, and more effective isotope separation may thus be achieved.

A number of applications of intercalated graphite fibers are based on the very weak interplanar bonding and high chemical stability of graphite fluoride and other acceptor compounds, making these materials attractive for high-temperature lubricants, low-friction coatings, protective wear-resistant coatings, lubrication for ball bearings and self-lubricating watches, sound-absorbing media, additives for abrasives, additives for polishing preparations, additives in combustion engine oil, and so on [8.150]. The chemical inertness of some

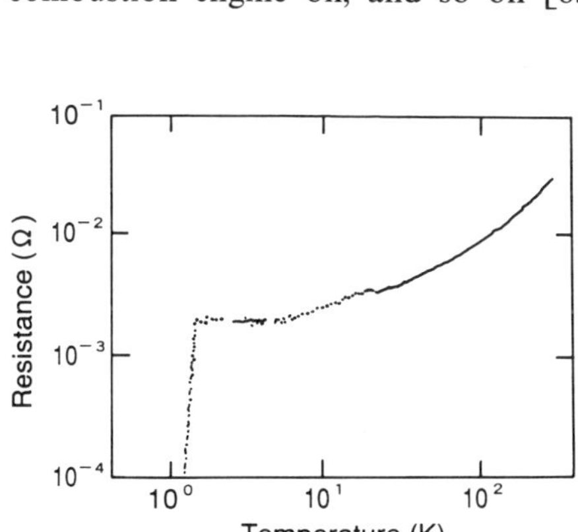

Fig. 8.45. Temperature dependence of the electrical resistance of KHg-intercalated (stage-1) continuous CVD fibers showing a transition to the superconducting state at low temperature [8.153]

acceptor intercalates has been discussed for potential applications as liners for auto exhaust tubes to prevent tar deposition [8.150]. The exfoliation property, which is exploited in Grafoil and vermicular graphite (used as high surface area materials and high-temperature seals), is under consideration in fibrous form for such applications as extinguishers for metal fires and as a fire retardant [8.150]. By enhancing the adhesion between fibers and the matrix materials, intercalation may lead to increased mechanical strength of carbon fiber-epoxy composites [8.155, 156].

Many applications of intercalated fibers have been considered using the intercalated fibers as catalysts, including the synthesis of acetylene, ammonia, cabonyl sulfide, and various polymerization reactions. For reactive GICs, such as alkali-metal GICs, the easy intercalation and deintercalation of these compounds has been used as a source and getter for metal vapors [8.150].

8.9 Summary and Conclusions

In this chapter we have shown that the special properties of fibrous host materials introduce a variety of novel physical properties, thereby allowing the study of new physical phenomena and the practical use of these fibers for a variety of applications. For carbon fiber science and technology, intercalation offers a rich variety of opportunities for properties modification, thereby also leading to new science and new applications.

Traditionally, pristine carbon fibers provide extremely high specific strength and modulus, and for this reason they have mainly been used for structural aerospace applications. Through the intercalation process we can introduce additional interesting properties to these pristine fibers with only minor degradation (or perhaps in some cases even an increase in modulus) of their mechanical properties. The availability of intercalated carbon fibers has made possible more accurate transport measurements on GIC materials because of their highly favorable aspect ratio, leading to a better understanding of low-temperature electrical transport behavior, of the thermal-conductivity mechanisms, and of weak localization and negative magnetoresistance effects. From a practical standpoint, intercalated carbon fibers have shown high potential (and in some instances commercial products) for a variety of practical applications, including primary batteries, secondary batteries, thermocells, EMI shielding devices, lightning protection for aircraft, and hydrogen isotope separators. Some of these applications exploit the high specific conductivity, strength, and high surface area that can be achieved and stabilized in fibrous intercalated materials.

Special attention has recently been given to applications exploiting the high specific electrical conductivity and the convenient geometry of intercalated fibers for applications. Intercalation is not only greatly expanding the range of technical applications for carbon fibers, but is also leading to new insights into our understanding of the physics of intercalation and carbon fiber science, often exploiting our ability to control the structure and properties of the intercalated fibers through control of the host-fiber characteristics and of the intercalation process.

Acknowledgements. The authors wish to thank Dr. G. Dresselhaus and their colleagues for many valuable discussions and critical reading of the manuscript. We give many thanks of Prof. J.-P. Issi and Dr. L. Piraux of the Catholic University of Louvain-la-Neuve for numerous discussions and for kindly supplying Figs. 8.28–30, and to Dr. J. Tsukamoto of the Research Association for Basic Polymer Technology for kindly supplying Figs. 8.8, 41, 44, 45. One of the authors (M.S.D.) wishes to acknowledge support by the Advanced Research Projects Agency of the Department of Defense, which was monitored by the Air Force Office of Scientific Research under Contract No. F49620-85-C0147. The other author (M.E.) offers thanks to Professor Emeritus T. Koyama of Shinshu University and Prof. M. Inagaki of University of Hokkaido for their sustained encouragement of this work, and to Dr. Roger Bacon of the Amoco Company for kindly supplying mesophase pitch fiber samples.

References

8.1 J.-P. Issi: In *Intercalation in Layered Materials*, volume 148 of NATO ASI Series B: Physics, ed. by M.S. Dresselhaus (Plenum, New York 1987) p. 347

8.2 F.L. Vogel, A. Hérold: Mater. Sci. Eng. **31** (1977), Proc. International Conference on Intercalation Compounds of Graphite

8.3 M.S. Dresselhaus, K. Sugihara, I.L. Spain, H.A. Goldberg: *Graphite Fibers and Filaments*, Springer Ser. Mater. Sci. Vol. 5 (Springer, Berlin, Heidelberg 1988)

8.4 M. Endo: J. Mater. Sci. **23**, 598 (1988)

8.5 T. Hamada, T. Nishida, Y. Sajiki, M. Matsumoto, M. Endo: J. Mater. Res. **2**, 850 (1987)

8.6 M. Guigon, A. Oberlin, G. Desarmot: Fibre Sci. Technol. **20**, 55 (1984)

8.7 R.J. Diefendorf, E.W. Tokarsky: J. Polymer Eng. Sci. **15**, 150 (1975)

8.8 A. Fourdeux, R. Perret, W. Ruland: In The International Conference on Carbon Fibers and Their Composites (The Plastics Institute, London 1971) p. 57

8.9 M.S. Dresselhaus: J. de Chimie Phys. **81**, 739 (1984)

8.10 M. Elzinga, D.T. Morelli, C. Uher: Phys. Rev. **B26**, 3321 (1982)

8.11 D.J. Johnson: Phil. Trans. Roy. Soc. **A294**, 443 (1980)

8.12 M.S. Dresselhaus, G. Dresselhaus: Adv. Phys. **30**, 139 (1981)

8.13 S.A. Solin: Adv. Chem. Phys. **49**, 455 (1982)

8.14 G. Kirczenow: In *Graphite Intercalation Compounds I*, ed. by S.A. Solin, H. Zabel, Springer Ser. Mater. Sci. Vol. 14 (Springer, Berlin, Heidelberg 1990), p. 59

8.15 R. Bacon: J. Appl. Phys. **31**, 283 (1960)

8.16 M. Endo: CHEMTECH Am. Chem. Soc. **18**, 568 (1988)

8.17 K. Matsumara, A. Takahashi, J. Tsukamoto: Synth. Met. **11**, 9 (1985)

8.18 J. Tsukamoto, K. Matsumura, T. Takahashi, K. Sakoda: Synth. Met. **13**, 255 (1986)

8.19 M. Endo, H. Yamanashi, G.L. Doll, M.S. Dresselhaus: J. Appl. Phys. **64**, 2995 (1988)

8.20 H. Menjo, B.S. Elman, G. Braunstein, M.S. Dresselhaus: J. de Chimie Phys. **81**, 835 (1984)

8.21 C. Meschi, J.P. Manceau, S. Flandrois, P. Delhaès, A. Ansert, L. Deschamps: Ann. de Phys. **11**, 199 (1986) colloq. 2, suppl. 2

8.22 Claude Meschi: Studies of the intercalation of chlorides into carbon fibers, in order to obtain electrical conductors, PhD thesis, L'Université de Bordeaux I (1988)

8.23 J. Shioya, H. Matsubara, S. Murakami: Synth. Met. **14**, 113 (1986)

8.24 H. Goldberg, I.L. Kalnin: Synth. Met. **3**, 169 (1981)

8.25 J.R. Gaier, M.E. Slabe: In Extended Abstracts of the 1986 MRS Symposium on Graphite Intercalation Compounds (MRS, Pittsburgh 1986) p. 216
8.26 M. Endo, M. Shikata, H. Touhara, K. Kadono, N. Watanabe: Transactions of the Institute of Electrical Engineers of Japan, **105-A**, 643 (1985) (in Japanese)
8.27 M. Endo, H. Ueno, M. Inagaki: Transactions of the Institute of Electrical Engineers of Japan, **105-A**, 329 (1985) (in Japanese)
8.28 M. Endo, H. Yamanashi, A. Sudou, M.S. Dresselhaus: International colloquium on layered compounds Pont-a-Mousson Conference (1988) p. 169
8.29 J.S. Speck, M. Endo, M.S. Dresselhaus: J. Crystal Growth **94**, 834 (1989)
8.30 S. Otani, Y. Kokubo, T. Koitobashi: Bull. Chem. Soc. (Japan) **43**, 3291 (1970)
8.31 M. Endo, T. Koyama, M. Inagaki: Synth. Met. **3**, 177 (1981)
8.32 W.D. Lee, G.P. Davis, F.L. Vogel: Carbon **23**, 731 (1985)
8.33 G. Hooley: In *Preparation and Crystal Growth of Materials with Layered Structures*, ed. by R.M.A. Leith (Reidel, Dordrecht 1977) p. 1
8.34 I.L. Kalnin, H.A. Goldberg: Synth. Met. **3**, 159 (1981)
8.35 M. Endo, T.C. Chieu, G. Timp, M.S. Dresselhaus, B.S. Elman: Phys. Rev. **B28**, 6982 (1983)
8.36 J.G. Hooley, V.R. Deitz: Carbon **16**, 251 (1978)
8.37 N. Watanabe: J. Fluorine Chemistry **22**, 205 (1983)
8.38 H. Touhara, K. Kadono, N. Watanabe, M. Endo: J. de Chim. Phys. **81**, 841 (1984)
8.39 N. Watanabe, T. Nakajima, H. Touhara: *Graphite Fluorides*, Studies in Inorganic Chemistry, Vol. 8 (Elsevier, Amsterdam 1988)
8.40 T. Nakajima, T. Ino, N. Watanabe, H. Takenaka: Carbon **26**, 397 (1988)
8.41 H. Touhara, K. Kadono, N. Watanabe: Tanso **117**, 998 (1984) (in Japanese)
8.42 M. Endo, J. Nakamura, H. Touhara, S. Morimoto: T. IEEE Japan **108A**, 81 (1988) (in Japanese)
8.43 T. Nakajima, A. Mabuchi, R. Hagiwara: Carbon **26**, 357 (1988)
8.44 J.R. Gaier, D.A. Jaworske: Synth. Met. **12**, 525 (1985)
8.45 J.R. Gaier: Stability of bromine intercalated graphite fibers, NASA TM-86859 (1985)
8.46 S.C. Moss, R. Moret: In *Graphite Intercalation Compounds I*, ed. by S.A. Solin, H. Zabel, Springer, Ser. Mater. Sci. Vol. 14 (Springer, Berlin, Heidelberg 1990), p. 5
8.47 A. Oberlin, M. Endo, T. Koyama: J. Cryst. Growth **32**, 335 (1976)
8.48 A. Oberlin, M. Endo, T. Koyama: Carbon **14**, 133 (1976)
8.49 T.C. Chieu, G. Timp, M.S. Dresselhaus, M. Endo, A.W. Moore: Phys. Rev. **B27**, 3686 (1983)
8.50 M. Endo, Y. Yamagishi, M. Inagaki: Synth. Met. **7**, 203 (1983)
8.51 P. Kwizera, M.S. Dresselhaus, G. Dresselhaus: Carbon **21**, 121 (1983)
8.52 L.E. McNeil, J. Steinbeck, L. Salamanca-Riba, G. Dresselhaus: Carbon **24**, 73 (1986)
8.53 L.E. McNeil, J. Steinbeck, L. Salamanca-Riba, G. Dresselhaus: Phys. Rev. **B31**, 2451 (1985)
8.54 X.W. Qian, S.A. Solin, J.R. Gaier: Phys. Rev. **B35**, 2436 (1987)
8.55 X.W. Qian, S.A. Solin, J.R. Gaier: In *Intercalation in Layered Materials*. Vol. 148 of NATO ASI Series B: Physics, ed. by M.S. Dresselhaus (Plenum, New York 1987) p. 477
8.56 M. Endo, T.C. Chieu, G. Timp, M.S. Dresselhaus, B.S. Elman: Synth. Met. **8**, 251 (1983)
8.57 G. Timp, M.S. Dresselhaus: J. Phys. C **17**, 2641 (1984)
8.58 P. Kwizera, M.S. Dresselhaus, D.R. Uhlmann, J.S. Perkins, C.R. Desper: Carbon **20**, 387 (1982)
8.59 H.A. Goldberg: Final report to US Army Res. Office, Contract # DAAE29-81-C-0016, (1985)
8.60 D.A. Jaworske, J.R. Gaier, C.C. Hung, B.A. Banks: SAMPE Quarterly (October 1986) p. 9
8.61 F. Tuinstra, J.L. Koenig: J. Chem. Phys. **53**, 1126 (1970)
8.62 S.A. Solin: In *Graphite Intercalation Compounds I*, ed. by S.A. Solin, H. Zabel, Springer Ser. Mater. Sci. Vol. 14 (Springer, Berlin, Heidelberg 1990), p. 157
8.63 P. Lespade, R. Al-Jishi, M.S. Dresselhaus: Carbon **20**, 427 (1982)
8.64 M.S. Dresselhaus, G. Dresselhaus: In *Light Scattering in Solids III*, ed. by M. Cardona, G. Güntherodt, Topics Appl. Phys. Vol. 51 (Springer, Berlin, Heidelberg 1982) p. 3
8.65 F.L. Vogel, G.M.T. Foley, C. Zeller, E.R. Falardeau, J. Gan: J. Mater. Sci. Eng. **31**, 264 (1977)

8.66 V. Natarajan, J.A. Woollam, A. Yavrouian: Synth. Met. **8**, 291 (1983)
8.67 V. Natarajan, J.A. Woollam: In Proceedings of the Symposium on Intercalated Graphite, Vol. 20, ed. by M.S. Dresselhaus, G. Dresselhaus, J.E. Fischer, and M.J. Moran (North-Holland, Amsterdam 1983) p. 45
8.68 C. Manini, J.-F. Marêché, E. McRae: Synth. Met. **8**, 261 (1983)
8.69 J.M. Murday, D.D. Dominguez, J.A. Moran, W.D. Lee, R. Eaton: Synth. Met. **9**, 397 (1984)
8.70 C. Manini, E. McRae, J.-F. Marêché, A. Hérold: Carbon **23**, 465 (1985)
8.71 I.L. Spain, A.R. Ubbelohde, D.A. Young: Phil. Trans. Roy. Soc. London A**262**, 345 (1967)
8.72 M.S. Dresselhaus: In *Intercalation in Layered Materials*, volume 148 of NATO ASI Series B: Physics, ed. by M.S. Dresselhaus (Plenum, New York 1987) p. 461
8.73 F.L. Vogel: Carbon **14**, 175 (1976)
8.74 T. Nakajima, N. Watanabe, I. Kameda, M. Endo: Carbon **24**, 343 (1986)
8.75 G.R. Bagga, Ph. Touzain, L. Bonnetain: In Extended Abstracts of the International Conf. on Carbon, Bordeaux, France (1984) p. 292
8.76 D.D. Dominguez, J.L. Lakshmanan, E.F. Barbano, J.S. Murday: In Proceedings of the Symposium on Intercalated Graphite, ed. by M.S. Dresselhaus, G. Dresselhaus, J.E. Fischer, M.J. Moran (North-Holland, New York 1983) p. 63
8.77 H. Oshima, J.A. Woollam, A. Yavrouian: J. Appl. Phys. **53**, 9220 (1982)
8.78 J.-P. Issi, B. Poulert, J. Heremans, M.S. Dresselhaus: Solid State Commun. **68**, 1065 (1988)
8.79 C.P. Davis, M. Endo, F.L. Vogel: In Extended Abstracts, 15th Biennial Conf. Carbon (1981) p. 363
8.80 D.A. Jaworske, J.D. Miller: NASA Technical Memorandum 87217 (1985)
8.81 D.A. Jaworske, J.R. Gaier, C. Maciag, M.E. Slabe: Carbon **25**, 779 (1987)
8.82 S.L. DiVittorio, M.S. Dresselhaus, M. Endo, T. Nakajima, B**43**, 12304 (1991)
8.83 J.M. Ziman: *Principles of the Theory of Solids*, 2nd edn. (Cambridge University Press, Cambridge 1972)
8.84 I.L. Kalnin, H.A. Goldberg: Synth. Met. **8**, 277 (1983)
8.85 M.Z. Tahar, M.S. Dresselhaus, M. Endo: Carbon **24**, 67 (1986)
8.86 L. Piraux, B. Nysten, J.-P. Issi, L. Salamanca-Riba, M.S. Dresselhaus: Solid State Commun **58**, 265 (1986)
8.87 L. Piraux, J.-P. Issi, L. Salamanca-Riba, M.S. Dresselhaus: Synth. Met. **16**, 93 (1986)
8.88 K. Sugihara: In Extended Abstracts of the Symposium on Intercalated Graphite at the Materials Research Society Meeting, Boston (1984) p. 60
8.89 K. Sugihara: Phys. Rev. B**37**, 7063 (1988)
8.90 Y.Y. Huang, Y.B. Fan, S.A. Solin, J.M. Zhang, P.C. Eklund, J. Heremans, G.G. Tibbetts: Solid State Commun. **64**, 443 (1987)
8.91 D.D. Dominguez, R.N. Bolster, J.S. Murday: In Extended Abstracts of the 15th Conf. on Carbon (American Carbon Society, University Park 1981) p. 365
8.92 A. Ansart, C. Meschi, S. Flandrois: Synth. Met. **23**, 455 (1988) Proc. 4th Int. Symp. on GICs, Jerusalem
8.93 J.A. Woollam, V. Natarajan, B. Brandt: Appl. Phys. Commun. **6**, 121 (1986)
8.94 J.A. Woollam, H. Chang, S. Nafis, D.J. Sellmyer: Private communication
8.95 J.A. Woollam, V. Natarajan, B. Brandt: Appl. Phys. Commun. **6**, 161 (1986)
8.96 J.W. McClure, W.J. Spry: Phys. Rev. **165**, 809 (1968)
8.97 L. Piraux, V. Bayot, J.-P. Michenaud, J.-P. Issi, J.F. Marêché, E. McRae: Solid State Commun. **59**, 711 (1986)
8.98 L.D. Woolf, J. Chin, Y.R. Lin-Liu, H. Ikezi: Phys. Rev. B**30**, 861 (1984)
8.99 L. Piraux, B. Nysten, J.-P. Issi, J.F. Marêché, E. McRae: Solid State Commun. **55**, 517 (1985)
8.100 L. Piraux, J.-P. Issi, J.-P. Michenaud, E. McRae, J.F. Marêché: Solid State Commun. **56**, 567 (1985)
8.101 Luc Piraux: Conduction, interaction and localization models in GICs, PhD thesis, Université Catholique de Louvain (1987)
8.102 L. Piraux, V. Bayot, X. Gonze, J.-P. Michenaud, J.-P. Issi: Phys. Rev. B**36**, 9045 (1987)

8.103 L. Piraux: In *Intercalation in Layered Materials*, Vol. 148 of NATO ASI Series B: Physics ed. by M.S. Dresselhaus (Plenum, New York 1987) p. 375
8.104 P.W. Anderson, E. Abrahams, T.V. Ramakrishnan: Phys. Rev. Lett. **43**, 718 (1979)
8.105 B.L. Altshuler, A.G. Aronov, P.A. Lee: Phys. Rev. Lett. **44**, 1288 (1980)
8.106 P.A. Lee, T.V. Ramakrishnan: Rev. Mod. Phys. **57**, 287 (1985)
8.107 G. Bergmann: Phys. Rep. **107**, 1 (1984)
8.108 L. Piraux, M. Kinany-Alaoui, J.-P. Issi, A. Perignon, P. Pernot, R. Vangelisti: Phys. Rev. B**38**, 4329 (1988)
8.109 L. Piraux, V. Bayot, J.-P. Issi, M.S. Dresselhaus, M. Endo, T. Nakajima: Phys. Rev. B**41**, 4961 (1990)
8.110 L. Piraux: J. Mat. Res. **5**, 1285 (1990)
8.111 V. Bayot, L. Piraux, J.-P. Michenand, J.-P. Issi, M. Le Laurain, A. Moore: Phys. Rev. B**41**, 11770 (1990).
8.112 A.A. Bright: Phys. Rev. B**20**, 5142 (1979)
8.113 S.L. DiVittorio, M.S. Dresselhaus, M. Endo, T. Nakajima: Phys. Rev. B**43**, 1313 (1991)
8.114 J. Heremans, M. Shayegan, M.S. Dresselhaus, J.-P. Issi: Phys. Rev. B**26**, 3338 (1982)
8.115 N. Daumas, A. Hérold: C.R. Acad. Sc. Paris **268**, 373 (1969)
8.116 L. Piraux, J.-P. Issi, J.F. Marêché, E. McRae: Synthetic Metals **30**, 245 (1989)
8.117 J. Tsukamoto, A. Takahashi, T. Tani, T. Ishiguro: Carbon **27**, 919 (1989)
8.118 J. Heremans, J.-P. Issi, I. Zabala-Martinez, M. Shayegan, M.S. Dresselhaus: Phys. Lett. **84A**, 387 (1981)
8.119 J.-P. Issi, J. Boxus, B. Poulaert, J. Heremans, M.S. Dresselhaus: Solid State Commun **44**, 449 (1982)
8.120 K. Sugihara: In Proceedings of the Symposium on Intercalated Graphite, ed. by M.S. Dresselhaus, G. Dresselhaus, J.E. Fischer, M.J. Moran, Vol. 20 (North-Holland, New York 1983) p. 157
8.121 K. Sugihara: Phys. Rev. B**28**, 2157 (1983)
8.122 M. Endo, I. Tamagawa, T. Koyama: Jpn. J. Appl. Phys. **16**, 1771 (1977)
8.123 T. Tsuzuku, K. Sugihara: In *Chemistry and Physics of Carbon* Vol. 12, ed. by P.L. Walker, Jr., P.A. Thrower (Dekker, New York 1975) p. 109
8.124 T. Takezawa, T. Tsuzuku, A. Ono, Y. Hishiyama: Philos. Mag. **19**, 623 (1969)
8.125 J.-P. Issi: Private communication
8.126 M. Kinany-Alaoui, L. Piraux, J.-P. Issi, P. Pernot, R. Vangelisti: Solid State Commun. **68**, 1065 (1988)
8.127 K. Sugihara: J. Phys. Soc. Japan **53**, 393 (1984)
8.128 K. Sugihara, K. Matsubara, T. Tsuzuku: J. Phys. Soc. Japan **53**, 795 (1984)
8.129 K. Sugihara: Phys. Rev. B**29**, 5872 (1984)
8.130 C. Herinckx, R. Perret, W. Ruland: Carbon **10**, 711 (1972)
8.131 T. Sugiura, T. Iijima, M. Sato, K. Fujimoto: Tanso **135**, 267 (1988) (in Japanese)
8.132 Ph. Touzain, G.R. Bagga: in International Symposium on Carbon (Toyohashi, Japan) (1984) p. 91
8.133 G.L. Hart, G. Pritchard, F.C. Stokes: In Reinforced Plastics/Composites Institute, The Society of the Plastics Industry, Inc., 32nd Annual Technical Conf. (1977) Sect. 9-D, p. 1
8.134. H. Oshima, J.A. Woollam, A. Yavrouian, M.B. Dowell: Synth. Met. **5**, 113 (1983)
8.135 M. Armand, P. Touzain: J. Mat. Sci. Eng. **31**, 319 (1977)
8.136 J.O. Besenhard, E. Theodoridou, H. Mohwald, J.J. Nickl: Synth. Met. **4**, 211 (1982)
8.137 J.O. Besenhard, H. Mohwald, J.J. Nickl: Revue de Chimie Minerale **19**, 588 (1982)
8.138 T. Nakajima, M. Kawaguchi, N. Watanabe: Synth. Met. **7**, 117 (1983)
8.139 I. Palchan, M. Talinaker: In *Intercalation in Layered Materials*, Vol. 148 of NATO ASI Series B: Physics, ed. by M.S. Dresselhaus (Plenum, New York 1987) p. 385
8.140 H. Touhara, H. Fujimoto, M. Endo, N. Watanabe, A. Tressaud: Solid State Ionics **14**, 163 (1984)
8.141 H. Touhara, K. Kadono, N. Watanabe, M. Endo: J. Chim. Phys. **81**, 841 (1984)

8.142 H. Touhara, H. Fujimoto, K. Kadono, N. Watanabe, M. Endo: Electrochimica Acta **32**, 293 (1987)
8.143 M. Endo, T. Momose, H. Touhara, N. Watanabe: J. Power Sources **20**, 99 (1987)
8.144 M. Endo, J. Nakamura, H. Touhara: ln Extended Abstracts of the 1988 MRS Symposium on Graphite Intercalation Compounds (MRS, Pittsburgh 1988) p. 157
8.145 J. M. Lalancette, R. Roussel: Canad. J. Chem. **54**, 3541 (1976)
8.146 M. Endo, T. Koyama, M. Inagaki: Ohyo Butsuri **49**, 563 (1980) (in Japanese)
8.147 M. Endo, T. Koyama: Tanso **111**, 59 (1980) (in Japanese)
8.148 M. Endo, T. Koyama: IEE of Japan **100A**, 633 (1980) (in Japanese)
8.149 Y. Maeda, H. Kitamura, E. Itoh, M. Inagaki: Synth. Met. **7**, 211 (1983)
8.150 R. Setton: Synth. Met. **23**, 467 (1988) Proc. 4th Int. Symp. on GICs, Jerusalem
8.151 S. Fukuda, J. Tsukamoto, A. Takahashi, K. Matsumura: Synth. Met. **18**, 491 (1987)
8.152 S. Fukuda, A. Takahashi, J. Tsukamoto: Synth. Met. **23**, 475 (1988) Proc. 4th Int. Symp. on GICs, Jerusalem
8.153 J. Tsukamoto, A. Takahashi, S. Fukuda, K. Kawasaki: Presented at Symposium on '88 Next Future Technology (1988)
8.154 T. Terai, Y. Takahashi: Carbon **22**, 91 (1984)
8.155 D.D.L. Chung, L.W. Wong: Carbon **24**, 639 (1986)
8.156 D.D.L. Chung, L.W. Wong: Carbon **24**, 443 (1986)

Subject Index

Page numbers in italics refer to the companion volume *Graphite Intercalation Compounds I*, ed. by Hartmut Zabel, Stuart A. Solin (Springer Series in Materials Science, Vol. 14)